MW00999072

Dairy Cattle:
Principles, Practices, Problems, Profits

Chapter 2 on Milk Marketing and Pricing by
Truman Graf, Ph.D.
Professor, Department of Agricultural Economics, University of Wisconsin, Madison, Wisconsin

Chapter 21 on Herd Health by
John S. Glenn, D.V.M., Ph.D.
*Extension Veterinarian, Cooperative Extension
University of California, Davis, California*

Dairy Cattle:

Principles, Practices, Problems, Profits

DONALD L. BATH, Ph.D.
Extension Dairy Nutritionist, Animal Science Extension,
University of California, Davis, California

FRANK N. DICKINSON, Ph.D.
Chief, Animal Improvement Programs Laboratory, Animal Science
Institute, Science and Education, Agricultural Research Service,
Beltsville Agricultural Research Center, U.S.D.A., Beltsville, Maryland

H. ALLEN TUCKER, Ph.D.
Professor of Physiology, Department of Animal Science,
Michigan State University, East Lansing, Michigan

ROBERT D. APPLEMAN, Ph.D.
Professor of Animal Science and Extension Animal Scientist, Dairy Management
University of Minnesota, St. Paul, Minnesota

THIRD EDITION

LEA & FEBIGER PHILADELPHIA
1985

Lea & Febiger
600 Washington Square
Philadelphia, PA. 19106
(215) 922-1330

First edition 1972
Reprinted 1973, 1975, 1977

Second edition 1978
Reprinted 1981

Library of Congress Cataloging in Publication Data
Main entry under title:

Dairy cattle.

Includes bibliographies and index.
1. Dairy cattle. 2. Dairy farming. I. Bath, Donald L.
SF208.D35 1985 636.2'14 84-20096
ISBN 0-8121-0955-4

PRINTED IN THE UNITED STATES OF AMERICA
Print Number 5 4 3 2 1

★ ★ ★ ★ ★

DR. RICHARD C. FOLEY *passed away on February 23, 1972, just as the first edition of this text was being readied for printing. He has been sorely missed by his former students, co-workers and friends. He conceived and was senior author of the first edition of this text, which occupied most of his work time for several years prior to his death. It is perhaps appropriate that he died while attending a meeting of dairymen—to the very last moment he dedicated his professional career to the dairymen and the great American dairy industry.*

Preface

The third edition of this text contains major updating as well as new material. The sections on breeding, feeding, reproduction, lactation, herd management, milk marketing, and herd health contain major revisions based on the latest research findings and changing conditions in the dairy industry.

This book is written for use as a text in senior dairy herd management courses at both the Associate degree (2-year) and Bachelor's degree (4-year) level. By stressing both the fundamental principles that determine why certain programs are more desirable than others and the detailed practices about how to breed, feed, and manage dairy cattle, we hope that this third edition will serve all segments of the dairy industry. It has been our intention also to produce a text that would be useful in all 50 states, in the major dairy areas of the world, and in those developing nations that are striving to improve their animal agriculture and to increase their supply of animal products.

Success in dairy farming, as in any other business, requires planned programs, based upon realistic standards of performance directed toward reasonable goals and executed with energy and enthusiasm. An effective dairy farm manager or operator must possess an understanding of the fundamental principles of economics, genetics, nutrition, physiology, and veterinary medicine, and a knowledge of desirable animal husbandry practices. The profitableness of the enterprise depends in large measure upon his ability through experience to solve the problems that are common to dairy farming everywhere and upon the soundness of his judgment in making decisions relative to all phases of breeding, feeding, and managing dairy cattle, and marketing dairy products.

Recognizing that the innumerable problems involved in managing dairy cattle and producing and marketing quality milk vary widely from place to place, some of the major problems have been presented as examples but no attempt has been made to discuss all of them or to present solutions that will apply universally.

In addition to planning this book as a text, we wanted a book that would be valuable to dairy farmers everywhere, to those individuals who sell to farmers or buy from farmers, as well as those who give them counsel at all levels. Finally in this era of intensive specialization this text may provide animal geneticists, nutritionists, physiologists, and veterinarians with a readily available source of information about those aspects of dairy herd management that are outside their respective fields of expertise. It may serve also as a useful reference for agricultural economists, agricultural engineers, dairy technologists, plant and soil scientists, and other individuals or groups who require current information about dairy cattle.

Davis, California
Beltsville, Maryland
East Lansing, Michigan
St. Paul, Minnesota

Donald L. Bath
Frank N. Dickinson
H. Allen Tucker
Robert D. Appleman

Acknowledgments

The authors are indebted to the many organizations and individuals who supplied charts, photographs, and tables. While, in each instance, the source is acknowledged, we wish it were possible to recognize each contributor individually.

We have utilized extensively the special knowledge of our colleagues in an effort to achieve the broadest possible in-depth coverage of the subject matter based on their experience and research. Many are cited in the list of references. We are especially grateful to the following individuals: D. W. Bates, University of Minnesota; B. G. Cassell, Virginia Tech University; R. L. Fogwell, Michigan State University; J. J. Ireland, Michigan State University; K. Huston, Wooster, OH; D. G. Johnson, University of Minnesota; R. P. Johnston, University of Wisconsin; H. H. Leipold, Manhattan, KA; G. D. Marx, University of Minnesota Technical College; B. T. McDaniel, Raleigh, NC; R. W. Mellenberger, Michigan State University; H. D. Norman, Beltsville, MD; D. E. Otterby, University of Minnesota; R. L. Powell, Beltsville, MD; K. H. Thomas, University of Minnesota; G. R. Wiggans, Beltsville, MD.

Our sincere thanks to our artists and typists, Gloria Bath, L. G. Dickinson, Nancy Feldman, Ann Tucker, and to the staff of Lea & Febiger.

D. L. B.
F. N. D.
H. A. T.
R. D. A.

Contents

I. *Profitable Dairy Farming*

III. *Nutritional Principles and Feeding Practices*

PART ONE

Profitable Dairy Farming

CHAPTER 1

Dairy Farming—U.S.A.

1.1 Introduction

Milk is defined as the "physiological secretion from the mammary gland of mammals." Since the dairy cow is the most commonly and extensively used source of milk in the United States, the term milk shall be understood to mean milk of the dairy cow unless otherwise specified.

Dairy cattle require more labor per animal and are influenced to a greater degree by the level of management they receive than any other class of farm animals. Tremendous advances have been made in recent years in the production, marketing, and processing of milk and dairy products. This progress, combined with the dairy cow's greater efficiency over other farm animals in converting feedstuffs into edible, nutritious human food, means that mankind should respect the cow's role in feeding the hungry world. Further, cattle utilize much more readily available feed (forages and by-product feeds) that cannot be directly consumed by humans.

This chapter briefly discusses why the United States has dairy cattle and a dairy industry. The discussion touches on the industry's development and current practices and projects of the near future. Included is a comparative analysis of dairy farming in the different regions of the United States and American producers versus producers in other parts of the world.

1.2 Why Dairy Cattle?

Most world food authorities believe that livestock production will retain its importance in the future. Phillips lists 5 factors that favor continuation of animal agriculture:

1. Food requirements of a rapidly expanding human population.
2. Nutritional merit or special qualities of animal food products.
3. Special ability of animals to transfer feedstuffs into edible food for humans.
4. Role of animals in maintaining soil fertility and in soil and water conservation.
5. Need for animals as a source of power.[1]

The authors feel that all of these except the last will continue to play an important role in U.S. agriculture.

Some people believe that animals no longer have a place in the world. They insist that only 10 to 30% of the energy fed is returned as food for humans. However, this figure holds true only if the entire diet of the animal is composed of food eaten by people.[2]

Ruminants, especially the dairy cow, play a significant role in maintaining a strong agricultural economy. The sale of dairy products ranks No. 1 in agricultural cash receipts in 7 states and is in the top four in 35 states. In 14 states, milk sales account for more than 20% of all farm cash receipts. Furthermore, dairy cows generate many times their farm value as income to allied businesses and industries concerned with manufacturing, processing, and distributing both the inputs (feed, equipment, and supplies) sold to dairymen and the outputs (milk and cattle) they market.

The ruminant does not compete for food essential to humans.[3,4] Ruminants utilize cellulose and non-protein sources of nitrogen, which humans cannot use directly. Furthermore, ruminants are scavengers; they consume crop residues and industrial by-products. More than 9 million tons of by-product feeds, left over after processing food grains for human products, are consumed annually.

Finally, food products from ruminant animals are in great demand. Combined, the dairy and beef industries supply two-thirds of the protein, 80% of the calcium, 62% of the phosphorus, and one-third of the energy the American people consume.

In addition to food, ruminants provide a wide array of other useful products, for example, leather, hair, essential enzymes (such as rennin) needed in manufacturing cheese, and pharmaceuticals such as insulin. Cash receipts for ruminants and their products account for 43% of the receipts from all U.S. agricultural commodities.

Dairy Cow Efficiency

Dairy cows are very efficient in converting feedstuffs into edible human food (Table 1.1). Another fact often overlooked is that dairy cattle, when offered an economical urea or ammonia supplement, can manufacture protein with an efficiency of conversion exceeding 100% if we consider only those quantities of plant products that cows consume and can also digest.[4]

In recent years, the dairy farmer has been more concerned with yield of milk per cow than with pounds of milk produced per pound of feed fed. Even though the efficiency with which dairy cows convert feed to milk decreases with high levels of feeding, it has been economically rewarding to breed and feed for high production. This is because proportionately less feed goes toward maintenance of the high producing cow, and most non-feed cash costs (for example, milking, housing, and waste handling) are essentially on a per head basis.[5]

TABLE 1.1 *Efficiency of Various Classes of Livestock in Converting Feed Nutrients to Edible Products*

	Efficiency of conversion of indicated nutrient, in %	
	Protein	*Energy*
Ruminants		
Dairy cattle	25	17
Beef cattle (edible cuts only)	4	3
Lambs (edible cuts only)	4	—(not reported)
Non-ruminants		
Hens (eggs)	26	18
Broilers	23	11
Turkeys	22	9
Swine	14	14

Source: Wedin, W. F., Hodgson, H. J., and Jacobson, N. L.: Utilizing plant and animal resources in producing human food. J. Animal Sci. *41*:667, 1975.

The feeding of concentrates to cows in DHIA herds increased from 3,000 lb per cow per year in 1950 to 6,000 lb in 1983, because (1) grain production increased significantly, (2) the U.S. population (human and livestock) could not consume all of the grain produced, and (3) price-depressing surpluses of grain accumulated.

In spite of the great amount of labor involved in producing milk, the cow is an efficient producer of energy and protein. Thus, the U.S. consumer can purchase protein and food solids at a lower cost per unit when they come from milk and dairy products rather than from meat and eggs (Table 1.2).

1.3 What is a Dairy Farmer?

Traditional dairy farmers in past years depended on the use of dairy cattle, land, and labor as major resources. The thrust in modern day dairying is the increased use of capital and management. The successful dairy farm operator must combine all of these resources into a productive and profitable unit.

Absolute rules that will assure a dairy farmer success cannot be written. This book does provide guidelines and important considerations, but in the final analy-

Table 1.2 *Retail Costs of Protein and Food Solids in Dairy Products, Red Meat, and Eggs*

Food	Price per pound $(U.S.)	Grams protein per pound	Cost of protein per pound $(U.S.)	Grams food solids per pound	Cost of food solids per pound $(U.S.)
Milk (instant dry nonfat)	.85	162.8	2.37	427.2	.90
Cottage cheese (creamed)	.50	61.7	3.68	96.4	2.35
Cheese (American)	1.20	113.4	4.80	278.1	1.96
Ice cream	.35	20.4	7.79	165.1	.96
Red meat (ground beef, lean)	1.10	93.9	5.32	140.3	3.56
Chicken (fryer, ready to cook)	.50	57.4	3.95	73.1	3.11
Eggs	.50	52.1	4.79	104.1	2.40

Source: Protein and food solids compositions. *Composition of Foods,* USDA Handbook No. 8, 1963.

sis, it is the expertise, judgment, and experience of the manager that determines his success or failure.

Most dairy farms consist of at least two enterprises, the dairy herd and the crop farm. Except in those areas of the United States where specialization has developed to the point that producers purchase all feedstuffs, the successful dairy farm operator must be both a good *crop farmer* and an outstanding *milk producer* (manager or herdsman).

The crop farmer produces feedstuffs that have a market value. Although this enter-

Table 1.3 *Rank of Major Competency Areas and Competencies Within Areas Necessary for Dairy Farm Herdsmen to Succeed*

Rank	Major area	Competencies
1	Record-keeping	A. Maintains breeding records
		B. Keeps animal identification up-to-date
		C. Analyzes and uses production records
		D. Keeps enterprise records and analyzes them at least annually
2	Milking	A. Insists on and uses correct milking procedures
		B. Maintains milking system in good condition
		C. Keeps mastitis losses at a minimum
3	Herd health	A. Minimizes calf losses
		B. Recognizes animal health problems and knows when to treat and/or seek assistance
4	Feeding	A. Utilizes basic principles of nutrition
		B. Uses technology needed to formulate the most profitable rations
		C. Provides an adequate and balanced ration to all groups of animals
5	Breeding	A. Detects cows in heat and determines best time for breeding
		B. Plans and follows constructive breeding programs
		C. Correctly inseminates cows if direct herd service is used
		D. Selects cows based on production and physical traits of economic importance, culling those not meeting standards
6	Business management	A. Maintains adequate and accurate farm records
		B. Obtains and uses credit wisely
7	Housing	A. Understands housing requirements, including ventilation needs
		B. Minimizes materials handling and labor requirements, especially manure and bulky feeds
8	Labor	A. Plans labor needs, including anticipation and preparation for peak work loads
		B. Uses labor efficiently, and recognizes circumstances requiring immediate attention

Source: Orth, R., Iowa State Univ., Ames.

prise, like the dairy industry, involves its own set of principles, practices, problems, and profits, it is not the purpose of this book to include a detailed discussion of this topic.

Dairy farmers, on the other hand, market these feedstuffs through their dairy cattle, with the expectation of profiting from more fully utilizing their available labor, capital, and management capabilities. Therefore, the producer makes daily decisions regarding the selection, breeding, feeding, managing, housing, and care of his dairy herd.

Many recent 4-year college and 2-year technical school graduates are seeking employment as herdsmen on dairy farms. A recent study identified and ranked the major areas in management that herdowners believe herdsmen must master to achieve success.[6] These areas of herdsmanship, and the more important competency items within each area (also ranked), are listed in Table 1.3.

These results, obtained from a survey of producers nationwide, suggest areas of study that students of dairying should definitely include in their curricula. It implies that students of dairying should make certain they become proficient in the competencies listed. It does not necessarily mean, for example, that milking procedures (No. 2A) are more important in achieving success on the dairy farm than, for instance, providing an adequate ration (No. 4C).

1.4 Composition and Nutritional Value of Milk

Chemically, milk is a complex mixture of fats, proteins, carbohydrates, minerals, vitamins, and other miscellaneous constituents dispersed in water. Table 1.4 lists these constituents along with the normal variation present.

Milk Fat

The most important milk constituent in determining the price received for milk is milk fat. Essentially all dairy products, except skim milk and those items made from skim milk, contain varying amounts of fat. Butter contains 80% or more fat, natural cheddar or American cheese 30 to 40%, and ice cream varies from 10 to 18% milk fat. The desirable qualities of smooth body and texture and rich mellow flavor of many dairy products are attributed to fat.

Table 1.4 *Gross Composition of Mixed Herd Milk*

Constituent	*Average content*	*Normal variation*
	%	
Water	87.2	82.4–90.7
Fat (milk fat)	3.7	2.5–6.0
Solids—not fat	9.1	6.8–11.6
Protein	3.5	2.7–4.8
Casein	2.8	2.3–4.0
Lactalbumins and lactoglobulins	0.7	0.4–0.8
Lactose (milk sugar)	4.9	3.5–6.0
Minerals	0.7	0.6–0.8
Total solids	12.8	9.3–17.6

Source: Compiled from various sources.[7–10]

In milk, fat is dispersed in the form of small globules of a true oil-in-water type emulsion. The globules range in size from about 0.5 to 20 microns (1 micron = 1/25,000 in.) with an average size of approximately 3 microns. Milk from the breeds of cattle that secrete a higher fat content generally contains globules of the larger size. The larger sized fat globules contribute to the easier formation of cream layers and are also more susceptible to partial churning during handling and transportation of milk. Milk fat aids in calcium absorption, and since milk contains abundant calcium, the complementary effect of fat in milk is especially important to the nutrition and health of humans.

Protein

Among the most complex of all organic compounds, proteins are essential to all forms of life. They are composed of a series of "building blocks" known as amino acids. Animals are able to synthesize proteins only from the proteins or amino acids they consume in their foods, although they can sometimes convert one amino acid to another. Those proteins that cannot be formed in the animal body and, therefore, must be present in the foods consumed are called *essential amino acids.* Thus, the nutritional or "biological value" of a given source of protein is measured in terms of the completeness with which it supplies essential amino acids. Milk, therefore, has a high biological value since it is an excellent source of the essential amino acids. In ruminants, proteins can be synthesized by ruminal bacteria, as described in Chapter 8.

Milk contains four classes of proteins—casein, α lactalbumin, β lactoglobulin, and immune globulins.

Casein constitutes about 80% of the total protein in milk and is unique in that it is found only in milk. In addition to amino acids, casein also contains phosphorus and exists in milk as a calcium salt known as calcium caseinate. Casein may be precipitated by acids or the enzyme rennin, and rennin-precipitated casein is the basis for curd formation in cheese. Casein is also used commercially for making glue, plastics, sizing for paper, and paint adhesives.

α lactalbumin and *β lactoglobulin* differ from casein in that they contain the sulfur-bearing amino acid, cysteine, and the amino acid, tryptophan, rather than phosphorus. They also differ in that they are coagulated easily by heat and are not precipitated by acids. Although a minor protein component of milk, they are important nutritionally since they complement the qualities of casein.

Immune globulins make up only about 0.1% of normal milk. However, their concentration increases greatly during the period of colostrum formation. These immune globulins act as an antibody carrier to protect the newborn calf from pathogenic organisms.

Milk provides man, by virtue of its amino acid composition, appreciable amounts of all *essential amino acids.* With the exception of the amino acids containing sulfur, the estimated average minimal daily requirement of essential amino acids for adult humans can be provided by 1 pint of milk. Furthermore, milk is extremely valuable in supplementing the diets of humans eating cereals, since cereals are especially low in the amino acid lysine, and 1 pint of milk contains twice the minimum adult daily requirement of that particular amino acid.

The average U.S. consumer obtains about 25% of his dietary protein from milk and milk products, but only 13% of his dietary energy (calories).

Lactose

Milk is the only source of the principal carbohydrate, lactose. The mammary gland is unique in being able to synthesize this sugar, which consists of two simple sugars: glucose and galactose. Different from fat and protein, lactose is in true solution in milk; thus it affects the freezing point, boiling point, and osmotic pressure of milk. Lactose is only one-sixth as sweet as sucrose (table sugar).

Certain bacteria have the ability to ferment lactose to produce lactic acid. This fermentation is responsible for the souring of milk and cream but is advantageous in the production of various types of cheese, buttermilk, and sour cream.

Favorable human nutritional features of lactose include its ability to suppress protein putrefaction in the intestine and thereby impede growth of many pathogenic organisms. The lactic acid produced is also of considerable value after the administration of antibiotics in that it assists in the re-establishment of desirable intestinal flora. Furthermore, the low solubility of lactose makes it less irritating than many other sugars; thus, it is valuable in the treatment of stomach ulcers.

Commercial uses of lactose, separated from cheese whey by crystallization, include pill coatings, an ingredient in baby foods, a filler in medicines, and a source of nutrients for antibiotic producing organisms.

Minerals

Milk is an excellent source of both calcium and phosphorus, major minerals necessary for increased skeletal growth. The diet of man is more often deficient in calcium than in any other nutrient, and researchers have concluded that addition of milk to the diet of school children markedly increases both body height and weight. On the average, a person living in the United States obtains nearly 80% of the calcium and about one-half of the phosphorus consumed daily from milk and milk products.

Milk is a poor source of iron and copper. It is fortunate, however, that milk is low in iron for iron would destroy certain vitamins. Furthermore, the presence of iron in larger amounts could result in oxidation of milk, causing a distinctive off-flavor.

Vitamins

All known vitamins are in milk, but it is an especially good source of riboflavin. About one-half of the U.S. consumer's daily intake of this vitamin is from milk and milk products. Most fluid milks sold to consumers are enriched with additional vitamin D so that milk provides man with a well balanced supply of the fat soluble vitamins (A, D, E, and K) as well as the water soluble vitamins (except vitamin C).

1.5 Development of the Dairy Industry

Milk is one of the oldest foods known to man. Historians tell us that records exist of cows being milked in 9000 B.C. The Bible mentions milk in many places, including Exodus 3:8, which describes the "land of milk and honey." Hippocrates, a famous physician 500 years before Christ, had his patients take milk as a medicine.

The first dairy cattle came to the West Indies with Christopher Columbus on his second voyage. In 1611, cows arrived at the Jamestown colony. From colonial times until about 1850, dairying in the United States changed little. Dual-purpose cattle were kept as family cows. Small herds produced raw milk, and farmers delivered butter in towns and cities close enough to be reached by horse-drawn wagons. Milk production was highly seasonal, reaching a peak in the spring and summer months; surplus milk was skimmed and churned into butter or processed for cheddar cheese on the farm. Storage was limited by the lack of refrigeration. The farmer plowing with oxen on an Illinois farm in 1833 and the housewife churning farm butter remind the reader how much American agriculture has progressed since that time.

Historical Milestones

The dairy industry is relatively young. The more important dates in the development of this industry in the United States are:

1841 First regular shipment of milk by rail—Orange County, New York, to New York City
1851 First cheese factory in Oneida, New York
1855 First college of agriculture, Michigan State University
1857 First condensery built by Gail Borden, at Burrville, Connecticut
1860s Establishment of several U.S. breed associations
1861 Mechanical refrigeration available
1871 First centrifugal separator, invented almost simultaneously by DeLaval, and Winthrop and Nielsen
1884 Milk bottle invented by H. D. Thatcher, Potsdam, New York
1885 Practical aspects of dairy farming communicated through popular publications, such as *Hoard's Dairyman*
1887 Establishment of agricultural experiment stations
1890 Babcock test for fat determination developed
1890 Tuberculin testing of dairy herds introduced
1895 Commercial pasteurizing machines introduced
1905 Production testing (DHIA) introduced in the United States
1908 First compulsory pasteurization law (Chicago) applying to all milk except that from tuberculin tested cows
1914 Establishment of the Agricultural Extension Service
1914 Tank trucks first used to transport milk
1919 Homogenized milk sold successfully in Torrington, Connecticut
1920s Classified pricing systems established in all major milk markets
1932 Addition of vitamin D to milk made practicable
1938 Bulk tanks began replacing milk cans on dairy farms
1939 Artificial insemination available commercially
1948 Ultra-high temperature pasteurization introduced
1948 First plastic-coated paper milk cartons introduced commercially
1968 Official acceptance of electronic testing for milk fat available commercially
1970s Automated milking unit detachers introduced.
1974 Nutrition labeling of fluid milk products

TABLE 1.5 *U.S. Milk Marketed as Fluid and Manufacturing Milk, by Selected Years*

Year	*Sold as whole milk*	*Sold as cream*	*Sold directly to consumers*	*Total marketed*	*Used on farm*	*Total produced*	*Grade A*	*Cash receipts*
	billion pounds						%	billion $
1960	103.9	7.9	2.1	114.0	9.2	123.1	67	—
1965	112.7	3.7	1.8	118.2	6.0	124.2	69	5.038
1970	110.0	1.2	1.7	113.0	4.0	117.0	74	6.525
1975	110.3	.4	1.5	112.2	3.2	115.5	80	9.866
1980	124.7	—	1.5	126.2	2.3	128.5	84	16.584
1982	132.1	—	1.3	133.5	2.3	135.8	85	18.274

Source: Economic Research Service, USDA: Dairy Situation, DS-360, May, 1976; and DS-393, June, 1983.

1980s UHT (ultra high temperature) milk gained national recognition
1980s Routine testing of cow's milk for components other than milk fat accepted
1980s Embryo transfer introduced and accepted
1980s Backflushing of milking units to sterilize units between cows developed
1980s Acceptance of somatic cell counts for monitoring subclinical mastitis adopted nationwide

1.6 Production and Consumption of Dairy Products

Dairy farmers marketed 133.5 billion pounds of milk in 1982, up 17.5% since 1975 (Table 1.5). Another 2.3 billion pounds were utilized on the farm, either consumed by the family or fed to calves. The five major dairy states of Wisconsin, California, New York, Minnesota, and Pennsylvania marketed over 50% of the total. Over 85% of the milk sold was eligible for fluid use (grade A). This amount is up 5% from 1975 and considerably higher than the 67% figure in 1960. Farmers are being encouraged to convert to grade A production by receiving considerably higher prices for fluid milk than for manufacturing milk and by smaller differences in sanitation standards between the two grades. Of the 19 billion pounds of manufacturing grade milk produced in 1982, over 11.5 billion pounds came from Wisconsin, Minnesota, and Iowa.

Higher prices and increased marketings brought cash receipts from farm marketed milk and cream to over $18.2 billion, up 85% from 1975.

Utilization

Although 85% of the U.S. milk supply is produced under Grade A standards and is eligible for fluid use, only 36.4% of the

TABLE 1.6 *U.S. Per Capita Civilian Consumption of Milk and Dairy Products, 1925–1982*

Year	*All products lb*[a]	*% change from 1945*	*Fluid milk and cream lb*[a]	*% change from 1945*
1925	800	+1	340	−15
1935	800	+1	330	−18
1945	790	—	400	—
1955	710	−11	350	−13
1965	620	−22	300	−25
1970	560	−29	280	−30
1975	540	−32	270	−33
1980	540	−32	250	−38
1982	560	−29	240	−40

[a]Figures rounded to nearest 10 pounds.
Source: Economic Research Service, USDA: Dairy Situation, DS-361, July, 1976; and DS-394, September, 1983.

1982 supply was consumed as fluid milk and cream. This proportion has been dropping for the past 10 years. Per capita civilian consumption of milk in all dairy products totaled 560 lb in 1982, up 3.7% from 1980 but down 10% from 1965 (Table 1.6).

The most noteworthy change in usage patterns has been that of cheese consumption (up 75% in the 17-year period since 1965). The phenomenal increase in sales of low fat fluid milk and yogurt has helped the dairy industry dispose of solids-not-fat (Table 1.7). Note, however, that these increased sales have been offset with marked declines in the consumption of fluid whole milk, cream, butter, evaporated and condensed milk, and nonfat dry milk. The net result has been a decrease in the amount of milk fat utilized.

Table 1.7 *U.S. Per Capita Civilian Consumption of Milk and Dairy Products, by Product, 1965, 1975, and 1982*

Item	Per capita sales[a] 1965	1975	1982	Percent change (1975 to 1982)
Fluid products				
Whole milk	264.0	195.0	133.0	−31.8
Low fat milk[b]	34.0	84.7	98.9	+16.8
Cream[c]	7.6	5.9	5.8	− 1.7
	305.6	285.6	237.7	
Cheese				
American cheese	5.7	7.9	11.4	+44.3
Cottage cheese	4.7	5.0	4.2	−16.0
Other	3.4	6.2	8.6	+38.7
	13.8	19.1	24.2	
Butter	5.8	4.4	4.5	+ 2.3
Frozen products				
Ice cream	18.8	18.8	17.7	− 5.9
Ice milk	6.6	7.6	6.7	−11.8
Sherbert	1.5	1.5	1.2	−20.0
	26.9	27.9	25.6	
Other				
Evaporated and condensed milk	10.8	5.2	4.0	−23.1
Nonfat dry milk	4.8	3.1	2.7	−12.9
Yogurt	0.3	2.0	0.6	−70.0
	15.9	10.3	7.3	

[a]Based on civilian population except fluid milk products, which are based on estimated population using fluid products from purchased sources.
[b]Includes skim milk, buttermilk, and flavored milk drinks.
[c]Includes milk and cream mixture.
Source: Economic Research Service, USDA: Dairy Situation, DS-359, March, 1976; DS-361, July, 1976; and DS-393, June, 1983.

1.7 Current Trends and Projections of the Dairy Industry

The last several years have been frustrating to most dairy farmers. They have found it difficult to adjust to erratic and rapidly increasing input costs. Since 1975, costs for farm machinery have doubled, building material costs increased 50%, and interest rates increased three-fold. Milk prices have increased only 55% (Table 1.8).

The net result for many producers has been delay in expenditures for equipment and buildings. Some dairymen with high debt loads have been forced out of business.

The trend toward larger and more highly mechanized dairy farms will continue.[11] High investments for buildings and equipment will encourage multiple man units, which provide the opportunity for more specialization and the spreading of overhead costs over more units of milk produced. Continued high costs for equipment, housing, and labor will also encourage development of more economical and labor efficient housing systems. Increased output of milk per cow, produced economically, will be an extremely important ingredient for profitable dairy farming. Milk produc-

Table 1.8 *Index of Prices Received for Milk and Prices Paid for Selected Inputs, 1975–1982.*

Item	Year 1975	1980	1981	1982
Prices received				
Milk	90	135	142	140
Prices paid				
Production items	91	138	148	149
Feed	100	123	134	122
Farm machinery	81	134	149	162
Building material	90	128	134	135
Fertilizer	120	134	144	144
Interest	77	173	211	233
Farm wages	85	126	137	141

U.S. annual average index, 1977 = 100
Source: Statistical Reporting Service, USDA: *Agricultural Prices,* Pr 1–3, June, 1983.

Table 1.9 *Top 30 Dairy States, Based on Milk Produced, 1982.*

	Total production in 1982 (million lbs.)	Percentage of U.S. production, 1982	Percent change in production, 1975–1982	Percent of all agricultural sales in state, 1982	Projected change 1990[a]
Top 10 states					
Wisconsin	22,230	17.1	+22.3	65.4	0
California	14,518	10.7	+33.8	20.3	+
New York	11,147	8.2	+12.6	61.8	0
Minnesota	10,341	7.6	+15.6	21.1	–
Pennsylvania	9,264	6.8	+29.7	47.5	0
Michigan	5,253	3.9	+18.5	27.0	–
Ohio	4,550	3.4	+ 7.0	18.8	– –
Iowa	4,301	3.2	+ 9.8	5.6	– –
Texas	3,770	2.8	+17.0	7.8	+
Washington	3,222	2.4	+38.8	16.3	+
Total, 10 states	89,596	66.1	+21.1	22.6	0
Second 10 states					
Missouri	2,905	2.2	– 3.8	10.2	– –
Illinois	2,657	2.0	+ 3.8	4.9	–
Kentucky	2,364	1.7	+ 1.9	13.2	–
Vermont	2,361	1.7	+17.5	90.0	0
Indiana	2,334	1.7	+ 5.6	7.2	–
Tennessee	2,326	1.7	+24.2	19.3	–
Idaho	2,253	1.7	+46.8	13.5	0
Florida	2,109	1.6	+ 7.7	9.2	+
Virginia	2,059	1.5	+17.5	18.7	0
South Dakota	1,762	1.3	+28.6	8.2	–
Total, 10 states	23,130	17.1	+12.2	10.5	–
Third 10 states					
North Carolina	1,686	1.2	+ 5.2	6.0	0
Maryland	1,581	1.2	+ 1.7	23.8	0
Georgia	1,412	1.0	+18.3	6.7	+
Nebraska	1,360	1.0	– 5.0	2.8	– –
Kansas	1,356	1.0	– 3.4	3.4	– –
Oregon	1,301	1.0	+31.4	11.0	+
Arizona	1,212	0.9	+44.3	10.9	+
Oklahoma	1,165	0.9	+ 9.9	6.2	0
Utah	1,162	0.9	+26.4	31.0	0
North Dakota	1,005	0.7	+ 8.8	4.4	–
Total, 10 states	13,240	9.8	+11.1	6.3	–
Remaining 20 states	9,829	7.0	+ 9.8	3.5	–
Total, all states	135,795	100.0	+17.6	13.8	–

[a]2 negative signs indicate a more pronounced drop in milk production is expected.
Source: Economic Research Service, USDA: *Dairy Situation,* DS-392, March, 1983; and DS-393, June, 1983; also *Agricultural Outlook,* AD-84, January–February, 1983.

ers producing less than 14,000 lb of milk per cow will find it difficult to survive.

Pounds of milk produced and value of dairy products sold are shown in Table 1.9. The top 30 states are listed in descending order. States with a high dependence on the dairy enterprise include Wisconsin, New York, Pennsylvania, and Vermont. These states have relatively limited alternative enterprises. This ordering does not imply that California (No. 2) and Minnesota (No. 4) do not have viable dairy industries; it simply means that total agricultural income is less dependent, proportionately, on dairying.

The top 10 states account for nearly two-thirds of all milk produced; the second 10 states produce 17%; the third 10 states,

10%. States that have shown the greatest reductions in relative quantities of milk produced have been in the Plains area, especially Kansas, Nebraska, and Missouri.

The location of major and secondary milk producing areas in the United States is shown in Figure 1.1. Dairying usually occurs near large urban centers or in areas suitable for forage production. An exception is Florida; milk production is dependent almost entirely on purchased concentrates and forages, which quite often include citrus pulp and other by-products of Florida's agriculture. Producers in many other states, particularly in the West and the Northeast, also purchase most of their concentrate needs.

Dairy farming in the Mountain and Pacific Coast states is located almost entirely in irrigated valleys, usually near urban centers. Milk production tends to be concentrated in small areas when land values are high, especially the Phoenix, Arizona area, the Chino and San Joaquin Valley areas of California, and western Washington.

A high percentage of the milk produced in Wisconsin, Minnesota, and South Dakota is utilized in the manufacture of cheese, butter, and powder; thus, the dairy farms need not be located in urban centers but, rather, are located where ample supplies of forage are available.

Urbanization and pollution control are two important reasons why dairy farmers move to new locations. Producers from the high cost areas in eastern New York, New Jersey, and other heavily populated northeastern states continue to move to western New York and Maine. The exodus of dairy farmers from the fringes of metropolitan Chicago, Cleveland, Detroit, and other major cities continues as a result of urban sprawl. Fewer relocations have occurred in the more rural states of Wisconsin and Minnesota.

Southern California milk producers in San Bernardino and Riverside counties are faced with severe pollution control measures, resulting in some drastic changes in the storage and handling of manure. All

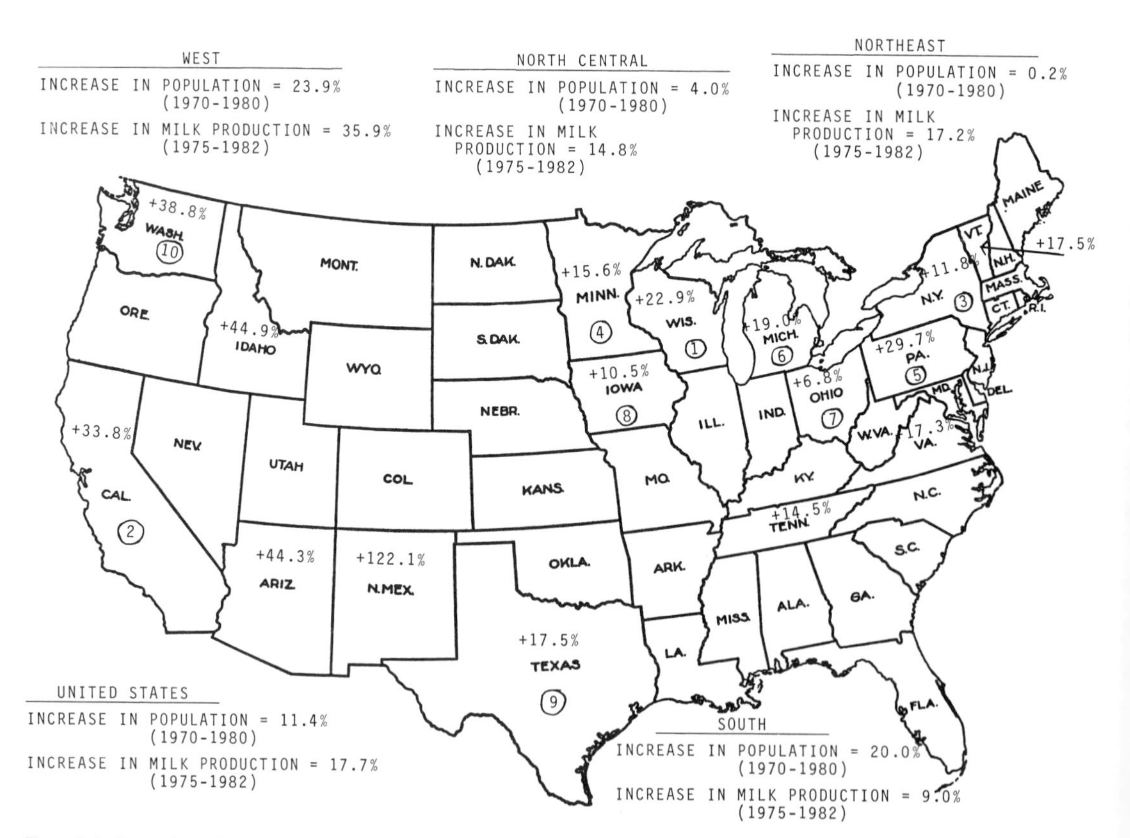

FIG. 1.1 Location of major dairy areas and those increasing rapidly in the United States. The percentage figure above selected states is the increase in total milk production since 1975. The circled number is the state rank.

run-off must be contained, and as much as 75% of the manure solids has had to be exported. Because of high costs, pollution control measures, and population pressures, many producers have been moving their operations to the San Joaquin Valley, and even to Oregon, Washington, Idaho, and New Mexico.

Of the top 20 milk producing states, Hoglund projects that relative milk production will continue to increase in Wisconsin, California, Texas, Washington, Florida, and North Carolina.[12] States expected to maintain present relative milk production levels include New York, Pennsylvania, Missouri, Vermont, Virginia, and Maryland. It is now estimated that relative milk production in Michigan and Minnesota will decline during the next 5 to 10 years.

Many factors, both on and off the farm, influence the viability and profitability of dairying. The individual producer plays an important role in determining the success or failure of the dairy enterprise. Management is the key to profits. However, individual producers must have access to adequate capital in developing viable operations. Too many dairy farmers are still mainly "labor" and not "management" oriented. For most, it requires a combination of both. A high percentage of the dairy operations in the Lake States, northern Corn Belt, and Northeast are not sufficiently large to require full-time management.

It is difficult to measure management skills. About all one can do is to measure the results of management. Important success factors include the following:

1. High, but economical milk production per cow.
2. High milk output per man.
3. High crop yields per acre.
4. Low investment per cow.
5. Timely and appropriate animal care and treatment practices.
6. Use of appropriate accounting and other management tools, especially dairy records.

Milk production will continue to be profitable on well-managed dairy farms. The majority of dairy farmers in the Lake States and Northeast will have 40 to 150 cows, average yearly production exceeding 14,000 lb per cow, and milk sales per man in the 600,000 to 800,000 lb per year range. Herds of 400 or more cows are not expected to be important in these areas but will continue to be increasingly important in the specialized milk producing areas in California, Washington, Arizona, Texas, and Florida.

1.8 Competitive Position for the U.S. Market

Most of the major supply regions of the world have emphasized increasing herd size. Average herd sizes vary from 105 cows in New Zealand to under 5 cows in Italy (U.S. average = 35.3 cows). Milk yield per cow is closely related to the amount of concentrates (grain) fed. In New Zealand, for example, production per cow is low, but dairying is a pasture based industry in which the dairy farm has no buildings except an open shed milking parlor. Cows are pastured year round, and no hay or grain storage or handling equipment exists.

Most of the economic and social forces operating in the United States are also evident abroad. Rapidly rising production costs in many countries have created concerns about the dairy farmer's ability to continue dairying. Inflation of land values and increasing production costs have threatened the "sharemilker" system of farm transfer in New Zealand. Traditionally, young sharemilkers would work on established dairy farms for several years, with the objective of saving enough money to start their own dairy farms. With the recent inflation, farm values have increased more than most sharemilkers can save.

1.9 Types of Dairy Farms

Climate, herd size, cost of labor and equipment are major factors affecting the organization of the dairy farm, type of housing, and the degree of mechanization built into the facilities.[11] Average herd size, by states, varied from a low of 35 cows in Minnesota (including only those farms with gross sales exceeding $10,000) to a high of 418 in Florida (1971 figures). To facilitate a discussion of the U.S. dairy industry and its variation east to west and north to south, the Bureau of Census, Department of Commerce divides the U.S. mainland into 10 districts. (Figure 1.2 offers a clear under-

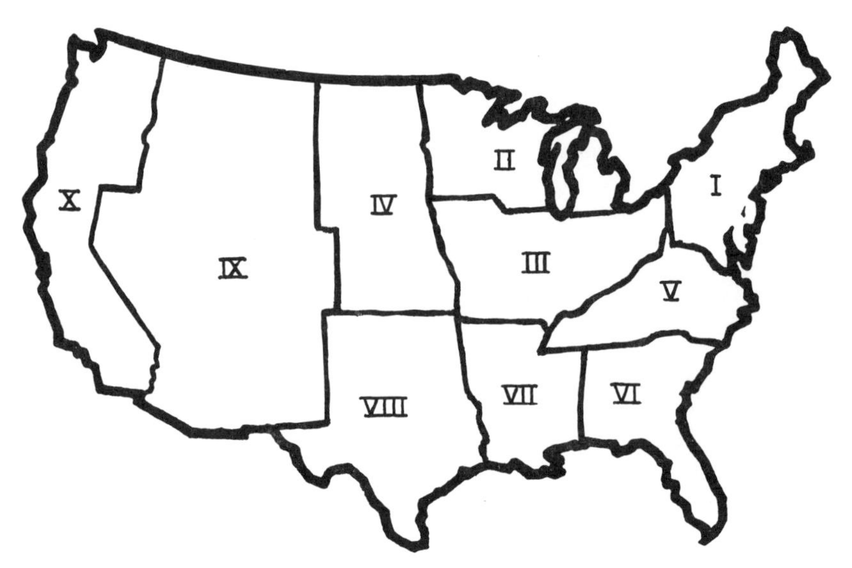

Fig. 1.2 Geographic regions of the United States.
I Northeast: Maine, New Hampshire, Vermont, Massachusetts, Rhode Island, Connecticut, New York, New Jersey, Pennsylvania, Delaware, Maryland
II Lake States: Michigan, Wisconsin, Minnesota
III Corn Belt: Ohio, Indiana, Illinois, Iowa, Missouri
IV Northern Plains: North Dakota, South Dakota, Nebraska, Kansas
V Appalachia: Virginia, West Virginia, North Carolina, Kentucky, Tennessee
VI Southeast: South Carolina, Georgia, Florida, Alabama
VII Delta States: Mississippi, Arkansas, Louisiana
VIII Southern Plains: Oklahoma, Texas
IX Mountain: Montana, Idaho, Wyoming, Colorado, New Mexico, Arizona, Utah, Nevada
X Pacific: Washington, Oregon, California

standing of these geographic regions discussed in the sections that follow.)

Housing

The Lake States lead in stanchion housing with 80% of the herds and 76% of the cows (Table 1.10). More than 50% of the new dairy barns being built in Wisconsin and Minnesota are of the tie-stall type. Nearly half of those being built in New York and Pennsylvania are stanchion barns, usually built for herds of fewer than 80 cows.

The Corn Belt, Northern Plains, the three southeastern areas, and Michigan shifted to loose housing systems about 15 or 20 years ago. While loose housing is still used extensively in most of these regions, the northern states have since installed free stalls.

Free-stall housing has gained in popularity in all but the southern Plains States. Even in California, this system is replacing or complementing corral-type systems. Dry lot or corral systems are most common in Arizona (90%), and in Florida, Texas, and California (50% or more). Frequently shades are provided when no other housing exists.

Milking System

The milking system selected is determined largely by the type of housing used (Table 1.10). In the Northeast and Lake States, more cows are milked in stanchion barns than are housed in them. The reason is that some producers have converted to loose or free-stall housing, but still milk in the old barn. The herringbone milking system has gained in popularity over the side-opening parlors. In the northern states, double-4 or 6 herringbone parlors are most common, while in the South and West, parlor sizes tend to be mostly in the double-6 to 10 size range.

Polygon and rotary parlors are increasing in numbers, scattered in many different regions, but they are more popular in large herd situations where the capital investment required can be spread over more cows.

TABLE 1.10 *Description of Typical Dairy Farms, by Regions in the United States*[a]

				Percentage distribution of housing type					Percentage distribution of type of milking system				
					Loose housing				Stanchion		Parlor		
Area	No. farms	Cows per farm	Lbs milk per cow	Stanchion	Loafing	Free-stall	Corral or dry lot	Other	Bucket	Pipeline	Herringbone	Side-opening	Other
Northeast	37,648	48.1	10,810	71	11	14	—	4	65	17	12	4	2
Lake States	60,561	37.9	11,110	80	6	11	—	3	71	15	9	3	2
Corn Belt	25,287	40.0	11,340	34	38	24	—	4	40	24	17	11	8
Appalachian	9,594	53.3	10,210	14	47	34	—	5	23	24	19	19	15
Delta	3,344	72.7	8,630	10	50	17	20	3	10	49	17	14	10
Southeast	2,240	157.3	9,390	15	36	22	24	3	12	36	28	14	10
Northern Plains	7,150	42.3	10,700	35	40	17	3	5	40	23	18	10	9
Southern Plains	3,985	81.7	9,470	13	43	7	33	4	11	44	22	13	10
Mountain	4,219	70.2	11,050	5	33	32	28	3	18	29	27	16	10
Pacific	5,410	144.6	12,030	17	17	35	28	2	12	40	22	17	9

	Percentage distribution of manure handling systems					Percentage distribution of silage crop storage system				Purchased feed—1969	
	Spread as:						Silage stored in				
Area	Solid	Liquid	Lagoon system	Apply on other farms	Other methods	Tons stored per cow	Conventional tower	Sealed tower	Bunker or trench	Tons commercial feed per animal	Percent of forages
Northeast	93	8	1	2	1	8.6	59	13	28	1.00	3
Lake States	94	7	—	2	1	8.4	78	12	10	.38	3
Corn Belt	92	8	2	2	2	7.5	67	15	18	.53	4
Appalachian	90	10	3	3	6	10.8	46	10	44	.49	6
Delta	66	8	15	6	11	7.7	32	12	56	.97	2
Southeast	68	11	16	7	10	12.7	42	10	48	1.68	6
Northern Plains	92	7	2	2	5	8.3	40	7	53	1.39	8
Southern Plains	72	10	8	9	10	4.1	23	9	68	1.15	35
Mountain	88	11	2	7	5	7.3	3	4	93	.67	31
Pacific	73	28	9	14	6	3.6	12	4	84	1.45	47

[a]No. of farms and purchased feed data obtained from census data. U.S. Dept. of Commerce, 1969 Census of Agriculture, Vol. V, Part 8, Special Reports. All other data obtained from Hoglund, C. R.: The U.S. Dairy Industry, Today and Tomorrow. Mich. Agr. Exp. Sta. Research Report 275, 1975.

Manure Handling System

Some combination of gutter cleaners, scrapers, stackers, and spreaders for handling manure in solid form is likely to remain the most popular method on dairy farms in the near future. This combination accounts for 90% or more of the systems used in the northern states, and at least two-thirds of the farms in the southern states and Pacific Coast region still use this system (Table 1.10). Liquid manure systems account for less than 10% of the systems in most areas, being most popular (28% of the farms) in the Pacific Coast region. Many producers who are considering installing a liquid system are continuing with their present spreader system until such time as pollution control regulations become more definitive.

Liquid holding ponds and lagoon systems are most common in Florida, Louisiana, Arizona, California, and New Mexico. Application of manure on land away from the dairy facility (often sold to another farmer) is common only in Arizona, California, and New Mexico, and to a lesser extent in Alabama, Florida, and Texas.

Feed Storage and Handling

The trend in recent years has been an increase in use of corn silage and the harvesting and feeding of more of the legume-grass mixtures as silage or haylage. In the Southwest, hay is still the major forage crop fed.

The heaviest feeding of silage occurs in the Appalachian and Southeast areas (Table 1.10). Sixty percent or more of the silage crop is stored in conventional tower silos in the Northeast, Lake States, and Corn Belt; more than 80% in Wisconsin. Interest in bunker-type storage increases as herd sizes expand and where large quantities of corn silage are fed. The break-even point in costs when comparing conventional tower silos and bunkers is approximately 1,000 to 1,200 tons. More than 70% of all silage crops in Maine, New Hampshire, Massachusetts, and Connecticut is stored in bunker silos.

Purchased Feed

Dairy farmers in the Lake States, Corn Belt, and Appalachian regions use mostly home-grown grains and buy comparatively little feed commercially. The South, West, and Northeast regions buy considerably more feed. It is only in the Pacific, Mountain, and Southern Plains regions that a high percentage of forages is purchased. Because of the heavy dependence on home grown feeds—both grains and forages—producers in the Lake States, Corn Belt, and Appalachian regions are better able to withstand periods of high feed costs.

1.10 Locating the Dairy Farm

Skilled producers can generally make a success of dairy farming and compete well with alternative agricultural enterprises wherever land is better suited to forage than to feed grains. A most important consideration in locating the dairy farm is the intensity or concentration of dairying present in the immediate area. Dairy farmers located in regions having concentrated cow populations are at a distinct advantage because:

1. More milk marketing alternatives are available.
2. More artificial insemination (AI) studs are available to service the area.
3. Services of competent and experienced veterinarians are more readily available.
4. Providing maintenance and repair of the milking system, feeding equipment, and other supplies are usually more convenient and reliable.
5. More likely to have strong Dairy Herd Improvement Associations (DHIAs) active in the community.
6. A supply of labor, interested and trained, is more likely to be available.
7. Credit and business management personnel are more familiar with, understanding of, and helpful to the producer's problems.

1.11 Choosing a Breed

The most important consideration in choosing a breed of dairy cattle to establish a dairy farm is the present and future milk market situation.

Market Considerations. Most areas of the United States favor breeds that produce the largest volume of milk. Undoubtedly, this preference explains why the Holstein breed has gained so much in popularity (47% of all dairy cattle in 1920; 90% now). In some areas, however, special milk markets have been developed, and Guernsey or Jersey breeds are quite common. Today's emphasis on protein, solids-not-fat, and cheese yield could change milk marketing patterns quite rapidly.

Personal preference. This aspect frequently plays a role in breed selection, especially if other members of the family have been in dairying. Personal satisfaction is extremely important to people, and most milk producers can be happy with good cattle of any breed.

Other factors to consider include:

Salvage value. The larger breed cows are worth considerably more when their useful life as milk producers has ended.

Suitability for meat. Half of the calves born are males; only a few are needed for breeding purposes. The selling of dairy calves for the veal or steer market, or the feeding of dairy beef, is of considerable economic importance on many dairy farms. The large size breeds are generally considered superior as meat producers although tenderness and taste evaluations of meat from the small breeds have been outstanding.

Calving difficulty. Although the newborn Jersey calf is quite large in relation to size of dam, this breed is generally recognized as having fewer problems with difficult calvings.

A summary of dairy breeds common in the United States, their estimated numbers, production, and percentage of animals bred artificially is shown in Table 1.11.

Registered or Grade Cattle

The producer with registered dairy cattle is engaged in, or hopes to be engaged in, two distinctly different enterprises: one, the production and selling of milk; two, the development and sale of breeding stock. To many, the challenge of developing a herd of superior animals, participating in breed association activities, and developing a reputation for being a "breeder" is what keeps them interested in dairying. The producer can find the purebred aspect a profitable and fascinating business, if he or she has the time (and finances) to carry out the exacting details necessary to achieve success. Not everyone should be a registered breeder.

The name, address, and Executive Secretary of the various breed associations are listed in Appendix Table 1-A.

Successful dairy farm operators, whether they own and cherish registered cattle or not, should have knowledge of and an interest in improving their skills in the breeding, feeding, and management of their cattle. These qualities are needed:

1. *Love for the dairy cow.* Unless people have a sincere love for a good dairy cow, they will never become great *breeders.* Unfortunately, too many have an obsession with the glamour activities, such as the show-ring and type classification but care little about the feeding, managing, and attention to the small details necessary for success.
2. *Ability to withstand disappointments.* Choice animals may be lost to the herd from injury, sterility, or death. The bull that had been counted on to develop the herd may not turn out as expected.
3. *Willingness to spend more time and finances.* Many time-consuming aspects to the registered business are not required of the grade cow producer, for example, (a) completing registrations and transfers carefully, (b) promptly answering inquiries, (c) developing a system for naming and identifying animals, (d) traveling and visiting other breeders' farms, and (e) assisting with many local, state, and national breed association activities.
4. *Advertising.* The registered breeder must work at establishing a name and informing his fellow milk producers that he has superior breeding stock available. Advertising is not limited to space in the breed magazine but includes hosting youth events, exhibiting at fairs and shows, enrolling in breed association sponsored "production testing" programs, and consigning superior animals to local or state sales.
5. *Integrity.* The registered business is based on the honesty of the breeder and the accuracy of the records. People whose word is as good as a "bond" are the ones making a success in breeding dairy cattle. Unfortunately, too many producers attempt to sell animals at the

TABLE 1.11 *Summary of Dairy Breeds in the United States*

Breed	Estimated % of total U.S. population	Estimated no. of living registered cows in U.S.	Avg. no. animals registered annually	Avg. production, U.S. Official DHI herds, 1981: No. herds / No. Cows			Proven sires available in AI January, 1984 (Pred. diff. '82)				Mature cow body weight (lbs)	Newborn calf weight (lbs)
				Milk	*%*	*Fat*	*No.*	*M*	*F*	*$*		
Ayrshire	1	85,000	10,500 females	427		18,456	16	+242	+7	+27	1,200	75
			475 males	12,226	3.9	481						
Brown Swiss	1	100,000	12,500 females	561		26,427	43	+383	+15	+50	1,500	95
			1,100 males	12,665	4.1	516						
Guernsey	2	164,000	20,500 females	985		54,954	33	+469	+15	+56	1,100	75
			750 males	11,222	4.7	523						
Holstein	90	2,824,000	353,000 females	32,404		2,781,387	543	+535	+18	+65	1,500	95
			29,000 males	15,480	3.6	562						
Jersey	5	370,000	46,200 females	1,486		105,193	58	+453	+18	+60	1,000	60
			2,000 males	10,608	4.9	516						
Milking Shorthorn	1	30,000	3,500 females	84		3,054	11	+408	+13	+48	1,250	75
			1,100 males	11,162	3.6	405						

Source: Figures compiled from various sources, including: 4-H Dairy: Calves & Heifers, Univ. of Minnesota Agr. Ext. Service Publ. B-10, 1975; various breed association annual reports; *DHI Letter*, ARS-USDA, *59*(1), 1983; and USDA-DHIA AI Sire Summary List.

"bottom of their herd" for foundation animals when really they should have been culled. Remember, a breeder is also a commercial milk producer, and both have cows that should be culled.

1.12 Summary

Success in dairy farming, as in any other business, depends upon the ability of the owner or manager to make the right decisions. An understanding of why we have a dairy industry, confidence in its future, and knowledge of what is required to become successful should encourage the student of dairying.

As U.S. dairy farms become fewer in number but larger in size, the level of management required to operate them profitably and successfully continues to rise. Successful dairy farm enterprises may be found in all regions of the United States, but it is important that the beginner select an area and breed of cattle that promise to be competitive 20 to 40 years later.

If the graduate of dairy science decides that farm ownership is not for him, many opportunities still exist within the industry to manage farms, to work in extension or research, or to become gainfully employed in agribusiness service or sales. Dairy farming and the support of the dairy farming industry are both a business and a way of life.

Review Questions

1. Are you optimistic or pessimistic about the future for dairy farming in your locality? List 5 reasons for your answer.
2. Study the major subject-matter areas and competencies indicated to be important in becoming a successful dairy farm herdsman, manager, or owner-operator (Table 1.3). List those in which you feel you already have achieved a satisfactory level of understanding. Make a separate list of those items in which you lack the knowledge needed to achieve success.
3. Compare your own consumption patterns of milk and dairy products with the national averages (Table 1.7). Then discuss with your parents or other elders how their utilization patterns have changed over the years. List the reasons why these changes have occurred.
4. Cite an example from your own experience, if possible, of a dairy farm similar to the majority within your region of the United States regarding herd size, production level, and type of housing, milking system, manure handling, and crop storage. Cite a second example of another farm quite dissimilar. Which do you prefer? Why?
5. Indicate your choice of breed of cattle, and state whether they would be registered or grade cattle, if you were to start dairy farming today. List your reasons for these decisions.

References

1. Phillips, R. W.: Factors favoring animal production. Proceedings—Second World Conference on Animal Production. American Dairy Science Assn., p. 15, 1969.
2. Mehren, G. L.: Keynote address. General Session. Proceedings—Second World Conference on Animal Production. American Dairy Science Assn., p. 10, 1969.
3. Reid, J. T.: A re-evaluation of the importance of forages. Mimeograph, Cornell Univ., 1975.
4. Oldfield, J. E., *et al.*: Ruminants as food producers, now and in the future. Council for Agricultural Science and Technology. Spec. Publ. No. 4, 1975.
5. Touchberry, R. W.: Efficiency of animal production potential of animal genetics. Proceedings—36th Minnesota Nutrition Conference, p. 12, 1975.
6. Orth, R.: Dairy farm competency listing. Master of Education thesis. University of Minnesota, 1975.
7. Nickerson, T. A.: Chemical composition of milk. J. Dairy Sci. *43:* 598, 1960.
8. U.S. Dept. of Agriculture: *Composition of Foods.* Agr. Handbook No. 8, 1963.
9. Laben, R. C.: Factors responsible for variation in milk composition. J. Dairy Sci. *46:* 1293, 1963.
10. Rook, J. A. F.: Variations in the chemical composition of the milk of the cow. Dairy Sci. Abstr. *23:* 251 and *23:* 303, 1961.
11. Hoglund, C. R.: The U.S. dairy industry, today and tomorrow. Michigan Agr. Exp. Sta. Res. Report 275, 1975.
12. Hoglund, C. R.: Some recent trends and projections for the dairy industry. Mimeograph. Michigan State University, 1976.

Suggested Additional Reading

Campbell, J. R. and Lasley, J. F.: *The Science of Animals That Serve Mankind.* New York: McGraw-Hill Book Co., 1975.

Campbell, J. R., and Marshall, R. T.: *The Science of Providing Milk for Man.* New York: McGraw-Hill Book Co., 1975.

CHAPTER 2

Milk Marketing and Pricing

2.1 Introduction

Dairy farming is a business, and the business does not stop at the farm gate. Success in dairy farming depends on profitable marketing, as much as it does on profitable production. Marketing is a critical component of the dairy farming business.

Fluid milk is unique among perishable foods in its combination of bulk (87% water), short storage life, high nutritional value, and centuries of traditional use in the home. As a result, an unlimited supply of pure, wholesome, fresh milk is required every day of the year to satisfy consumer demand. However, widely varying supply, demand, and competitive conditions exist between markets and over time, which in turn have caused widely fluctuating farm milk prices.

Due to this unique combination of supply-demand conditions, the dairy marketing industry has become one of the most regulated and controlled in the entire economy. Milk haulers, plants, distributors, and retailers are subject to a wide variety of regulations and government programs, including environmental regulation, regulation of hauling rates and procedures, Federal Milk Marketing Orders regulating about 81% of U.S. grade A farm milk prices, State Milk Marketing Orders regulating trade practices and farm wholesale and retail milk prices, USDA price support programs undergirding farm milk prices, and import quotas and duties regulating quantities of dairy products that can be imported. In addition, more than 15,000 health and sanitation jurisdictions in the United States check dairy plants, farms, and stores for compliance with sanitation and quality regulations.[1]

This combination of local, state, and federal sanitary regulations, and state and federal price controls makes milk marketing and pricing a complex but fascinating subject for study.

2.2 Supply

From a previous all-time high of 127 billion pounds in 1964, U.S. milk production decreased to 115.5 billion pounds annually in the period 1973–75, but then increased to a new all-time high of 138.9 billion pounds in 1983. In 1973–75 increased milk per cow almost exactly offset the decline in the number of milk cows, resulting in static milk production. However, production per cow increased over 9% in the 1979–83 period, and the number of milk cows increased over 3% resulting in a 15.5 billion pound increase in milk production. Average milk production per cow was 12,531 lb. in 1983, and the nation had 11.1 million milk cows (Table 2.1 and Figure 2.1).

TABLE 2.1 *Milk Production, Number of Cows, and Milk Per Cow, 1950–83*[a]

	Milk cows		*Milk production per cow*		*Milk production*	
Year	*Number million*	*1967 = 100 percent*	*Amount pounds*	*1967 = 100 percent*	*Amount billion pounds*	*1967 = 100 percent*
1950	21.9	163	5,314	60	116.6	98
1960	17.5	131	7,029	79	123.1	104
1970	12.0	90	9,747	110	117.0	99
1975	11.2	84	10,354	117	115.5	97
1979	10.7	80	11,488	130	123.4	104
1981	10.9	82	12,177	137	133.0	113
1983	11.1	83	12,531	142	138.9	117

[a]Source: Economic Research Service, USDA: *Dairy Situation*, DS-359, March, 1976, and DS-395, January, 1984: Shaw, C. N.: *Outlook for Dairy*, Economic Research Service, USDA, November, 1976; Statistical Reporting Service, USDA; *Milk Production*, February, 1983.

Future milk production will depend on the milk-feed ratio (pounds of concentrate ration equal in value to 1 lb of milk sold to plants), cull cow prices, the state of the general economy, and dairy legislation.

Seasonality of Milk Production

Recently, milk production has been lowest in February and highest in May (Figure 2.2). However, the seasonal spread between the low and high points has been decreasing—dropping from 54% in 1950 to only 18% in 1983.

The seasonal "swing" in 1983 average farm milk prices was $.70 per hundredweight, with a "low" price of $13.20 per hundredweight in June and July, and a "high" price of $13.90 per hundredweight in November. Seasonal price variations, improved feeding practices, and a more uniform monthly freshening schedule help account for decreasing seasonal variation in production. Since the monthly demand for dairy products is relatively constant, seasonal "swings" in milk production result in increased processing, storage, and marketing costs.

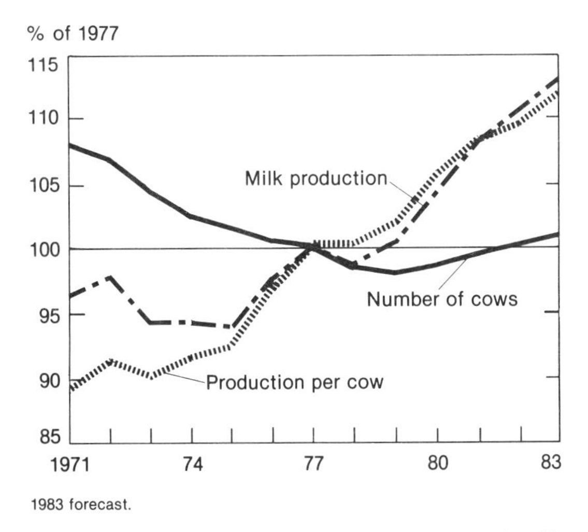

FIG. 2.1 Milk production, number of cows, and milk per cow. (Courtesy of USDA: *1983 Handbook of Agricultural Charts,* Agricultural Handbook No. 619, December, 1983)

2.3 Dairy Plants

Over three-fourths of the dairy manufacturing plants and seven-eighths of the fluid processing plants have closed in the past third century. Meanwhile, output per manufacturing plant has more than quintupled. All phases of the dairy processing industry

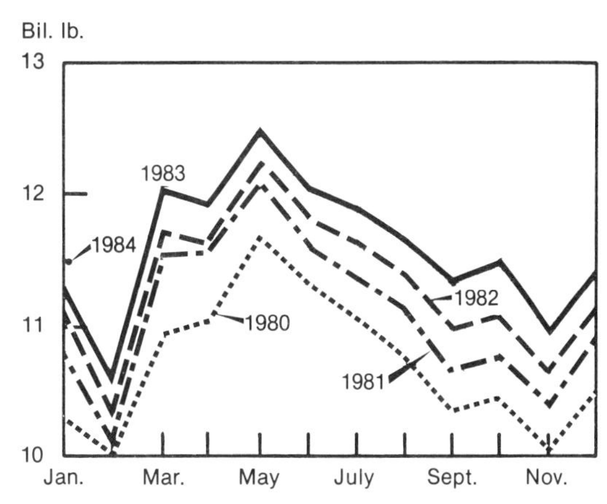

FIG. 2.2 U.S. milk production by months. (Courtesy of Economic Research Service, USDA: *Dairy Outlook and Situation,* DS-395, January, 1984)

have experienced these dramatic plant changes. Mergers, acquisitions, and "closing" are resulting in fewer and larger dairy processing plants.

The economic impact of environmental controls also affects dairy plant operations. A recent government study cited 12% closures of dairy plants due solely to pollution controls, between 1972 and 1977—almost half (49%) of total plant closures in that period. The report also cited consumer cost increases of 1.9 to 2.7 cents per pound of butter, 0.6 to 4.4 cents per pound of cheese, 0.8 to 4.7 cents per half gallon of ice cream, 0.4 to 5.5 cents per half gallon of milk, and 0.6 to 0.7 cents per can of evaporated milk to cover the cost of pollution controls.[2]

Many small milk distributors have ceased processing to become distributors for large companies, or have merged with other dealers to achieve larger volume and greater efficiency. The trend toward fewer and larger plants has generally resulted in lower unit costs, which helps increase farm milk prices. However, when larger plants result in less competition between plants and less bargaining power for dairy farmers, farm milk prices are hurt rather than helped. Farmers have substantially increased the size of their dairy cooperatives in recent years, in an attempt to deal with processing and marketing concentration.

2.4 Demand

Contrary to popular opinion, the United States ranks low among nations in per capita consumption of dairy products. In 1982 the United States ranked seventeenth out of 35 major countries in per capita commercial consumption of fluid milk and cream (217 pounds) and twenty-second in per capita consumption of butter (3.8 pounds). These amounts were only 44% and 11% respectively of the per capita consumption in the top country. The United States ranks near the bottom (28th) among nations in per capita consumption of nonfat dry milk, and 12th in per capita consumption of cheese.

Since 1925, U.S. per capita consumption of butterfat has dropped over one-third, compared with an increase of over one-sixth in per capita disappearance of milk solids-not-fat. This resulted in a 263

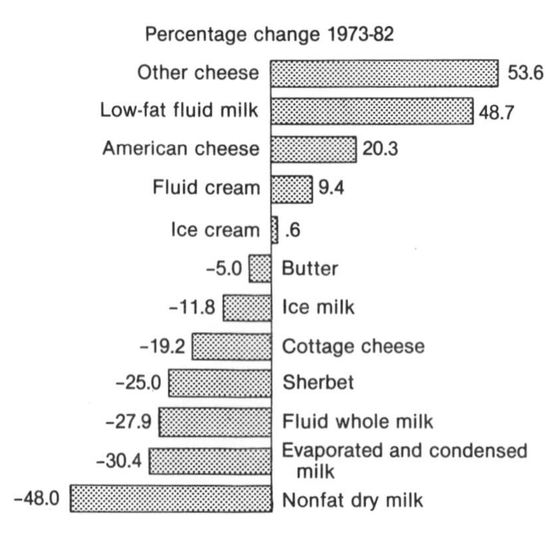

Fig. 2.3 Changes in per capita dairy product sales, 1973–1982. (Courtesy of USDA: *1983 Handbook of Agricultural Charts,* Agricultural Handbook No. 619, December, 1983)

pounds (32%) decline in per capita consumption of milk equivalent between 1925 and 1982—from 819 pounds to 556 pounds. In 1982, U.S. per capita consumption was lower for fluid whole milk, nonfat dry milk, cottage cheese, sherbert, ice milk, butter, and evaporated and condensed milk, than they were a decade before. Meanwhile, per capita sales were up approximately one-half for cheese, and low fat milk, and 5% for fluid cream (Figure 2.3).

This increase in cheese sales has been industry wide; this growth includes processed and various foreign types (Figures 2.4 and 2.5). As examples, per capita consump-

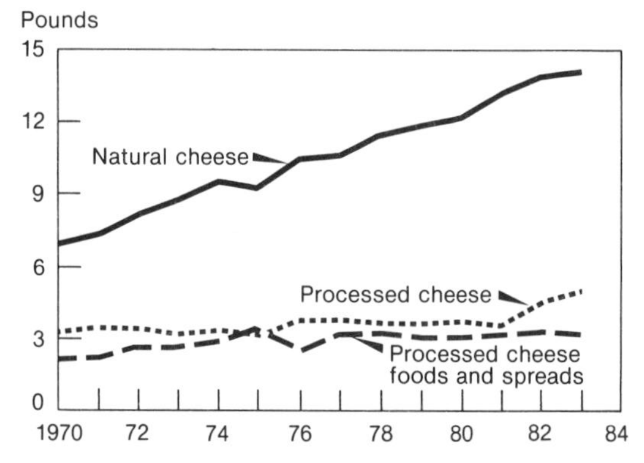

Fig. 2.4 Per capita consumption of processed and natural cheese. (Courtesy of Economic Research Service, USDA: *Dairy Outlook and Situation,* DS-397, June, 1984)

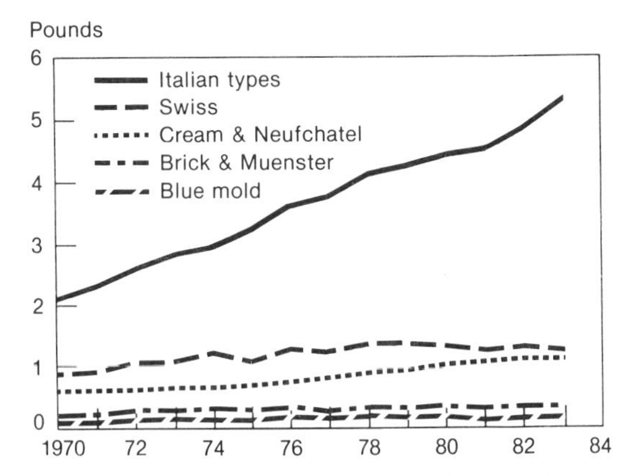

FIG. 2.5 Per capita consumption of other cheese. (Courtesy of Economic Research Service, USDA: *Dairy Outlook and Situation*, DS-397, June, 1984)

tion of processed cheese increased approximately one-third and consumption of Italian cheese three-fourths in the decade 1973–1982.

Cheese consumption will likely continue to increase because of:

1. Higher meat prices, which encourage cheese consumption as an alternate protein source.
2. Higher incomes; that is, a 10% increase in consumer incomes is associated with a 2 to 5% increase in cheese purchases.
3. Continued popularity of pizza, which stimulates use of Italian cheeses.
4. Increased "dining out," resulting in more drive-in restaurant and supper club business, all big users of cheese in foods such as cheeseburgers, snacks, dressings, and fondues.
5. More home patio entertaining, which involves a heavy use of cheese.
6. Continued increase in use of cheese as an ingredient in other foods.
7. Popularity of cheese foods and spreads.
8. Continued shift in consumer tastes to "exotic" foreign and specialty type cheeses.

However, the situation is serious with respect to butter, which had a 74% decline in per capita consumption in the 1940–1982 period, while margarine consumption increased 183%. Per capita sales of margarine are now more than 2½ times butter sales.

Based on current trends in consumer demand, the increase in per capita sales of low-fat and high solids nonfat dairy products, and the decrease in per capita sales of high fat products appear likely to continue because of increasing consumer infatuation with "weight watching," "low calorie diets," and "cholesterol levels." Substantial increases in promotion, merchandising, and nutrition information expenditures will be needed to reverse the downtrend in per capita consumption of high fat dairy products, if indeed it can be reversed.

Factors Influencing Demand

Price. The demand for dairy products is generally inelastic; that is, an increase in price is accompanied by a less than proportional decrease in consumption.

The demand for fluid milk items is more inelastic than the demand for manufactured products. For example, a national study involving household purchasing patterns for dairy products reveals the following short run decreases in purchases by consumers, with a 10% increase in price: natural cheese—8.5%, butter—7.3%, ice cream—6.9%, nonfat dry milk powder—4.5%, cottage cheese—4.3%, total fluid milk—1.4%, and liquid skim milk—1.2%.[3]

Increased dairy marketing costs result in relatively high prices for manufactured dairy products. This increase in price causes decreased consumer purchases, or the substitution of another food item that costs less (for example, margarine for butter). This situation in turn results in increased dairy surpluses and therefore lower farm milk prices. The decreases in consumer purchases associated with increased prices emphasize the importance of maximum marketing efficiency.

Age and Sex. Per capita consumption, and the proportion of people drinking milk, decrease with age. Under 6 years of age, 95% of both sexes drink an average of 1.5 pints of milk daily. Among teenagers, boys continue to consume an average of 1.5 pints of milk daily, but girls drink only 0.93 of a pint. Average daily milk consumption for people over 35 years old is less than 0.5 of a pint.[4] Convincing teenage and older females and post-teenage males that continuation of their earlier milk drinking habits is nutritionally advantageous would immensely increase milk consumption.

Level of Income. Milk consumption increases with income. One nationwide study reported that a $5,000 increase in annual income was associated with a 10% increase in the number of people drinking milk, and an increase of 3½ ounces per day in per capita milk consumption.[4] Inflation, unemployment, and reduced per capita disposable income discourage consumers from purchasing dairy products, and encourage the purchasing of lower priced substitutes.

Carbonated Beverages. An increase in the consumption of carbonated beverages normally results in decreased consumption of fluid milk items. One-third of the children, and one-half of the teenagers of both sexes drink carbonated beverages regularly at mealtime and between meals. Convincing children and teenagers that milk is "better for them" and also a more "fun drink" could help dairymen increase sales of their products, and hence their incomes. Competing with carbonated beverages is a major challenge confronting dairymen.

Distribution Method. Home delivery of milk has all but ceased. Only 2% of fluid milk sold in Federal Order markets was home delivered in 1982, compared with 10% a decade earlier. Greater emphasis by dairies on large volume home delivery customers could help reduce unit delivery costs, and would enable "home delivery" to compete more effectively with "store sales." This could help increase consumption of dairy products.

Type and Size of Container. Glass bottles have been largely replaced by paper and plastic containers, with 39 and 60%, respectively, of 1982 Federal Order milk packaged in these containers. Plastic containers in particular have had a rapid rise—only 25% of the Federal Order milk was packaged in this type of container a decade previously.

In 1982, over three-fourths (82%) of Federal Order milk was packaged in gallon and half gallon containers. Only 5% of Federal Order milk was still packaged in quart containers—down almost one-half in the decade.

The shift to larger sized paper and plastic containers has increased sales of fluid milk because of greater consumer convenience associated with these containers.

Promotion of Milk and Dairy Products. The dairy industry ranks low among American industries in product promotion. Market research indicates that 90% of all fluid milk items is consumed by only 45% of the population.[4] A larger potential market for milk therefore exists among more than one-half the population.

New York, which initiated a mandatory dairy promotion "check-off" on farm milk in 1972, found that the increase in farm value of milk sold exceeded advertising costs by 10.5 cents per capita in New York City and by 2.8 cents in Albany. The net loss from advertising for Syracuse was 3.8 cents per capita.[5] Demographic and institutional variations had a bearing on the effectiveness of dairy advertising, but overall it was very profitable. In another study conducted in New York City from July 1976 to June 1977, milk sales rose 4.9% at 8.5¢ per capita advertising, compared to what they would have been with no advertising. Dairy farmers therefore earned $2.69 for each dollar invested in dairy promotion—169% net return. The study also determined that in 10 major U.S. markets during 1979, 10.4¢ per capita advertising, increased milk sales 3.6% compared to what they would have been with no advertising. Dairy farmers therefore received $2.20 for each dollar invested in promotion—a 120% net return. On this basis dairy promotion could result in a net increase to farmers of $400 million annually.[6]

With returns of this magnitude, annual expenditures of about $31 million in 1982 by dairy farmers for milk promotion seems small when viewed against 1982 farm milk sales of $18.3 billion, and approximately $689 million spent by the carbonated and noncarbonated drink industry, and $1.03 billion spent in the alcoholic industry promoting its products in 1982. Milk was therefore outspent in promotion 55 to 1 by other beverages.

Dairy promotion assessments on farm milk were first authorized in Federal Milk Order markets in 1971. As of January 1984, assessments were in effect in 6 order markets. The assessment rate in the 6 markets varied from 10 to 14 cents per hundredweight, and approximately $13 million was collected for dairy promotion in 1982. Nineteen State Milk Orders, representing about two-thirds of U.S. milk production also had dairy promotion assessments. In addition to Federal and State Milk Order assessments, the 1983 Dairy Law provided for a 15 cents per hundredweight dairy promotion assessment on farm milk during 1984, and January-September 1985, with a farm referendum option for continuation after then, and with credits for most State and Federal Milk Order assessments. Together these various dairy promotion assessments total about $200 million annually—including approximately $140 million new money annually from the 15 cents per hundredweight promotion assessment mandated by the 1983 Dairy Law. This is a substantial increase over previous dairy promotion funding, but still substantially less than competitive products are spending on promotion. The dairy industry seemingly still has a long way to go in maximizing returns from dairy promotion.

Filled and Imitation Products. Filled milk is prepared by blending a vegetable fat or oil with skim milk; the vegetable fat or oil serves as a replacement for milk fat. Imitation milk contains no dairy products, and it is prepared by blending together water, vegetable fat or oil, vegetable protein or sodium caseinate (interpreted by the Food and Drug Administration (FDA) as nondairy), sugars, minerals, flavoring, some vitamins, emulsifiers, and stabilizers. Filled milk sales are legal in a large number of states. Filled milks can also be sold in interstate commerce, since the FDA did not challenge a federal district court ruling approving its movement in interstate commerce. Sales of properly labeled imitation milk are not prohibited by federal law, but some states have laws that prohibit its sale within their borders.

In the late sixties, filled and imitation milk appeared to present a major challenge to fluid whole milk consumption especially in the Southwest, on the Pacific Coast, and in Hawaii. Sales of these substitute products decreased in the early seventies, and were negligible or nonexistent in U.S. markets in the early eighties. However, consumption of filled and imitation milk may increase because of improved product stability and taste, increased emphasis on the products by the trade, retail price increases for milk, and recent court rulings.

Changes in FDA standards of identity are increasing the competitiveness of vegetable fat ice cream. Federal standards of identity for vegetable fat ice cream now permit the name "mellorine" on interstate sales, which is probably more appealing to consumers than the former name "imitation ice cream," and therefore more competitive with ice cream—and milk fat. A federal standard of identity has also been established for filled evaporated milk increasing the competitiveness of this product.

A federal court ruling in the early 1970s legalized the interstate shipment of filled cheese. In 1974 President Ford signed a bill repealing the Filled Cheese Act of 1896, which had imposed various taxes on manufacturers and distributors of this product. These developments will likely increase competitiveness of this substitute dairy product.

Both the Carter and Reagan Administrations refused to impose import quotas on subsidized imported casein, which was selling for about one-half the price of domestic nonfat solids of comparable protein content, and in turn used in the manufacture of imitation cheese.

As a result of these various actions or inactions, imitation cheese production was about 5% of total cheese production in the early eighties, with 16–32% lower average retail prices for imitation cheese than natural cheese. Also average retail prices for pizza with imitation cheese were 14% lower than pizza with only natural cheese, and pizza with imitation cheese claimed over one-half (57%) of the shelf space in major supermarkets.[7]

2.5 Milk Prices

Grade A farm milk prices have been averaging in excess of $1.00 per hundredweight more than manufacturing grade prices (Table 2.2). With differentials of this magnitude, rapid conversion of the 15% grade B milk in 1982, to grade A farm milk is likely to occur.

Butterfat Differentials

Butterfat differentials are the amount by which the price of farm milk is increased or decreased for each "point" (0.1%) of butterfat test. The procedure used to calculate butterfat differentials in most Federal Milk Order markets (and generally copied in non-order markets) is to multiply the aver-

TABLE 2.2 *Farm Prices for Milk of Average Butterfat Test, 1960–1983[a]*

Year	*Grade A*	*Manufacturing grade*	*Difference*
1965	$ 4.63	$ 3.34	$1.29
1970	6.05	4.70	1.35
1975	9.02	7.63	1.39
1980	13.21	12.05	1.16
1981	13.94	12.73	1.21
1982	13.73	12.66	1.07
1983	13.72	12.62	1.10

[a]Source: Economic Research Service, USDA: *Dairy Outlook and Situation*, DS-395, January, 1984.

age wholesale price for 92 score butter in Chicago by 0.115 (overrun allowance). For example, the USDA support price for U.S. grade A or higher butter at Chicago effective December 1, 1983, was $1.4325 cents per pound. At this price the butterfat differential for farm milk would be approximately 16.5 cents (1.4325 × 0.115).

With this procedure, differentials are based solely on the value of butterfat, with no allowance for the value of solids-not-fat. Use of butterfat differentials to price farm milk assumes fat and solids-not-fat prices and content fluctuate together. Since these two assumptions are frequently not valid, criticism of butterfat differentials has been voiced by many in the dairy industry.

Component Pricing

Milk fat and protein are the two major variable constituents in milk. Lactose remains fairly constant at about 5%, and minerals at about 0.7%, but protein and fat vary considerably seasonally and between breeds and herds. On the average, farm milk contains about 3.7% butterfat and 8.55% solids-not-fat, including about 3.2% protein. A one point (0.1%) change in milk fat test is normally associated with a 0.4 point (0.04%) change in solids-not-fat, and in protein, since protein is the major variable constituent within the solids-not-fat portion. However, considerable variation from this average relationship does occur.

Component pricing takes into consideration the value of variations in solids-not-fat as well as fat; therefore some advocate it to correct what they feel is an inequity to dairy farmers under current butterfat differential pricing. Their contention is that milk has been priced on the basis of its other major variable constituent fat, for over 80 years, so why not also solids-not-fat, or protein?

Major issues that advocates see with respect to component pricing are:

1. Without component pricing, farmers who shipped milk with higher nonfat solids and protein did not get full credit for the added value of nonfat constituents in the higher fat test milk. For example, the skim milk portion of fluid milk constitutes over one-half of the total value of the milk (56% in the Chicago Federal Milk Order Market in December 1983). Yet the skim milk portion receives no credit in computing butterfat differentials.
2. Additional protein results in higher yields of skim milk powder, cheese, and other high protein dairy products. These added yields provide the extra money needed for component pricing, with little or no new net outlay by plants.
3. Consumers are becoming more protein conscious, and milk and dairy products provide consumers with relatively low cost protein. Emphasizing this aspect can offset the decline in milk equivalent sales attributable to a decline in consumer demand for butterfat. While many other components in dairy products are valuable, protein appears to have more "glamour," is readily recognizable by consumers, and is therefore the most marketable.

Because of these valid criticisms of "butterfat differential" pricing, various forms of "component pricing" were being formulated and adopted by early 1984. These component pricing plans continued to give price credit for butterfat in farm milk, but in addition gave credit for solids-not-fat including protein. Some also involved "end product" pricing in which farm milk prices are established on the basis of the yield and market value of cheese or other dairy products that can be manufactured from the milk. Most also established maximum somatic cell counts at which "component" premiums would be paid, because of decreasing product yields with increasing somatic cell counts, even though solids-not-fat and protein content continued to increase.

The National Milk Producers Federation representing virtually all U.S. dairy cooperatives developed a component pricing plan in early 1984, for consideration by its members. Also virtually all Wisconsin dairy cooperatives which market or represent about 90% of Wisconsin milk, had either adopted or formulated component pricing plans by early 1984, paying a premium of about 10 cents for each point (.1%) of protein above average, and either a "quality" premium of 6 to 12 cents per hundredweight of milk for minimum somatic cell counts, or establishing maximum somatic cell counts at which component premiums would be paid. As a result of these "component" and "quality" premiums, many farmers were receiving an additional 30 cents or more per hundredweight for their milk.

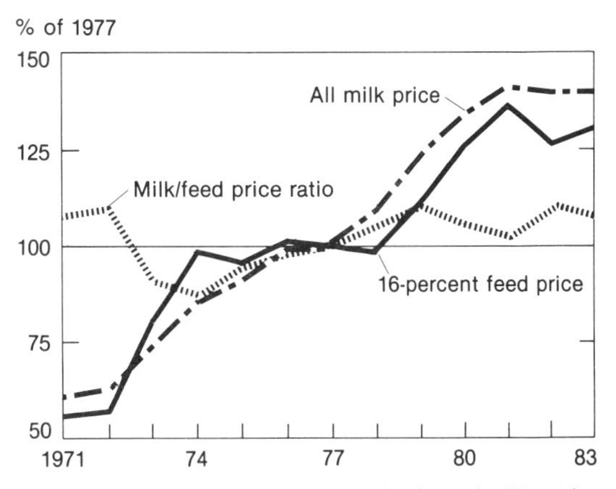

1983 forecast. 16% feed price is the average paid by farmers for 16% protein dairy concentrate. Milk/feed price ratio is the pounds of 16% feed equal in value to 1 pound of milk sold to plants.

Fig. 2.7 Milk-feed price relationships. (Courtesy of USDA: *1983 Handbook of Agricultural Charts,* Agricultural Handbook No. 619, December, 1983)

Milk-Feed Price Relationships

Average farm milk prices have more than doubled since the early 1970s, increasing from an average of $5.71 per hundredweight in 1970, to over $13.00 per hundredweight in 1983 (Figures 2.6 and 2.7). However, this increase does not necessarily mean that dairy farming is now more profitable. The milk-feed ratio is a key measure of the relationship between changes in farm milk prices and farm costs, since it represents the pounds of 16% protein dairy concentrate equal in value to 1 pound of milk sold to plants.

This ratio did not change appreciably during the period in which farm milk prices were making their spectacular rise (Figure 2.7). For example, in 1983 1 lb of farm milk bought 1.45 lb of concentrate—only 5% below the 1970 ratio of 1.53 (Table 2.3). The decline in the milk-feed ratio during the period in which milk prices were rising sharply emphasizes the importance of highly efficient milk production practices, regardless of farm milk prices.

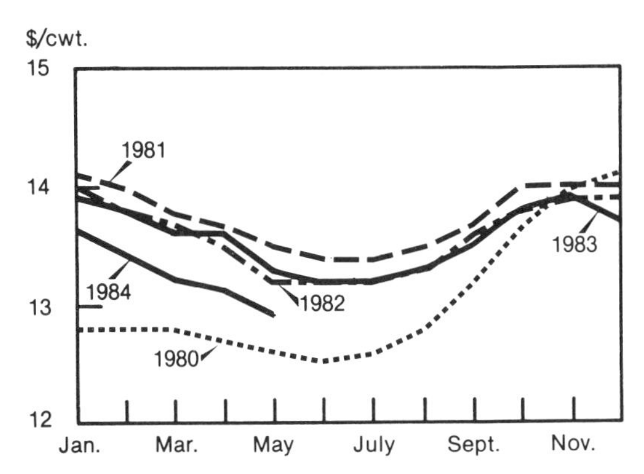

Fig. 2.6 Milk prices. (U.S. average price received by farmers for deliveries to plants and dealers.) (Courtesy of Economic Research Service, USDA: *Dairy Outlook and Situation,* DS-397, June, 1984)

Dairy–Nondairy Price Relationships

Retail dairy prices lag price rises for other foods and consumer products. In the 1967–July 1983 period, retail prices for all consumer items rose 197% and for all foods 192%—47 and 42 percentage points more than the dairy price rise of 150%. Retail dairy prices consistently lagged other consumer items throughout the 16

Table 2.3 *Milk-Feed Price Ratios, 1970–1983*[a]

Year	*Pounds*	*Milk-feed price ratio*[b] *1970 = 100*
1970	1.53	100
1975	1.31	85.6
1980	1.48	96.7
1981	1.44	94.1
1982	1.53	100
1983	1.45	94.8

[a]Source: Economic Research Service, USDA: *Dairy Situation* DS-374, March, 1979 and DS-395, January, 1984.
[b]Pounds of 16% protein dairy concentrate equal to one pound of milk sold to plants.

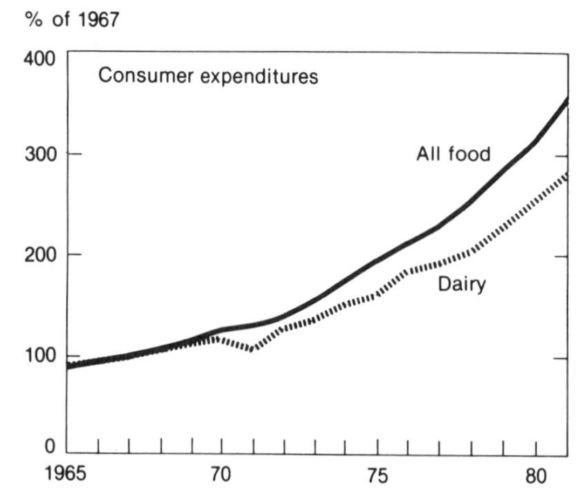

FIG. 2.8 Retail price indices. (Courtesy of USDA: *1982 Handbook of Agricultural Charts,* Agricultural Handbook No. 609, November, 1982)

year period (Figure 2.8). Dairy farmers have not been as successful in achieving higher prices for their products as have producers and processors of other commodities and products.

Wages are going up faster than milk prices; for example it took 15.8 minutes at average wage rates to earn enough to buy a half gallon of milk at grocery stores in 1950, compared with 7.9 minutes in 1982. Therefore, the "real" price of milk in terms of the amount of work required to earn the money to buy it decreased 50% between 1950 and 1982. The decrease in the "real price" of other dairy products during the same period of time was: American Cheese 21%, evaporated milk 29%, butter 51%, and ice cream 58%.

Marketing Margins

Marketing fluid milk is an expensive complicated process because of processing, sanitation, refrigeration, packaging, transportation, and distribution costs, and the many processes to which milk is subjected. As a result, dairy farmers receive only about one-half (52% in 1982) of the "consumer's dollar" for fluid milk. Assembling milk took 4 cents of the consumer's dollar, and processing, wholesaling-transportation, and retailing almost equally divided the other 45 cents (Table 2.4).

The farmer's share of the consumer's dollar for fluid milk has remained relatively low, and shown little change over the years. In 1975 it was 54%, and in the 3-year period 1979–81 it averaged 53%, even though the farm to retail price spread increased from 35.7 cents in 1975, to 52.4 cents in 1981.

TABLE 2.4 *What the Consumer's Fluid Whole Milk Dollar Pays For*[a,b]

Production	52¢
Assembly	4¢
Processing	13¢
Wholesaling and transportation	16¢
Retailing	15¢

[a]Source: Economic Research Service, USDA: *Dairy Outlook and Situation,* DS-392, March, 1983.
[b]Based on 1982 prices, costs, and margins.

The farmer's share of the consumer's dollar for butter has also been quite low and static over the years, averaging 65% in 1975, and 66% in the three year period 1979–81, even though the farm to retail price spread increased from 36.2 cents in 1975, to 66.8 cents in 1981 (Table 2.5).

The farmer's share of the consumer's overall dairy dollar was 52 cents in 1982, compared to 50 cents for meat products, 51 cents for poultry products and 63 cents for eggs (Figure 2.9).

The milk dealer assembles, processes, and distributes fluid milk to homes or retail stores, for which he received 33 cents of the consumer's milk dollar in 1982 (Table 2.4). Raw materials accounted for 65% of the milk dealer's cost, delivery and sales expenses took 15%, and processing costs also took 15%. Net profit after taxes was less than 2% of the milk dealer's dollar (Table 2.6).

2.6 Price Supports

The farm prices for milk, and the wholesale and retail prices for all dairy products are greatly affected by the government dairy price support program, which until recently have been based on the parity principle. The term parity comes from the Latin word "paritas," meaning equal. As used in government agricultural programs, parity refers to the relationship between the prices received by a producer for his products and the prices he has to pay for equipment, labor, and supplies to produce those products, with 1910–14 farm product

TABLE 2.5 *Marketing Margins and Farmer's Share of the Consumer's Dollar, for Various Dairy Products, 1975–81*[a]

	Farm to retail spread (¢)				Farmer's share of consumer's dollar (%)			
DAIRY PRODUCTS	*1975*	*1979*	*1980*	*1981*	*1975*	*1979*	*1980*	*1981*
Butter	36.2	57.0	62.3	66.8	65	66	67	66
Fluid Whole Milk (1/2 gal)	35.7	44.0	49.1	52.4	54	54	53	53

[a]Source: Economic Research Service, USDA: *Developments in Farm to Retail Price Spreads For Food Products,* Agricultural Economics Report, No. 449, March, 1980, and No. 488, September, 1982, and No. 500, May, 1983.

prices representing 100% of parity. Thus, the percent of parity farmers receive at any later date depends on whether prices they pay have gone up more (or less) since 1910–14 than prices they receive. If prices paid since 1910–14 went up more than the prices received, farmers got less than 100% of parity, and vice versa. Thus, "parity prices" give a unit of a farm commodity (such as milk) the same purchasing power in terms of the goods and services farmers buy as it had in the 1910–14 base period.

The Agricultural Act of 1949, although amended several times since then, still provides the permanent authority for the milk price support program. This law requires the Secretary of Agriculture to support the price of milk at between 75 and 90% of parity as of the beginning of the marketing year (October 1). The Secretary cannot decrease the support price during a marketing year, but can increase it up to the 90% of parity maximum if he feels it is necessary to assure adequate milk supplies and farm income. However Congress temporarily amended the permanent legislation several times in the early 1980's in an attempt to encourage a reduction in milk production. The most substantive revision covered the period December 1, 1983–October 1, 1985. This revision reduced the support price $.50 to $12.60 per hundredweight for milk of average butterfat test, irrespective of parity levels, for the period December 1, 1983 to April 1, 1985. The legislation also permitted another $.50 per hundredweight decrease in support prices during April 1, 1985–July 1, 1985; and a further $.50 per hundredweight decrease during July 1–October 1, 1985 if annual price support purchases were projected to exceed 6 and 5 billion pounds of milk equivalent respectively commencing April 1 and July 1, 1985. The legislation also provided for $10 per hundredweight payments during January 1, 1984–April 1, 1985 to farmers who reduced milk production 5 to 30% below their base (either their 1981–82 or 1982 average production). The $10 per hundredweight payments were largely financed by a $.50 per hundredweight "assessment" on all farm milk

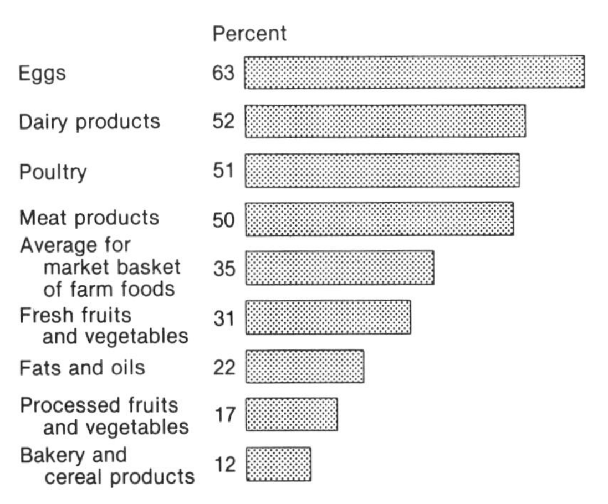

FIG. 2.9 Farmer's share of the market-basket dollar, by food group in 1982. (Courtesy of USDA: *1983 Handbook of Agricultural Charts,* Agricultural Handbook No. 619, December, 1983)

TABLE 2.6 *What becomes of the Milk Dealers' Gross Income Dollar?*[a]

Item	*Percent*
Raw product costs	64.5
Delivery and sales expense	14.5
Processing costs	14.6
Administrative expenses	3.4
Income taxes	1.5
Net profit[b]	1.5

[a]Source: Milk Industry Foundation: *Milk Facts,* 1983 ed., Washington, D.C., July, 1983.
[b]Profit after income taxes.

sales. A $.15 per hundredweight "dairy promotion assessment" was also authorized during the 1984–October 1, 1985 period, with continuation after that dependent on an affirmative vote by dairy farmers in a referendum.

Nevertheless unless additional new legislation is enacted prior to October 1, 1985, milk support prices will again be based on parity after that date, since permanent legislation—the Agricultural Act of 1949, requires this.

Calculation of Milk Parity and Support Price

The Secretary of Agriculture announced an 80% of parity support price of $13.10 per hundredweight for manufacturing grade milk of average butterfat test, for the marketing year commencing October 1, 1980. This was the last time support prices were based on parity at least through the first half of the 1980s. Thereafter during this period support prices were set administratively without regard to parity.

The parity procedure and calculations used to arrive at the $13.10 per hundredweight support price are:

1. Compute the most recent 10 calendar year (1970–1979) average of "all milk wholesale" prices ($8.40 per hundredweight), and divide it by the most recent 10 calendar year (1970–1979) average index of prices received by farmers (439-base period 1910–14 = 100) to obtain the Adjusted Base Price ($1.91 per hundredweight). The Adjusted Base Price is what farmers received for milk in the 1910–14 period adjusted for changes in prices received for all farm commodities since then.

$8.40 per hundredweight ÷ 439
= $1.91 per hundredweight
(Adjusted Base Price)

2. Multiply the Adjusted Base Price ($1.91 per hundredweight) by the current index of prices paid by farmers for goods and services (972-base period 1910–14 = 100) to obtain the 100% of parity price ($18.60 per hundredweight) for all milk wholesale as of October 1, 1980.

(1.91 per hundredweight × 972
= $18.60 per hundredweight
(100% parity for all milk wholesale)

3. Multiply the 100% of parity price for milk wholesale ($18.60 per hundredweight) by the most recent 10 calendar year (1970–1979) average ratio of "manufacturing grade milk prices" to "all milk wholesale prices" (88%), to obtain the 100% parity equivalent price ($16.37 per hundredweight) for manufacturing milk. It is necessary to calculate the 100% parity equivalent for manufacturing milk, since only manufacturing milk prices are supported.

$18.60 per hundredweight × 88%
= $16.37 per hundredweight
(100% parity equivalent price for manufacturing milk)

4. Multiply the 100% parity equivalent price for manufacturing milk ($16.37 per hundredweight) by 80% (the percent of parity support announced by the Secretary of Agriculture) to obtain the support price ($13.10 per hundredweight) for manufacturing milk.

$16.37 per hundredweight × 80%
= $13.10 per hundredweight
(80% of parity support price for manufacturing milk on October 1, 1980)

By October 1, 1983 the $13.10 per hundredweight support price was only 65% of parity—15 percentage points less than it had been 3 years earlier. This decrease in the "percent of parity" for the $13.10 per hundredweight support price reflects the increased costs of production during the 3-year period October 1, 1980–October 1, 1983.

Product Prices

Milk price supports are implemented through USDA open market purchases of butter, cheddar cheese, and nonfat dry milk, at prices that, with average yields and processing-marketing costs for these products, enable dairy plants to pay farmers the support price. By providing an outlet for excess quantities of these 3 major dairy products, market prices of all manufactured dairy products are maintained, because of competition among plants for milk supplies.

The USDA purchase prices for butter, nonfat dry milk, and cheddar cheese are calculated from the support price for milk. The procedure used to determine the purchase prices for the three products of $1.49 per pound for butter, $.94 per pound for nonfat dry milk, and $1.395 per pound for cheddar cheese, during the period October 1, 1980—December 1, 1983 was as follows:

One hundred pounds of milk were estimated to yield 4.48 pounds of butter, and 8.13 pounds of nonfat dry milk.

Adding the USDA's estimate of an appropriate manufacturing allowance of $1.22 per hundredweight of milk for the cost of turning milk into butter and nonfat dry milk, to the farm milk support price of $13.10, equates to a gross price to plants of $14.32 per hundredweight of milk.

USDA purchase prices for butter were set at $1.49 per pound.

4.48 lbs butter × $1.49
= $6.68 butter value in 100 lbs. milk.

$14.32 − $6.68 = $7.64 ÷ 8.13
= $.94 per pound USDA purchase price for nonfat dry milk.

One hundred pounds of milk were estimated to yield 10.1 pounds of cheddar cheese, and ¼ pound of butterfat (from the whey cream).

Adding the USDA's estimate of an appropriate manufacturing allowance of $1.37 per hundredweight of milk for the cost of turning milk into cheese, to the farm milk support price of $13.10, equates to a gross price to plants of $14.47 per hundredweight of milk. Subtracting $.38 for the butterfat in the whey cream, equates to a gross price to plants of $14.09 per hundredweight of milk for cheese.

$14.09 ÷ 10.1 = $1.395 per pound USDA purchase price for cheddar cheese.

Support and Farm Milk Prices Compared

Dairy plants are not legally required to pay farmers the support price for manufacturing milk. However, USDA purchase prices are set at levels that are estimated to permit plants with average costs and yields to pay farmers at least the support price. As a result, farm prices for manufacturing milk were at or above the announced support level in 13 of the 15 years between 1961 and 1976, and above the minimum 75% of parity support levels in all 15 years. Manufacturing milk prices paid to farmers were only 2 cents per hundredweight below support prices in each of the 2 years they failed to reach support levels. Since 1976 the situation reversed, with farm prices for manufacturing milk being below support prices in 6 of 7 years from 1977–1983. Plants contended manufacturing allowances used in calculating USDA purchase prices for butter, nonfat dry milk, and cheese, during that time were generally less than actual costs, and therefore plants could not afford to pay farmers the support prices. Controversy therefore existed throughout the period with respect to achievement of support levels.

Volume of Price Support Purchases

Price support levels for manufacturing milk have been achieved in the decade 1967–1976 with USDA purchases of less than 7% of the milk fat and less than 8% of the solids-not-fat produced in any 1 year. In the 3 years 1974–76, USDA purchases were less than 2% of milkfat and less than 5% of the solids-not-fat produced in either year. However, in the early 1980s, substantial USDA purchases were necessary to maintain farm milk prices at the price support level, reaching 10.5% of milk production in 1982, and 12% in 1983 (Figure 2.10). USDA price support purchases were 14.3 billion pounds milk equivalent in 1982, and 16.8 billion pounds in 1983.

Net government expenditures on dairy price support purchases averaged $231 million annually in the quarter century 1950–1975—approximately 4.2% of average gross farm income from the sale of milk during that period of time. However, net price support costs escalated sharply in the early 1980s because of large milk surpluses, reaching $2.2 billion in fiscal year 1982, and $2.6 billion in fiscal year 1983—in excess of 13% of farm cash receipts from the sale of milk (Figure 2.11). These substantial price support expenditures in turn brought on the reduced support prices, farmer assessments, and payments for reducing milk production, discussed in previous sections.

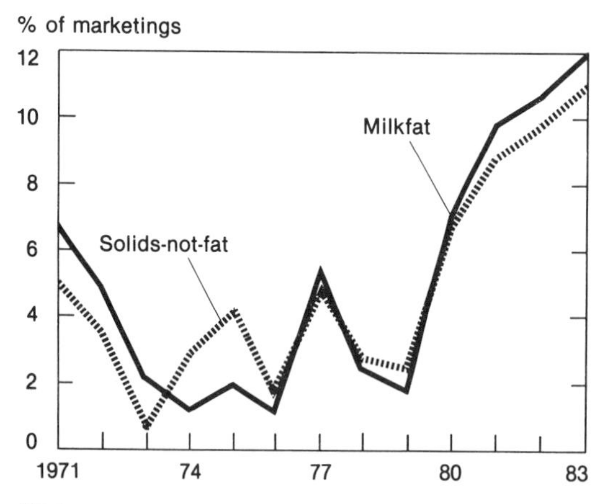

Fig. 2.10 Milk solids removed from the market by CCC programs. (Courtesy of USDA: *1983 Handbook of Agricultural Charts,* Agricultural Handbook No. 619, December, 1983)

2.7 Federal Milk Marketing Orders

These orders are legal instruments authorized by the Agricultural Marketing Agreement Act of 1937, and issued by the U.S. Secretary of Agriculture at the request of a majority of the dairy producers. They establish minimum grade A farm milk prices within specific geographic areas designated as "markets," which may cover portions of several states, therefore involving interstate commerce. Federal Milk Orders regulate only grade A milk prices—they do not regulate grade B milk prices. Federal Milk Orders establish minimum not maximum farm milk prices.

The administrator of each order: (1) announces minimum prices to be paid to producers by handlers, (2) collects reports from handlers showing milk receipts and quantities used in each type of use or dairy product, and (3) receives from each handler reports of all payments to producers. In addition, the administrator verifies handlers' reports by regular audits and, in most markets, verifies milk weights and butterfat tests. Administrative expenses are covered by an assessment generally ranging from 2 to 5 cents per hundredweight of milk, which is paid by handlers.

Milk handled in Federal Order markets increased fivefold, and average deliveries per producer increased sixfold, in the period 1950–1982. Producer deliveries to Federal Milk Order markets in 1982 were 92 billion pounds—81% of all grade A milk. Gross value of this milk was $12.6 billion—69% of farm cash receipts from dairy products.

Improved transportation, sanitation, and refrigeration facilities that increase the mobility of milk, and pressures for more geographic balance of farm milk prices, have encouraged mergers of Federal Milk Orders. As these trends continue, more Federal Orders will cover several states, and even regions in the future. The author and others predict the number of Federal Milk Orders will eventually decline to 5 to 10 regional orders.[8] The number of Federal Milk Orders declined from a high of 83 in 1962, to 45 as of January 1, 1984.

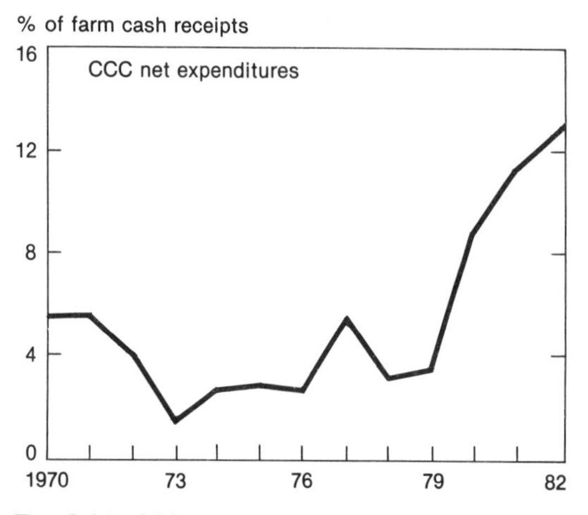

Fig. 2.11 CCC net expenditures as % of farm cash receipts from marketings of milk and cream. (Courtesy of USDA: *Handbook of Agricultural Charts,* Agricultural Handbook No. 619, December, 1983)

Classified-Use System

Federal Milk Orders use the "classified-use" system in determining "blend" prices to be paid producers. Dealers are obligated to the market administrators "pool" for the class prices on the quantities of milk they utilize in the various classes. Dairy producers in turn receive a blend price, which is a weighted average of class prices and the volume of milk used in the various classes.

Class I Milk. In addition to fluid whole milk, most markets categorize the following as class I: low-fat milk, skim milk, buttermilk, flavored milk drinks, and other uses for which grade A milk is required by health departments.

Class II Milk. Class II milk usually includes frozen desserts, dips, cottage cheese, fluid cream products, sour cream, and yogurt.

Class III. Milk in this class is used in "hard" manufactured products such as butter, cheese, and nonfat dry milk. However the U.S. Supreme Court agreed to review a lawsuit in the mid 1980's involving the issue of whether fluid milk reconstituted from butterfat and nonfat dry milk should be classified as a class III rather than class I product. Proponents contend this would reduce the price of the product, (since class III prices are considerably less than class I prices in all Federal Milk Order markets), and therefore increase dairy sales. Most dairy farm groups oppose the proposal on the basis it would weaken if not destroy the classified pricing system. The Supreme Court ruled against the consumer proponents in mid-1984, on the basis that consumers do not have the right to challenge milk marketing orders.

Class I Pricing

The class I price in all 45 Federal Order markets as of January 1, 1984, was based on the Minnesota–Wisconsin (M–W) manufacturing milk price for the second previous month, plus a specified differential. The second preceding month M–W price is used to give milk dealers time to price their packaged fluid products. The class I price differential was lowest in the Upper Midwest (Minneapolis–St. Paul area) Milk Marketing Order ($1.12 per hundredweight); and was highest in the Southeastern Florida Order ($3.15 per hundredweight). The wide variation between markets in class I differentials is based on estimated differences in supply and production costs.

Class II and III Pricing

Class III prices in all of the Federal Order markets with a class III category as of January 1, 1984, were based on the M–W price series for the current month.

Class II prices were generally determined by adding 10 to 15 cents per hundredweight to the M–W price for the current month. A few markets used some combination of the M–W price, and the butter-power formula price.

Blend Prices

The pounds of farm milk assigned to each class in the pool, multiplied by appropriate class prices, determine total pooled value of farm milk. Total pooled value of milk divided by pounds of milk in the pool determines blend prices paid producers.

Example:

	Pounds		Price per cwt	Total pooled value
Class I	500	×	$13.76	$ 68.80
Class II	200	×	12.60	25.20
Class III	300	×	12.50	37.50
TOTAL	1,000			$131.50

Blend price = $13.15 per cwt.
($131.50 ÷ 1,000 lb)

August, 1983, Class I utilization in the Chicago Regional Order was 21%—22 percentage points below the national average, and class I prices in Chicago were 89 cents per hundredweight below the national average. As a result, Chicago's blend price was 63 cents per hundredweight below the national average. The Upper Midwest Order (Minneapolis–St. Paul area) had an even lower class I price and utilization, and hence lower blend price than did Chicago (Table 2.7).

The Chicago–Minneapolis area has the largest volume of reserve (grade A) milk, and also the largest volume of grade B milk converting to grade A. This area therefore has the lowest class I utilization—and blend prices.

Other Pricing Factors

Location Differentials. Class I and blend prices are subject to adjustments depending on the location of producers or the plant to which they sell their milk. Typical deductions from a major city basing point such as in the Chicago Regional Federal Milk Order are: −12 cents per hundredweight for the first 100 miles, and −2.3 cents for each additional 15 miles, with a maximum deduct of 36 cents over 250 miles. Thus the milk price at a dairy

TABLE 2.7 *Average Class I Prices and Differentials, and Blend Prices, in Selected Federal Milk Orders, August, 1983*[a]

Federal Milk Order market	Class I price[b] differential	Class I[c] price	Blend[c] price	% Class I utilization
	$ per cwt			
Upper Midwest (Minneapolis–St. Paul)	1.12	13.62	12.72	15
Chicago Regional	1.26	13.76	12.95	21
Nebraska–Western Iowa	1.60	14.10	13.09	37
Puget Sound	1.85	14.35	13.07	30
Texas Panhandle	2.25	14.75	14.40	83
Central Arizona	2.52	15.02	14.04	60
Southeastern Florida	3.15	15.65	15.38	90
Average all Federal Order markets	—	14.65	13.58	43

[a]Source: Agricultural Marketing Service, Dairy Division, USDA: *Federal Milk Order Market Statistics*, August, 1983.
[b]Amount to be added to the Minnesota–Wisconsin manufacturing grade milk (3.5% butterfat) price for the second preceding month.
[c]3.5% butterfat.

plant 275 miles from a major city basing point would be lowered 36 cents per hundredweight to reflect the cost of hauling the milk to the consuming area. A plant 115 miles from the basing point would have a location differential of 14.3 cents per hundredweight.

Louisville Plan. This plan is designed to encourage constant production throughout the year, by rewarding those producers who deliver a high proportion of their annual milk production during the August–December low production fall months. Also known as the "take-out and pay-back plan," part of the payments due producers are withheld during the spring months and repaid during the months of short supplies.

Seven of 46 Federal Orders included this plan as of January 1, 1983. Deductions during the period March through July 1982 were 20 cents to 50 cents per hundredweight, with payments of 21 cents to 50 cents per hundredweight in August–December 1982.

Seasonal-Base Plans. Another method used to bring production of fluid milk into balance with demand is the base-excess plan. Producers establish a production base during the months of low production. The producer receives a lower price for milk produced above his established base in the "base paying" months. Nine of 46 Orders had seasonal-base plans as of January 1, 1983. The most common "base-forming period" is September through January, with the "base-paying period" from March through July.

Class I Base Plans. The Food and Agriculture Act of 1973 authorized "class I base plans" in Federal Milk Orders. Under this program, producers receive the class I price for their "base" milk and the class III price for deliveries in excess of base (instead of the blend price for all their milk as is otherwise the case in Federal Order marketing). Thus, the intent of the class I base program was to bring production closer to class I utilization, thereby increasing farm milk prices. "Bases" would normally be determined by multiplying the market class I utilization percentage times the individual producer's milk volume during the base-forming period.

A separate producer referendum was required for base plans—other provisions of Federal Milk Orders would not be voted on when base plans were being considered. Bloc voting by cooperatives was prohibited—each producer was to vote individually on base plans.

However, only two Federal Milk Marketing Orders adopted class I base plans (Puget Sound and Georgia), and they had only limited success in reducing milk production. Continued authorization for class I base plans in Federal Milk Orders was therefore not renewed in the 1981 Agricultural Act. Reinstitution of Class I base plans in Federal Milk Orders in the future would therefore require new legislation.

Formula Pricing. As the proportion of grade B milk continues to decline from the 15% of total marketings in 1982, it will be necessary eventually to use something other than the price of manufacturing milk to determine class prices in Federal Milk Marketing Orders. Class II and III milk will probably have to be priced by product formulas, using average yields and costs for products such as butter, nonfat dry milk powder, cheese, and whey powder. Such formulas were used before the adoption of the M–W manufacturing milk price series in the early 1960s.

The difficulty with these product formulas is in determining representative and equitable manufacturing costs for various products. Errors on the low side cause financial difficulties for milk plants, and errors on the high side create supply and price distortions within order markets.

Economic Formulas. Economic formulas for pricing class I milk are receiving increasing attention as an alternative to the M–W manufacturing milk price. Economic formulas use factors such as cost of producing milk, cost of living, consumer income, gross national product, and other general economic indicators in determining class I prices.

An economic formula was used in various new England markets to price class I Federal Order milk from the late 1940s until the early 1970s. However, as of 1976, no Federal Order markets were using economic formulas to price class I milk.

A national economic formula was adopted by the National Milk Producers Federation in June 1969, and presented at Federal Milk Order Public Hearings in early 1970. The reasons advanced by the proponents of the formula were:

1. National hearings are cumbersome, expensive, and time consuming. Differences among the representatives from various regions may make it difficult to reach an agreement promptly. Competition among various cooperatives without a formula as a guide may result in class I prices that are not realistic in relation to supply and demand, income, and public interest considerations.
2. A national formula, which is impersonal and understood by consumers, handlers, and producers, will be more acceptable than administrative decisions.
3. Annual hearings and a national class I pricing formula could effect prompt and equitable class I price changes.

However, the USDA concluded that the proposed formula would not: (1) reflect accurately needed changes in fluid milk prices in relation to supply and demand; (2) maintain appropriate relationships among markets and uses of milk, particularly between fluid and manufacturing milk prices; and (3) be compatible with the responsibilities of the Secretary of Agriculture for other programs, especially the price-support program. Therefore the USDA terminated the proceedings.[9] Since many producers and dairy leaders favor economic pricing formulas, national pricing on revised economic formulas will likely come up again.

Super Pool Pricing. "Super pool" premiums above Federal Order minimum class I prices were being received in 34 markets in the 45 Federal Milk Orders as of January 1984. The average class I premium was $.50 per hundredweight. Dairy farmers justify these premiums because of "high" production costs. Super pool premiums are the result of negotiations between cooperatives and dairy plants, and are therefore independent of Federal Orders. They are administered solely by employees of cooperatives.

Super pool premiums originated in upper Midwest markets, which have lower class I and blend prices than other parts of the country. The Chicago Regional Order market, which was one of the first to implement super pool pricing (in 1958), generally typifies super pool programs throughout the country. Fourteen Wisconsin and Illinois dairy cooperatives have been supplying Chicago Regional Order market with approximately 90% of its milk requirements. These 14 cooperatives announce monthly "super pool" class I prices based on their interpretation of supply, demand, and price conditions. They attempt to set super pool prices at levels that will satisfy farmer members but not encourage dealers to buy milk elsewhere.

Since class I super pool prices and Federal Order class I prices both vary from month to month, Class I super pool premiums also vary. For example, in December

1982 the Chicago Federal Order class I price was $13.82 per hundredweight, and the class I super pool price $14.58 per hundredweight, resulting in a class I super pool premium of 76¢ per hundredweight. In December, 1983, the Chicago Federal Order price decreased to $13.78 per hundredweight, but the class I super pool price rose to $14.64 per hundredweight. Therefore, the class I super pool premium increased to $.86 per hundredweight.

Super pool premiums resulted in $2.8 million in additional income to Chicago area dairy farmers in 1983 ($170 per producer). This increase amounted to 2.3 cents per hundredweight higher "blend" prices.

Dairy farmers prefer increased Federal Order class I prices, rather than constantly "negotiating" super pool premiums. Dairy farmer interest in economic formulas for pricing class I milk also stems in part from their uneasiness over privately negotiated super pool premiums. As of early 1984, the USDA had declined both, contending that additional quantities of milk would be attracted, lowering class I utilization and eventually blend prices.

Standby Pool Pricing. The "standby pool" was created in 1970 to maximize blend prices by carrying grade A reserve milk supplies outside Federal Milk Orders. It was also independent of Federal Orders, and was administered by seventeen cooperatives who paid an average of approximately 2½ cents per hundredweight, from their class I sales, into the standby pool in the 1970s. This money made possible payments of approximately 25 cents per hundredweight to about a dozen Minnesota and Wisconsin nonorder grade A milk plants, which agreed to ship milk from the reserve supply area to deficit markets in the South and East North Central regions of the country at prevailing prices plus transportation costs. If the milk was not needed for class I purposes, the nonorder Wisconsin-Minnesota plants could process it into manufactured dairy products at their normal margin, plus approximately 25 cents per hundredweight standby pool payments.

The standby pool program reflected the large volume of upper Midwest reserve milk available for Order markets, potentially depressing class I utilization percentages and blend prices. Because of increased transportation costs, sales to deficit areas slackened, and the standby pool ceased operation in 1981. However, the format for it remains in place, and reactivation of it could occur if circumstances favored its reinstitution.

Legislative Status. The following Federal Milk Order proposals by dairy farmers were rejected when the 1973 Agricultural Act was passed.

1. Authorization to extend negotiated super pool premiums in Federal Milk Order markets to all milk in these markets, provided the premiums covered at least two-thirds of the order milk.
2. Assessing "nonmembers" for services performed by cooperatives.
3. Federal Order standby pools.
4. Assigning class I bases to cooperatives rather than individual farmers.
5. Assessing producers and cooperatives for handler services.
6. Prohibition of producer "checkoff" refunds if two-thirds of the producers voting in an individual producer referendum favor a promotion program. (Under present legislation producers may request refunds.)
7. Authorizing manufacturing milk orders for checking butterfat tests and weights—but without minimum price provisions.

Proposals incorporated into the new law were: (1) requiring Federal Milk Order hearings when requested by at least one-third of the farmers in the market; (2) requiring the Secretary of Agriculture to consider the level of farm income needed to assure adequate production when setting Federal Milk Order prices; and (3) Authorization for class I base plans—later rescinded in the 1981 Agricultural Act.

Until new legislation is enacted these Federal Milk Order mandates and prohibitions will continue to apply.

2.8 State Milk Control Programs

These programs were developed at about the same time as Federal Order programs. Both have similar objectives of ensuring an adequate milk supply at fair and reasonable prices to farmers and consumers, marketed in an orderly stabilized fashion. Both price only grade A milk. Nei-

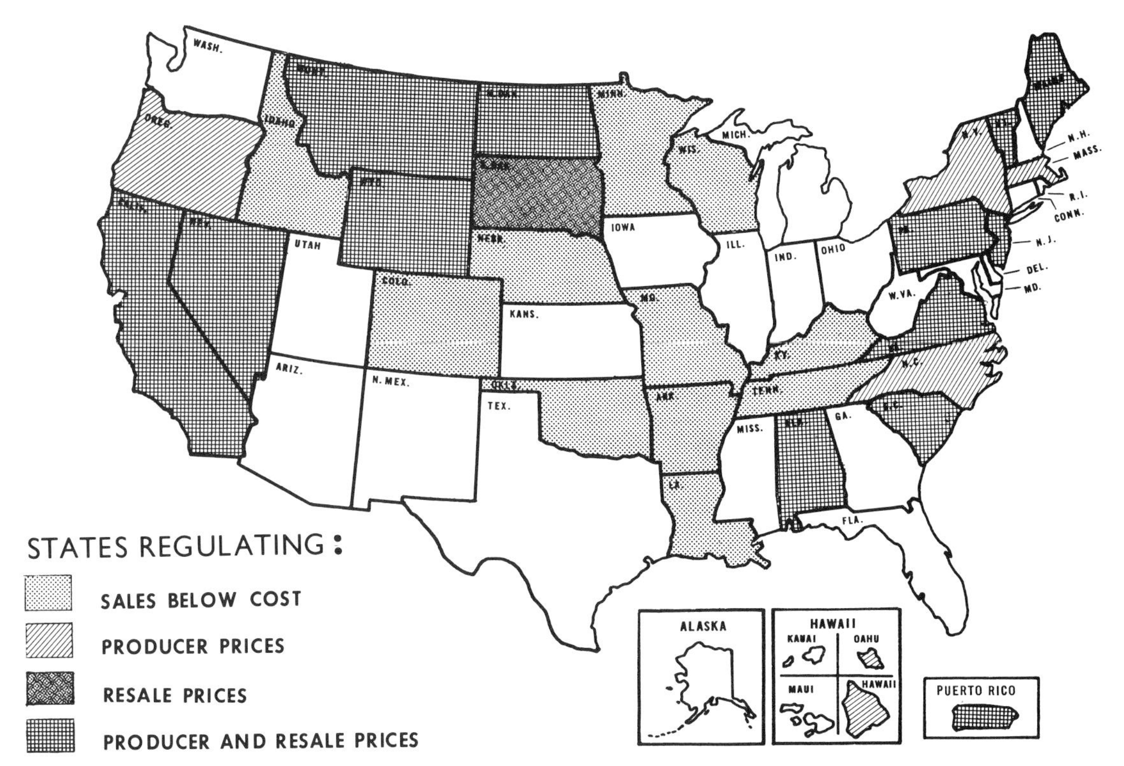

FIG. 2.12 States regulating fluid milk prices, January 1, 1977. (Courtesy of Economics, Statistics, and Cooperative Service. USDA: *Government's Role in Pricing Fluid Milk in the United States,* Ag. Econ. Report No. 397, March, 1978)

ther prices grade B milk. State Orders price milk only within the particular state, whereas Federal Orders can cover several states. Federal Milk Orders establish only farm milk prices, while State Orders establish prices at various stages in the marketing system—farm, wholesale, retail—as well as regulating trade practices (Figure 2.12). In 1982, approximately 14% of all grade A farm milk was priced under State Orders in the 17 states that controlled producer milk prices.

Some states set maximum prices at either wholesale, retail, or at both levels. Some set both minimum and maximum wholesale and retail milk prices. Several states also require a minimum markup particularly by retailers, while others require that milk prices be filed with the state. Most states directly specify class I prices in their Orders.

All state control agencies also have the authority to require milk distributors to be licensed, and to require licenses for selling milk in specific marketing areas. Control agencies also have the authority to inspect and investigate operations of milk dealers, audit their records, and require regular reports by them.

Trade practices prohibited or controlled by state Orders include: (1) free service or equipment, (2) discounts or rebates to some but not to all, (3) sales below cost, and (4) price discrimination.

State milk promotion activities include advertising, merchandising, research, and education. Most states with promotion programs directly assess producers for most or all of the operational costs, while a few use appropriated funds.

2.9 Dairy Imports and Exports

A recent USDA study concludes that additional imports of 500 million pounds of milk equivalent (0.4% of U.S. production) would reduce U.S. farm milk prices about 9 cents per hundredweight.[10] On this basis, farm milk prices were reduced approximately 50 cents per hundredweight during 1973, the peak year for dairy imports.

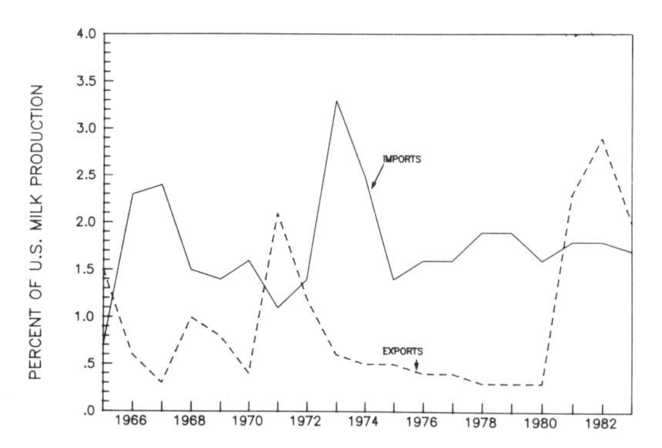

Fig. 2.13 U.S. Dairy imports and exports as % of U.S. Milk Production. (Courtesy of Economic Research Service, USDA: *Dairy Outlook and Situation*, DS-380, May, 1980, DS-393, June, 1983, and DS-395, January, 1984) 1983 data January–September.

During the decade 1973–1982, dairy imports ranged from 1.4 to 3.3% of U.S. milk production. Meanwhile exports ranged from 0.3 to 2.9% of milk production. Dairy imports exceeded exports in 8 of the 10 years, and import volume averaged over 2 times more than export volume (Figure 2.13). Butter imports hit a high of 5.6% of U.S. civilian consumption in 1973, cheese imports a high of 10.3% of consumption in 1974, and nonfat dry milk a high of 24% of consumption in 1973.

The United States has been a net importer of dairy products, even though U.S. milk production has consistently exceeded consumption.

A large proportion of U.S. dairy exports are noncommercial, and financed under the "Food for Peace" program (the Agricultural Trade Development and Assistance Act of 1954–Public Law 480). Under this government financed export program, USDA purchases of manufactured dairy products are distributed to foreign nations, to help improve the diets of people, especially children in developing nations.

Import Quotas

Under Section 22 of the Agricultural Adjustment Act of 1937, the President has authority to establish import quotas on foreign dairy products that "interfere with USDA price support programs." Dairy import quotas were 1.6% of U.S. milk production in 1983, and were in effect on virtually every manufactured dairy product.

The pressure for "protectionism" by the U.S. dairy industry is likely to increase in coming years, as world milk production continues to increase—over 3% in 1983 (over one-fifth of U.S. annual milk production) with projected continued foreign production increases in later years. Meanwhile, world per capita consumption of dairy products is declining as consumer prices increase. As a result, the imbalance between supplies and commercial demand in developed countries increases, and pressures to "export" intensify. This situation increases both U.S. export pressures, and requests for more import "protection" by the U.S. dairy industry.

Subsidized Import Protection

The Trade Act of 1979 requires the U.S. Secretary of the Treasury to act within 60 days on complaints involving export subsidies by other countries. Export subsidies are payments by foreign countries to exporters, enabling them to sell their products at lower prices in the United States, hence making them more competitive with U.S. domestic dairy products.

The "subsidized import" provision of the 1979 Trade Act permits embargoes against dairy imports which are priced lower than U.S. domestic dairy products because of export subsidies by other countries. The Trade Act has been used several times to get European Common Market countries to reduce their export subsidies, which were averaging approximately 17% of U.S. cheddar cheese wholesale prices, 20% of nonfat dry milk prices, and 35% of butter prices in the early 1980s.

Free Trade

A USDA study concludes that the elimination of all trade barriers would not seriously damage the U.S. dairy industry, since New Zealand and Australia would be the major world exporters of dairy products under a "free trade" situation. Since these 2 countries produce less than 5% of the world's milk supply and do not possess the resources to increase their milk output significantly, their exports "could not seriously affect the U.S. dairy industry."[10]

The study also concludes that "free trade" in dairy products is unlikely, since it would involve such great adjustments for high priced EEC countries as to make free trade politically unacceptable.

2.10 Dairy Industry Concentration

Fluid Milk

Substantial concentration exists in the fluid milk industry. Twelve companies, representing less than 2% of all firms engaged in processing fluid milk products, accounted for more than one-third of the total sales of all fluid milk products in 1980. Four companies representing less than .5% of the firms, accounted for over one-sixth of fluid milk sales. These four large firms distributed milk in an average of 25 Federal Milk Order markets, averaging about one-tenth of total sales in these markets. One of the four firms operated in all 45 Federal Milk Order markets. These four large firms averaged more than two-thirds of the sales by firms in the upper half of the size distribution, in the markets in which they sold.[11]

Large scale vertically integrated plants accounted for over one-fourth (26%) of U.S. fluid milk sales in 1979. Firms operating vertically integrated supermarkets alone accounted for over 16.8% of U.S. fluid sales. The share of total U.S. fluid milk sales by these integrated supermarkets increased from 3.3% in 1964, to 9.0% in 1970, to 16.8% in 1979.[12]

Even greater concentration is likely in the future because an increasing proportion of fluid milk products is being processed and sold by chain stores that have their own processing facilities, or by milk processors who operate chains of retail stores. Advantages claimed for this vertical integration are: (1) better quality control, (2) lower distribution costs, and (3) lower unit processing costs because of large volume operations. This increased concentration in the fluid milk industry will have a substantial effect on milk procurement, distribution, marketing efficiency, cooperatives, and dairy farmer bargaining.

Cooperatives are rather minor factors in the distribution of fluid milk and fluid milk products. All dairy cooperatives combined accounted for only 16% of fluid milk and fluid milk products distributed by U.S. fluid milk handlers in 1980—up only slightly from 12% in 1973.

Butter

Construction of large butter-skim milk powder plants in the 1960s and 1970s increased competition in butter manufacture. The number of butter plants decreased 28%, and volume per plant increased 61% in the 5-year period 1977–82. Cooperatives manufactured 64% of total production in 1980. Land O'Lakes is the primary distributor of butter, with over one-fourth of U.S. retail sales. Geographically Land O'Lakes butter comprises approximately one-third of total retail butter sales east of the Rocky mountains, and one-half of retail sales east of the Appalacian mountains.

A relatively small group of primary receivers also purchase butter in bulk from creameries. They store, print, package, and distribute it to wholesale and retail outlets.

USDA purchases for price support have been large in the early 1980s, reaching approximately one-third of total U.S. production in 1982 and 1983.

Cheese

Many plants with 1 million pounds a day or more milk volume were constructed in the 1970s and early 1980s, and many small plants were closed. In the 5-year period 1977–82, the number of cheese plants declined 12%, while production per plant increased 54%. This trend will likely continue.

Cooperatives manufactured 47% of U.S. cheese production in 1980. Assemblers, brokers, cooperatives, reprocessors, wholesalers, and retail food store chains are important buyers of cheese from cheese plants. However USDA has been the largest single buyer in recent years, with price support purchases reaching about one-fourth of cheddar cheese production in 1982 and 1983.

Nonfat Dry Milk

Cooperatives processed about 85% of total nonfat dry milk production in 1980. The USDA purchases nonfat dry milk from any seller who meets its specifications, as part of its price support operations—which totaled about two-thirds of U.S. production in 1982 and 1983.

Almost 90% of the "powder" used commercially is purchased by bakeries, by the dairy industry for use in cottage cheese, fortified skim milk, ice cream, and prepared dry mixes, and by consumers for use in the home.

Evaporated Milk

Economical and efficient production, packaging, and distribution of canned, evaporated and condensed whole milk for retail sale to consumers require an abundant supply of surplus milk of high quality, large processing plants, and a national distribution system. Multiunit corporations dominate the field, with cooperatives and individual proprietors being relatively unimportant.

Ice Cream

Multiunit corporations operating 25% of the ice cream plants account for about 75% of total sales. This trend to fewer but larger plants will continue. Medium-sized firms, in order to compete, are establishing their own outlets in specialized ice cream parlors or convenience-type dairy stores. Smaller companies face stiff competition from larger firms and retail food chains that produce their own ice cream. Cooperatives manufactured only 14% of total U.S. production in 1980.

Cottage Cheese

Per capita consumption of cottage cheese in 1973 was approximately double the 1945 level, but declined about one-fifth the following decade.

Concentration is increasing dramatically in the cottage cheese industry, and attrition of cottage cheese plants is rapidly increasing. In the 5-year period 1977–82, the number of cottage cheese plants declined 24%, while production per plant increased 20%. Cooperatives manufactured 22% of U.S. production of cottage cheese in 1980.

Environmental and sanitation requirements have resulted in substantial sewage costs, which many smaller plants have been unable to meet. Larger plants have almost completely taken over production, with one firm distributing to most of the states east of the Mississippi River out of just one plant.

Some cottage cheese sales by cottage cheese plants are to fluid milk dairies, who in turn distribute the product. The trend in the future will be for more direct sales to warehouses, which in turn will distribute the product to increasingly dominant food store chains.

Cooperatives

The Capper–Volstedt Act of 1922 permits farmers to band together lawfully to market their products collectively, provided they do not engage in predatory price discrimination, and other illegal practices. It was this Act that provided the impetus for the substantial growth of dairy cooperatives, thereby giving dairymen greater bargaining power in the marketing of their milk and dairy products.

However, cooperatives have recently come under attack for allegedly having too much marketing power, achieving too much "concentration," and unduly enhancing milk prices.

Examples:

1. The Federal Trade Commission (FTC) in a September 30, 1975, Staff Paper Report concluded that dairy cooperatives "dominate" certain markets, and engage in "anticompetitive activities and unfair practices." The report also charges that Market Orders create entry barriers, interfere with movement of commodities to market, and allocate markets to keep prices above what the market would otherwise establish, thereby contributing to the economic power of cooperatives.
2. The U.S. Justice Department has instituted well-publicized suits against large dairy cooperatives, and spokesmen for the Department have offered vocal opposition to Capper–Volstedt cooperative exemption from antitrust prosecution.
3. The Council of Economic Advisors and the Office of Management and Budget, in the Economic Report of the President have argued that the Capper–Volstedt Act be reevaluated in light of "recent cooperative abuses."

4. Specific legislative proposals have been created to (a) eliminate taxation exemption privileges of agricultural cooperatives, and (b) tax both cooperatives and farmer members for patronage refunds.
5. A 1976 FTC decision stipulated that the Capper–Volstedt Act covers only processing cooperatives—not bargaining cooperatives.

Cooperatives dispute "concentration" charges because:

1. In the 1960–1970 decade private agribusiness firms grew far faster than agricultural cooperatives—sales 8 times faster, assets 18 times faster, net worth 25 times faster, and net assets 24 times faster.
2. In 1982 the largest private corporation Exxon had 17 times more dollar sales than all 6,211 agricultural cooperatives combined.
3. Only 3 cooperatives had over $1 billion assets in 1981 and 1982, compared to 99 of Fortune's 100 largest industrial corporations. Exxon alone had more than double the assets of all 6,211 farmer-owned cooperatives in 1982.
4. Only 9 of Fortune's current "500" in 1983 were cooperatives—493 were private corporations.

At the time of this writing, the outcome of the "anticooperative" agitation was unsettled. However, cooperatives contend that the April 20, 1975, U.S. Justice Department Associated Milk Producers, Inc. (AMPI) consent decree, and the June, 1976, U.S. Department of Justice proposed consent judgment to settle its antitrust suit against Mid-America Dairymen (MID AM), do not appear to support charges of cooperative monopolization. Nevertheless, it is likely that dairy cooperatives will be under closer scrutiny with respect to their marketing and bargaining practices in the future than they have been in the past.

2.11 Summary

Success in dairy farming depends as much on profitable marketing as on profitable production. Marketing is therefore a critical component of the dairy farming business.

The dairy industry is one of the most regulated and controlled in the entire economy, through a wide variety of legislation and government programs. The kinds of regulation include hauling rates, Federal Milk Marketing Orders regulating farm milk prices, State Milk Orders regulating trade practices and farm wholesale and retail milk prices, USDA price support programs undergirding farm milk prices, import quotas and duties regulating quantities of dairy products that can be imported, 1,500 health and sanitation jurisdictions regularly conducting sanitation and quality inspections,[1] and environmental safeguards. This combination of local, state and federal regulations and programs, plus varying seasonal and geographic supply–demand and competitive conditions, make milk marketing and pricing a complex, but fascinating subject for study.

Dramatic changes are continuously occurring in milk marketing, and these changes profoundly affect farm milk prices. Dairying is becoming more concentrated geographically, seasonality of production is decreasing, production per cow is increasing dramatically, and sales of whole milk by farmers are increasing steadily, as farmers use less milk for feed and home consumption, and sell less farm separated cream. Almost 7 out of every 8 quarts of farm milk sold now meet grade A standards, but only about 2 out of every 5 quarts are sold in fluid form. Over half is sold for use in lower priced manufactured products, mostly in midwestern states, resulting in lower farm milk prices there. This situation has created pressures for better geographic balance of farm milk prices. These pressures will likely intensify with the continued conversion to grade A milk in the upper Midwest, where most of the grade B milk is located.

A reduction of over three-fourths in the number of dairy manufacturing plants and seven-eights in the number of fluid processing plants has occurred in the past quarter century. Meanwhile the output per manufacturing plant has more than quintupled, and many small milk distributors have ceased processing to become distributors for large companies, or have merged with other dealers to achieve larger volume and greater efficiency. This trend toward fewer and larger plants has generally resulted in lower unit costs, which helps farm milk prices, but can also result in less competition between plants, which reduces farm milk prices.

Home delivery sales are small and decreasing—only 2% of fluid milk sold in Federal Order markets was home delivered in 1982. Glass milk bottles are virtually extinct, with 99% of 1982 Federal Milk Order milk packaged in paper and plastic containers. Pint and quart containers are also less available—over four-fifths (82%) of Federal Order milk was packaged in gallon and half gallon containers in 1982.

Farm milk prices are still largely based on weight and butterfat percentage. However, increasing interest and attention are being directed toward "component" pricing, i.e., determining the price of milk based on both its fat and one or more of the nonfat solids components.

The United States ranks near the bottom among major nations in per capita consumption of dairy products, with a one-third decline in per capita consumption of milk equivalent occurring between 1925 and 1982. In 1982 per capita consumption was lower for fluid whole milk, nonfat dry milk, cottage cheese, sherbert, ice milk, butter, and evaporated and condensed milk than it was a decade before. Stepped-up dairy promotion programs, although helping to increase some dairy sales, have not yet increased overall dairy product consumption.

The USDA supports the farm price of manufacturing milk by offering to buy butter, skim milk powder, and cheddar cheese at preannounced purchase prices. Under this procedure farm prices for manufacturing milk were generally at or above support levels in the 1960s and 1970s. However, in the early 1980s milk surpluses depressed farm milk prices below support levels.

The Secretary of Agriculture has been required by law to support the price of milk at between 75 and 90% of parity since the late 1940s and the early 1980s. A freeze in the support price of farm milk in 1982 and 1983, and a reduction in the support price in 1984 because of milk surpluses, resulted in farm milk prices of only slightly more than 60% of parity by early 1984. Farmers reducing milk production below their bases (1981–82 or 1982 average daily production) were eligible for $10 per hundredweight on reduced production in 1984, and in January–March 1985.

Federal Milk Orders price about 81% of U.S. grade A milk production, and State Orders price about 14%. Both Federal and State Orders price milk according to use, with class I (fluid) prices highest and class III (manufactured products) lowest. Thus, a higher class I utilization will increase farm milk (blend) prices. A peak of 83 Federal Orders was reached in 1962, but the number declined to only 45 as of January, 1984. An eventual decline to 5 to 10 regional Orders is projected, as pressures for geographic farm price equalization intensify, plants and markets merge, and transportation and sanitation facilities improve.

The United States has quotas and duties regulating quantities of dairy products that can be imported. Nevertheless, dairy imports averaged over 2 times more than exports, and exceeded exports in 8 of 10 years between 1973 and 1982, even though U.S. milk production consistently exceeded consumption. This import situation has depressed farm milk prices by an average of 9 cents per hundredweight for each 500 million pounds of milk equivalent imported, and has brought on pressures for dairy import restrictions. Dairy import quotas were only 1.6% of U.S. milk production in 1983.

The U.S. Trade Act of 1979 contains "subsidized import provisions" permitting import embargoes against government subsidized exports from foreign countries, which undercut U.S. domestic prices. This legislation will improve the competitiveness of domestic U.S. dairy products with imported dairy products, and will likely reduce dairy imports, thereby improving farm milk prices.

Rapid changes have occurred in dairying during the 1970s and early 1980s. Dairy marketing and milk pricing are expected to continue this pace during the late 1980s and 1990s.

Review Questions

1. Are you optimistic or pessimistic about the future marketing of milk from U.S. dairy farms? List 5 reasons for your answer.
2. Define parity and explain how it ties into the dairy price support program. Describe and detail the mechanics and operation of the

dairy price support program and its effects on farm milk prices.

3. Describe and detail mechanics and operation of the Federal Milk Order program, its effects on farm milk prices, and major differences between Federal Milk Orders and state milk control programs.
4. Discuss and detail changes in demand that have been occurring for various dairy products, and factors that affect demand for dairy products. What has been the impact of both of the above on farm milk prices?
5. Have dairy imports affected farm milk prices? Why or why not? Discuss government programs and policies that affect dairy imports and exports.
6. Are you for or against component pricing of fluid whole milk? State 3 reasons why. Are you for or against the geographic variation in class I fluid and farm milk blend prices (with the midwest lowest), that has historically existed in Federal Milk Orders? State 3 reasons why.
7. List 3 reasons why you are for or against each of the following: (1) Louisville plan, and other base-excess plans, (b) super pool pricing, (c) $10 per hundredweight USDA payment not to produce milk.
8. Do you feel that the dairy producer is getting his fair share of the retail price for fluid whole milk? Give 3 reasons for your answer.

References

1. Milk Industry Foundation: *Milk Facts,* 1982 ed., Washington, D.C., 1982.
2. US-EPA: *Economic Analysis of Proposed Effluent Guidelines for the Dairy Processing Industry,* November, 1973.
3. Boehm, W. T., and Babb E. M.; *Household Consumption of Beverage Milk Products.* Sta. Bull. 75. March, 1975; *Household Consumption of Storable Manufactured Products.* Sta. Bull. 85. June, 1975; and *Household Consumption of Perishable Manufactured Dairy Products: Frozen Desserts and Specialty Products.* Sta. Bull. 105. West Lafayette, Indiana: Purdue Univ., Agr. Exp. Sta., September, 1975.
4. American Dairy Association: *Milk Beverage Consumption Patterns National Survey,* Chicago, 1962.
5. Thompson, S. R., and Eiler, D. A.: *Producer Returns from Increased Milk Advertising.* Amer. J. Agr. Econ., Vol. 57, No. 3, August, 1975.
6. United Dairy Industry Association: *10-Market Study Show Non-Brand Advertising Pays,* Chicago, 1980.
7. Graf, T. F.: *Economic Impact of Imitation Cheese.* Agr. Econ. Staff Paper Series No. 208, University of Wisconsin–Madison, August, 1982.
8. Graf, T. F., and Jacobson, R. E.: *Resolving Grade B Milk Conversion and Low Class I Utilization Pricing and Pooling Problems.* Agr. Exp. Sta. Res. Report R 2503, University of Wisconsin–Madison, June, 1973.
9. Consumer and Marketing Service, USDA: *Decisions and Order to Terminate Proceeding on Proposed Amendments to Marketing Agreements and Orders.* Docket No. A0-10-A41 et al. 7 CFR Parts 1001–1138, January, 1971.
10. Economic Research Service, USDA: *The Impact of Dairy Imports on the U.S. Dairy Industry.* Agr. Econ. Report No. 278, January, 1975.
11. Rourke, J. P.: *Marketing Study.* Dairy Field, Vol. 165, No. 5, May, 1981.
12. Lough, H. W.: *Fluid Milk Processing Market Structure.* ESS Staff Report, No. AGESS810415, USDA, April, 1981.

CHAPTER 3

Dairy Record Keeping

3.1 Introduction

Complete and accurate records are the backbone of a highly profitable dairy farm operation. The use of adequate records to make management decisions could turn many deficit operations into profitable ones and will almost certainly make an already profitable operation even more profitable. Probably the one single deficiency of management that costs dairymen the most in lost income is the lack of good records.

The primary purpose of dairy records is to give the dairyman detailed information on individual cows as well as on the entire herd, for day-to-day decisionmaking, evaluation of past management practices, and long-range planning. The desirable characteristics of a dairy record keeping system are that it be simple, complete, accurate, up-to-date, and understandable, and that it require a minimum of time to keep.

Two basic kinds of organized record keeping programs are available to dairymen. One kind, generally referred to as farm business records or farm accounting records, gives dairymen information on the financial status of the farm operation. These records are discussed in Chapter 23. The second kind, dairy records available through the National Cooperative Dairy Herd Improvement Program (NCDHIP) is the subject of this chapter. This program is referred to by various names: the DHIA Program, NCDHIP, or the DHIA System, where DHIA is an acronym for Dairy Herd Improvement Association.

Whether or not a dairyman should rely on farm-kept records or join the DHIA Program depends on the direct and indirect costs of each record keeping system compared to its benefits. On most farms, the cost-benefit advantage should favor the DHIA Program over farm-kept records for a variety of reasons.

Hand-kept records are the least sophisticated and least valuable of the record keeping options available to dairymen. The direct costs are minimal but the indirect costs may be high in terms of time spent on record keeping relative to the benefits received. Perhaps the greatest disadvantage of hand-kept records is the difficulty of obtaining the many projections of statuses and events that are so valuable for planning purposes.

Farm computers offer an alternative for farm-kept records that is superior to hand-kept records on many but not necessarily all farms. The indirect costs of record keeping on a farm computer may be quite high due to the time required for data entry and checking. The main advantage of a farm computer over hand-kept records is the ease of obtaining summaries and projections. But, the pool of data available is usually considerably less than through the DHIA Program.

On the average, participation in the DHIA Program costs about $1.50 per cow per month. Nationwide, about 25 different dairy record keeping plans are offered in the DHIA Program (Section 3.4), so every dairyman has his choice of a variety of dairy record keeping plans that should satisfy his needs. In some plans, a DHIA Supervisor, an employee of the local or state Dairy Herd Improvement Association, visits each farm at approximately monthly intervals and collects and records all the information needed. In other plans farm employees record the information. Therefore, these plans usually cost less. In both cases, pre-printed barn sheets are provided to simplify data recording. Data from the almost 70,000 dairy farms participating in the DHIA Program are analyzed and summarized at regional Dairy Records Processing Centers (DRPC). The DRPC have developed highly complex computing systems to process each farm's records and provide dairymen with a wealth of management information (Section 3.5) which is the primary reason-for-being of the DHIA Program. The DHIA Program provides benefits that are usually impossible to obtain through farm records including: tests on each cow's milk for fat, protein, and/or solids-not-fat content as well as a somatic cell count; sophisticated computerized data base systems at DRPCs that can be accessed on-line from farm terminals to generate a wide variety of reports the dairyman can customize to his own needs, and; an interface to the national research program on genetic improvement of dairy cattle that includes genetic evaluations of bulls and cows and the various breed associations' improvement programs. Research has shown that continuous enrollment in the DHIA Program increases profits on both a farm and a per-cow basis and that the increases are greater in the larger herds.[1]

3.2 Farm-Kept Records

Farm-kept records should include the following data on each cow no matter whether hand-kept or on a farm computer:

1. Birth data and ancestry (at least parents, preferably grandparents also).
2. Periodic milk yield (weekly or monthly) and an estimate of total lactation production.
3. Reproduction information (heats, breedings, service sires, and calvings).
4. Health and veterinary data especially on reproduction and mastitis.

In addition, the following data would be desirable:

1. Periodic tests and lactation yield for milk fat and other components included in the milk pricing formula.
2. Consumption of concentrates and special feeds.

Farm-kept records should also include accurate estimates of cost and income factors. However, these can probably be obtained best from the farm accounting information which will be discussed in Chapter 23.

Many alternatives exist for maintaining hand-kept records. Permanent-page books, loose-leaf books, and separate folders for each cow are the most common methods used. The permanent-page book is the least flexible method. A loose-leaf book has the advantage of utilizing printed or photocopied forms which may be put in different groupings, such as heifers, milking cows, dry cows, dead cows, or different strings. A separate folder for each cow has the added advantage that other material can be filed quickly (such as breeding slips or veterinary receipts). A few states may still have forms for hand-kept dairy records obtainable from the extension service. However, in most states the supplying of such forms has been abandoned because of the advantages of the DHIA Program.

Many alternatives also exist for farm computers but for practical purposes the choice may be limited in any given area. Adequate computer hardware is available in many different brands of microcomputers. Computer software is gradually becoming more widely available. Dairymen planning to acquire a microcomputing system for farm records should be certain of the availability of continued hardware and software support to protect their investment and maximize benefits from their farm computer.

3.3 Organization of the DHIA Program

Organized dairy record keeping in the United States had its beginning in Michigan in 1905 when a dairyman named Helmer Rabild called a meeting to propose production testing to local dairymen. Mr. Rabild's proposal apparently was stimulated by a dairy record keeping program in Denmark. The idea caught on. The first DHIA was established in Mr. Rabild's area in 1906. The program was so successful that the U.S. Department of Agriculture became involved as a sponsor in the early 1920s and the program soon became national in scope. For many decades, the program was run by the Extension Service in the counties and states. In the mid-1960s dairymen expressed a desire to exert more leadership and formed a National DHIA while Extension expressed a desire to change its role in the program to education. Thus began a decade of rapid change, improvement and expansion that led to the present DHIA Program.

The DHIA Program is governed by a Policy Board composed of representatives of the major participating groups: National DHIA; Agricultural Research Service and Extension Service, U.S. Department of Agriculture; State Cooperative Extension Services; Purebred Dairy Cattle Association; National Association of Animal Breeders, and; Extension Committee on Organization and Policy. The Policy Board determines policy under which the DHIA Program operates and coordinates and approves operational aspects of the program which are the individual responsibilities of the participating groups. All these groups work to make the record keeping plans more valuable to dairymen whose herds are enrolled in them. The public sector and private sector participants in the DHIA Program and the purpose and membership of the NCDHIP Policy Board are shown in Appendix I-B.

A dairyman joins his local DHIA when he wants to participate in the program. The local DHIAs provide the working framework whereby the business of dairy record keeping is accomplished on the farm. In most of the states the local DHIAs are organized into a state Dairy Herd Improvement Association. The state DHIAs usually constitute the legal entity of dairy record keeping in their state for employing DHIA Supervisors, obtaining insurance for dairymen and DHIA Supervisors, enforcing the Official DHI Rules, and providing liaison and direct representation to the National DHIA. Most state DHIAs are members of National DHIA. The effectiveness and influence of the National DHIA is a splendid example of the benefits that can accrue to dairymen from working together to strengthen the industry.

3.4 Dairy Record Keeping Plans in the DHIA Program

Dairymen have their choice of approximately 25 dairy record keeping plans to obtain management information for their herds. Not all plans are offered in all parts of the country. The dairy record keeping plans available to each dairyman are those that are offered by the DRPC that serves his area and that are sanctioned by his State DHIA.

Six official dairy record keeping plans are available. The official plans offer special benefits to participants including collection and reporting of data by a DHIA Supervisor, production records of a high degree of accuracy and integrity that can be used for merchandising cattle, and participation in breed association programs and the national genetic improvement program. Participation in official plans also entails special responsibilities including adherence to the Official Dairy Herd Improvement Rules that prescribe conditions and procedures for obtaining certain data at the farm each test day.

Dairymen also can participate in any of about 18 different non-official dairy record keeping plans depending where they are located. The non-official plans generally offer the same information for cow and herd management as the official plans (with a few important exceptions). But, data either are reported by the dairyman rather than a DHIA Supervisor or if reported by a DHIA Supervisor not all the Official Rules are followed. Therefore, production data from the non-official plans do not have the same degree of authenticity as data from the official plans that are reported by a DHIA Supervisor according to the Official Rules.

Each of the dairy record keeping plans has attributes that make it more attractive to some dairymen than to others.

Official DHI Plans

The oldest and most widely used plan is the Official DHI Plan with approximately 2.5 million cows enrolled from 31,000 herds. Typical information obtained by dairymen from this plan is described in Section 3.5. A DHIA Supervisor visits each enrolled herd approximately monthly. The DHIA Supervisor records a variety of information on each cow and on the herd. For each cow he records dates of important events since the last test day such as breedings or calvings. If the cow is milking, he records milk yield at each milking on test day to measure 24-hour yield and takes a sample of milk from each milking. The milk samples are composited for later analysis. The samples may be tested by the DHIA Supervisor for milk fat content in which case the Supervisor records all data on the DRPCs pre-printed barn sheets and sends the sheets to the DRPC for analysis and summarization of the data. Many states have central milk testing laboratories operated by the State DHIA where the milk samples may be tested for protein or solids-not-fat content and somatic cell count as well as for milk fat content. In the states with central milk testing laboratories, the DHIA Supervisors may send the barn sheets to the central laboratory along with the milk samples or may send the barn sheets directly to the DRPC because the central laboratory's milk test results are teleprocessed from the laboratory to the DRPC. After the data are computer-processed at the DRPC, the information is returned to the farm for the dairyman's use in managing his herd, usually within a week after the DHIA Supervisor's visit.

Two variations of the Official DHI Plan are available, Official DHI AM/PM with Monitoring Device and Official DHI with Alternate AM/PM Component Sampling. In Official DHI AM/PM with Monitoring Device, milk weights and samples are taken at only one milking each test day and that milking must be alternated from AM to PM each succeeding test day. Thus far, a device for recording the time of start of each milking has been required for participation in this plan but elimination of that requirement is being considered. In the Official DHI with Alternate AM/PM Component Sampling Plan, milk weights for each cow are taken at both milkings on test day but samples for component testing are taken from only one milking. For several years these plans were offered only to herds on twice a day milking but they are now available for herds that milk three times a day.

Official DHIR Plans

Official DHIR (Dairy Herd Improvement Registry) Plans are dairy record keeping plans similar to the three Official DHI Plans but with an additional set of rules imposed by the sponsoring breed associations. These plans are for herds with registered cows and the participating dairymen receive additional benefits from special breed association programs.

Owner–Sampler Plans

The Owner–Sampler Plans and all the remaining plans that will be described are in the non-official category. As the name Owner–Sampler implies, all data in this group of plans are reported by the herd owner with no DHIA Supervisor involvement. There are no national rules for the Owner–Sampler Plans but participating dairymen must adhere to operational procedures required by the DRPCs for reporting data if the data are computer processed. Three of the Owner–Sampler Plans are parallel in name and procedures to the three Official DHI Plans. Additional plans in the Owner–Sampler category include: a milk-only plan in which no milk component information is provided; an AM/PM milk-only plan; plans in which either breed average or milk plant herd average component tests are used to generate milk component information for each cow, and; plans in which trimonthly milk weights or component tests are utilized. The Owner–Sampler Plans generally offer about the same information for within herd management as the equivalent Official DHI Plans but participating dairymen do not receive the benefits from authenticity of information in the official plans.

Table 3.1 *Superior Milk Yield (pounds) of Cows Enrolled in the DHIA Program*

Year	*Yield per cow, official DHI*	*Yield per cow, non-official plans*	*Yield per cow, non-DHIA*	*Superiority of official DHI cows*	*Superiority of non-official plan cows*
1979	14,786	13,969	9,531	5,255	4,438
1980	14,960	14,080	9,907	5,053	4,173
1981	15,137	14,177	10,085	5,052	4,092
1982	15,280	14,213	10,231	5,049	3,982
1983	15,521	14,544	10,517	5,004	4,027

Source: DHI Letter, USDA, *60:* 1, 1984.

Commercial Plans

Three dairy record keeping plans are offered in the Commercial category that parallel the three Official DHI Plans. The Commercial Plans are used by large herds mostly in the western states. A DHIA Supervisor usually collects all the data but the herds cannot participate in the official plans because facilities or some other constraint make it impossible for them to adhere to all the Official Rules.

Supervised Nonofficial Plans

This group of plans offers an opportunity to dairymen to benefit from the services of a DHIA Supervisor without having to follow all the Official Rules. Plans offered parallel the Official DHI Plans plus milk only plans. The majority of herds enrolled in Supervised Nonofficial Plans participate in the supervised Sampling AM/PM Plan and do not have a monitoring device to record the starting times of milkings as required in the Official DHI AM/PM with Monitoring Device Plan. Some of the Supervised Nonofficial Plans do not have clearly distinctive characteristics from the Commercial Plans. They are just offered in different parts of the country under different names.

Basic Plans

A Basic Management Information Plan is offered in one region for herds that are newly interested in the DHIA Program and others that want only simple, basic information for management purposes.

Dairy record keeping programs exist to satisfy the needs for management information of virtually every dairy herd in the country. There is scarcely a herd that would not benefit from participation in the DHIA Program, regardless of location or size. Table 3.1 shows the average benefits that accrue to participating herds in terms of extra milk yield per cow. Milk yield per cow has the greatest single influence on profitability of most herds.

3.5 Typical Management Information Available from the DHI Program

The information available to dairymen from the DHI Program differs according to the DRPC that processes the data. A core set of essential information is common to all DRPCs but is presented differently by each one. In addition, each DRPC generates special information that particularly appeals to the dairymen that it serves. A typical array of DHIA information will be shown in this section to demonstrate the valuable management information that dairymen can obtain from the DHI Program.

Most of the management information is calculated by the DRPC from the input data entered on the Cow Barn Sheet (Figure 3.1). After a herd joins the DHI Program, the DRPC sends a preprinted barn sheet each month on which to enter the following test day data. Barn sheets contain identification information for each cow in the herd that has ever calved and previous test day data including status codes and dates to help in entering accurate data next test day. Blank columns are provided to enter important events and dates that occurred since the last test day and for milk weights, component percentages and somatic cell counts or other mastitis screening test score. One of the major responsibilities of the DHIA Supervisor is to insure the accu-

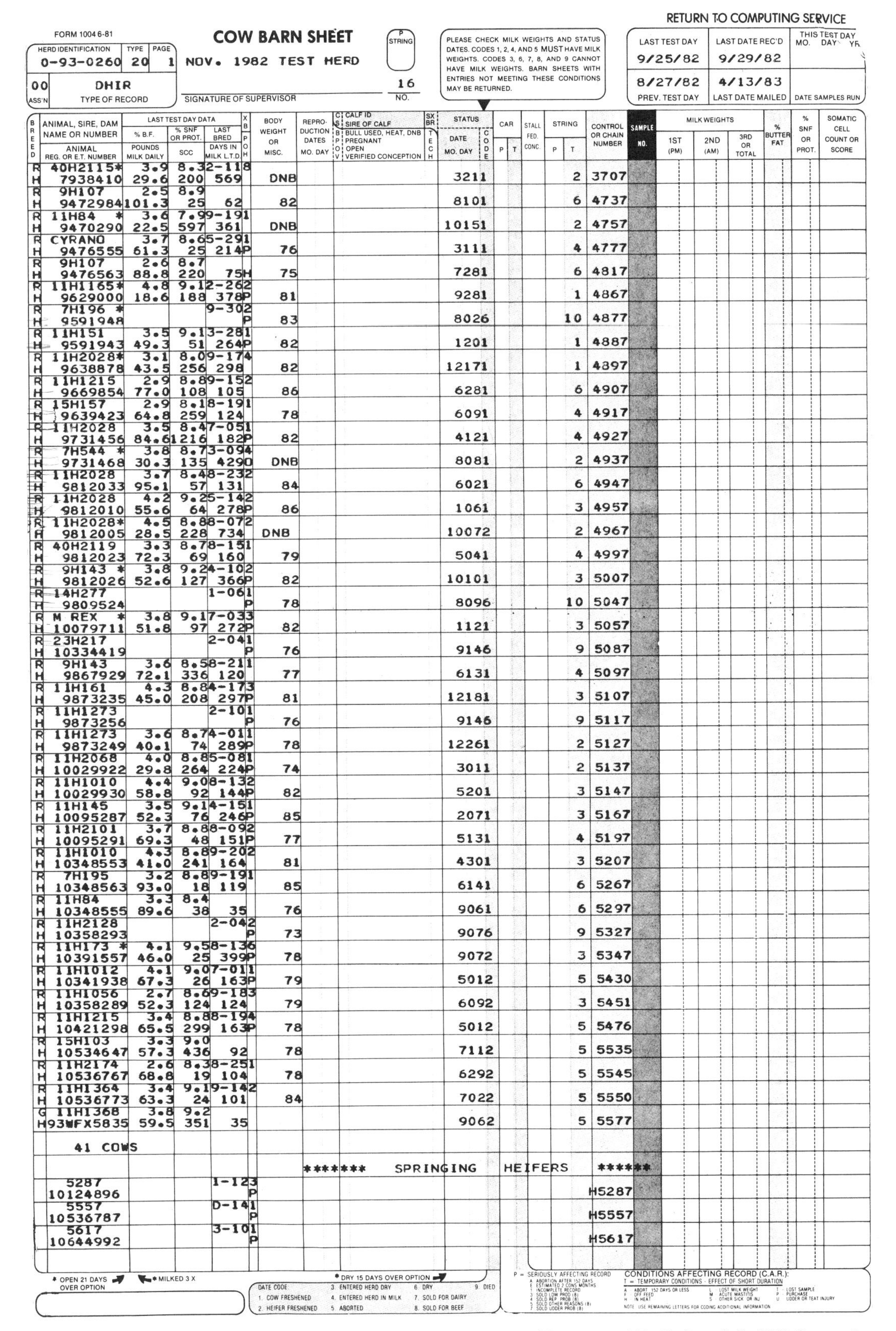

RETURN TO COMPUTING SERVICE

FORM 1004 6-81

COW BARN SHEET

HERD IDENTIFICATION	TYPE	PAGE
0-93-0260	20	1

NOV. 1982 TEST HERD

ASS'N	TYPE OF RECORD
00	DHIR

P STRING

SIGNATURE OF SUPERVISOR — NO. 16

PLEASE CHECK MILK WEIGHTS AND STATUS DATES. CODES 1, 2, 4, AND 5 MUST HAVE MILK WEIGHTS. CODES 3, 6, 7, 8, AND 9 CANNOT HAVE MILK WEIGHTS. BARN SHEETS WITH ENTRIES NOT MEETING THESE CONDITIONS MAY BE RETURNED.

LAST TEST DAY	LAST DATE REC'D	THIS TEST DAY MO. DAY YR.
9/25/82	9/29/82	
8/27/82	4/13/83	
PREV. TEST DAY	LAST DATE MAILED	DATE SAMPLES RUN

BREED	ANIMAL, SIRE, DAM NAME OR NUMBER / ANIMAL REG. OR E.T. NUMBER	% B.F. / POUNDS MILK DAILY	% SNF OR PROT. / SCC	LAST BRED / DAYS IN MILK L.T.D.	XB / POH	BODY WEIGHT OR MISC.	REPRODUCTION DATES MO. DAY	CS / BPOV: CALF ID, SIRE OF CALF, BULL USED, HEAT, DNB, PREGNANT, OPEN, VERIFIED CONCEPTION	SX BR / TECH	STATUS DATE MO. DAY / CODE	CAR P T	STALL FED. CONC.	STRING P	STRING T	CONTROL OR CHAIN NUMBER	SAMPLE NO.	MILK WEIGHTS 1ST (PM)	2ND (AM)	3RD OR TOTAL	% BUTTER FAT	% SNF OR PROT.	SOMATIC CELL COUNT OR SCORE
R	40H2115*	3.9	8.3	2-11	8																	
H	7938410	29.6	200	569		DNB				3211				2	3707							
R	9H107	2.5	8.9																			
H	9472984	101.3	25	62		82				8101				6	4737							
R	11H84 *	3.6	7.9	9-19	1																	
H	9470290	22.5	597	361		DNB				10151				2	4757							
R	CYRANO	3.7	8.6	5-29	1																	
H	9476555	61.3	25	214	P	76				3111				4	4777							
R	9H107	2.6	8.7																			
H	9476563	88.8	220	75	H	75				7281				6	4817							
R	11H1165*	4.8	9.1	2-26	2																	
H	9629000	18.6	188	378	P	81				9281				1	4867							
R	7H196 *			9-30	2																	
H	9591948				P	83				8026				10	4877							
R	11H151	3.5	9.1	3-28	1																	
H	9591943	49.3	51	264	P	82				1201				1	4887							
R	11H2028*	3.1	8.0	9-17	4																	
H	9638878	43.5	256	298		82				12171				1	4897							
R	11H1215	2.9	8.8	9-15	2																	
H	9669854	77.0	108	105		86				6281				6	4907							
R	15H157	2.9	8.1	8-19	1																	
H	9639423	64.8	259	124		78				6091				4	4917							
R	11H2028	3.5	8.4	7-05	1																	
H	9731456	84.6	1216	182	P	82				4121				4	4927							
R	7H544 *	3.8	8.7	3-09	4																	
H	9731468	30.3	135	429	O	DNB				8081				2	4937							
R	11H2028	3.7	8.4	8-23	2																	
H	9812033	95.1	57	131		84				6021				6	4947							
R	11H2028	4.2	9.2	5-14	2																	
H	9812010	55.6	64	278	P	86				1061				3	4957							
R	11H2028*	4.5	8.8	8-07	2																	
H	9812005	28.5	228	734		DNB				10072				2	4967							
R	40H2119	3.3	8.7	8-15	1																	
H	9812023	72.3	69	160		79				5041				4	4997							
R	9H143 *	3.8	9.2	4-10	2																	
H	9812026	52.6	127	366	P	82				10101				3	5007							
R	14H277			1-06	1																	
H	9809524				P	78				8096				10	5047							
R	M REX *	3.8	9.1	7-03	3																	
H	10079711	51.8	97	272	P	82				1121				3	5057							
R	23H217			2-04	1																	
H	10334419				P	76				9146				9	5087							
R	9H143	3.6	8.5	8-21	1																	
H	9867929	72.1	336	120		77				6131				4	5097							
R	11H161	4.3	8.8	4-17	3																	
H	9873235	45.0	208	297	P	81				12181				3	5107							
R	11H1273			2-10	1																	
H	9873256				P	76				9146				9	5117							
R	11H1273	3.6	8.7	4-01	1																	
H	9873249	40.1	74	289	P	78				12261				2	5127							
R	11H2068	4.0	8.8	5-08	1																	
H	10029922	29.8	264	224	P	74				3011				2	5137							
R	11H1010	4.4	9.0	8-13	2																	
H	10029930	58.8	92	144	P	82				5201				3	5147							
R	11H145	3.5	9.1	4-15	1																	
H	10095287	52.3	76	246	P	85				2071				3	5167							
R	11H2101	3.7	8.8	8-09	2																	
H	10095291	69.3	48	151	P	77				5131				4	5197							
R	11H1010	4.3	8.8	9-20	2																	
H	10348553	41.0	241	164		81				4301				3	5207							
R	7H195	3.2	8.8	9-19	1																	
H	10348563	93.0	18	119		85				6141				6	5267							
R	11H84	3.3	8.4																			
H	10348555	89.6	38	35		76				9061				6	5297							
R	11H2128			2-04	2																	
H	10358293				P	73				9076				9	5327							
R	11H173 *	4.1	9.5	8-13	6																	
H	10391557	46.0	25	399	P	78				9072				3	5347							
R	11H1012	4.1	9.0	7-01	1																	
H	10341938	67.3	26	163	P	79				5012				5	5430							
R	11H1056	2.7	8.6	9-18	3																	
H	10358289	52.3	124	124		79				6092				3	5451							
R	11H1215	3.4	8.8	8-19	4																	
H	10421298	65.5	299	163	P	78				5012				5	5476							
R	15H103	3.3	9.0																			
H	10534647	57.3	436	92		78				7112				5	5535							
R	11H2174	2.6	8.3	8-25	1																	
H	10536767	68.8	19	104		78				6292				5	5545							
R	11H1364	3.4	9.1	9-14	2																	
H	10536773	63.3	24	101		84				7022				5	5550							
G	11H1368	3.8	9.2																			
H	93WFX5835	59.5	351	35						9062				5	5577							
	41 COWS																					
							******* SPRINGING HEIFERS *******															
	5287			1-12	3																	
	10124896				P										H5287							
	5557			D-14	1																	
	10536787				P										H5557							
	5617			3-10	1																	
	10644992				P										H5617							

* OPEN 21 DAYS OVER OPTION — * MILKED 3 X — * DRY 15 DAYS OVER OPTION

DATE CODE: 1. COW FRESHENED 2. HEIFER FRESHENED 3. ENTERED HERD DRY 4. ENTERED HERD IN MILK 5. ABORTED 6. DRY 7. SOLD FOR DAIRY 8. SOLD FOR BEEF 9. DIED

P = SERIOUSLY AFFECTING RECORD: A ABORTION AFTER 152 DAYS; E ESTIMATED 2 CONS MONTHS; 1 INCOMPLETE RECORD; 3 SOLD LOW PROD (8); 4 SOLD REP. PROB (8); 5 SOLD OTHER REASONS (8); 7 SOLD UDDER PROB (8)

CONDITIONS AFFECTING RECORD (C.A.R.):
T = TEMPORARY CONDITIONS - EFFECT OF SHORT DURATION
A - ABORT 152 DAYS OR LESS; F - OFF FEED; H - IN HEAT; L - LOST MILK WEIGHT; M - ACUTE MASTITIS; S - OTHER SICK OR INJ; T - LOST SAMPLE; P - PURCHASE; U - UDDER OR TEAT INJURY
NOTE: USE REMAINING LETTERS FOR CODING ADDITIONAL INFORMATION

FIG. 3.1 A typical Cow Barn Sheet used in the DHIA Program. (Courtesy of B. H. Crandall, DHI Computing Service, Provo, Utah)

racy and completeness of information on the barn sheet that is essential to the DRPC being able to computer-process and summarize information for the management reports. The reader should examine the preprinted data and blank fields in Figure 3.1 in detail for a few cows to understand better the information reported on each cow on test day.

A detailed set of information on each cow in the herd is prepared each test day by the DRPC and reported on the Sample Day Lactation Report (Figure 3.2). This report is divided into three sections, identification, sample day data, and lactation to date. The best way for the reader to appreciate the scope of information for each cow on this report is to examine each field of data for a few cows, interpreting the codes and the dates along with the data. Two lines of data are given for each cow. In the lactation to date section, the top line shows actual yields of milk, milkfat, and protein, while the bottom line shows the standardized yields (Section 5.3) as signified by the code P in the Conditions Affecting Record column. If you make a list of all the information you can acquire from this report for just one cow, you will have an impressive set of information that may surprise you in just how much you know about the cow's present status and past history. The codes at the bottom of the report will be very helpful.

In addition to the information for each cow on the Sample Day Lactation Report, most DRPCs will periodically provide an Individual Cow Record (Figure 3.3), usually when each cow completes her lactation. Individual Cow Pages are even available every month as a special option from some of the DRPCs. The Individual Cow Page gives detailed histories of production, breedings, calvings, each sample day from

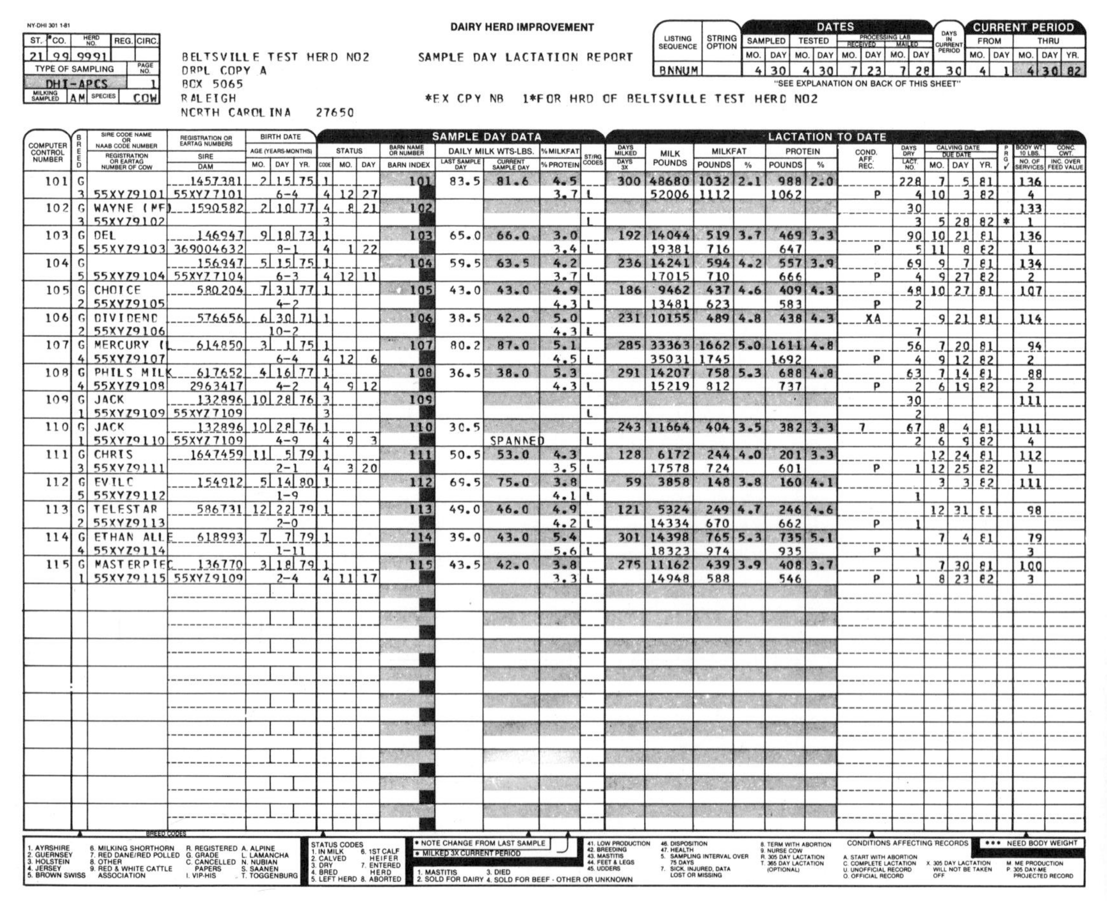

NY-DHI 301 1-81

DAIRY HERD IMPROVEMENT

SAMPLE DAY LACTATION REPORT

ST.	CO.	HERD NO.	REG.	CIRC.
21	99	9991		

TYPE OF SAMPLING: DHI-APCS — PAGE NO.: 1
MILKING SAMPLED: AM — SPECIES: COW

BELTSVILLE TEST HERD NO2
DRPL COPY A
BOX 5065
RALEIGH
NORTH CAROLINA 27650

*EX CPY NB 1*FOR HRD OF BELTSVILLE TEST HERD NO2

LISTING SEQUENCE	STRING OPTION	DATES: SAMPLED MO.	DAY	TESTED MO.	DAY	PROCESSING LAB RECEIVED MO.	DAY	MAILED MO.	DAY	DAYS IN CURRENT PERIOD	CURRENT PERIOD FROM MO.	DAY	THRU MO.	DAY	YR.
BNNUM		4	30	4	30	7	23	7	28	30	4	1	4	30	82

"SEE EXPLANATION ON BACK OF THIS SHEET"

COMPUTER CONTROL NUMBER	BREED	SIRE CODE NAME OR NAAB CODE NUMBER / REGISTRATION OR EARTAG NUMBER OF COW	REGISTRATION OR EARTAG NUMBERS: SIRE / DAM	BIRTH DATE MO.	DAY (AGE YEARS-MONTHS)	YR.	STATUS CODE	STATUS MO.	STATUS DAY	BARN NAME OR NUMBER / BARN INDEX	DAILY MILK WTS-LBS. LAST SAMPLE DAY	CURRENT SAMPLE DAY	% MILKFAT / % PROTEIN	ST/RG CODES	DAYS MILKED / DAYS 3X	MILK POUNDS	MILKFAT POUNDS	MILKFAT %	PROTEIN POUNDS	PROTEIN %	COND. AFF. REC.	DAYS DRY / LACT. NO.	CALVING DATE / DUE DATE MO.	DAY	YR.	PRG ✓	BODY WT. 10 LBS. / NO. OF SERVICES	CONC. CWT. / INC. OVER FEED VALUE
101	G		1457381	2	15	75	1			101	83.5	81.6	4.5		300	48680	1032	2.1	988	2.0		228	7	5	81		136	
	3	55XYZ9101	55XYZ7101		6-4		4	12	27				3.7	L		52006	1112		1062		P	4	10	3	82		4	
102	G	WAYNE (MF	1590582	2	10	77	4	8	21	102												30					133	
	3	55XYZ9102					3							L								3	5	28	82	*	1	
103	G	DEL	146947	9	18	73	1			103	65.0	66.0	3.0		192	14044	519	3.7	469	3.3		90	10	21	81		136	
	5	55XYZ9103	369004632		8-1		4	1	22				3.4	L		19381	716		647		P	5	11	8	82		1	
104	G		156947	5	15	75	1			104	59.5	63.5	4.2		236	14241	594	4.2	557	3.9		69	9	7	81		134	
	5	55XYZ9104	55XYZ7104		6-3		4	12	11				3.7	L		17015	710		666		P	4	9	27	82		2	
105	G	CHOICE	580204	7	31	77	1			105	43.0	43.0	4.9		186	9462	437	4.6	409	4.3		48	10	27	81		107	
	2	55XYZ9105			4-2								4.3	L		13481	623		583		P	2						
106	G	DIVIDEND	576656	6	30	71	1			106	38.5	42.0	5.0		231	10155	489	4.8	438	4.3	XA		9	21	81		114	
	2	55XYZ9106			10-2								4.3	L								7						
107	G	MERCURY (L	614850	3	1	75	1			107	80.2	87.0	5.1		285	33363	1662	5.0	1611	4.8		56	7	20	81		94	
	4	55XYZ9107			6-4		4	12	6				4.5	L		35031	1745		1692		P	4	9	12	82		2	
108	G	PHILS MILK	617652	4	16	77	1			108	36.5	38.0	5.3		291	14207	758	5.3	688	4.8		63	7	14	81		88	
	4	55XYZ9108	2963417		4-2		4	9	12				4.3	L		15219	812		737		P	2	6	19	82		2	
109	G	JACK	132896	10	28	76	3			109												30					111	
	1	55XYZ9109	55XYZ7109				3							L								2						
110	G	JACK	132896	10	28	76	1			110	30.5				243	11664	404	3.5	382	3.3	7	67	8	4	81		111	
	1	55XYZ9110	55XYZ7109		4-9		4	9	3			SPANNED		L								2	6	9	82		4	
111	G	CHRIS	1647459	11	5	79	1			111	50.5	53.0	4.3		128	6172	244	4.0	201	3.3			12	24	81		112	
	3	55XYZ9111			2-1		4	3	20				3.5	L		17578	724		601		P	1	12	25	82		1	
112	G	EVILC	154912	5	14	80	1			112	69.5	75.0	3.8		59	3858	148	3.8	160	4.1			3	3	82		111	
	5	55XYZ9112			1-9								4.1	L								1						
113	G	TELESTAR	586731	12	22	79	1			113	49.0	46.0	4.9		121	5324	249	4.7	246	4.6			12	31	81		98	
	2	55XYZ9113			2-0								4.2	L		14334	670		662		P	1						
114	G	ETHAN ALLE	618993	7	7	79	1			114	39.0	43.0	5.4		301	14398	765	5.3	735	5.1			7	4	81		79	
	4	55XYZ9114			1-11								5.6	L		18323	974		935		P	1					3	
115	G	MASTERPIEC	136770	3	18	79	1			115	43.5	42.0	3.8		275	11162	439	3.9	408	3.7			7	30	81		100	
	1	55XYZ9115	55XYZ9109		2-4		4	11	17				3.3	L		14948	588		546		P	1	8	23	82		3	

BREED CODES: 1. AYRSHIRE 2. GUERNSEY 3. HOLSTEIN 4. JERSEY 5. BROWN SWISS 6. MILKING SHORTHORN 7. RED DANE/RED POLLED 8. OTHER 9. RED & WHITE CATTLE ASSOCIATION R. REGISTERED G. GRADE C. CANCELLED PAPERS I. VIP-HIS A. ALPINE L. LAMANCHA N. NUBIAN S. SAANEN T. TOGGENBURG

STATUS CODES: 1. IN MILK 2. CALVED 3. DRY 4. BRED 5. LEFT HERD 6. 1ST CALF HEIFER 7. ENTERED HERD 8. ABORTED

* NOTE CHANGE FROM LAST SAMPLE
* MILKED 3X CURRENT PERIOD

1. MASTITIS 2. SOLD FOR DAIRY 3. DIED 4. SOLD FOR BEEF - OTHER OR UNKNOWN

41. LOW PRODUCTION 42. BREEDING 43. MASTITIS 44. FEET & LEGS 45. UDDERS 46. DISPOSITION 47. HEALTH 5. SAMPLING INTERVAL OVER 75 DAYS 7. SICK, INJURED, DATA LOST OR MISSING 8. TERM WITH ABORTION 9. NURSE COW R. 305 DAY LACTATION T. 365 DAY LACTATION (OPTIONAL)

CONDITIONS AFFECTING RECORDS: A. START WITH ABORTION C. COMPLETE LACTATION U. UNOFFICIAL RECORD O. OFFICIAL RECORD X. 305 DAY LACTATION WILL NOT BE TAKEN OFF M. ME PRODUCTION P. 305 DAY-ME PROJECTED RECORD

*** NEED BODY WEIGHT

FIG. 3.2 A typical Sample Day Lactation Report used in the DHIA Program. (Courtesy of L. H. Wadell, New York DRPC, Ithaca, New York)

INDIVIDUAL COW RECORD

NY-DHI 309 1-80

ST.	CO.	HERD NO.	PAGE NO.
21	67	0963	1
REG.	CIRC.	DATE PROCESSED	
9	4	05-02-79	
CCNUM		YEAR	

FARM NAME
DOE, JOHN
ANY STREET
ANY CITY
ANY STATE ZIP C

COMPUTER CONTROL NUMBER	BIRTH DATE MO.	DAY	YR.	BR	COW REGISTRATION OR EARTAG	BARN INDEX	BARN NAME OR NUMBER
142	1	28	73	03	8390195		PRNCES
	SPECIES	COW					

NAME

ETA VALUES
± MILK
± % TEST
+ 300
- .05

ETA CALC 79-01

BR	SIRE REGISTRATION OR EARTAG	CODE NAME	NEAISC 79-05 SIRE COMPARISON NO. DAUG.	MILK	TEST	MILKFAT	MILK CONFID.
03	1458744	ASTRONAUT	6028	850	- .02	28	40
	NAAB CODE NUMBER	40H 96					

NAME

COMPUTER CONTROL NUMBER	BR	DAM REGISTRATION OR EARTAG	BARN INDEX	BARN NAME OR NUMBER
	03	7707623		

NAME SUSAN

LACTATION PRODUCTION SUMMARY

LACT. NO.	CALVING DATE MO.	DAY	YR.	AGE AT CALVING YR.	MO.	DAYS DRY	CODES	305 DAY, (OR LESS) ACTUAL PRODUCTION DAYS / DAYS 3X	MILK	% MILKFAT / % PROTEIN	LBS. MILKFAT / LBS. PROTEIN	305 DAY-M.E. SEASON PRODUCTION MILK	MILKFAT / PROTEIN	COMPLETE LACTATION DAYS / PROTEIN DAYS	MILK	MILKFAT / PROTEIN	VALUE OF PRODUCT	TYPE OF RECORD
1	4	01	75	2	02			305	11,621	3.4	391	15,340	516	576	17,629	610	1576	DHI
2	1	30	77	4	00	94		305	13,084	3.4	442	13,607	460	327 239	13,363	455 272	1243	DHI
3	3	09	78	5	01	76		305	15,346	3.4 3.1	517 479	15,499	522	317 317	15,459	522 485	1532	DHI
4	4	19	79	6	02	89		7	457	2.8	13 16			7	LACT. TO DATE			DHI

2. SOLD FOR DAIRY
3. DIED
4. SOLD FOR BEEF
5. SAMPLING INTERVAL OVER 75 DAYS
7. SICK, INJURED, DATA LOST OR MISSING
8. TERM WITH ABORTION
9. NURSE COW

▲ CODES
A. START WITH ABORTION
T. TWINS
3X. 3X MILKING ONE OR MORE PERIODS
F. DATA FROM FIELD CORRECTION
H. PART OR ALL OF RECORD NOT MADE IN THIS HERD.
I. MASTITIS

AVAILABLE LIFETIME PRODUCTION	MILK	MILKFAT	PROTEIN
	46,451	1,587	757

SAMPLE DAY DATA

(Upper figure in each cell: % MILKFAT / % PROTEIN; lower figure: MILK / LTD DAYS)

LACT. NO.	SAMPLE 1 % MILKFAT / MILK	% PROTEIN / LTD DAYS	SAMPLE 2 % MILKFAT / MILK	% PROTEIN / LTD DAYS	SAMPLE 3 % MILKFAT / MILK	% PROTEIN / LTD DAYS	SAMPLE 4 % MILKFAT / MILK	% PROTEIN / LTD DAYS	SAMPLE 5 % MILKFAT / MILK	% PROTEIN / LTD DAYS	SAMPLE 6 % MILKFAT / MILK	% PROTEIN / LTD DAYS	SAMPLE 7 % MILKFAT / MILK	% PROTEIN / LTD DAYS	SAMPLE 8 % MILKFAT / MILK	% PROTEIN / LTD DAYS	SAMPLE 9 % MILKFAT / MILK	% PROTEIN / LTD DAYS	SAMPLE 10 % MILKFAT / MILK	% PROTEIN / LTD DAYS	SAMPLE 11 % MILKFAT / MILK	% PROTEIN / LTD DAYS	MAST. COUNT
1	4.0 49	18	3.4 46	51	3.0 46	79	3.4 37	107	3.2 41	143	3.4 38	178	2.9 38	208	3.0 33	239	3.9 22	264	3.4 33	297	3.4 31	331	
2	4.0 57	28	3.3 51	56	3.5 49	88	3.3 53	117	3.3 55	3.0 146	2.9 45	2.8 174	3.0 38	3.1 207	3.2 33	3.2 236	3.3 27	2.8 266	3.7 18	3.5 297			
3	4.0 74	3.3 15	3.5 71	3.6 45	3.3 75	2.8 77	2.9 68	3.1 108	3.5 57	2.7 136	3.0 50	3.0 170	3.1 42	3.0 197	3.5 29	3.3 231	3.4 26	3.1 259	3.5 16	3.8 290			
4	2.8 73	3.6 7																					
5																							

COW BREEDING RECORD DAYS OPEN	NO. BREED-INGS	LAST SUCCESSFUL BRED DATE MO.	DAY	YR.	SIRE IDENTITY CODE NAME OR NAAB CODE NO. REGISTRATION OR EARTAG	BR	LACT. NO.	COW BODY WT. 10 LBS.	CALF INFORMATION COMPUTER CONTROL NUMBER	BIRTH DATE MO.	DAY	YR.	CALF EARTAG	BARN NAME OR NUMBER / BARN INDEX	EST. PRODUCING ABILITY MILK	% FAT	PROB	DISP	SEX
							1	105	36	4	01	75	21WJK5218						F
381	4	4	16	76	1684030	03	2	124		1	30	77			+ 299	- .10		4	M
120	2	5	30	77	AMERICUS 1669134	03	3	140		3	09	78			+ 322	- .03		4	M
125	3	7	12	78	1763499	03	4	153		4	19	79			+ 300	- .05		4	M
							5	153											

FIG. 3.3 A typical Individual Cow Record used in the DHIA Program. (Courtesy of L. H. Wadell, New York DRPC, Ithaca, New York)

past lactations, genetic merit of the cow's sire and dam, estimated transmitting ability (ETA) of the cow herself and an Estimated Producing Ability of her calves. This information is valuable for a wide variety of decisions.

A variety of other reports for individual cow management is available from all the DRPCs.

Action Sheets are pocket-size lists of cows requiring specific attention on prescribed dates such as breed, dry off, due to calve, check if pregnant, candidate for culling, and recommended for dry cow treatment to prevent mastitis.

Calf Identification Sheet has the calf identification information reported by the dairyman plus space to log the calf's vaccination, health and breeding record until she enters the milking herd, and also to sketch her color markings for positive identification and registration.

Annual Herd Register constitutes an inventory of the herd including complete identification of each animal and space for disposal information.

Special Listings are produced by most of the DRPCs showing individual cow information for especially important management factors like somatic cell counts for each test day in the current lactation, test day production values for the current lactation, and each cow's recent reproduction history.

Management of the entire dairy herd as a business unit has become more important as herds have increased in size. DRPCs have responded to the growing need to monitor the effectiveness of past management practices and for future planning by producing sophisticated summaries of important management factors for the entire herd and for designated management units within the herd, called management strings. An especially valuable report for herd management is the Dairy Herd Profile, the first page of which is shown in Figure 3.4. Many pages would be required to explain the usefulness of the many sections of information on the Dairy Herd Profile. As with the individual cow information shown in Figures 3.1 through 3.3, a detailed examination of a few sections of the Dairy Herd Profile will reveal the wealth of information provided to dairymen. For example, note in the upper left corner how statistics are summarized separately for cows in different lactations as well as for the entire herd. This breakdown enables dairymen to evaluate management trends by comparing groups of cows in different parities. Improvement and decline in management can be detected by comparing the status of various management factors for 1st lactation versus 4th lactation or all cows.

DAIRY HERD PROFILE PAGE 1

NYDHI 334 4/78

ST.	CO.	HERD NO.	REG.	CIRC.
21	67	0963	99	99

TYPE OF SAMPLING: DHI — PAGE NO. 1 — SPECIES: COW

FARM NAME
DOE, JOHN
ANY STREET
ANY CITY
ANY STATE ZIP C

DATE LAST SAMPLED (MONTH DAY YEAR): MAR 27 1979
DATE PROCESSED: 04-23-79
HERD OWNERS TELEPHONE NUMBER: 607-256-2126

LACTATION NUMBER	NUMBER OF COWS	AVERAGE AGE AT CALVING YRS. MOS.	AVERAGE BODY WEIGHT AT CALVING	LTD DAYS IN MILK: AVERAGE DAYS	≤ 90	91-180	181-305	> 305	DRY	DAYS DRY BEFORE CALVING: AVERAGE DAYS	< 42	42-55	56-60	> 60	DAYS OPEN: AVERAGE DAYS	≤ 100	101-150	> 150	AVERAGE DAYS IN MILK: FIRST BREEDING	LAST BREEDING	SERVICES PER CONCEPTION: AVERAGE NUMBER	1	2	3	4+	NUMBER NOT GOING TO BREED	CALVING INTERVAL: PREVIOUS	PROJECTED	LACTATION NUMBER
1	45	2- 3	116	159	20	56	16	7	2						93	71	19	10	74	95	2.0	50	25	8	17			12.4	1
2	45	3- 4	139	199	16	31	22	11	20	56	8	36	36	19	130	52	19	29	78	125	2.2	40	45		15		13.0	13.4	2
3	19	4- 5	147	209	22	17	28	17	17	70	13	13	7	67	94	65	18	18	84	100	2.3	67	8		25		11.4	12.5	3
4	19	5- 8	154	184	28	22	28		22	69	14		29	57	91	78	11	11	43	99	1.7	55	27	9	9		12.2	12.5	4
5 & OVER	37	8- 2	156	162	35	22	24	3	16	88	3	19	10	68	120	55	23	23	71	128	2.0	42	50		8		13.9	13.5	5 & OVER
ALL	165	4- 6	139	179	23	33	22	7	14	70	8	22	22	48	109	63	19	19	73	111	2.0	49	33	3	15		12.9	12.9	ALL
PREGNANT	68														108	65	15	20											PREGNANT
PREGNANT (?)	63														115	51	29	21											PREG. (?)
OPEN	21																												OPEN

LACT. NO.	COWS ENTERING HERD — FIRST CALF HEIFERS: % RAISED	% PURCHASED	% ALL	NUMBER ALL	OTHER: %	NUMBER	% NEW ANIMALS	COWS LEAVING HERD — NUMBER OF COW BY REASON: DAIRY PURPOSES	DIED	LOW PROD.	BREEDING PROBLEMS	MASTITIS	FEET AND LEGS	UDDERS	OTHER OR UNKNOWN	% ANIMALS LEAVING	AVERAGE ETA'S: MILK	MILKFAT %	PROTEIN %	PRODUCT VALUE	SOMATIC CELL COUNTS (000's/ml): NUMBER OF COWS	PERCENT OF COWS < 300	300-500	500 TO 750: NO.	%	> 750: NO.	%	AVERAGE COUNT	LACTATION NUMBER
1									1						6	4	- 245	+ .02		- 24	44	91	5			2	5	186	1
2								1	2						5	5	+ 450	- .03		40	35	71	11	2	6	4	11	287	2
3								1	1						6	5	+ 75	- .03		1	16	44	19	2	13	4	25	938	3
4															4	2	- 314	- .01		- 38	15	53	13			5	33	630	4
5 & OVER									1						7	5	+ 280	+ .06		39	30	43	7	1	3	14	47	1049	5 & OVER
ALL	92	8	93	39	7	3	25	2	5						28	21	+ 101	+ .00		9	140	66	9	5	4	29	21	530	ALL

ETA DISTRIBUTION AND AVERAGES

LACT. NO.	NUMBER ETA AVAILABLE	PERCENT OF COWS: < -500	-500 THRU 0	1-500	> 500	NUMBER > 500	MILK LBS.	MILKFAT %	PROTEIN %	PRODUCT VALUE
1	29	28	48	21	3	1	- 267	- .04		-36
2	45	11	16	38	36	16	+ 316	- .03		25
3	18	11	22	44	22	4	+ 216	+ .00		20
4	18	6	11	33	50	9	+ 544	- .03		51
5&OVER	37	8	24	43	24	9	+ 247	+ .00		24
ALL	147	13	24	36	27	39	+ 199	- .02		15

DRY PERIOD EVALUATION

4 % < 42 DAYS 23 % 42-55 DAYS 73 % > 55 DAYS

LACT. NO.	PERCENT OF COWS: < 30	30-39	40-49	50-59	60-69	70-79	80-89	90-99	≥ 100
1									
2	6	3	14	58	11		6		3
3				23	23	15	8	8	23
4				25	33	17	8		17
5&OVER		3	6	23	16	3	6	13	29
ALL	2	2	8	37	17	5	7	5	16

A.I. SERVICE SIRE ANALYSIS

SIRE IDENTITY	NO. OF COWS BRED	PREDICTED DIFFERENCES: MILK	MILKFAT %	PROTEIN %	PRODUCT VALUE
STREPHON	11	1750	-.14		154
COMMANDER	11	794	-.03		77
TRIPLE THREAT	8	50	.16		32
JUPITER	8	1500	-.01		155
WAYNE	6	1855	-.10		174
PERSUADER	5	1547	-.03		156
ELATE	4	1300	-.02		132
STAR	4	878	-.06		79
ELEVATION	3	1400	-.04		138
LEWIE	3	1000	-.06		92
CONDUCTOR	2	1663	-.02		170
TIPPY	2	493	-.14		24
BOLD C	2	650	.30		126
BARRETT	2	1150	-.04		112
MIGHTY	2	579	-.13		34
ASTRO MINER	2	1507	-.06		146
JASPER	1	450	-.28		8
IVANHOE STAR	1	500	.04		58
CAVALIER	1	1550	-.22		116
GLENDELL	1	1777	-.10		166
NEIL	1	1500	.00		157
KENNEDY	1	400	-.08		25
MONARCH	1	1451	-.32		86
REMAINING 1 AI SIRES	1	549	-.16		26
27 %COWS BRED TO > 1200	45	1603	-.07		152
5 %COWS BRED TO 800-1200	9	979	-.06		91
AVERAGE WEIGHTED P.D.	83	1150	-.04		111

WEIGHTED PRED. DIFF.

(0 %) COWS BRED-AI PROOF NOT AVAILABLE
(25 %) COWS BRED-NON AI SIRES
(4 %) COWS BRED-SIRE NOT IDENTIFIED
(21 %) COWS NOT BRED

RELATIONSHIP OF DAYS DRY AND PROJECTED 305 DAY PRODUCTION

DAYS DRY BEFORE CALVING	NUMBER OF COWS	PERCENT OF COWS	305 DAY ACTUAL PROJECTED RECORDS: MILK	MILKFAT POUNDS	MILKFAT %	PROTEIN POUNDS	PROTEIN %	ME - SEASON ADJUSTED: MILK	MILKFAT	PROTEIN
<30	2	2	17184	626	3.6	576	3.3	19728	720	661
30-39	2	2	16444	562	3.4	513	3.1	17049	583	532
40-49	7	8	19310	689	3.6	658	3.4	22238	795	757
50-59	34	37	18497	696	3.8	619	3.4	20193	758	676
60-69	16	17	19362	723	3.7	635	3.3	20366	759	670
≥ 70	31	34	18490	701	3.8	605	3.3	19093	724	625

EPA INFORMATION ON YOUNG STOCK

50 ANIMALS LESS THAN 12 MONTHS OF AGE	AVERAGE MILK EPA + 917 PERCENT > ZERO 82 AVERAGE MILKFAT % EPA - .09 AVERAGE PROTEIN % EPA
62 ANIMALS 12 MONTHS OR GREATER IN AGE	AVERAGE MILK EPA + 512 PERCENT > ZERO 77 AVERAGE MILKFAT % EPA +.00 AVERAGE PROTEIN % EPA

PEAK PRODUCTION FIRST 95 DAYS; % DROP S. D. PRODUCTION FOR 30 DAY PERIOD AT MID-LACTATION

LACT. NO.	PERCENT OF COWS: ≤ 39	40-49	50-59	60-69	70-79	80-89	90-99	100-119	≥ 120	AVERAGE LBS.	% DROP — PERCENT OF COWS: < 8 %	8-10 %	11-12 %	> 12 %	AVERAGE %
1		9	34	45	5	5	2			62	66	7		28	- 5
2 & OVER		3	1	7	17	28	24	16	4	87	54	11	2	33	- 9
ALL		5	11	19	13	21	17	11	3	79	58	9	1	31	- 8

PRODUCTION PER DAY OF LIFE SINCE 2 YEARS OF AGE

CATEG.	1ST LACT.	2ND & 3RD LACT.	4 + LACT.	ALL
MILK	31	37	36	36
% MILKFAT	3.6	3.7	3.7	3.7
% PROTEIN	3.3	3.3	3.2	3.3

FRESH COW PRODUCTION AND PERSISTENCY ANALYSIS

CALVING MONTH-YR.	NO. OF COWS	1ST SAMPLE	2ND SAMPLE	3RD SAMPLE	4TH SAMPLE	5TH SAMPLE	6TH SAMPLE	7TH SAMPLE	8TH SAMPLE
8-78	10	75	75	68	62	56	54	46	42
9-78	12	75	76	68	62	58	54	51	
10-78	27	67	71	66	62	60	60		
11-78	12	75	76	75	68	66			
12-78	17	70	76	72	70				
1-79	19	76	87	77					
2-79	6	70	81						
3-79	10	80							

MILK PRODUCTION

COOPERATIVE PROGRAM-U.S. DEPARTMENT OF AGRICULTURE-STATE AGRICULTURE EXTENSION SERVICES AND STATE DHI COOPERATIVES IN CONNECTICUT, DELAWARE, MAINE, MASSACHUSETTS, NEW HAMPSHIRE, NEW JERSEY, NEW YORK, RHODE ISLAND, VERMONT and WEST VIRGINIA

Fig. 3.4 A typical Dairy Herd Profile sheet for herd management in the DHIA Program. (Courtesy of L. H. Wadell, New York DRPC, Ithaca, New York)

Other valuable herd management reports are also provided by the DRPCs.

A *Herd Summary* shows annual data on herd size, production, feeding, income, and labor efficiency for the past 10 years.

A *Herd Summary and Management Report* details the herd's status for the past month compared to the past year for a variety of management factors.

Genetic and Environmental Herd Profile shows long term trends, up to 15 years, in production, feeding, genetic merit of cows in the herd, genetic merit of cows culled, top ETA cows, and sires used.

Ladder of Progress Sheet compares change in status between the past two six month periods in production, feeding, breeding, income, turnover, and miscellaneous categories to aid the dairyman to assess progress attained in various aspects of herd management and how well his goals were met.

On-Farm Computing Equipment

The present array of DHIA information results from several decades of evolution in the dairy industry and management needs of dairymen. As herds and investments have grown larger, management decisions have greater economic impact and require more extensive information on a real-time basis. Traditional monthly feedback of management information on paper from the DRPC no longer suffices in many herds. Therefore on-farm computing equipment, ranging from dumb terminals to microcomputers with many megabytes of storage and communications capability, has proliferated as a source of management information. Obtaining adequate computer hardware is easy and relatively inexpensive. But, obtaining, maintaining, and improving software to generate information for management decisions can be much more difficult for most dairymen.

At least 4 DRPCs have responded to this problem with on-line data storage and software packages so participating dairymen can input data and obtain reports at any time while having to acquire a minimum of computer expertise. Dairymen can customize their reports for their own needs. Thus they can obtain individualized reports at relatively little cost in dollars or time invested in learning computer programming and data processing procedures. Some farms have microcomputers that can be used on a stand-alone basis for some purposes and can also act as terminals for communicating with a DRPC. This microcomputer configuration takes advantage of the DRPCs sophistication of summarization as well as permitting use of the computer for stand-alone work when appropriate. Critical factors in deciding whether a microcomputer should be used to communicate with a DRPC or used on a stand-alone basis are the acquisition of software to produce the management information desired, maintenance and improvement of software and the expense of data entry and verification. When a dairyman joins the DHI Program he can virtually eliminate the software generation and maintenance problems and a large part of the work of data entry and verification. He would pay the DHIA Supervisor and the DRPC for those services. He would enter some data into the DRPC system through his on-farm computer if he chooses but the DHIA Supervisor and DRPC can take care of all data entry and verification.

On the other hand, automated equipment is available for milking parlors that permits electronic data collection. This equipment can be connected to on-farm computing equipment to permit on-farm data analysis that generates all the DHIA information essential for official records. The DHI Program has provisions for dairymen to utilize automated equipment of various levels of sophistication including taking on all the responsibilities of a DRPC. Four configurations of on-farm automated equipment are sanctioned for integration into the DHI Program.

Configuration A: The traditional DHIA data reporting system involving a DHIA Supervisor completing the barn sheets is used with the possible involvement of some automated data collection devices and/or an on-farm computer.

Configuration B: A dairy records collection unit is utilized including an automated data capture system; data are transmitted to a DRPC through a terminal or microcomputer; the DRPC computes all the DHIA information.

Configuration C: The farm has a dairy records processing system with automated data capture; all DHIA records are computed on the farm and transmitted to a DRPC.

Configuration D: The farm assumes all responsibility of a DRPC with automated data capture and computation of all DHIA information on the farm; appropriate DHIA information is transmitted to required recipients such as USDA, Extension and breed associations.

Advantages and disadvantages exist for each type of on-farm computer configuration for obtaining dairy records. Relatively few dairymen have the work time and expertise available on their farms to utilize an on-farm computer to full advantage without a significant amount of outside help. Dairymen should weigh all factors carefully before deciding on the type of computer hardware and software and automated equipment that will pay a reasonable return for their investment in it. The DHI Program offers many alternatives and a high degree of excellence in variety and accuracy of information. Dairymen should be sure that some other alternative will be superior on a long term basis before rejecting the DHI Program as a primary source of dairy records.

3.6 The Broad Spectrum of Benefits from the DHIA Program

The dairymen whose herds are enrolled in the DHIA Program benefit the most from it. The exact extent of each dairyman's benefits depends largely on the completeness of the information he reports and his own ability to take advantage of the valuable management information he receives. Average monetary benefits for cows in herds participating in the DHIA Program can be estimated from the superiority in yields shown in Table 3.1. No matter how the yield superiority figures in Table 3.1 are converted into profit from the DHIA Program, the $18.00 or so annual investment for each cow's annual data yields a handsome return.

A significant additional advantage of the DHIA Program is that the data from the program are used for teaching and research for the benefit of all dairymen and the general public, in exchange for the public and private support that is given the program. Some of the most important ways in which all dairymen and the general public benefit from the DHIA Program are described in the remainder of this section.

Teaching and Research

Information obtained from the DHIA Program is used widely in dairy courses in high schools and universities as well as in dairy extension education programs. It is also a primary source of data for research on many aspects of dairy cattle genetics and management.

Genetic Improvement

Perhaps the greatest single benefit to accrue to the dairy industry as a whole from the DHIA Program has been the information to evaluate genetic merit of dairy cattle. The dairy industry has used DHIA information to achieve a rate of genetic improvement that far surpasses that of other livestock industries in the United States. For over 40 years, a national program for genetic evaluation of dairy bulls (and more recently of cows also) has been based on information from the DHIA Program. Without this information it would have been impossible for the AI industry to contribute as much genetic progress as it has. First, the cost would have been prohibitive to obtain sufficient information on enough cows to calculate accurate estimates of the genetic merit of their sires. Second, without some sort of regulation on the accuracy and integrity of information, many different kinds of data and methods would probably have been used to estimate genetic merit, and these might have varied considerably in accuracy. Third, the teaching and research support of scientists at the land-grant universities and research support of scientists in USDA using DHIA records has permitted the rapid progress in methodology of genetic evaluation, which would not have been possible otherwise.

Animal Identification

Animal identification has become a major problem in recent years as herds have grown larger. The uniform national eartag system has enabled unique, permanent identification for each cow. However, the eartags are small and not readable except within arm's length. Practical methods of identification that are inexpensive, permanent, nondamaging to the animal, and readable at a distance of at least 100 to 150 feet are needed.

Methods of identification of cows can be divided into 2 categories, permanent and nonpermanent.[2] Permanent identification includes ear and udder tattoos, sketches, photographs, brands (acid, caustic, hot iron, and freeze brands), and electronic implants. Nonpermanent types of identification include neck straps, eartags, anklet straps, brisket tags, neck chains with tags, flank tags, tail tags, and marking paint and crayon, and external electronic devices attached to animals in various ways. Examples of nonpermanent types of identification are shown in Figure 3.5. None of these methods is ideal under all conditions. Many types of tags have the inherent disadvantage of requiring the skin to be pierced for

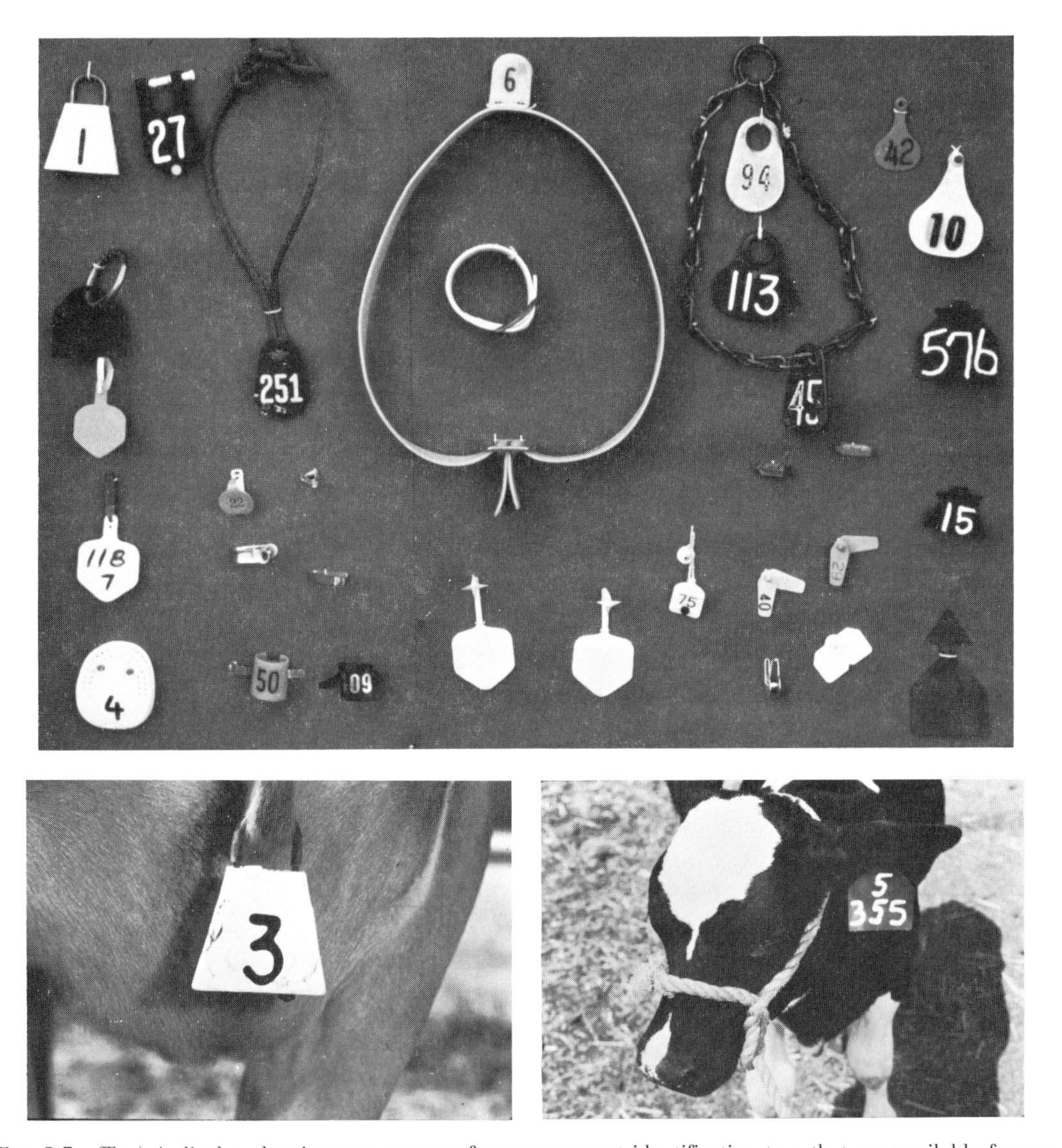

Fig. 3.5 *(Top)* A display showing many types of nonpermanent identification tags that are available for use on dairy cattle. Examples of a brisket tag *(bottom left)* and one type of eartag *(bottom right)*. (Courtesy of N. W. Hooven, Jr.)

the tag to be attached to the animal. With eartags this is not particularly harmful to the animal; however, it is a serious disadvantage with brisket and flank tags where the risk of infection is fairly high. Tail tags are difficult to apply so they will be permanently retained. If they are not tight enough they slide off, especially during the fly season; if they are too tight, blood circulation is reduced and part of the animal's tail may slough off.

Freeze branding is a practical method of permanent identification. The application of extreme cold selectively destroys the melanocytes or pigment-producing cells on the skin, resulting in the growth of white hair in the branded area. This method can also be used on white animals since a longer application of the extreme cold will destroy the hair completely, leaving a bare skin brand. Examples of freeze brands are shown in Figure 3.6. When performed properly, freeze branding has all the desirable results of hot branding with few of the disadvantages. Discomfort to the animal is less, skin damage and therefore hide damage is greatly reduced, and the risk of infection is virtually eliminated. The equipment needed, procedure used, and the results of freeze branding are described in Appendix I-C.

Another method that has promise for large herds is electronic identification. A small digital circuit called a transponder would contain the animal's unique identification. It could be implanted subcutaneously or hung on an animal in various ways. The transponder can be passive, reacting only when an interrogator energizes it or it can be active, that is, have its own power supply (battery). An electronic interrogator then senses a code in the transponder and converts it to the animal's ID. Such devices could be part of automatic milking, feeding, and data recording systems to provide automated real-time cow control and data analysis systems.

A major failing of dairymen nationwide

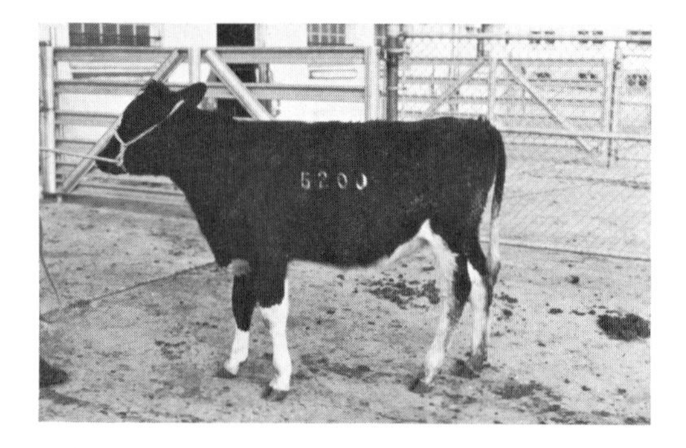

FIG. **3.6** Examples of freeze brands. A heifer branded on her side *(top left)* and a closeup of the brand *(top right)*. *(Bottom left)* A heifer on which various exposure times were tested for liquid nitrogen brands. *(Bottom right)* Heifers which have been freeze branded on the rump. (Courtesy of N. W. Hooven, Jr.)

is the failure to identify their cows sufficiently for profitable within-herd management. The problem is almost exclusively with non-registered cows. This deficiency is reflected in the DHIA Program where about 40% of all official lactation records cannot be used for genetic evaluations due to lack of sire identification. Identification needed to maximize benefits from DHIA data includes an acceptable identification number for each cow, her sire and her dam and the cow's birth date. Acceptable identification for DHIA purposes includes registration numbers, uniform series eartag numbers and VIP numbers. National DHIA sponsors the Verified Identification Program (VIP), to identify DHIA cows. VIP cows are issued identification certificates after their identification information has been verified by the DHIA Supervisor. The primary purpose of VIP is to increase the identification of nonregistered cows for the benefit of the dairymen who own them and the dairy industry as a whole. The Holstein Association sponsors a grade identification program. The other breed associations all have programs whereby grades or their offspring can be registered if they meet specified qualifications.

Milk Weighing and Sampling Devices

Under the leadership of the NCDHIP Policy Board, a technical committee has established standards of accuracy for devices for weighing and sampling milk and has tested numerous devices for accuracy. Testing and approval of such devices are imperative to allow dairymen and DHIA Supervisors to obtain accurate, unbiased estimates of milk and component yields for the dairy record keeping plans. Any company that wishes to market a weighing and sampling device for use in the DHIA Program must submit that device to the technical committee for testing and approval. Therefore, it is now possible for dairymen to purchase a variety of different weighing and sampling devices with confidence that they are accurate. Many of the weighing and sampling devices that have been approved for DHIA use are shown in Figure 3.7.

Fig. 3.7 A display of milk weighing and sampling devices including many which have been approved for Official DHI use. *1.* De Laval Calibrated Receiver Jar (A.); *2.* Milk-O-Meter (A.); *3.* Surge Calibrated Receiver Jar (A.); *4.* Milk Scales (A.); *5.* Chore-Boy Roll-O-Measure (A.); *6.* Milko-Scope (N. A.); *7.* Bodmin Milk Meter (N. A.); *8.* Meterite (N. A.); *9.* Tru-Test (A.); *10.* Ross-Holm Meter (N. A.); *11.* Waikato Weight and Rate Meter (A.); *12.* Sta-Rite Calibrated Receiver Jar (A.); *13.* Dari-Kool Calibrated Receiver Jar (A.). (A = Approved; N. A. = Not Approved.) (Courtesy of J. W. Smith, J. Dairy Sci. *52:* 1, pp. 129–130, 1969, and American Dairy Science Association.)

3.7 Future Dairy Record Systems

In the future we can expect the evolution of the DHIA Program to continue, in the direction of business management systems for herds of all sizes. This continued development will include a proliferation of on-farm terminals and computers, a decrease in the traditional periodic (monthly) collection of all the data on each herd, major revisions in the computer hardware and software employed by the DRPC, and adoption of advanced data collection, transmission, storage, and analysis techniques using the latest in computer and communications technology. These developments will better serve the business management needs of larger dairy herds where the economic importance of milk production per cow may be paralleled or even surpassed by the efficiency of labor management and the management of capital resources. Real-time system updating and interrogation will be a must. Up-to-the-minute management information will be available to the manager whenever he chooses to ask for it.

3.8 Summary

This chapters offers many compelling arguments for dairymen to join the DHIA Program and utilize the wealth of valuable information therein. The most persuasive reason to join the DHIA Program is the financial advantage. Where else can a dairyman invest $18.00 per cow to obtain a return of 1,000% 2,000%, or 3,000%? Just joining the program is not enough. Dairymen must report accurate data, study the management information carefully, look for weaknesses in their operation that are decreasing profits, and consult expert advisors on difficult problems. The DHIA Program exists to serve the participating dairymen above all else.

Review Questions

1. What are the advantages and disadvantages of each of the dairy record keeping plans that are available to dairymen?
2. List the advantages and disadvantages of participating in each of the dairy record keeping plans depending on whether you had a herd of 50 cows, 100 cows or 500 cows or larger. How important is each advantage and disadvantage insofar as it affects a herd's profit picture?
3. List at least 10 ways in which dairy record keeping information would enable you to turn your investment in records into a profit.
4. List some advantages of having a DHIA Supervisor visit your farm to obtain dairy record keeping data versus participating in a plan in which a DHIA Supervisor does not visit your farm.
5. What are the accepted methods of permanent identification for dairy cows?
6. Name as many methods for temporary identification of dairy cows as you can.
7. How will accurate identification of all animals in your herd (i.e., sire, dam, and birth date as the minimum) help you make more profitable management decisions from DHIA management information?
8. Describe the primary components of an automated cow identification and data recording, reporting, and retrieval system envisioning electronic transponders in the milking herd as one end of the system and the DRPC as the other. Do this with and without inclusion of an on-farm computer.

References

1. McCaffree, J. D., Everett, R. W., Ainslie, H. R., and McDaniel, B. T.: *Economic Value of Dairy Herd Improvement Programs.* Search Agriculture, Vol. 4, No. 1, Cornell University, 1974.
2. Hooven, N. W., Jr.: Freeze branding for animal identification. J. Dairy Sci. *51:* 146, 1968.

Suggested Additional Reading

Agricultural Research Service, USDA: Dairy Herd Improvement Letters.
DHIA Handbook, Extension Service, USDA.
DHIA Supervisors' Manual from Dairy Records Processing Center that serves your state.
Dickinson, F. N.: *Future Dairy Record Systems.* Proceedings of Symposium of Large Herd Management, University of Florida Press, 1978.
Literature from your state and local extension services or state DHIA on the Dairy Herd Improvement Program.
Minutes of the NCDHIP Policy Board, Official DHI rules and Official DHIR rules.

PART TWO

Breeding Better Dairy Cattle

CHAPTER 4

Genetic Basis for Improvement

4.1 Introduction

The first documented genetic improvement of cattle was accomplished by Robert Bakewell during the period between 1760 and 1795 in Leicestershire, England. Prior to that time, natural selection primarily determined the evolution of cattle. However, Bakewell was first, so far as is known, to introduce man's intellect as the determining force in the evolution of the bovine species. The Colling brothers were next to gain fame. They did so through the application of Bakewell's principles of careful selection based on individual performance, progeny testing, and inbreeding, especially line breeding to favored sires. The Colling brothers' insight, learned apparently at least in part from observing Bakewell's success, led to the formation of the Shorthorn breed and the establishment of the first herd book for cattle in 1822. This latter event was another important milestone; i.e., the formation and recording of pure breeds of livestock whose genetic constitution would be protected from outside "contamination" by certain rules.

During the remainder of the 19th century many new breeds and herd books were formed. This formative period lasted for at least 150 years after Bakewell. Ideals were established for each breed, and breeders directed their efforts towards matching these ideals. These ideals were specified largely in terms of color markings or type characteristics with little direct emphasis placed on the improvement of production traits. This occurred partly because it was thought that the ideals of color and conformation that were adopted by the breeds were closely associated genetically with superiority for production traits.

Science Enters Genetic Improvement

The opportunity to make substantial genetic progress for milk yield on a scientifically sound basis became feasible through 2 important developments near the turn of the 20th century. The first of these was the formation of the first cow testing association, which occurred in Denmark in 1895. So far as is known, this was the first organized program to record the yield of dairy cows and to provide information on their producing ability. These associations spread rapidly throughout Europe and into the United States, where the first cow testing association commenced operation in Michigan in 1906. The second great development was the rediscovery of Mendel's work in 1900. Mendel's research was the foundation of modern genetic knowledge. Understandably it took a while for this basic knowledge to be built on and expanded. Several decades were spent by dairy cattle researchers and breeders (as well as most other plant and animal researchers and breeders, for that matter) attempting to explain the inheritance of all traits in terms of Mendel's principles. It wasn't until the mid-1930s that methodology evolved to the point that breeders could really begin to make genetic improvement in traits such as milk yield. At that point in time the compatibility between biometry and genetics was finally well recognized and accepted—and modern genetic practices were on the way. The men who were instrumental in making the transition possible were William Bateson (1861–1926), English geneticist who coined the term "genetics"; Francis Galton (1822–1911), English statistician who is considered the founder of biometry; Ronald A. Fisher (1890–1962), English biometrician, and Sewall Wright (1889–   ), American geneticist, who between them largely laid the foundation for the science of population genetics; and Jay L. Lush (1896–1982) who did so much himself and through his many outstanding students to translate theory into practice and to enhance the adoption and utilization of scientific breeding principles.

The rapid progress in genetic improvement of dairy cattle over the past 20 years has been made possible by three major developments; processes for diluting and preserving bull semen, statistical procedures for genetic evaluation of traits affected by large numbers of genes, and high speed electronic computers for analyzing large amounts of data.

4.2 The Cell, Carrier of the Genetic Material

Dairy cattle, as well as all other living organisms, are made up of cells. Among the various types of materials in cells, all of which contribute to the amazing process known as life, is the "genetic material." Most of this genetic material is in the 30 pairs of cattle chromosomes which are found in the nuclei of cells (Chapter 18). It is these chromosomes and the genes of which they are constructed that determine the animal's genotype, or genetic makeup.

Basically there are 2 types of cells in the body, sex cells and somatic cells. Sex cells (spermatozoa in the male and ova in the female) are those cells whose special behavior during cell division makes sexual reproduction possible. Somatic cells are those cells that constitute all the rest of the body. All the different kinds of somatic cells divide and reproduce themselves by the process known as mitosis. During mitotic division each parent cell becomes 2 offspring cells, each of which is essentially the same as the original parent cell. On the contrary, sex cells undergo a special division known as meiosis, or reduction division, which reduces the number of chromosomes in the egg and sperm to one-half the normal number. This is necessary so that when a sperm and an egg unite to form a zygote that will develop into a calf, that calf will have the same number of chromosomes in its body cells as did each of its parents. Meiosis is also the process that results in a parent transmitting only one of each pair of its own genes to an offspring. This in turn insures the continuance of the sampling nature of inheritance and the resultant variability of genotypes within the species.

4.3 Qualitative Traits

The most simply inherited traits are frequently referred to as "mendelian traits." Usually just one locus or a few loci will have major effects on such traits. Frequently there will be additional loci which exert minor effects through genes called modifying genes. These traits are more accurately termed "qualitative traits" because the phenotypes (what the animal is or does, its physical traits and performance traits) tend to fall into discrete categories rather than being measured on a more or less continuous scale.* Environment usually plays a relatively minor roll in influencing the phenotypic category. For many qualitative traits of cattle it is known how many alleles are present at the major loci and the relation among those alleles for additivity, dominance, and epistasis. However, even for these, the most simply inherited traits, it is not known which chromosomes con-

*The authors are indebted to H. W. Leipold and Keith Huston for many helpful comments on qualitative traits in this section, Table 4.1, and Suggested Additional Reading.

tain the major locus or loci that control them. In addition, most qualitative traits are probably affected by modifying genes. These genes cause minor variations in qualitative traits so that some variation in phenotype exists within the major categories.

Examples of qualitative traits in dairy cattle are hair color, inherited abnormalities, horned vs polled, and blood antigens.

Hair Color

It appears that there are at least 9 major loci which affect coat colors and patterns in cattle.[1] Four of these have at least 3 alleles each. It is probable that there are other major loci, or at least modifying genes, which also affect color and pattern. The colors and patterns caused by the genes at these loci are very important to the breeder. All of the breeds have color requirements which are considered desirable in order to distinguish animals of that breed and to maintain a uniformity of appearance. In recent years there has been a tendency in most of the breed organizations to relax color or color-pattern requirements. This has been done by introducing provisional registration systems. This is a very progressive and beneficial development and will enable the retention in the breeds of animals with superior genotypes for yield that would have previously been lost due to unacceptable coloring.

In general, black is dominant to the other colors, although in some cases the dominance is incomplete. For instance, black is completely dominant to red in Holsteins. But crossing Holsteins with other breeds usually results in a mixture of black and colored hairs on the nonwhite areas of the offspring. Environment has little effect on hair color except for extreme circumstances such as molybdenum toxicity (Chapter 9), long exposure to tropical sun, or freeze branding (Chapter 3).

Hereditary Defects

The gross hereditary defects of cattle are usually categorized as lethal, semilethal, or subvital, depending on the severity of the abnormality and the proportion of afflicted offspring that die. Most of these abnormalities are simple mendelian recessives. There are over 40 different loci known to contain genes that cause gross abnormalities or reduced vitality in cattle. The phenotypic expression of 4 of these gross abnormalities is shown in Figure 4.1. Three of these are lethal or semilethal and the fourth, mule foot or syndactylism, seriously limits an animal's usefulness. Most of the more common gross hereditary defects found in American dairy cattle are listed in Table 4.1. In addition to the genes that cause the defects described in Table 4.1, there are probably many more genes that reduce fertility in either males or females by causing fetal death, especially early in gestation. It is very difficult to prove the existence of genes in these latter categories. Many factors in the environment reduce fertility, so a single-gene effect is difficult to identify especially if it simply reduces fertility rather than rendering an animal sterile. Also, early fetal death is frequently indistinguishable from failure to conceive or from missed heats. However, most of the defects listed in Table 4.1 are readily identifiable, especially by a veterinarian. The parents of the defective calf, therefore, can be identified as carriers of defective genes.

The genes causing most of these defects probably originated as mutations. Since most of these genes are recessive to the normal allele, they are virtually impossible to eliminate from a population of cattle. This is because most of the recessive defect genes are masked in the phenotype by the dominant normal allele. Carrier animals are very difficult to identify and usually become known only when they are mated to a carrier and produce a calf homozygous for the defective allele. The probability of a calf born of 2 carrier parents being defective is only 1 in 4 (Figure 4.2).

Horned vs Polled

Two alleles at a single locus control the presence or absence of horns, but there are additional modifying genes that result in an in-between condition, referred to as the presence of "scurs." The allele for polled is dominant to the allele for horns, although the dominance is apparently not complete. Therefore, a naturally horned animal is homozygous for the horned allele whereas a naturally polled animal is either hetero-

Fig. 4.1 Examples of genetic abnormalities in dairy cattle. *(Top left)* A Holstein with the single-toe condition called mule foot (syndactylism). *(Top right)* A Jersey calf homozygous for the recessive abnormality, limber legs. *(Bottom left)* A Guernsey calf which has inherited the bulldog condition (achondroplasia). *(Bottom right)* A Holstein calf with a brain hernia. These genetic abnormalities are described in Table 4.1. (Top left, bottom left, and bottom right, courtesy of K. A. Huston; reprinted with permission from *Breeding and Improvement of Farm Animals.* Top right courtesy of the American Jersey Cattle Club.)

Table 4.1 *Some of the More Common Hereditary Defects Known to be Present in Dairy Cattle*

Name	*Description of abnormality and breeds in which identified*	*Relation of abnormal gene to normal allele*
Albinism, complete	Lack of pigment in hair, skin, and eyes (Guernsey)	Recessive
Albinism, incomplete	Lack of pigment in hair and skin; blue or glass eyes (Guernsey)	Dominant
Albinism, incomplete	Lack of pigment in hair and skin; eyes pigmented, ghost pattern (Guernsey, Holstein)	Recessive
Amputated	One or more legs, or parts of legs missing; apparently some modifying genes (Brown Swiss)	Recessive
Arthrogryposis	Permanent joint contracture, present at birth (probably all breeds)	Recessive
Blindness	Usually caused by small eyes or development of cataracts in fetus (Jersey). Nonhereditary types also occur.	Recessive
Bulldog (Achondroplasia, 3 types)	Very short bones, short heads with rounded foreheads and protruding mandible; aborted, born dead or die quickly (Guernsey, Jersey, Holstein)	Recessive

TABLE 4.1 *Some of the More Common Hereditary Defects Known to be Present in Dairy Cattle (Continued)*

Name	*Description of abnormality and breeds in which identified*	*Relation of abnormal gene to normal allele*
Cerebral hernia	Brain protrudes through opening in skull; calves stillborn or die shortly after birth (Holstein)	Recessive
Closed anus	No anal opening or anal muscle development (Holstein)	Recessive
Congenital dropsy	Abnormal fluid retention in fetus causes heavy bloated calf at birth; calf stillborn or dies soon after birth (Ayrshire)	Recessive
Congenital eye defect	Opaque body beneath enlarged and distorted cornea impairs vision (Holstein, Jersey)	Recessive
Congenital spasms	Calves show muscle spasms of head and neck, die shortly after birth (Jersey)	Recessive
Crampy (Spastic Syndrome)	Spastic contracture of muscles in hind legs (probably all breeds)	Recessive with incomplete penetrance
Curly hair	Curled hair gives woolly appearance (Ayrshire)	Recessive
Curved limbs	Fore and rear limbs curved anteriorly; calves stillborn or die soon after birth (Guernsey)	Recessive
Dwarfism	Latent lethal, seldom live to puberty (Jersey, Holstein)	Recessive
Flexed pasterns	Feet turned back so calf walks on toes, front feet usually only ones affected, disappears within a few weeks of birth if calf survives (Jersey). There is a similar non-hereditary condition.	Recessive
Fused or missing teats	Teats on same side of udder partially fused (Guernsey)	Recessive
Hairless	Almost no hair on body; calves die soon after birth due to lack of heat regulation; 2 other similar but less severe conditions are also caused by recessive genes (Guernsey, Holstein)	Recessive
Hydrocephalus	Abnormalities of bones and internal organs; bulging forehead (Ayrshire, Holstein, Jersey). Non-hereditary types also occur.	Recessive
Hypoplasia of cerebellum	Calves have no sense of balance; usually die or are killed (Holstein)	Recessive
Impacted molars	Lower jaw short and narrow with impacted molar teeth, jaw frequently fractured (Jersey, Milking Shorthorn)	Recessive
Imperfect skin	Defective skin on lower legs, around eyes, and some mucous membranes permits bacteria to enter; fatal infection develops soon after birth (Ayrshire, Holstein, Jersey)	Recessive
Limber legs	Legs non-functional, joints move in all directions (Jersey)	Recessive
Marble bone disease (Osteopetrosis)	Abnormal bone formation in fetus. Born dead. (Brown Swiss)	Recessive
Misshapen feet	2 to 4 months after birth, toes on front feet spread wide apart, and feet become sore causing lameness (Jersey)	Recessive

TABLE 4.1 *Some of the More Common Hereditary Defects Known to be Present in Dairy Cattle (Continued)*

Name	*Description of abnormality and breeds in which identified*	*Relation of abnormal gene to normal allele*
Mulefoot (Syndactylism)	Only one toe on affected feet, become sore and cause lameness (Holstein)	Recessive
Mummification	Calves die or are aborted around 8th month of gestation; have short necks, stiff legs, and prominent joints (Guernsey, Holstein, Jersey, Red Dane)	Recessive
Muscle contracture	Birth very difficult or impossible due to rigid muscles, head drawn back to shoulder, legs folded (Holstein)	Recessive
Notched ears (Cropped ears)	Symmetrical notches in both ears (Ayrshire)	Incomplete dominant
Paralyzed hindquarters	Calves cannot stand and usually die within a few weeks after birth (Red Dane)	Two complementary genes
Parrot mouth	Abnormal shortening of lower jaw; degree varies (Guernsey, Holstein, Jersey, probably other breeds also)	Several genes, some dominant, some recessive
Pink tooth (Congenital porphyria)	Pink-grey cast to teeth, pale mucous membranes, body sunburns easily (Ayrshire, Holstein, Shorthorn)	Recessive
Prolonged gestation I	Calves carried 1 to 3 months beyond normal term; usually very heavy, born dead, or die soon after birth; dam often dies also (Holstein)	Recessive
Prolonged gestation II	Calves carried up to 500 days; calves born dead, incompletely developed with no pituitary gland (Guernsey)	Recessive
Rectal-vaginal constriction (RVC)	Fertile but difficult or impossible calving as name implies (Jersey)	Recessive
Screw tail	Tail shortened and twisted like corkscrew (Red Poll)	Recessive
Sex-linked lethal	Females transmit sex-linked recessive causing death of male zygotes (Holstein); 2 other genes known; 1 causes deaths in the ratio of 3 males:1 female; other causes all females to be born dead but does not affect males.	Sex-linked recessive
Spastic lameness	Feet may miss ground when walking, difficulty in rising, poor balance after rising (Holstein)	Recessive
Umbilical hernia	Failure of umbilical ring to close after birth (virtually all dairy breeds)	Incomplete dominant
Weaver	Calves begin losing control of rear legs at about six months of age (Brown Swiss)	Recessive
White heifer disease	Reproductive tract fails to develop normally. Affected animals usually sterile (Milking Shorthorn)	Recessive, probably several genes involved
Wry face	Asymmetric development of nose gives face a twisted appearance (Jersey)	Recessive
Wry tail	Sacral vertebrae set off-center (virtually all dairy breeds)	Recessive

Source: Information in this table was taken from many sources but primarily from references on hereditary defects cited at the end of this chapter and from personal communications with H. W. Leipold and Keith Huston.

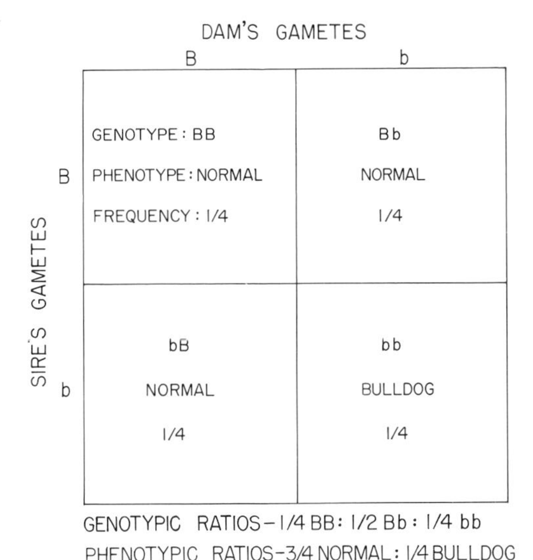

Fig. 4.2 Inheritance of the mendelian recessive lethal gene causing the bulldog condition in cattle.

zygous (one horned gene and one polled gene) or homozygous for the polled allele. Most dairy cattle are naturally horned. Of course, we seldom see horns on American dairy cattle, because they have been dehorned with mechanical devices, heat, or chemicals to change their phenotype. The inheritance of the horned-polled trait follows the same principles as the trait in Figure 4.2.

Blood Antigens

Approximately a dozen polymorphic antigenic systems have been defined in the erythrocytes (red blood cells) of cattle. Most of these systems contain multiple alleles. In fact, the B blood group system in cattle is known to contain over 300 alleles at a single locus and is the most complex allelic series that has been studied in any animal species. The genotypes for these blood systems can be determined very accurately from the phenotype by laboratory analysis of an animal's blood. Stormont[2] has pointed out that blood type records of identification on dairy cattle are tantamount to using fingerprints for identifying humans. Based on studies in his laboratory, he has estimated that approximately 2 trillion $[2(10^{12})]$ blood types are theoretically possible in cattle. Although blood typing cannot prove parentage conclusively, it can disprove it conclusively. Blood typing can resolve over 90% of all cases of questionable parentage involving 2 dairy bulls. This capability of blood typing has made possible an exotic technique used in multiple ovulation and embryo transfer, the mixing of semen from 2 bulls to breed the donor cow. Which bull is the sire of each calf can be affirmed by blood typing each calf, the donor dam and the 2 bulls. For over 30 years a rule has been in effect that the blood types of all bulls used in artificial insemination must be officially recorded. So far the primary practical benefit gained from blood typing has been an improvement in the accuracy of identification. This has cleared up many cases of questionable parentage and decreased the incidence of fraudulent parentage. Another important benefit of this work has been several important contributions to our basic understanding of gene action in multiple allelic series. It is possible that blood typing, and also the study of milk protein types, will gain in importance in the future. Evidence is accumulating that some genotypes of blood and milk proteins of dairy cattle are weakly associated with producing ability and fertility. Such associations could be used as a partial basis for selection of animals of superior genetic merit.

Qualitative traits should not be considered less important than the category of traits to be discussed in the next section. However, no one should be misled into the mistaken belief that most or even many qualitative traits have important influences on yield or other traits that are indicative of productivity. The fact is that few qualitative traits are known to be directly associated with yield to any important degree at the present time. The obviously debilitating effects of genes causing gross abnormalities affect the animal's entire constitution and ability to survive but are not associated directly with its genetic ability for production.

In the remainder of this section on breeding and genetics the discussion will be devoted primarily to quantitative traits and their genetic improvement.

4.4 Quantitative Traits

Quantitative traits are characterized by 2 important basic differences from qualitative traits: (1) they are influenced by many pairs of genes; (2) the phenotypic expression is frequently strongly dependent on the environment. These 2 effects combine to cause the phenotypes of quantitative characters to vary in an essentially continuous manner. For example, a cow's body weight does not fall into discrete classes such as the qualitative traits cited earlier. Her weight varies almost constantly by very small amounts. We are limited in attempting to place body weights on a truly continuous distribution by our ability and need to measure them accurately enough. For most purposes it is quite satisfactory to measure body weight to the nearest pound or even the nearest 10 pounds. But this is strictly for our convenience and does not change the fact that body weight is a continuous variable with an infinite number of units.

The basic nature of quantitative traits, the combined influence of large numbers of genes and changes in environment, makes it much more difficult to determine the genotype accurately than is the case for most qualitative traits. Sometimes the animal's phenotype tells us very little about the genotype. For example, environmental changes, especially from one herd to another, can cause sizable changes in the phenotype for milk yield.

Therefore, the genetic improvement of quantitative traits poses quite a challenge. But the meeting of this challenge is very important because many of the economically important traits of dairy cattle such as milk yield and composition, conformation, efficiency, and disease resistance are inherited quantitatively. The genetic ability of his cows for production will be largely dependent on the dairyman's ability to make the best use of the formidable scientific developments available to dairy cattle breeders. The profitability of the herd will, in turn, depend to an important degree on the producing ability of the cows.

4.5 Genetic Improvement of Quantitative Traits

An animal's phenotype is the result of its genotype acting in and reacting to an environment. The goal of genetic improvement of dairy cattle is to produce the best possible genotype which will operate at top efficiency in the environment to which it is subjected, in order to bring the greatest possible profit to the dairyman. In working towards this goal, the genetic improvement of dairy cattle is really a 2-phased process. Phase 1 is devoted to estimating each animal's breeding value or transmitting ability which is one-half breeding value (i.e., bridging the gap between the phenotype and genotype) so that animals with superior genotypes can be identified. Phase 2 is making the best possible utilization for breeding purposes of those animals with superior genotypes, so the rate of genetic progress will be maximized. Chapter 5 will be largely devoted to Phase 1 of this process. Various approaches to Phase 2 will be discussed in Chapter 6.

4.6 Transmitting Ability vs. Breeding Value

Transmitting ability is the average value of the gametes produced by an animal. The average value of the gametes is the average genetic merit for a trait that an animal passes on to its offspring. Since breeding value is an animal's genetic worth for a trait, transmitting ability is equal to one-half breeding value. It is important to understand that transmitting ability is not an absolute quantity but is always expressed relative to some genetic base. The role and importance of genetic bases will be discussed in Chapter 7. In the United States, an animal's genetic merit for each trait is expressed in terms of transmitting ability. However, some countries express genetic merit as breeding value.

4.7 Basic Concepts in the Study of Quantitative Inheritance

Familiarity with the following concepts will make the subsequent discussion more meaningful. These are tools that are used in genetic work. At first, they may seem somewhat abstract and difficult to grasp fully, but as the discussion progresses the significance, the meaning, and the value of these concepts should become clear.

The Normal Curve

Estimation procedures used in quantitative genetics generally involve the use of fairly large numbers of measurements (called data). Obviously techniques are required to summarize such large numbers of data into a few meaningful and readily understandable numbers, or statistics. Many of these estimation procedures are based on the phenomenon that is characteristic of biological data known as the normal distribution, or normal curve. The shape of the normal curve is shown in Figure 4.3. This normal curve is a "frequency distribution" where the height of the curve at any point on the base line indicates the "frequency" or relative proportion of animals in the population whose phenotype has that particular value. This is analogous to piling up animals, placing each animal on the pile right where its phenotypic value occurs on the base line. Then when the pile is completed, a line is drawn along the top of the pile from one end to the other. The variability of most biological traits as they occur in their natural state follows the normal distribution rather closely. Therefore, knowing this fact about the distribution of these traits gives a very workable starting point for estimation procedures. The most important characteristics to know about the normal distribution are the center point, called the average or the mean, and the manner in which the data are distributed around this center point, called the variance. A measure of variance that is commonly used in conjunction with the mean to describe a particular distribution is called the standard deviation.

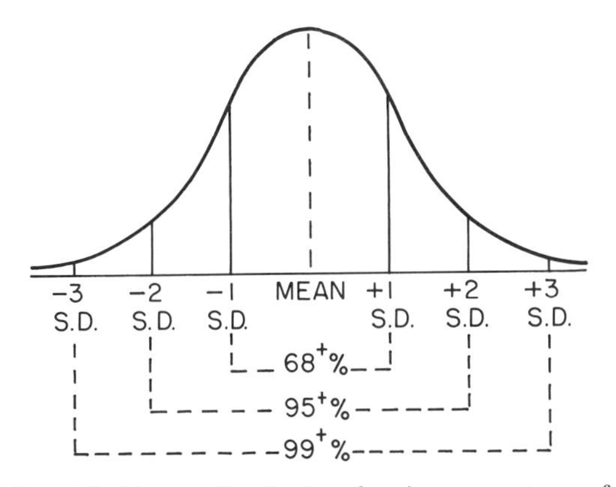

Fig. 4.3 Normal distribution showing percentages of data within 1, 2, and 3 standard deviations above and below mean.

The Mean

The mean is the average of all values being measured. It is calculated by summing the measurements and dividing the sum by the number of measurements. In Figure 4.3 the mean is the center point of the distribution. Knowing the mean of a group of measurements gives only part of the picture. The manner in which they vary around the mean is also needed to complete the picture.

The Variance

By definition, the variance is the average squared deviation of the individual measurements from the mean. A slight departure from the usual procedure of calculating averages occurs here. The average of the square deviations is obtained by dividing by one less than the number of measurements. This is necessary to obtain an unbiased estimate of the population variance. The actual definition is based on mathematical properties of the normal distribution. The important thought to grasp thoroughly is the concept of variability of biological data. The variance is the commonly used measure of variability for data that follow the normal distribution. In mathematical formulas the variance is represented by the Greek letter sigma (σ) with a superscript "2." Subscripts, usually letters, are used to identify the type of variance. Thus σ_P^2 means the phenotypic variance, where the subscript "P" indicates phenotypic. In quantitative genetic work it is possible to partition the phenotypic variance into portions due to inheritance and to environmental effects. This is a powerful analytical device and is the backbone of much of our knowledge of inheritance of quantitative traits in dairy cattle.

Standard Deviation

The standard deviation is the most common function of the variance that is used along with the mean to describe a set of measurements. Mathematically, the standard deviation is the square root of the variance. In terms of the normal distribution, the area between one standard deviation above and below the mean (Figure 4.3) contains 68.26% of the measurements; the area between 2 standard deviations contains 95.46%; and the area between 3 standard deviations contains 99.74%. These unusual percentages are the result of the mathematical formulation for the normal distribution. For most study and discussion purposes the student need remember only the whole percentages of 68%, 95%, and 99%. The standard deviation has important practical uses in dairy cattle selection because the relative merit of animals for different traits are sometimes expressed in terms of standard deviation units, that is, each animal's deviation from the mean for a trait divided by the standard deviation for the trait.

Regression

Regression is very useful in explaining the variability of 2 traits when there is a cause-and-effect relationship between them. Regression estimates how many units of change in the dependent variable (the effect) result from or are caused by each unit of change in the independent variable (the cause). When stating a regression (cause-and-effect) relationship the units of both the independent and dependent variables must be used. For example, the regression of an animal's breeding value (or transmitting ability) on the breeding value (or transmitting ability) of either parent will be shown to be 0.5 (Section 5.2) because the animal received one-half of its genes from each parent (one-half of its genotype was caused by each parent).

Correlation

The correlation between 2 traits measures their tendency to vary in the same direction (positive correlation) or in opposite directions (negative correlation). In other words, the correlation estimates the incidental variation in 2 traits when both traits are affected by outside sources of variation. This tendency is measured in terms of the correlation coefficient. The correlation coefficient is not dependent on units of measurement. The maximum possible positive correlation is 1.0 indicating both traits vary in the same direction consistently. A correlation of 0 means that the 2 traits are completely independent. The largest possible negative correlation is −1.0 indicating that the 2 traits vary consistently in opposite directions. There are 2 types of correlations, phenotypic and genetic. Both are important in the genetic improvement of dairy cattle.

The phenotypic correlation, as the name implies, is simply the correlation between phenotypes. For example, there is a positive phenotypic correlation between body size and milk yield because, on the average, larger cows give more milk than smaller cows. On the other hand, cows with higher yields tend to have a lower milk fat percentage. Therefore, there is a negative phenotypic correlation between milk yield and milk fat percent. The phenotypic correlations are usually influenced by both inheritance and environment.

The genetic correlation is that portion of the correlation between 2 traits that is due to inheritance. The causes of genetic correlations are still not entirely clear. They appear to be due mainly either to pleiotropy (a single gene or group of genes influencing both traits at the same time) or to linkage between 2 loci, each of which affects one of the traits that are correlated. Statistical techniques have been developed to estimate genetic correlations, particularly using relatives. Genetic correlations are very important and useful in the genetic improvement of dairy cattle. The ability to measure another trait that is correlated genetically with yield strengthens the estimate of transmitting ability for yield. The use of genetically correlated traits is discussed in detail in Chapter 6.

4.8 Partitioning the Phenotypic Variance

The goal of all methods of estimating breeding value is to make some function of the phenotype as accurate an indicator of the genotype as possible. It has been found that this can be done best by studying the variability of phenotypes (specifically the variance) and attempting to determine the relative importance of various factors that cause this variability.

The phenotypic variance can be represented by the following mathematical relationship: $\sigma_P^2 = \sigma_H^2 + \sigma_E^2 + \sigma_{HE}^2$. This equation means that the phenotypic variance σ_P^2 is composed of three parts, the hereditary variance σ_H^2 plus the environmental variance σ_E^2 plus the covariance or interaction between inheritance and environment, σ_{HE}^2. The phenotypic variance σ_P^2 is what can be observed and measured. The hereditary variance σ_H^2 is the part that is to be estimated as accurately as possible. Therefore, the goal is to remove as much as possible of the environmental variance and the covariance between heredity and environment from the phenotypic variance. Mathematically, this would mean making $\sigma_P^2 - \sigma_E^2 - \sigma_{HE}^2 = \sigma_H^2$. This shows that, if all σ_E^2 and σ_{HE}^2 could be removed from σ_P^2, the variance remaining would equal σ_H^2. For milk yield, expressed as a deviation from herdmates, $\sigma_H^2 = 0.25$, or 25% of the phenotypic variance, on the average. Therefore 25% of the phenotypic variance (σ_P^2) is due to σ_H^2 and 75% is due to $\sigma_E^2 + \sigma_{HE}^2$. This is depicted graphically in Figure 4.4. One should be able to clearly relate Figure 4.4 to the equation for the phenotypic variance using both percentages and mathematical symbols. Figure 4.4 shows that the environmental variance plus the covariance between heredity and environment comprise about 75% of the phenotypic variance. Research has rather uniformly indicated that the covariance between heredity and environment is usually small and of relatively little importance in most cattle breeding work except for the correlation within herds of the mean and the variance of lactation records. Therefore, most of the 75% of the non-hereditary phenotypic variance is actually due to the effects of environment. These environmental effects are usually put into 2 categories for purposes of discussion: permanent environmental effects and temporary environmental effects. Typical permanent environmental effects would be inter-herd differences such as different feeding regimes, housing, milking facilities and techniques, care of heifers and dry cows, and disease control especially on mastitis and reproductive problems. Temporary environmental effects are those that last for only relatively short periods of time and are subject to frequent change. The hereditary variance, σ_H^2, can be partitioned into portions due to additive effects, dominance, and epistasis. This concept is important because these portions of the hereditary variance are the bases for the various systems of mating (Section 6.5). Mathematically the hereditary variance can be defined: $\sigma_H^2 = \sigma_A^2 + \sigma_D^2 + \sigma_I^2$, where σ_H^2 is the hereditary variance, σ_A^2 is the additive portion of the hereditary variance, σ_D^2 is that portion of the hereditary variance that is due to dominance, and σ_I^2 is that portion of the hereditary variance due to epistatic effects. This equation shows that the hereditary variance is the sum of the additive, dominance, and epistatic variances. Actually this definition is just an approximation because interaction effects are possible among σ_A^2, σ_D^2, and σ_I^2.

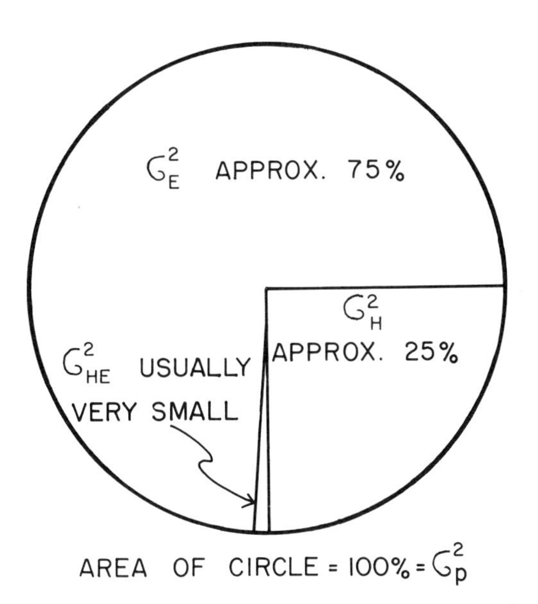

Fig. 4.4 Components of the phenotypic variance for milk yield.

Heritability

Heritability is the percentage of differences among animals for a particular trait that is due to the genes that they inherited. Heritability can be defined two different ways in terms of additive, dominance and epistatic genetic effects. Heritability in the broad sense of the word is the fraction of the phenotypic variance that is caused by genotypic differences among individuals. In terms of the equation for the phenotypic variance, heritability $= \sigma_H^2/\sigma_P^2 = \sigma_H^2/$

($\sigma_H^2 + \sigma_E^2 + \sigma_{HE}^2$). Heritability in the narrow sense is that fraction of the phenotypic variance (σ_P^2) that is due to additive genetic variation (σ_A^2). Therefore, heritability in the narrow sense is defined as σ_A^2/σ_P^2. This latter definition is more appropriate for the improvement of dairy cattle since most genetic improvement of dairy cattle is the result of selection, and selection primarily makes use of the additive genetic variation. Heritability is usually represented by the symbol h^2 for reasons that will be explained in Section 5.2. The greater the genetic influence on a trait relative to the environmental influence, the higher is the heritability. In general, the higher the heritability of a trait, the greater is the genetic progress that can be made, although many other factors are also important as will be shown later. Heritabilities of many important traits in dairy cattle are shown in Chapter 6.

4.9 Summary

Much progress has been made since scientists have become interested in genetic improvement of dairy cattle. The biggest single breakthrough after the rediscovery of Mendel's work was the acceptance of quantitative inheritance as a reality. Since then the pursuit of genetic excellence has been greatly hastened by artificial insemination and electronic computers. Some qualitative traits are, and will continue to be, important. However, most of the effort expended on improving economically important traits is in the realm of quantitative genetics. The basis for much of this work is the normal distribution. Knowledge of the properties of this distribution has made possible the development of most of the statistical procedures that have proved effective. In order to utilize these procedures, one must understand the concept of variance. The partitioning of the phenotypic variance into its component parts, heredity, environment, and the interaction between the two, is basic to estimating breeding value.

Review Questions

1. Name several phenotypic traits of dairy cows which are almost entirely genetically determined and several which are usually affected by changes in the environment. What sort of environmental changes would be necessary to change the phenotype of the first group vs the second group?
2. How can you describe to someone the comparative characteristics of 2 groups of cows, one group with a mean lactation milk yield of 13,000 lbs and the other with 14,000 lbs?
3. Now, how can you compare the 2 groups if you also know that the group that averages 14,000 lbs has a standard deviation of 1500 lbs and the 13,000-lb group has a standard deviation of 3000 lbs?
4. In which group of cows (from question 3) would you be likely to find the several cows with the highest milk yields? Why? How about the lowest milk yields and why? Hint: If there were cows in each group 3 standard deviations above their group mean, which group would have the highest producing cows? Use the same reasoning to determine the group with the lowest producers.
5. If you mated a group of polled bulls to a group of horned cows, what proportion of the calves would be polled? Explain your answer.
6. Should you be concerned about hereditary defects? How would you protect your herd?
7. How much genetic progress would you expect to make in a quantitative trait such as milk yield if you thought of it strictly in qualitative terms? Explain.
8. What other terms can you think of to use in lieu of "breeding value"?
9. Why is the normal distribution important to quantitative genetic improvement?
10. If your herd of 300 milking cows had a mean lactation milk yield of 14,000 lbs and a standard deviation of 2000 lbs, how many cows would you expect to be 1, 2, and 3 standard deviations, respectively, above the herd average?
11. Make separate lists of permanent and temporary environmental effects that you think would affect yield; classify each as major, intermediate, or minor in relation to the influence it would have on lactation milk yield.

References

1. Lauvergne, J. J.: Genetique de la couleur du pelage, des bovins domestiques. Bibliographia Genetica *20:* 1, 1966.
2. Stormont, C.: Contribution of blood typing to dairy science progress. J. Dairy Sci. *50:* 253, 1967.

Suggested Additional Reading

Butler, J. E.: Bovine immunoglobulins: a review. J. Dairy Sci. *52:* 1895, 1969.

Johansson, I.: *Genetic Aspects of Dairy Cattle Breeding.* Urbana, Ill.: University of Illinois Press, 1961.

Sinnott, E. W., Dunn, L. C., and Dobzhansky, T.: *Principles of Genetics.* 4th ed. New York: McGraw-Hill Book Co., Inc., 1960.

Snedecor, G. W.: *Statistical Methods.* 5th ed. Ames, Iowa: Iowa State College Press, 1956.

For information about hereditary defects, see the following and their bibliographies:

Fechheimer, N. S.: Prenatal and postnatal mortality in cattle: genetic aspects of calf losses. Pub. 1685. National Academy of Sciences, 1968.

Gilmore, L. O.: *Dairy Cattle Breeding.* Philadelphia: J. B. Lippincott Co., 1952, Chapter 9.

Green, H. J., Leipold, H. W., Huston, K., and Dennis, S. M.: Congenital defects in cattle. Ir. Vet. J. *27:* 37, 1973.

Johansson, I., and Rendel, J.: *Genetics and Animal Breeding.* San Francisco: W. H. Freeman and Co., 1968, Chapter 8.

Jolly, R. D., and Leipold, H. W.: Inherited diseases of cattle—a perspective. N. Z. Vet. J. *21:* 147, 1973.

Lasley, J. F.: *Genetics of Livestock Improvement.* Englewood Cliffs, N.J.: Prentice-Hall, Inc., 1963, Chapter 11.

Lauvergne, J. J.: Genetique de la couleur du pelage des bovins domestiques. Bibliographia Genetica *20:* 1, 1966.

Leipold, H. W.: Genetics and disease in cattle. Bovine Pract. Baltimore, Maryland, Proceedings, 1978.

Leipold, H. W., Adrian, R. W., Huston, K., Trotter, D. M., and Dennis, S. M.: Anatomy of hereditary bovine syndactylism. I. Osteology. J. Dairy Sci. *52:* 1422, 1969.

Leipold, H. W., Dennis, S. M., and Huston, K.: Congenital defects in cattle: nature, cause and effect. Adv. Vet. Sci. Comp. Med. *16:* 103, 1972.

Leipold, H. W., Ojo, S. A., and Huston, K.: Genetic defects of the skeletal system in cattle. Proc. World Congress Genetics Appl. Anim. Prod. Madrid, Spain, *3:* 35, 1974.

Ojo, S. A., Leipold, H. W., and Hibbs, C. M.: Bovine congenital defects: Syndactyly in cattle. J.A.V.M.A. *166:* 607, 1975.

Warwick, E. J., and Legates, J. E. *Breeding and Improvement of Farm Animals.* 7th ed. New York: McGraw-Hill Book Co., Inc., 1979.

Stormont, C.: Genetics and disease. Advances in Veterinary Science *4:* 137, 1958.

Winters, L. M.: *Animal Breeding.* 5th ed. New York: John Wiley and Sons, Inc., 1954, Chapter 9.

CHAPTER 5

Estimating Breeding Value

5.1 Introduction

The first phase of making genetic improvement is to identify each animal's genetic merit for the trait or traits to be improved. As discussed in Section 4.6, genetic merit may be expressed as breeding value, an estimate of the merit of an animal's full complement of genes, or as transmitting ability, an estimate of the merit of a random gamete or one-half an animal's full complement of genes. Both terms will be used or referred to in Chapters 5, 6 and 7 so the fact that transmitting ability = one-half breeding value should be remembered. Initially, discussion will center on breeding value and then gradually turn to transmitting ability as genetic evaluations are explained. Breeding value is defined separately for each trait. For example, a cow's breeding value for milk yield might be very high but at the same time her breeding value for milk fat yield might be much lower because she transmits genes for low fat test to her offspring. Naturally, if the genes each animal possessed were known, breeding value could be determined easily. From the discussion in Chapter 4 it is obvious that this is not true for quantitative traits. Therefore, sources of indirect information about each animal's genotype must be sought. No matter what the source of information is, it is based on phenotypes. It will be shown in this chapter that phenotypes by themselves may be rather poor estimators of breeding value, but their usefulness can be greatly improved by putting them on a comparative basis. Phenotypes for some traits other than milk and component yields can be useful for improving yield or profitability, but the emphasis of traits that have no relation to yield or profitability should be avoided.

5.2 Sources of Information on Breeding Value

Modern techniques for estimating genetic merit based on DHIA data make possible accurate estimates of breeding value. The bull and the cow shown in Figures 5.1 and 5.2 are examples of animals with superior breeding values.

There are 4 basic sources of information on breeding value. These are: (1) the animal itself; (2) the animal's progeny; (3) the animal's ancestors; and (4) collateral relatives. All of these sources provide information on genetic merit because all of the individuals are related to the animal either by descent or through common ancestors. Therefore, they all have some genes in common with the animal. Figure 5.3 shows diagrammatically how each of these sources of information is related to the animal's breeding value. In each case where a line connects two breeding values or a breeding value and a phenotype, that line indicates a direct relationship. These lines, or paths, are pathways of biological influence. The arrow on each line indicates the direction in which influence is exerted. Each path has a value and that value defines the degree of relation. For example, the path from the sire's breeding value to the animal's breeding value has a value of 0.5. This means that on the average one-half of the variation in the animal's breeding value is caused by the breeding value of the sire. Similarly, the path from each animal's breeding value to its phenotype has a value of h. The letter "h" in this case is equal to the square root of heritability, which was defined in Chapter 4 as h^2 just so its square root could be expressed simply as h. This means that for each trait a portion of the variation in the animal's phenotype is caused by the animal's breeding value or genotype, and that portion is equal to the square root of the heritability for that trait.

These 4 sources of information and how each relates to the animal's breeding value will be considered separately. For the time being the accuracy with which each phenotype is known will be ignored. This will be discussed later in the chapter. It is also possible that for some traits an animal may not have a phenotype. Obviously, bulls do not have a phenotype for milk or component yields.

The Animal Itself

The first source to be considered is the animal's own phenotype. As indicated earlier, the relation between an animal's phenotype and its breeding value is equal to the square root of heritability. This relation is shown diagrammatically in the center of Figure 5.3. A general rule to keep in mind is that, other things being equal, the higher

FIG. 5.1 "Round Oak Rag Apple Elevation," a Holstein-Friesian bull that was one of the greatest bulls of his time in genetic merit for yield and type with several generations of ancestors also with outstanding genetic merit. (Courtesy of Select Sires, Inc., Jack Remsberg, Middletown, Maryland, and the Holstein-Friesian Association of America)

FIG. 5.2 "Beecher Arlinda Ellen," a Holstein-Friesian cow with world's record production and an outstanding pedigree of genetic superiority for yield. (Courtesy of Jim Miller, Cary, Illinois, and the Holstein-Friesian Association of America)

the heritability of a trait the more valuable is the animal's phenotype as an estimator of its genotype. Therefore, the higher the heritability, the less reason there is to consider any information other than the animal's own phenotype. However, heritability is rather low for many of the quantitative traits of dairy cattle that are economically important. Therefore, in most cases sources of information in addition to the animal's own phenotype must be taken into account. In the case of bulls, which have no phenotype for yield traits, these other sources of information are the only ones from which estimates of breeding value can be made.

Progeny

A second source of information about an animal's breeding value is its progeny. Diagrammatically, the animal's progeny constitute the block of information at the left side of Figure 5.3. A sample one-half of the animal's genes has descended to each of its progeny. However, because of the sampling nature of inheritance, it is never known which genes were passed to a particular offspring or what the overall merit of that sample of genes really was. Therefore, the larger the number of progeny that are available with measurements of their phenotypes the more valuable is the source of information as an aid in estimating an animal's breeding value. The path from each progeny's breeding value to its phenotype is, as usual, equal to the square root of heritability. In order to calculate the relationship between a progeny's phenotype and the animal's breeding value 2 paths must be traversed. These are the path from the progeny phenotype to the progeny breeding value and the path from the progeny breeding value to the animal's breeding value. When traversing multiple paths, the total value of all paths is equal to the product of the values of the individual paths. Therefore, the relationships between the animal's genotype and the phenotype of either of the progeny in Figure 5.3 equals 0.5h. In other words, the relationship is one-half the square root of heritability.

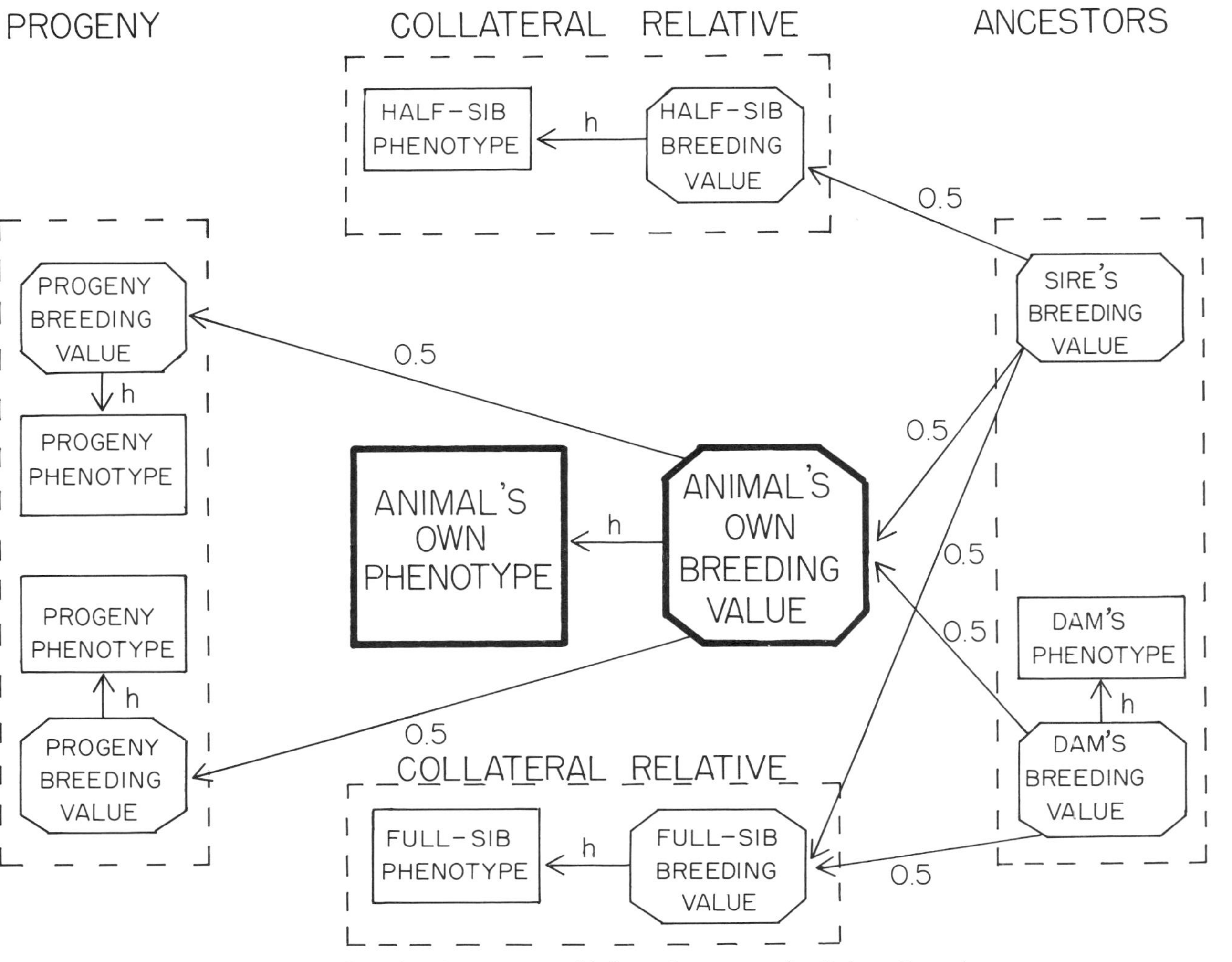

Fig. 5.3 Diagram showing relationship of various sources of information to an animal's breeding value.

Since there are 2 progeny, twice as much information is available as if there had been only 1 progeny.

Ancestors

A third source of information on an animal's breeding value is its ancestors. The block on the right side of Figure 5.3 shows ancestors of the animal, in this case the sire and dam. Ancestors are animals in previous generations which are related to an animal by descent. The calculation of the relationships in the case of ancestors follows exactly the same principles as were described for progeny. For example, the path from the dam's phenotype to the animal's breeding value is equal to 0.5h. One new element is introduced here. All the paths that were traversed did not exert biological influence in the same direction, as has been the case until now. Starting at the dam's phenotype the first path (h) was traversed against the direction of biological influence. Then the path (0.5) between the breeding values of the dam and the animal was traversed in the direction of biological influence. This demonstrates the principle that it is permissible to change direction of biological influence once when traversing a set of paths. Traversing a change in the direction of biological influence can be done only at a source of variation that is itself exerting influence in multiple directions. The only way such a change would make sense biologically would be if such a source of variation were approached against the direction of biological influence. For example, the sire's breeding value cannot be estimated by going from the dam's breeding value in the direction of biological influence to the animal's breeding value and then against the direction of biological influence to the sire's breeding value. Obviously, the animal's breeding value does not exert any biological influence on the breeding value of either the sire or the dam because the animal did not pass on genes either to its sire or dam.

Collateral Relatives

The fourth source of information is from collateral relatives. Two examples of collateral relatives are shown in Figure 5.3, at the top and bottom center. The collateral relative at the top is a half sib and the one at the bottom is a full sib. A sib is an animal that is related to another animal through common parentage; i.e., either full brothers or sisters or half brothers or sisters. When using the full sib to estimate the animal's breeding value another principle is introduced. There are 2 distinct sets of paths going from the full sib's phenotype to the animal's breeding value. One set goes through the sire's breeding value and the other through the dam's breeding value. The value of each set of paths is calculated according to the usual rules, namely the product of the individual component paths. After the value of each set of paths has been calculated the sets are added. In this case, each set of paths has a value of 0.25h. The sum of these 2 sets of paths equals a total value of 0.5h. Thus, strictly on the basis of relationships, one full sib is worth as much for estimating an animal's breeding value as one progeny. In each case the relation between the phenotype and the animal's breeding value is 0.5h.

All sources of information that are useful for estimating an animal's breeding value will fall in one of these 4 categories. The use of these relationships will be discussed in detail later in this chapter in Section 5.6. Before that, however, attention should be given to the all-important phenotype of yield, which is the primary goal of genetic improvement, as well as the main tool for making that improvement.

5.3 A Standard Definition for Phenotypes

In Chapter 4 a phenotype was defined as "what an animal is or does." Henceforth, the word phenotype will carry an implied reference to a specific trait or group of traits. In order to utilize traits or phenotypes for genetic improvement uniform bases for their definition and evaluation are necessary.

The most commonly evaluated traits of dairy cows are the yield traits. To compare the yields of different cows, or if several persons are to evaluate the yield of a single cow, one of the first things that must be done is to put yield on a standard basis so that it will mean the same thing to everyone. At the present time, lactation records in the United States are standardized on 3 bases: the number of days in the lactation, the number of times the cow was milked per day, and her age and month of the year at calving.

Length of Lactation

The accepted standard length for a lactation record is 305 days. The logic for this particular length is that, on the average, a cow milking for 10 months would total 305 days of lactation. A 10-month lactation with calving at 12-month intervals has been considered some sort of an "ideal" for many years. When cows milk longer than 305 days their yield for the first 305 days usually is taken as the lactation yield. In some cases cows do not milk for the full 305 days. They may go dry of their own accord, or their lactation may be terminated for a variety of reasons. For many years incomplete lactations were defined as those which were terminated in less than 305 days due to causes that were entirely environmental and had no relation to the cow's genetic ability to complete a normal-length lactation.

Research[1] has showed that projecting all lactation records to 305 days in milk, including lactations where the cow was reported to have gone dry, resulted in higher heritabilities and predictive values in the resultant records. The rationale for these results is that any cows that conceive early in lactation are forced dry before 305 days in milk while still milking well to allow for a sufficient dry period. Therefore, they are unfairly penalized by not being permitted to complete 305 days. Even if some cows' records do not deserve projection because they go dry naturally, projection of the records will result in credit for little additional yield because last test day production will be quite low. It is impossible to determine with 100% accuracy in the DHIA Program which cows' records deserve projection and which do not. Therefore, based on the latest scientific evidence all lactation records with less than 305 days in milk used in USDA-DHIA genetic evaluations are being projected to 305 days in milk.

The procedure for projecting lactations records of less than 305 days in milk to 305 days[2] is based on the sum of production for the number of days the cow actually milked plus an estimate of what the cow would have produced in the remainder of the 305 day lactation derived from the last available test day production data. An example of how to project a milk production record to 305 days is shown in Appendix Table II-A-1. The same procedure is used for other lactations, breeds, regions, and for fat production, but additional tables of factors are required. Similar procedures are being developed to project protein and solids-not-fat records.

A lactation record that has been projected from less than 305 days in milk should not be given quite as much emphasis as one that was for the full 305 days, because shorter records tend to be more variable and the projection factors do not account for this greater variability. Therefore, a projected record is less reliable than one that was for the full 305 days. Sets of factors for use in assigning relative weighting to incomplete records of various lengths is given in Appendix II-A (Table II-A-2).[3] Projected records that are used in genetic evaluations should be weighted according to these factors.

Number of Milkings per Day

Records are standardized to a 2-times-a-day milking basis, usually referred to as "2X." Factors for converting 3X or 4X records to a 2X basis are shown in Appendix II-B.[4] These factors are intended for use on age-corrected records (described below). The same factors are used for all breeds and for both milk and fat. However, factors are given for 3 different age groupings.

Age and Month of the Year at Calving

The third adjustment made to standardize lactation records is for age and month of the year at calving. Records have traditionally been standardized to a Mature Equivalent (ME) basis. The correct interpretation of an ME record is: the amount of milk or components that the same cow would have produced if she had calved in an environmentally average month and been of mature age. An ME record does *not* predict what a cow will produce in the future. Obviously, such a prediction could not be made with much accuracy since many things may happen to a cow before she reaches mature age, and environmental conditions affecting her future records may be quite different. Neither should ME factors be taken as an indication that heifers mature slower in some regions of the country or that cows are genetically superior in other regions. Breed and regional differences in ME factors are caused by all of the multitude of genetic and environmental effects that determine the yields of cows of different ages.

The age and month of calving adjustment factors used in the United States were published in 1974.[5] These factors were developed from a national set of NCDHIP lactation records using a statistical procedure that estimated the effects of both age and month of the year at calving on the amount of milk and milk fat that cows produced. Factors were calculated on a regional basis for each breed and were also pooled on a national basis. The complete array of factors includes 74 tables of factors plus explanatory material.[5] A partial set of the national Holstein factors for milk is shown in Appendix II-C. These factors can be used for practice calculations. The table of factors should be used to standardize milk yield to an ME basis by multiplying yield for the first 305 days of lactation by the ME factor corresponding to the cow's age and month of the year at calving for the appropriate breed, region of the country, and month of the year.

5.4 Genetic Evaluation of Dairy Cattle in the Past

Genetic improvement of dairy cattle centered on identifying genetically superior cows from the mid-1700s to the early part of the 20th century, with relatively little effort expended on the genetic evaluation of bulls. This course of events was caused by lack of knowledge of genetic principles and statistical procedures. Many methods were used to evaluate the genotypes of cows. Each method has its advantages and disadvantages. Discussion will be centered primarily on milk yield. In most cases the same statements will also apply to yield of milk components.

Using Lactation Yield Directly

In general, just the knowledge of a cow's lactation yield tells little about her breeding value for yield. One of the most important influences on a cow's yield is the herd environment. A cow's production could easily vary by as much as 50% if she were in a well-managed herd as opposed to a poorly managed herd. Such herd differences are the so-called permanent environmental effects. On the average, about 80% of the differences between herds is due to environment and about 20% is genetic. In addition, there is a wide variety of intraherd effects that act constantly on each cow as mentioned in Section 4.8. Standardization of records for length, times milked per day, and age and month at calving has helped to reduce the nongenetic variation from these 3 sources. Furthermore these corrections put records on a standard basis so that comparisons among records can be made more accurately. But, it must be realized that these adjustments do not remove all the variation due to these sources of variation.

Early in the development of dairy cattle genetics, it was realized that multiple lactation records on a cow were useful to obtain a better estimate of her genotype for yield. However, multiple records are not proportionately more valuable as their numbers increase because there is a relationship among succeeding records of a cow, which is termed the "repeatability" of records. The relationship among a cow's lactation records or their repeatability results from the fact that the cow had the same genotype in both lactations and usually had many of the same general environmental effects.

In general terms, repeatability is a measure of the tendency of repeated expressions of a trait to be similar. Repeatability can vary from 0, where there is no relation among repeated expressions of a trait, to 1.0 where repeated expressions of a trait are precisely the same. The higher the repeatability the less the need for repeated measurements of a trait. A single measurement on a qualitative trait such as coat color is usually sufficient. Little is to be gained by observing coat color again at a later time. However, with a quantitative trait such as milk yield, information on more than one lactation can be valuable because repeated measurements tend to average out temporary environmental differences which affect an individual lactation. Unfortunately, repeated lactations on a cow do little to reduce permanent environmental differences.

The repeatability of milk yield for succeeding lactations in the same cow is approximately 0.5. The value of information on succeeding lactations for a cow compared to just one lactation can be calculated from the formula:

$$\frac{\dfrac{nr}{1 + (n - 1)r}}{r},$$

where n is the number of lactations and r is the repeatability. Values obtained from this formula assuming the repeatability is 0.5 are shown in Table 5.1. Two records are worth about ⅓ more than one record and 3 records are worth about 50% more. While the information on additional lactations is worth more and should be used if already available, the merits of waiting the additional time necessary to obtain later lactations before making important decisions is questionable.

The advantage of using an animal's own performance, including standardized records and multiple records, is that this method is relatively simple to measure and understand. The disadvantage of using an animal's own performance is that many sources of non-genetic variability arising from permanent and temporary environmental effects are present even in standardized records. Thus, these records frequently deviate widely from the animal's true breeding value.

TABLE 5.1 *Value of a cow's succeeding lactations compared to having a single record*

Number of Lactations	*Value of Information*
2	1.33
3	1.50
4	1.60
5	1.67
6	1.71
8	1.78
10	1.82
20	1.90

The Daughter–Dam Comparison

The first widely used method for putting yield on a relative basis was the daughter–dam comparison. Basically, this method expressed each cow's yield as a deviation from the yield of her dam (daughter's yield − dam's yield). The records used were expressed on the standardized (305-2X-ME) basis. If there were multiple records on either the daughter or the dam the mean of those records was used. This was also the case when the daughter–dam comparison was used to evaluate the breeding value of dairy sires. The daughter–dam comparison, or modifications thereof, was used for about three decades, starting in the early 1930s, as the standard basis for genetic evaluation of the yield of dairy cattle in the United States. At least a dozen indices which were all variations of the basic daughter–dam comparison were used by various segments of the dairy industry during this period. The two primary indices which evolved from the daughter–dam comparison were the Equal Parent Index and the Regression Index.

The Equal Parent Index was based on the assumption that the daughter's production should be midway between that of the sire (if he could express production) and that of the dam. The formula by which the Equal Parent Index was calculated is: Equal Parent Index = daughter average + (daughter average − dam's average). For example, if the average of the daughter's records was 11,000 lbs and the average yield of the dam was 8,000 lbs the Equal Parent Index would be 11,000 + (11,000 − 8,000) = 14,000 lbs.

The Regression Index was based on the fact that daughters of superior sires and dams did not usually produce as much as their superior parents. This is an example of the widely known tendency of progeny to regress towards the population mean. In order to take this phenomenon into account the following Regression Index was devised:

$$\text{Regression Index} = \frac{\text{Equal Parent Index} + \text{Breed Average}}{2}$$

Therefore, the Regression Index gave a

value that was the mean of the Equal Parent Index and the breed average.

The advantages of the daughter–dam comparison over the use of the daughter's records were that it attempted to take into account both the sampling nature of inheritance and the fact that environmental effects were important in determining the yield of the daughters and the dams. The primary disadvantage of this method was the time lag of over 2½ years between the time that the dams and the daughters made their first records. Obviously, during this time interval many environmental changes were likely to occur even within a single herd. Therefore, the daughter–dam comparison was, in many cases, severely biased by those environmental changes.

Even more serious biases could be introduced into the comparison if the dam and her daughter made their records in different herds.

In the early 1960s, the daughter–dam comparison was largely abandoned in favor of the herdmate comparison which was superior in eliminating environmental effects from estimates of breeding value.

The Herdmate Comparison

The herdmate comparison, also referred to as the stablemate comparison, was the first genetic evaluation procedure developed specifically for evaluating bulls. The herdmate comparison compared each cow's record with the records of other cows milking in the same herd at the same time. Thus, all records used in the comparison were subjected to approximately the same permanent environmental effects as well as many of the same temporary environmental effects. This method largely overcame the primary weakness of the daughter–dam comparison, i.e., the opportunity for widely different environmental effects to act on records used for genetic evaluation. However, the herdmate comparison was based on some important assumptions. If these assumptions were not met, biased genetic evaluations resulted. But, in most cases genetic evaluations of bulls and cows were still considerably more accurate than those made by the daughter–dam comparison indices. These assumptions were:

1) All animals used in genetic evaluations were random samples of one genetic population in each breed.
2) There was no genetic trend in each breed.
3) The herdmates of all cows were subjected to the same severity of culling.
4) Each cow and her herdmates received the same level of treatment with no preferential treatment being given to any animal.

The herdmate comparison method was utilized on a national basis for over a decade to calculate USDA-DHIA genetic evaluations. It was replaced in 1974 by a more accurate procedure to be described shortly.

The Contemporary Comparison

The contemporary comparison is based on the same principles as the herdmate comparison except that only first records are used. The name stems from the fact that all animals used are "contemporaries"; that is, they commenced their first records at approximately the same time.

Proponents of the contemporary comparison claimed 2 primary advantages over the herdmate comparison. First, the use of first records minimized many of the possible inaccuracies that might arise from the use of age-correction factors that were inappropriate for a given herd. Second, the use of first records avoided bias that might arise from the inclusion of later records, which were made only by cows that escaped being culled and therefore were a selected group or received preferential treatment. In addition to these 2 primary advantages, the contemporary comparison could be calculated with simpler statistical procedures than the herdmate comparison and was less expensive to run on a computer. The primary disadvantage of the contemporary comparison, and one that may have counterbalanced all the advantages, was that the amount of information available for estimating breeding value was usually considerably less than for the herdmate comparison.[6] In each herd there would normally be many more cows that would qualify as herdmates than would qualify as true contemporaries.

5.5 Genetic Evaluation of Dairy Cattle at Present

Most of the developed countries with significant dairy industries have well-organized research programs for the genetic improvement of dairy cattle. These programs are sponsored by the government in many countries where the payback to the public from objective, unbiased research, free from commercial influences is recognized. The U.S. dairy industry and the taxpayers have benefitted greatly from the national research program on genetic improvement of dairy cattle conducted by the Animal Improvement Programs Laboratory (AIPL), Agricultural Research Service, U.S. Department of Agriculture and research at the land-grant universities. Genetic evaluations of bulls (Sire Summaries) and cows (Cow Indexes) emanating from this research have been the primary source of information for identifying animals with superior genetic merit for yield.

The primary objective of a genetic evaluation program should be accurate, useful evaluations that will enable dairymen and others in the field to maximize the rate of genetic improvement for economically important traits. Scientists agree that the most accurate genetic evaluations are those with Best Linear Unbiased Prediction (BLUP) properties. However, scientists do not agree on the best way to obtain BLUP results. In fact, attributes considered an advantage of one method over another by one scientist may be viewed just the opposite by another scientist. Therefore, space will not be taken here discussing the advantages and disadvantages of different methodologies. Rather, the attributes of the more commonly used methodologies will be discussed to provide information that can be used as a basis for comparison.

Genetic Evaluation Procedures in the United States

Since 1974, national genetic evaluations of bulls and cows for milk and fat have been calculated by AIPL-USDA using a procedure called the Modified Contemporary Comparison. National Sire Summaries for protein and solids-not-fat have been calculated by a Mixed Model Method since 1976. Research on genetic evaluations of type traits using a Mixed Model Method was started by AIPL-USDA with the Jersey breed in 1977 and now includes genetic evaluations of Uniform Linear Type Traits for all breeds except Holstein (the Holstein Association calculates type evaluations for Holsteins). Since 1976 research has been conducted by AIPL-USDA that has yielded Sire Summaries (and later Cow Indexes) for milk yield for Holsteins in Mexico in cooperation with Holstein Friesian de Mexico. In early 1983, research on Sire Summary procedures for dairy goats by AIPL-USDA culminated in the first national Buck Summary. Since the early 1970s, regional Sire Summaries (and more recently Cow Indexes) for yield traits have been calculated in the northeast by Cornell University in cooperation with the Eastern A.I. Cooperative.

Modified Contemporary Comparison

The model for the Modified Contemporary Comparison Sire Summaries can be shown diagrammatically as follows:

$$\underset{1}{PD82} = \underset{2}{R}(\underset{3}{DAU - MCA} + \underset{4}{SPD}) + \underset{5}{(1 - R)\,AM}$$

The equation can be partitioned into five parts for explanation.

1 PD82 represents a bull's Predicted Difference, that is his transmitting ability, for a trait expressed as a deviation from the 1982 genetic base (which will be discussed in detail in chapter 7).

2 Represents repeatability which is used in the calculation of the Modified Contemporary Comparison to weight daughter information (parts 3 and 4 of the formula) versus pedigree information (part 5 of the formula), and also as an indicator of reliability of the PD.

3 DAU − MCA represents the deviation of daughter records from the average of the modified contemporaries. 'Modified Contemporaries' was a new concept introduced with the Modified Contemporary Comparison. Lactation records of other cows calving in the same herd, year, and season as a particular daughter record are divided into two contemporary groups.

First lactations are one group and second and later lactations constitute the other group. Daughter first records are compared primarily with first lactation contemporary records with 2nd and later noncontemporary records counting as one additional contemporary record for comparison. Second and later daughter records are compared with second and later contemporaries, with first lactation non-contemporaries counting as one additional contemporary record for comparison. Therefore, this method uses all available lactation records but emphasizes the comparisons that are most accurate, 1st versus 1st and later versus later.

4 SPD represents the adjustment of the daughter-modified contemporary deviation for the PD of the sires of the modified contemporaries. This adjusts for the genetic merit of the competition to which a bull's daughters are subjected. The beneficial impact of this adjustment is enhanced by an iteration procedure that is incorporated into the Modified Contemporary Comparison.

5 (1 − R) AM represents the incorporation of ancestor merit directly into the calculation of the PD. Ancestor merit is weighted in comparison to daughter information by the use of (1 − R) and R as weighting factors.

The model for the Modified Contemporary Comparison Cow Indexes can be shown diagrammatically as follows:

$$CI82 = .5\,[w(\overline{MCD}_{cow} + ADC) + (1 - w)(PD82_{sire} + CI82_{dam})],$$

where CI82 represents a cow's Cow Index, that is her transmitting ability, for a trait expressed as a deviation from the 1982 genetic base, .5 is a factor to express CI82 as transmitting ability (one-half breeding value), w represents a weighting factor dependent on the amount of information in the cow's average Modified Contemporary Deviation ($\overline{MCD}$) and the repeatability of her sire's PD82 and her dam's CI82, ADC is an adjustment for the average genetic merit of the dams of contemporaries based on the cow's year of birth, and the other terms have been defined.

Important Attributes of the Modified Contemporary Comparison

All available lactation records are used in both Sire Summaries and Cow Indexes. Scientists have argued for many years over what records should be used in genetic evaluations. Many scientists have maintained that first lactations are sufficient and that later lactations are not needed. Most other countries have historically used only first lactations. However, research (7, 8) has shown that the use of later lactations is important for the correct ranking of bulls on lifetime productivity and profitability of their daughters.

Records-in-progress (RIPs) with 40 days or more in milk are used in both Sire Summaries and Cow Indexes. However, a cow must have at least 100 days in milk to qualify for the Elite Cow Index list on a single lactation record. Research on the use of RIPs showed that they ranked bulls accurately and permitted the first summary on a bull to be calculated about 6 months earlier in each bull's life.

All lactation records are standardized to a twice a day milking, 305 days in milk, mature equivalent basis (2X-305 day-ME) (Section 5.3). Lactation records used in genetic evaluations have been standardized in the U.S. for many decades, but standardization of records is still not practiced in some countries where only first lactations are used. Instead of standardizing records, some countries limit the records that are used in genetic evaluations to a minimum days in milk and to certain ages at calving. The disadvantage of selecting records in this manner is that the data used for genetic evaluations may be biased in unknown ways and no genetic evaluation procedure will account for those biases. Therefore biased genetic evaluations could result, that is, bulls and cows would not be ranked correctly for genetic merit.

Records are weighted for length of lactation according to the correlation between records of different lengths and 305 day records.

The daughter-modified contemporary deviations are calculated on a within herd-year-season basis using 5 month moving seasons centered on the daughter's month of calving. Research many years ago showed that moving seasons were superior to fixed seasons for calculating genetic evaluations on a within herd-year-season basis. However, since the statistical properties of moving seasons are difficult to define, their use in the Modified Contemporary Comparison interferes with that procedure producing results with true BLUP properties.

The within-herd environmental correlation among each bull's daughters (also referred to as the sire by herd interaction) is accounted for in the Modified Contemporary Comparison. This attribute is important due to the summarization of all bulls including those with most of their daughters in a single herd.

Cow Index procedures are an integral part of the Modified Contemporary Comparison Sire Summary calculations. Most of the calculations needed for Cow Indexes are accomplished during computations for the Sire Summaries, so highly accurate Cow Indexes are obtained relatively inexpensively.

The Modified Contemporary Comparison is more expensive to calculate on small data sets but less expensive on large data sets than the Mixed Model Methodology (to be described next). An important reason that AIPL-USDA continues to use the Modified Contemporary Comparison for national genetic evaluations is the prohibitive cost of the Mixed Model Method with the millions of lactation records that are used in each Sire Summary run.

Mixed Model Method

The term Mixed Model refers to the statistical properties of the effects in these models where some of the effects are random and some are fixed. This procedure also carries other labels including BLUP Procedure, Linear Model Method and Direct Comparison. The basic purpose of the Mixed Model Methodology is to produce genetic evaluations with BLUP properties, but, technically, in many cases the procedures used yield close approximations to BLUP results due to limitations in statistical or computer technology.

The Mixed Model Method used for national genetic evaluations for protein and solids-not-fat is represented by the following equation:

$$y_{ijklm} = ht_{ij} + g_k + s_{kl} + hs_{ikl} + c_{iklm} + e_{ijklm},$$

where y_{ijklm} is a daughter's yield for the trait being evaluated, ht_{ij} is a fixed effect due to herd, year and season, g_k is a fixed effect common to sires in the kth genetic group, s_{kl} is a random effect common to daughters of the lth sire in the kth genetic group, hs_{ikl} is a random effect common to daughters of the lth sire in the kth genetic group in the ith herd (the herd by sire interaction), c_{iklm} is a random cow effect of the mth daughter of the lth sire in the kth genetic group in the ith herd, and e_{ijklm} is unexplained variation.

Attributes of the Mixed Model Method

The attributes described for the Modified Contemporary Comparison regarding use of all records, RIPs, standardization and weighting of records, and the sire by herd interaction are also attributes of the Mixed Model Method used by AIPL-USDA.

The Mixed Model Methods all use fixed seasons. AIPL-USDA uses two 6-month seasons, January through June and July through December. The use of fixed seasons in the Mixed Model Methods enhances definition of the statistical properties of the models and the theoretical conformity of the results to BLUP properties, but, from a practical viewpoint the use of moving seasons may provide a sounder basis for calculation of within season comparisons.

Cow Index techniques are available for Mixed Model Methods but are much more costly to calculate than those for the Modified Contemporary Comparison and are of questionable accuracy for daughters of non-AI bulls with most or all daughters in a single herd.

An example of USDA-DHIA Sire Summary information is shown in Figure 7.1.

5.6 Pedigree Estimates of Breeding Value

A pedigree is a record of ancestry. Pedigrees should contain genetic evaluations and information on the performance of ancestors. Typical information would include genetic evaluations for yield and type characteristics for both sexes and lactation records of cows. Useful estimates of breeding value or transmitting ability (one-half breeding value) can be calculated for an animal from pedigree information alone. In essence, the accuracy of pedigree estimates of transmitting ability is based on 2 factors: (1) the relationship of each item of information to the animal's breeding value and (2) the relative accuracy of the various items of information. The relation of various sources of information to the animal's breeding value was discussed in Section 5.2.

A pedigree may contain information on as many as 4 or 5 generations including that of the animal itself. As shown in Figure 5.4 approximately one-eighth of the genes from each great grandparent are passed on to an animal. Therefore, most of the time, there is little advantage in tracing a pedigree beyond the grandparents except to check for inbreeding (Chapter 6).

Pedigree information is most valuable for evaluating young bulls and cows that have neither performance information nor progeny and for estimating the prospective merit of animals that would result from planned mating. As animals grow older, their progeny or performance records should be given more emphasis in relation to the information on ancestors. This is especially true for bulls where an accurate progeny test is by far the most valuable single source of information on transmitting ability. It is true to a lesser extent for cows that are daughters of sires with accurate summaries. In fact the sire's transmitting ability is a more accurate single piece of information on a cow's transmitting ability, and has a higher correlation with her transmitting ability than does a single lactation record on the cow with no information on the environmental conditions under which that record was made.

Unfortunately, establishing arbitrary weights or rules for weighting different items of information in pedigrees is diffi-

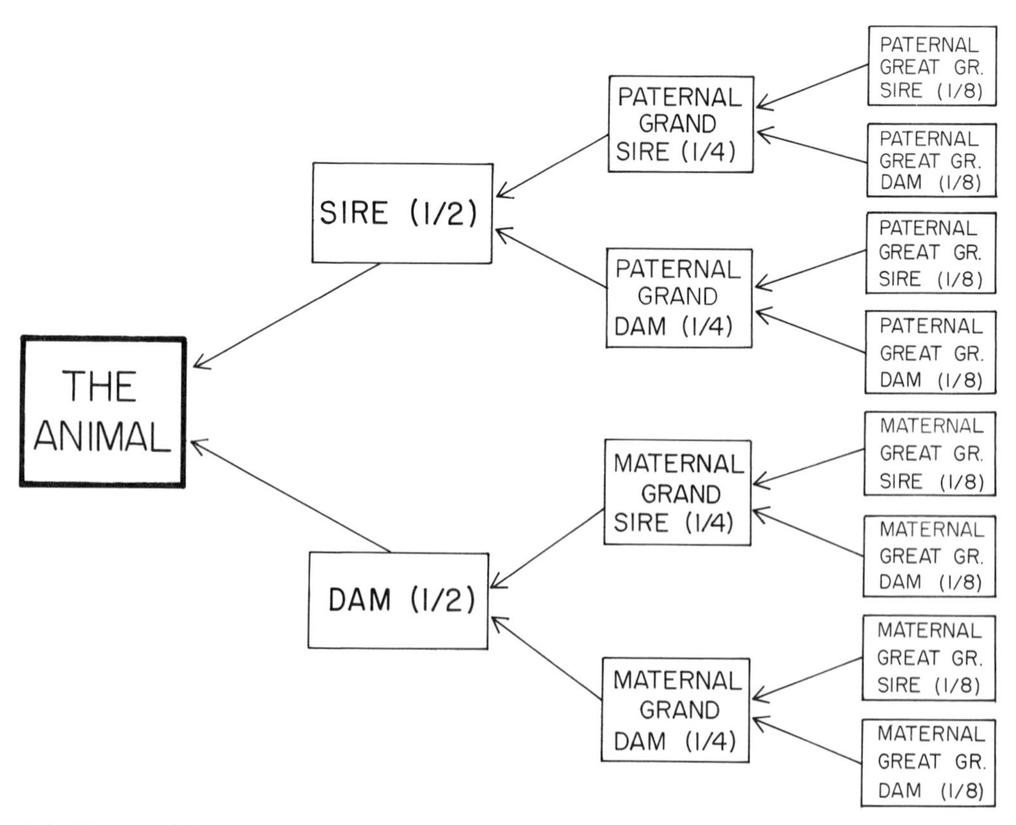

FIG. 5.4 Diagram of the proportion of genes passed to animal from each ancestor.

cult. Therefore, the proper evaluation of a pedigree requires considerable good judgment.

For bulls, the most accurate estimates of transmitting ability are Predicted Differences with high repeatabilities or genetic evaluations calculated by a mixed model procedure with a large number of daughters in many herds. Herdmate Comparisons are next in accuracy followed by daughter–dam comparisons, which may still be found in some pedigrees and daughter averages. At the bottom of the list is highly selected information such as an occasional high record made by one of the daughters or selected show ring winnings of the daughters. This kind of information may be a worthwhile addition to a pedigree for advertising individual animals or for arousing interest, but it is of limited value as a basis for estimating an animal's transmitting ability.

On the female side of the pedigree the various kinds of information can be ranked on the same hierarchy of accuracy. The most accurate information available on a national basis is the USDA-DHIA Cow Index Value. Most DRPCs also estimate transmitting ability on cows that are processed through their center.

Usually the key information in a pedigree is obtained from the animal, the sire, the dam, the maternal grandsire, and the paternal grandsire, in that order. The more information that is available on the animal itself and its close ancestors, the less the importance attached to remote ancestors. A pedigree should be evaluated by starting with the information on the animal itself and working backwards. If the animal being evaluated is an AI bull with a PD that has a high Repeatability, information on ancestors will usually add little to the estimate of transmitting ability for yield. This is true because of the close relation between the progeny test information and the bull's breeding value (Figure 5.3) and the accuracy of such information.

On the other hand, for a young bull that has no progeny, the best information is to be found on the parents. If the sire has a PD with a high Repeatability, then that is the most accurate single item of information that you can have on the parents. The next most accurate information is usually a reliable cow index value on the dam. If accurate information is not available on the sire, the paternal grandsire's information should be evaluated. If his PD has a high Repeatability, that constitutes the best information on the probable transmitting ability of the sire. If a cow index is not available on the dam, the maternal grandsire should be considered carefully.

Figure 5.5 shows a 3-generation pedigree of the Holstein bull Rockalli Son of Bova, a son of Round Oak Rag Apple Elevation (see Figure 5.1). Rockalli Son of Bova was the highest bull for PD milk and the 3rd highest for PD $ on the January, 1984 USDA-DHIA Sire Summary run. His pedigree has an array of information that would be desirable in all pedigrees.

5.7 Summary

Obtaining the most accurate estimate of the breeding value or transmitting ability of the animals that are to be used for breeding is extremely important in a genetic improvement program. Basically there are 4 sources where such information can be obtained: the animal itself; its progeny; its ancestors; and collateral relatives. In order to obtain estimates of breeding value or transmitting ability that will be generally understood it is necessary to have a standardized expression of phenotypes. This has been done according to prescribed rules and procedures. The development of more refined methods for estimating the breeding value or transmitting ability of cows and bulls has evolved to a point where the present estimates are by far the most accurate that have ever been available. Estimates of transmitting ability for yield are now available for all cows in herds enrolled in the Official DHIA recordkeeping plans and for bulls with enough daughters to be eligible for a USDA-DHIA Sire Summary.

The concept of weighting different kinds of estimates of transmitting ability enters into pedigree evaluation and is an important first step in estimating transmitting ability, especially for animals with no progeny or phenotype. The ranking of estimates of transmitting ability of bulls in descending order of accuracy is:

1) Modified Contemporary Comparison or Mixed Model Method.

NAME ROCKALLI SON OF BOVA   REG. NO. 1665634
SEX MALE   BORN 3/12/74

SELECT SIRES, INC.
11740 U. S. 42
PLAIN CITY, OH   43064

88-6Y   TPI +756 GM 7/83
6-01 88 V+E 2431231 00000 0
PD +1792M +35F 95% REP 1/84
490 DAUS AVG 19604M 3.42% 670F
CONTEMPS 18091M 641F

PDT +1.76 ±.23 1/84 92 HRD
.89R 126 DAU
EFT DAU/HRD 2.1
ALL DAU AVE ACT SC 79.5

EFT DAU AVE AASC 81.9
STD DEV AASC 3.2

TIDY BURKE ELEVATION
1271810   TPI +241 GM 1/73

PD -320M -20F 99% REP 10/74
2314 DAUS AVG 14995M 3.50% 525F
CONTEMPS 14383M 517F

PDT +1.68 ±.10 1/84 331 HRD
.98R 1129 DAU
EFT DAU/HRD 11.3
ALL DAU AVE ACT SC 81.5
EFT DAU AVE AASC 81.3
STD DEV AASC 3.3

2ND GET/SIRE   S W REGNL 1959

ROUND OAK RAG APPLE ELEVATION
1491007 96-10Y   TPI +522 GM 1/84
10-11 96 EEE 1211111 00000 0
ALL-AMERICAN GET/SIRE 1978
ALL-AMERICAN GET/SIRE 1977
H.M.ALL-AMER GET/SIRE 1981
H.M.ALL-AMER GET/SIRE 1980
PD +383M +23F 99% REP 1/84
53763 DAUS AVG 18407M 3.65% 672F
CONTEMPS 17821M 644F
PDT +2.21 ±.07 1/84 7612 HRD
.99R 40062 DAU
EFT DAU/HRD 16.7
ALL DAU AVE ACT SC 82.8
EFT DAU AVE AASC 83.0
STD DEV AASC 4.1

ROUND OAK IVANHOE EVE
5749758 94-12Y94-10Y93-9Y   4
COW INDEX -61M +15F 47% REP
MC DEV +4378 +217 9 REC

	Age		Days	Milk	%	Fat
DHR	2-01	2X	305D	13070M	4.2%	554
DHR			311	13260	4.2	562
DHR	3-01	2X	305	17680	4.2	743
DHR			353	19430	4.2	821
DHR	4-03	2X	305	22640	4.2	949
DHR			345	24440	4.2	1024
DHR	5-03	2X	305	19930	4.0	802
DHR			351	20810	4.0	838
DHR	6-06	2X	305	19620	3.8	754
DHR	7-08	2X	305	24050	3.9	936
DHR			351	25870	3.9	1013
DHR	8-11	2X	305	23640	4.1	960
DHR	11-00	2X	181	14950	4.2	628
DHR	13-09	2X	305	20070	4.3	856
LIFETIME			3317	196030	4.1	8070

ROCKALLI MAGNET DARCY
7945156 85-3Y
3-05 85 +VVV 2422322 21102 0

COW INDEX +211M -6F 49% REP
MC DEV +1757 +17 2 REC

	Age		Days	Milk	%	Fat
DHR	3-02	2X	305D	19860M	3.5%	699F
DHR			365	22520	3.6	807
DHR	4-10	2X	305	23760	3.6	865
DHR			365	27000	3.7	998

C ROMANDALE SHALIMAR MAGNET*RC
1560362 90-3Y   TPI +357 GM 5/7

PD +603M -9F 99% REP 1/8
2364 DAUS AVG 17724M 3.44% 609
CONTEMPS 16562M 602

PDT +.78 ±.07 1/84 930 HRD
.99R 1593 DAU
EFT DAU/HRD 3.4
ALL DAU AVE ACT SC 80.9
EFT DAU AVE AASC 81.2
STD DEV AASC 4.6

5TH AGED BULL   E NAT SHOW 1968

ROCKALLI HE-MAN DENISE
7772145 90-7Y 90-5Y 88-4Y GMD 2
COW INDEX -266M -4F 52% REP
MC DEV +4182 +127 7 REC

	Age		Days	Milk	%	Fat
DHR	2-04	2X	362D	21780M	3.7%	796
DHR	3-05	2X	305	24290	3.6	870
DHR			356	26940	3.6	980
DHR	4-06	2X	305	21410	3.8	813
DHR			365	23690	3.8	900
DHR	5-08	2X	305	27940	3.7	1025
DHR			351	30970	3.7	1159
DHR	6-09	2X	305	27100	3.5	954
DHR			331	28450	3.6	101[illegible]
DHR	7-09	2X	305	27120	3.5	947
DHR			365	30650	3.6	110[illegible]
DHR	10-10	2X	348	18020	3.5	634
LIFETIME			3112	210500	3.7	775[illegible]

*RC CARRIER RED HAIR COLOR

60 3246.0 02/16/84
Holstein Association
OFFICIAL PEDIGREE PREPARED BY
HOLSTEIN-FRIESIAN ASSOCIATION OF AMERICA
BRATTLEBORO, VERMONT 05301

All completed lactation and classification data for animals, will be provided upon request, where space limitations prevent printing all performance records.

FIG. 5.5 Pedigree of the Holstein bull Rockalli Son of Bova showing genetic merit for yield and type. (Courtesy of Select Sires and the Holstein Association)

2) Herdmate comparison.
3) Daughter–dam comparison.
4) Daughter average—no selection of information.
5) Selected information on outstanding records, show ring winnings, or distant relatives.

Accuracy of various types of phenotypic information for estimating the transmitting ability of cows in descending order:

1) Modified Contemporary Comparison or Mixed Model Cow Index or an estimate of transmitting ability.
2) Herdmate Comparison Cow Index.
3) Daughter–dam comparison.
4) Lactation records, average or listing of individual records.
5) Selected information such as highest records.

In Chapter 6 methods for making genetic improvement in dairy cattle will be discussed, that is, ways to utilize the genotypes that are available in order to make the fastest possible genetic progress.

Review Questions

1. For many years some dairy cattle breeders argued that a cow's genotype should be judged on the basis of her highest record. What merit does such a proposal have in light of today's scientific knowledge of dairy cattle genetics?
2. What are some arguments for and against changing the Mature Equivalent procedure to a procedure that standardizes lactation records to the age of average yield?
3. Make a list of environmental effects that are likely to influence the amount of milk a cow will produce in a lactation.
4. Describe the difference between a regression relationship vs a correlation relationship between 2 traits, and give an example of each.
5. How would genetic trend affect the comparison of bulls' PDs over time if this trend were not accounted for in the genetic evaluation procedure?
6. Considering past improvements in methods for evaluating the transmitting ability of cows and bulls, what improvements would you say are still needed, and how could these be accomplished?
7. How could pedigree information be made more valuable for estimating transmitting ability of: (1) cows; (2) bulls?
8. Using the pedigree in Figure 5.5, list the ancestors for which you feel you have a reliable estimate of transmitting ability vs those for which the estimate is of questionable reliability.
9. Using the pedigree in Figure 5.5 assume you had no information on the bull himself, and then no information on various ancestors in turn, and decide how that would affect your estimate of his transmitting ability and your weighting of the remaining information.

References

1. Norman, H. D., Dickinson, F. N., and Wright, J. R.: Merit of extending completed records of less than 305 days. J. Dairy Sci. *68:* In press. 1985.
2. Wiggans, G. R., and Powell, R. L.; Projection factors for milk and fat lactation records. Dairy Herd Improvement Letter, ARS-56-1, March, 1980.
3. Wiggins, G. R.: Length weights for AP and Standard DHI. Mimeograph, 1984.
4. Kendrick, J. R.: Standardizing Dairy Herd Improvement Association records in proving sires. ARS-52-1, January 1955.
5. Norman, H. D., Miller, P. D., McDaniel, B. T., Dickinson, F. N., and Henderson, C. R.: USDA-DHIA factors for standardizing 305-day lactation records for age and month of calving. ARS-NE-40, September 1974.
6. McDaniel, B. T., Norman, H. D., and Dickinson, F. N.: Herdmates versus contemporaries for evaluating progeny tests of dairy bulls. J. Dairy Sci. *56:* 1545, 1973.
7. Cassell, G. B., McDaniel, B. T., and Norman, H. D.: Modified Contemporary Comparison sire evaluations from first, all, and later lactations. J. Dairy Sci. *66:* 140, 1983.
8. Cassell, B. G., McDaniel, B. T., and Norman, H. D.: Modified Contemporary Comparisons for first and second lactations in the same and different herds. J. Dairy Sci. *66:* 315, 1983.
9. The USDA-DHIA modified contemporary comparison sire summary and cow index procedures. USDA Production Research Report No. 165, March, 1976.
10. Dickinson, F. N., Norman, H. D., Keown, J. F., and Waite, L. G.: Revisions to USDA methodology for sire summaries and cow indexes. J. Dairy Sci. *57:* 977, 1974.

Suggested Additional Reading

Johansson, I.: *Genetic Aspects of Dairy Cattle Breeding*. Urbana, Ill.: University of Illinois Press, 1961.

Johansson, I., and Rendel, J.: *Genetics and Ani-*

mal Breeding. San Francisco: W. H. Freeman and Co., 1968.

Lush, J. L.: *Animal Breeding Plans*. Ames, Iowa: Iowa State College Press, 1945.

Warwick, E. J., and Legates, J. E.: *Breeding and Improvement of Farm Animals,* 7th ed. New York: McGraw-Hill Book Co., Inc., 1979.

Schmidt, G. H., and Van Vleck, L. D.: *Principles of Dairy Science*. San Francisco: W. H. Freeman and Co., 1974.

Journal of Dairy Science.

Agricultural Research Service, USDA: Dairy Herd Improvement Letters.

DHIA Handbook, Extension Service, USDA.

CHAPTER 6

Methods for Genetic Improvement

6.1 Introduction

The first phase of making genetic improvement, estimating breeding value, was discussed in Chapter 5. The second phase is the combining of genotypes to produce offspring with superior genetic merit. There are 2 basic methods for combining genotypes to cause genetic improvement: (1) increase the proportion or frequency of genes that cause desirable genotypes, and (2) recombine genes into the most advantageous combinations without changing their frequency.

Most genetic improvement in dairy cattle results primarily from the first method. At present, the national program of genetic improvement of dairy cattle in the United States is based primarily on changing gene frequency (method 1). Recombining genes through systems of mating (method 2) also plays a role in genetic improvement in some circumstances.

6.2 Gene Frequency

The "frequency" of a gene is the proportion of loci in the population on which the gene is present in its allelic series. Gene frequency can range from 0 where the gene is completely absent to 1.0 where it is the only allele present in the population. For example, the frequency of the gene causing red color is fairly low in Holsteins, probably less than 10%. Therefore, the frequency of the allele causing black color is at least 90%. The red gene in Holsteins has been forced down to a relatively low frequency because for many years red Holsteins were undesirable and could not be registered. In addition many dairymen with registered Holsteins did not want their neighbors to know that they had any cows that were carriers of the red factor. In spite of many decades of discrimination against the red allele, it was not eliminated, because the rarer a recessive gene becomes, the more difficult it is to decrease its frequency any further. The problems of changing the frequency of genes that affect quantitative traits are inherently more complex than those affecting qualitative traits because many more genes affect quantitative traits.

Methods of Changing Gene Frequency

Basically there are 4 ways to change gene frequency: mutation, chance, selection, and migration (bringing in genetic material (animals, semen or embryos) with different gene frequencies). Mutation and chance are of little practical importance in dairy cattle breeding. Dairymen cannot cause mutations. Those that do occur spontaneously have a small probability of being beneficial, as well as only a remote possibility of being recognized. This is particularly true if their primary effect is on a quantitative trait. Chance changes in gene frequency (sometimes called random drift) are of little practical importance to the dairy cattle breeder. The most important single force for changing gene frequency and making genetic improvement is selection (Section 6.3). Migration usually refers to the introduction of new genetic material across country lines, i.e., from outside the native population of cattle. Migration has been an important factor in genetic improvement of dairy cattle in many countries. In recent years the term "gene migration" has been used relative to new concepts for improving some U.S. breeds.

6.3 Selection

Selection is the process of causing a differential rate of reproduction among animals. This is done so that those animals with the most desirable genotypes will leave the most offspring. As selection is practiced generation after generation the more desirable genes become more frequent and the less desirable alleles become rarer. Thus, the overall genetic merit of the population increases, as the more desirable genes gradually replace their less desirable alleles. The type of selection used and the effectiveness with which it is carried out largely determine the amount of genetic progress that a dairyman makes in his herd. The opportunity for rapid genetic progress by selection is the greatest today that it has ever been.

The discussion will be primarily concerned with selection for additive gene effects. Additive effects (Sections 4.8 and 6.5) are by far the most important ones for making genetic improvement in dairy cattle because they constitute most of the genetic variation. Non-additive effects will be discussed in later sections under systems of mating. The rate of genetic progress is determined by: (1) intensity of selection; (2) accuracy of selection; (3) genetic variation or gene frequencies on which selection pressure is exerted; and, (4) generation length. Intensity of selection, or selection pressure, refers to the proportion of animals in the population that are selected for breeding purposes. The intensity and accuracy of selection determine the selection differential which is the average genetic superiority of the selected animals over the average merit of the population from which they were selected. Traits included in selection programs should have all the following characteristics: economic importance, sufficient genetic control, sufficient genetic variation, and accurate measures of merit in individual animals.

Single Trait Selection

Obviously, selection can be for a single trait or for multiple traits. Single trait selection is when all selection pressure is exerted on a single trait, and all other traits are ignored. There is little place for single trait selection in dairy cattle. It would be inadvisable to select for milk yield alone ignoring all other traits that are associated with yield or that contribute to a strong healthy cow capable of high yield. In fact, if such a selection program were followed, other important traits might deteriorate to the point where little or no progress would be made for yield. In dairy cattle, there are no advantages to a single trait selection that are not greatly outweighed by accompanying disadvantages.

Selection in dairy cattle should be on a multiple trait basis. But the more traits selected for, the less genetic progress that can be made in the most important ones. Table 6.1 shows the relative amount of genetic progress that can be made in a trait of primary importance when also selecting

TABLE 6.1 *Relative Progress in a Primary Trait When Selecting for Various Numbers of Independent Traits.*

Number of traits selected for	*Relative progress in primary trait (%)*
1	100
2	71
3	58
4	50
6	41
8	35
10	32

for various numbers of unrelated traits. If the trait of primary importance is positively correlated to other traits then greater progress can be made than shown in Table 6.1, but if the primary trait is negatively correlated to other traits less progress will be made.

Selection for multiple traits can be accomplished basically in 3 different ways. These are the tandem method, the method of independent culling levels, and the selection index method.

Tandem Method of Selection

The tandem method of selection means selection first for one trait and then for a second one, and so on. Usually selection will later be brought to bear on the first trait again and then on the others. There is little place for this method of selection as a primary method for genetic improvement in modern dairy herd management. Occasionally dairymen may resort to this method because they have not had proper training in methods of genetic improvement. They keep changing the goals of their selection program, placing too much emphasis on single traits and changing degrees of emphasis among traits. Because of the interrelationships of so many important traits in dairy cattle the tandem method is likely to be wasteful, inefficient, and result in little genetic progress. The tandem method will usually result in much less progress than the most efficient method, the selection index method. Also, when the same information is used for both methods, the tandem method cannot be more efficient than the method of independent culling levels.[1]

Independent Culling Levels

In the method of independent culling levels several traits are considered simultaneously, and a minimum acceptable phenotypic level is set for each trait. When an individual falls below the minimum phenotypic value in any trait it is eliminated regardless of its phenotypic merit in other traits. Independent culling is frequently forced on dairymen when they must resort to culling cows due to debilitating disease or physical or physiological breakdown. Animals that are culled for cases of severe mastitis and serious reproductive problems are examples of this method. Dairymen may use independent culling levels to select bulls for their breeding programs but this method reduces the average genetic superiority of bulls selected compared to methods that will be recommended in Chapter 7 based on selection indexes.

Selection Index Method

The selection index method is the most efficient method for making genetic improvement in dairy cattle. Every dairyman should understand clearly the principles involved in this method. When the selection index method is based on accurate unbiased information it is almost always more efficient and will result in more rapid genetic improvement than either the tandem method or independent culling levels.

An example of how to construct a simplified selection index is shown in Appendix II-D. This example should be studied in conjunction with the discussion in this section. The procedures shown in Appendix II-D will help greatly to guide your thinking on selection to make decisions on selection much more profitable. Dairymen who orient their thinking to the selection index method have a tremendous advantage over their neighbors who do not.

In a selection index the worth and importance of several traits are considered simultaneously. The animals chosen for breeding are those with the highest composite scores based on all the traits for which they are being selected. Those with the lowest composite scores are eliminated. The basic principles for the selection index method were given by L. N. Hazel in 1943.[2] The advantages of the selection index method are: (1) it accounts for several important traits simultaneously; (2) it accounts for the heritability of each trait; (3) it includes the known genetic and phenotypic relationships among the traits; (4) decisions are based in part on the relative economic importance of each trait.

The information needed to construct a selection index is:

1) Heritability of each trait to be included in the index (Section 4.8),
2) Relative economic importance of each trait,

3) The genetic and phenotypic correlations among the traits (Section 4.7),
4) The phenotypic standard deviation of each trait (Section 4.7).

The combining of all this information permits a marked superiority in one trait to balance out deficiencies in one or more other traits so that the economic result of the selection process will be the greatest possible economic progress. The heritability of each trait is included in the index to provide a weighting factor based on the genetic influence that is exerted on each trait. The relative economic importance gives the opportunity to assign an economic worth to each of the traits. The genetic correlations and phenotypic correlations, when they are available, are a valuable addition to the index because selection for one trait is sometimes antagonistic to selection for another trait or 2 traits may be so closely related that selection for one will suffice to improve both. When selection indexes are calculated on computers the values of each trait are expressed in terms of standard deviation units to put all traits on a standard scale of measurement.

A simplified selection index formula is given below ignoring the genetic and phenotypic correlations.

$I = v_1h_1{}^2P_1 + v_2h_2{}^2P_2 + \ldots + v_nh_n{}^2P_n$, where

I = the index value to be calculated on each animal,

v = the relative economic value assigned to one standard deviation of change in each trait,

h^2 = the heritability of each trait,

P = the phenotypic value for each trait expressed as a deviation from the mean for that trait in standard deviation units (Section 4.7). The subscripts, 1, 2, . . . , n refer to the traits that are included in the index, trait 1, trait 2, and so on to trait n, which is the last one.

The relative economic value of traits differs in every herd. A significant advantage of the selection index concept is the inclusion of economic values of various traits when dairymen plan their selection program. Average heritabilities of most traits of interest to dairymen are shown in Table 6.2. Phenotypic and genetic correlations of

TABLE 6.2 *Approximate Heritability of Some Traits in Dairy Cattle*

Trait	*Approximate heritability*
Yield:	
M.E. milk	.30
M.E. fat	.25
Milk (deviation from herdmates)	.25
Fat (deviation from herdmates)	.25
Protein	.25
Solids-not-fat	.25
Fat percent	.50
Protein percent	.50
Solids-not-fat percent	.50
Linear Type Appraisal Traits:	
Final type score	.30
Stature	.40
Chest and body (strength)	.20
Dairy character	.20
Foot angle	.10
Rear legs (side view)	.15
Rear legs (rear view)	.10
Pelvic angle (rump side view)	.20
Rump width	.25
Fore udder attachment	.20
Rear udder height	.15
Rear udder width	.15
Udder depth	.25
Suspensory ligament	.15
Teat placement (rear view)	.20
Disease Susceptibility:	
Mastitis	.10
Milk fever	.05
Ketosis	.05
Breeding problems	.05
Cystic ovaries	.05
Intensity of edema	.05
Persistency of edema	.10
Milking Characteristics:	
Milking speed	.30
Milk leak	.20
Body Characteristics:	
Body weight	.35
Height of tail setting	.25
Height of thurls	.25
Depth of body	.25
Tightness of shoulders	.25
Straightness of hocks	.20
Strength of pasterns	.15
Typical head	.15
Strength of head	.45
Arch of back	.15
Udder Characteristics:	
Rear udder length	.15

Table 6.2 *Approximate Heritability of Some Traits in Dairy Cattle (Continued)*

Trait	Approximate heritability
Udder Characteristics:	
Rear udder bulginess	.10
Rear udder funnelness	.10
Fore udder length	.15
Fore udder bulginess	.10
Fore udder funnelness	.10
Udder quality	.05
Forward slope of udder	.10
Strength of rear udder attachment	.15
Strength of fore udder attachment	.15
Udder quartering	.10
Rear teats forward	.10
Rear teats sideways	.30
Fore teats forward	.25
Fore teats sideways	.15
Rear teat spacing	.25
Fore teat spacing	.25
Rear to fore teat spacing	.30
Miscellaneous:	
Excitability	.25
Feeding speed	.15
Calving ease	.05
Liveability	.10

Source: Research by H. D. Norman, R. L. Powell, L. D. Van Vleck, J. M. White, W. E. Vinson, J. R. Thompson, L. P. Johnson, A. E. Freeman, K. L. Lee, J. R. Wright, and other sources.

various traits with first lactation milk yield are shown in Table 6.3. Values in both these tables should be studied carefully by dairymen having a tendency to place heavy emphasis on non-yield traits in their selection programs. Methods for practicing selection for important traits will be discussed in detail in Chapter 7.

6.4 Migration

Migration is the term used by geneticists to describe the process of introducing new genetic material into a population. This process is not limited to wild animal or wild bird migration but also includes the introduction of animals, semen and embryos into cattle populations through careful planning. The primary objective of migration in terms of genetic improvement is to bring in germ plasm with a higher frequency of superior genes and thereby raise the average genetic merit of the population.

Migration can change gene frequencies and raise the average genetic merit of a population of cows rapidly if it is practiced correctly. The impact of migration is dependent on 2 factors: (1) the proportion that the new germ plasm constitutes of the entire pool of germ plasm; and (2) the difference in genetic merit (or frequency of superior genes) between the new germ plasm and the original pool of germ plasm.

The term gene migration has been popular in the U.S. in recent years. It signifies the introduction of new genes into breeds perceived as lagging behind in economic competitiveness. The principles of gene

Table 6.3 *Phenotypic and Genetic Correlations of Various Traits with First Lactation Milk Yield*

Trait	Phenotypic correlation[1]	Genetic correlation[1]
Fat yield (first lactation)	.85	.70
SNF yield	.85	.90
Protein yield	.85	.90
Fat percent (first lactation)	−.35	−.35
SNF percent	−.30	−.25
Protein percent	−.35	−.30
Lifetime milk yield	.35	.80
Length productive life	.25	.75
Mastitis	−.05	.10
Breeding problems	.05	.00
Milking speed	.05	.00
Excitability	.00	.05
Feeding speed	.15	.45
Linear Type Traits		
Final score	.29	.00
Stature	.11	−.01
Strength	.12	.07
Dairy character	.50	.68
Foot angle	.00	−.24
Rear legs (side view)	.02	.14
Pelvic angle	.04	.19
Thurl width	.10	−.11
Fore udder attachment	−.09	−.47
Rear udder height	.12	−.13
Rear udder width	.16	.09
Udder depth	−.27	−.64
Suspensory ligament	.14	.12
Front teat placement	.02	−.12

[1]Some correlations are based partially on lactations other than first lactations.

Source: Studies by H. D. Norman, J. M. White, and other sources.

migration across breed lines are exactly the same as across country lines. In both situations gene migration results in a loss of purity of characteristics of the breed or country into which genes are introduced. That fact may cause some resistance to gene migration among breeders regardless of the potential for economic or sociological benefits.

6.5 Systems of Mating

Systems of mating constitute breeding plans that are designed to combine the genes in a population into the most advantageous genotypic combinations. This is done without changing gene frequencies, since a system of mating by itself will not change gene frequency. For this reason a system of mating should seldom if ever be used without selection. The combination of these 2 procedures, a system of mating accompanied by effective selection may be a powerful tool for genetic improvement in special circumstances, for example in developing countries or when trying to improve traits influenced strongly by nonadditive genetic effects. In a system of mating, the individual matings are usually planned on the basis of the relationships of the animals involved, in order to accomplish a specific purpose in terms of genetic variation. The purpose of systems of mating can be explained using the formula for the phenotypic variance which was discussed in Section 4.8. The phenotypic variance was defined as $\sigma_P^2 = \sigma_H^2 + \sigma_E^2 + \sigma_{HE}^2$ where σ_H^2 represented the hereditary variance and σ_H^2 was partitioned into $\sigma_A^2 + \sigma_D^2 + \sigma_I^2$, the additive, dominance, and epistatic variances.

Inbreeding

Inbreeding is the mating of animals that are more closely related to each other than the average relationship in a population. Inbreeding tends to increase the degree of genetic homozygosity. Therefore, its main purpose is to make the inbred animals more homozygous for the superior genes at a high proportion of their loci. Obviously, undesirable genes tend to become homozygous at about the same rate as the desirable genes during inbreeding unless such genes are deleterious to the extent that there is considerable natural selection against them. Therefore, inbreeding must usually be accompanied by severe selection. Lack of sufficiently intense selection is probably the reason that inbreeding in large animals has frequently done more harm than good.

Inbreeding results from one or more ancestors having contributed genes on both the male and female sides of an animal's pedigree. An example of inbreeding is shown in Figure 6.1, the schematic diagram of a pedigree. In this case the same bull "H" sired both the paternal grandsire and paternal grand-dam, and the paternal grand-dam was also the maternal great grand-dam on the grand-dam's side. For simplicity, a letter is assigned to each animal in the pedigree. A path to show inheritance is drawn from each ancestor to its offspring. Each path has a value of ½. An animal is inbred if you can start at the animal whose pedigree you are studying (A) and trace paths to some ancestor and back to the animal itself without having traversed the same path twice. This can be done in Figure 6.1. In fact, 2 sets of paths can be traced. One set of paths goes back to animal E via the route A-B-E-G-C-A. The

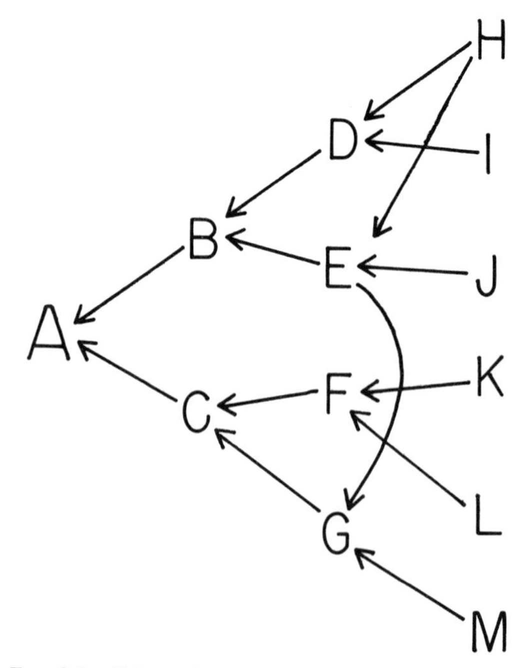

FIG. 6.1 Schematic representation of a pedigree showing inbreeding.

other set of paths goes back to animal H via the route A-B-D-H-E-G-C-A. It is permissible to use the same path in 2 different sets of paths, but no path can be used twice within the same set. For example, you cannot trace inbreeding through the route A-B-D-H-E-B-A because the path A-B would be traversed twice. This does show, however, that animal B is inbred.

The degree to which an animal is inbred (its inbreeding coefficient) is equal to one-half the relationship between its parents. When the common ancestors of the parents are not themselves inbred, an animal's inbreeding coefficient can be calculated from the formula:

$$F_A = \frac{\Sigma[(\frac{1}{2})^{(GS+GD)}]}{2},$$

where F_A indicates the inbreeding coefficient of animal A, Σ indicates summation of sets of paths if there is more than one common ancestor, GS is the number of generations from the animal's sire to the common ancestor, and GD is the number of generations from the animal's dam to the common ancestor.

The numerator of the inbreeding coefficient formula is the relationship between an animal's sire and dam. Therefore, if the parents are unrelated an animal's inbreeding coefficient is zero. In Figure 6.1 the parents of A (B and C) are related because they have two common ancestors (E and H). Therefore A is inbred. In addition, D and E are related (half sibs) through H, meaning that B is also inbred. On the basis of this information the inbreeding coefficient of B can be calculated from the formula above. In this case,

$$F_B = \frac{(0.5)^{(1+1)}}{2} = \frac{0.5^2}{2} = 0.125.$$

This shows that the inbreeding coefficient of animal B (F_B) is one-eighth or 0.125, and from the numerator that the relationship between D and E is 0.25, as usual for half sibs.

A more elaborate formula must be used to calculate the inbreeding coefficient of animal A since one of its parents (B) is also inbred. The principle is exactly the same as above; that is, the inbreeding coefficient is equal to one-half the relationship between the parents. The general formula for the relationship between 2 animals (X and Y) when either or both parents or the common ancestor (Z) may be inbred is:

$$R_{XY} = \frac{\Sigma(\frac{1}{2})^{(GX+GY)}(1 + F_Z)}{\sqrt{(1 + F_X)(1 + F_Y)}}.$$

Here, X and Y represent the parents. Z represents a common ancestor and F indicates inbreeding as before. The relationship between the parents (B and C) of animal A in Figure 6.1 is:

$$R_{BC} = \frac{0.5^{(2+3)}(1.0) + 0.5^{(1+2)}(1.0)}{\sqrt{1 + .125)\ (1 + 0)}}$$

$$= \frac{0.5^5 + 0.5^3}{\sqrt{1.125}} = \frac{.031 + .125}{\sqrt{1.125}}$$

$$= \frac{.156}{1.06} = 0.15.$$

Therefore, the relationship between B and C is 15%. The inbreeding of A is one half the relationship or 7.5%.

The inbreeding coefficients of offspring from some matings that might be made among relatives are shown in Table 6.4. These are minimum inbreeding coefficients that would result from these matings assuming that neither the parents nor any common ancestors were themselves inbred.

The consequences of inbreeding without selection are: (1) increase in homozygosity; (2) decrease in genetic and phenotypic variation; (3) decrease in average phenotypic merit; and (4) decrease in adaptability. Perhaps the classic example of indiscrimi-

TABLE 6.4 *Inbreeding Coefficients in Offspring Resulting from Some Matings of Related Animals*

Mating	*Inbreeding of offspring (%)*
Sire x daughter	25.0
Sire x dam	25.0
Full sibs	25.0
Half sibs	12.5
Sire x granddaughter	12.5
Son of a sire x granddaughter of the sire	6.25
Grandson of a sire x granddaughter of the sire	3.13

nate inbreeding without adequate selection in cattle occurred in the Duchess family of Shorthorns in the 1800s. The Duchess family was maintained for 40 years with an inbreeding coefficient of approximately 40%. Many animals of the family were imported by American breeders, who through additional inbreeding, with almost no selection on individual merit and a fanatical attempt to exploit the Duchess name for profit, bred the family practically into extinction in the late 1800s due to low fertility.

Most experiments with inbreeding dairy cattle have resulted in a general loss of vigor and producing ability as inbreeding increased.[3-11] Average decreases in lactation milk yield resulting from each additional percent inbreeding have ranged from 33 lb to 210 lb with approximately 50 lb the most commonly found decrease. Other deficiencies resulting from inbreeding dairy cattle have been sterility, lack of vigor, decrease in birth weight, higher calf mortality, and lower milk fat yield.

Average effects of 3 levels of inbreeding on some economically important traits are shown in Table 6.5. None of these effects can be considered desirable except possibly the increase in fat percent. However, this increase is probably due to the drop in milk yield and the negative genetic correlation between the 2.

Intentional inbreeding should never be practiced by a dairyman unless he is prepared (and can afford) to cull the inbreds very heavily. The most successful application of inbreeding has been in plant genetics where great masses of individuals can be produced rapidly and cheaply, and all can be culled if necessary. In fact, many of the genetic breakthroughs in plant species have resulted from intense inbreeding. However, in most cases this inbreeding was accompanied by extremely intensive selection, such as culling over 99% of all inbreds for many consecutive generations. Obviously, such practices have little applicability to dairy cattle breeding. These procedures could not be widely used in dairy cattle for many reasons: high monetary value of each individual, long generation life, high cost of rearing individuals and relatively low reproductive rate especially of females. Inbreeding in dairy cattle has been found to be little problem in the U.S.[12] Dairymen can protect their herds from the undesirable effects of accidental inbreeding by recording sire identification on all cows in their herd and avoiding matings to a cow's sire or a son of her sire.

TABLE 6.5 *Average Effects of Three Levels of Inbreeding on Some Economically Important Traits of Dairy Cattle*

Trait	*Level of inbreeding* 6.25%	12.5%	25.0%
Milk yield	−300 lb	−600 lb	−1200 lb
Fat yield	−9 lb	−18 lb	−35 lb
Fat %	.02%	.04%	.12%
Birth weight	−1.5 lb	−3.0 lb	−6.0 lb
Yearling weight	−10 lb	−25 lb	−60 lb
Two-year weight	−20 lb	−40 lb	−60 lb
Yearling height	−0.6 cm	−1.2 cm	−2.4 cm
Two-year height	−0.4 cm	−0.8 cm	−2.4 cm
Yearling heart girth	−1.0 cm	−2.0 cm	−4.0 cm
Two-year heart girth	−1.2 cm	−2.4 cm	−4.8 cm
Mature heart girth	−0.8 cm	−1.6 cm	−3.2 cm
Mortality (% of outbred)	112%	125%	150%

Source: Young, C. W.: Inbreeding and the gene pool. J. Dairy Sci. *67:* 472, 1984.

Line Breeding. Line breeding is the mating of several generations of offspring to a particular animal or to that animal's descendants. It is an attempt to concentrate the genes of a superior animal in later generations in order to reconstruct the superior animal's genotype as closely as possible in the hopes of producing others just like him. In practice, line breeding has been most common with superior sires, mainly because bulls have a much greater reproductive rate than do cows. An example of line breeding would be the use of a single bull as the sire of an animal as well as its maternal grandsire and also its great grandsire.

Close Breeding. Intense inbreeding by mating full or half sibs, cousins, or parents to offspring is close breeding. Matings of this type are rather rare in dairy cattle in the United States, because most dairymen who milk or breed cows for a livelihood cannot afford the great risks involved in mating such close relatives.

Crossbreeding

Crossbreeding is the mating of animals of different breeds. There are 2 distinct purposes for which crossbreeding may be practiced. The first is the crossing of superior animals from 2 breeds in an attempt to produce a new generation that is phenotypically superior on the average to either of the parent breeds because of hybrid vigor or heterosis. In this case an attempt is made to maximize the effect of the dominance and epistatic portions of the hereditary variance to a point beyond that which can be done by mating animals within a breed. The second purpose is to cross breeds of widely different characteristics and adaptability to produce a new breed with many of the best characteristics of each of the parent breeds. Crossbreeding for the second purpose is done primarily to make use of the additive genetic variation in the parent breeds, and any heterosis that results is of secondary importance. In neither case does crossbreeding create any new genes or change gene frequency from that which existed in the parental populations. However, in both cases crossbreeding does change the genotypic frequencies, hopefully resulting in animals with genotypes that are superior to either of the parent breeds.

Crossbreeding for Hybrid Vigor. Hereditary variance was defined in Section 4.8 as $\sigma_H^2 = \sigma_A^2 + \sigma_D^2 + \sigma_I^2$.

If all the hereditary variance were strictly additive then $\sigma_H^2 = \sigma_A^2$. In a constant environment and in the absence of selection, the mean phenotypic merit of the offspring would equal the mean of the parents except for the effect of chance segregation of genes. In such a situation there would be no opportunity for heterosis. However, for most traits there is some dominance and epistatic variance so there is at least an opportunity to make use of heterosis to develop superior phenotypes.

The effects of crossbreeding for heterosis are increased heterozygosity, improved hardiness and adaptability, and animals that may have superior phenotypes for specific economically important traits. All 3 of these are advantages in animals that are important for commercial purposes (milk or meat production). However, these animals may not be nearly as valuable for breeding purposes as their phenotypes would indicate they should be. They cannot transmit heterozygosity because during meiosis only one allele of each pair is transmitted to each zygote. Therefore, an animal cannot transmit any heterotic advantage that is due to dominance. In addition it can transmit only about one-half of its own heterotic superiority that is due to epistatic effects.

One of the major reasons that crossbreeding for heterosis has not been utilized more widely in dairy cattle is that in most cases each female is both a breeding animal and a producing animal. In the meat species, however, it is possible to keep parent stocks primarily for breeding purposes to produce the commercial stocks. This has not yet proved feasible in dairy cattle but may someday through embryo transfer, at least on a limited scale. Even so this does not preclude the possibility of making some use of heterosis in dairy cattle because artificial insemination makes the superior bulls of different breeds equally accessible to all commercial dairymen no matter what breed they may have in their herd.

Heterosis can be measured and interpreted in 2 different ways. The first of these is the strictly scientific interpretation where heterosis is the amount by which offspring exceed the mean breeding value of the parents. If the parents are widely different in their breeding values for a trait it is possible to have considerable heterosis in offspring that are not as good phenotypically as the superior parent. The second interpretation of heterosis is economic heterosis. This terminology is frequently applied to the amount by which the offspring exceed the superior parent. This is the type of heterosis that could be profitable for some dairymen.

While heterosis is frequently measured and reported on individual traits, the true effects of heterosis should not be judged this way any more than a selection program should be based on a single trait. The economic advantages of heterosis from crossbreeding should not be judged just on how much milk or fat is produced by crossbreds vs purebreds. They should be measured on some basis of overall economic merit such as how much milk and components are produced for each mating that is made. This would take into account the cumulative effects of heterosis for many traits such as reproductive efficiency, livability of calves, growth rate, disease resistance, and producing ability.

Two major dairy cattle crossbreeding

studies have been conducted in the United States in the 1950s and 1960s. These showed varying amounts of heterosis for yield and other traits depending on the specific crosses. In general, economic heterosis was not found for individual traits with the exception of yield in some crosses[13] in one study and livabilty[14] in the other. However, both experiments showed evidence of economic heterosis in the crossbreds resulting from the cumulative effect of heterosis in many economically important traits.[15,16]

It is doubtful if crossbreeding for heterosis will be a major part of dairy cattle breeding programs in the United States in the foreseeable future. Selection based on additive genetic variation appears to offer a far greater and safer opportunity for the vast majority of dairymen to make genetic improvement in their herds.

Crossbreeding to Create New Breeds. A second purpose of crossbreeding is to produce offspring superior to the parental breeds because they are more adaptable to a particular environment or otherwise possess special characteristics not present in the parent breeds. This special adaptability is not due primarily to heterosis. In fact, any heterotic effects would be almost incidental to the primary purpose of this type of crossbreeding which is to utilize the additive genetic variation present in the parent breeds. This type of crossbreeding is especially useful in instances where breeds with widely different characteristics are crossed to make rapid changes in the phenotype which is typical of one of the parents, usually a "native" breed. There are several obvious instances where this is very desirable. For example, crossbreeding dairy bulls to cows of the traditional beef breeds produces a generation of cows with milking capabilities that are usually far superior to the beef cows in the parental generation. Also, the level of milk and fat yield can be changed quite drastically by crossing cows of one dairy breed to bulls of another breed with differing levels of milk yield and fat content. Cows that are well adapted to some areas of the tropics and produce a great deal more milk than the native cattle can be developed by crossing the native cows to bulls of the European breeds. This mating system should be accompanied by proper selection in order for genetic progress to reach its maximum effectiveness. Crossbreeding of this type has not worked well in the Southern United States apparently because the climate and disease and feeding problems are not severe enough to give sufficient advantage to European-Zebu crosses over the pure European breeds. However, this system may be useful in the truly tropical parts of the world where it is necessary for cattle to possess many of the characteristics of the Zebu and other indigenous cattle in order to survive.

Other Systems of Mating

Several other systems of mating have been utilized in dairy cattle breeding. Most of these are ramifications or combinations of inbreeding or crossbreeding systems.

Grading Up. Grading up has been the name traditionally attached to the system by which purebred sires from a single breed are used on inferior herds of mixed or unknown breeding to raise their genetic and phenotypic merit. This system is perhaps not so common any more because there are fewer of these inferior herds left, and there is less of a difference between purebred and grade herds than years ago.

Outbreeding. Outbreeding is the opposite of inbreeding in that it is the mating of animals less closely related than the average relationship within the breed. Outbreeding tends to increase the heterozygosity of the offspring. Because of this, outbred animals may tend to have higher phenotypic merit for the same reasons as crossbreds. But like crossbreds, their breeding value may tend to be lower than their phenotypic value because some of their phenotypic superiority is due to dominance and epistatic effects that cannot be transmitted to their offspring.

Outcrossing. Outcrossing is similar to outbreeding except that animals from inbred lines have been crossed to produce the outcrosses. Once the outcross has been made the breeding plan reverts back to inbreeding the outcrosses.

6.6 Summary

In this chapter 2 basic methods for the second and last phase of making genetic improvement were discussed: (1) increasing the frequency of genes that cause superior phenotypes; and (2) recombining the genes present in the population so that the most favorable genotypic combinations are formed. The 4 methods by which gene frequency can be changed are mutation, chance, selection, and migration. Chance and mutation are of little practical importance to dairymen. Selection is described as the primary method for making genetic improvement. Migration is also important as a secondary method. In most herds genetic progress will be determined largely by the effectiveness with which selection is carried out. Continued selection for a single trait without attention to others is the least effective method of selection and can cause many problems. The tandem method of selection is somewhat better than single trait selection but less effective than the method of independent culling levels, in which animals are culled on the basis of a phenotypic value assigned to each trait independently. The most effective method of selection is through the use of selection indexes where several traits are considered simultaneously. The importance of each trait is weighted according to its economic value and heritability. Index procedures that can be used by all dairymen for selection will be described in Chapter 7.

The methodology and advantages and disadvantages of the 2 main systems of mating in dairy cattle, crossbreeding and inbreeding, were described. Inbreeding is a risky procedure and should be attempted only by dairymen with a thorough knowledge of the potential consequences as well as the financial resources to cover losses that may be incurred. Crossbreeding may be pursued with 2 goals in mind: the maximizing of potential benefits due to heterosis, or hybrid vigor; and crossing breeds of widely different backgrounds and characteristics to create a new breed with greater adaptability to a particular environment than the parent breeds.

Application of these principles and those described in Chapter 5, in the manner that is most effective for a particular herd situation, is the essence of success in making rapid genetic improvement and breeding cows for greater profitability.

Review Questions

1. Construct a pedigree showing linebreeding to a bull using letter designations for animals as in Figure 6.1.
2. Construct a pedigree showing 2 different kinds of closebreeding.
3. Make a list of reasons why the method of independent culling levels results in faster genetic progress for several traits than the tandem method.
4. List reasons why the selection index method causes more rapid genetic improvement than other selection methods.
5. Can you explain why selection changes gene frequency but systems of mating do not? What is the importance of this genetic phenomenon in the genetic improvement of dairy cattle?
6. List all the traits which you might want to include in a selection index and for which you can make an objective estimate of economic worth; list the basis for each estimate of economic worth.
7. List at least 5 traits (for instance, various type traits) that you did not list in question 6, state how you might determine an economic value of each, and try to give a reasonable economic value for each trait. What problems did you encounter doing this?
8. What factors enable dairymen to make faster genetic progress in their herds now than was possible in the past?

References

1. Hazel, L. N., and Lush, J. L.: The efficiency of three methods of selection. J. Heredity *33:* 393, 1942.
2. Hazel, L. N.: The genetic basis for constructing selection indexes. Genetics *28:* 476, 1943.
3. Tyler, W. J., Chapman, A. B., and Dickerson, G. E.: Growth and production of inbred and outbred Holstein-Friesian cattle. J. Dairy Sci. *32:* 247, 1949.
4. Davis, H. P., and Plum, M.: Influence of inbreeding on production. J. Animal Sci. *11:* 739, 1952.
5. Swett, W. W., Matthews, C. A., and Fohrman, N. H.: Effect of inbreeding on body size, anatomy and producing capacity of grade Holstein cows. USDA Tech. Bull. 990, 1949.
6. Laben, R. C., Cupps, P. T., Mead, S. W., and Regan, W. M.: Some effects of inbreeding and evidence of heterosis through outcrossing in a Holstein-Friesian herd. J. Dairy Sci. *38:* 525, 1955.
7. Von Krosigk, C. M., and Lush, J. L.: Effect of inbreeding on production in Holsteins. J. Dairy Sci., *41:* 105, 1958.
8. Hillers, J., and Freeman, A. E.: Effects of inbreeding and selection in a closed Guernsey herd. J. Dairy Sci. *47:* 894, 1964.
9. Gaalaas, R. F., Harvey, W. R., and Plowman, R. D.: Effect of inbreeding on production in

different lactations. J. Dairy Sci. *45:* 781, 1962.
10. Young, C. W.: Watch for inbreeding when selecting sires. Hoard's Dairyman *114:* 1181, 1969.
11. Young, C. W., Tyler, W. J., Freeman, A. E., Voelker, H. H., McGilliard, L. D., and Ludwick, T. M.: Inbreeding investigations with dairy cattle in the north central region of the United States. North Central Regional Research Publication 191, 1969.
12. Young, C. W.: Inbreeding and the gene pool. J. Dairy Sci. *67:* 472, 1984.
13. McDowell, R. F., and McDaniel, B. T.: Interbreed matings of dairy cattle. I. Yield traits, feed efficiency, type and rate of milking. J. Dairy Sci. *51:* 767, 1968.
14. Dickinson, F. N., and Touchberry, R. W.: Livability of purebred vs. crossbred dairy cattle. J. Dairy Sci. *44:* 879, 1961.
15. McDowell, R. E., and McDaniel, B. T.: Interbreed matings of dairy cattle. III. Economic aspects. J. Dairy Sci. *51:* 1649, 1968.
16. Touchberry, R. W.: A comparison of the general performance of crossbred and purebred dairy cattle. J. Animal Sci. *31:*169, 1970.

Suggested Additional Reading

Johansson, I.: *Genetic Aspects of Dairy Cattle Breeding.* Urbana, Ill.: University of Illinois Press, 1961.

Johansson, I., and Rendel, J.: *Genetics and Animal Breeding.* San Francisco: W. H. Freeman and Co., 1968.

Lush, J. L.: *Animal Breeding Plans.* Ames, Iowa: Iowa State College Press, 1945.

Proceedings of National Workshop on Genetic Improvement of Dairy Cattle. Sponsored by Extension Service, USDA. April, 1976.

Warwick, E. J., and Legates, J. E.: *Breeding and Improvement of Farm Animals.* 7th ed. New York: McGraw-Hill Book Co., Inc., 1979.

Schmidt, G. H., and Van Vleck, L. D.: *Principles of Dairy Science.* San Francisco: W. H. Freeman and Company, 1974.

CHAPTER 7

The Science and the Art of a Profitable Breeding Program

7.1 Introduction

The primary goal of a breeding program should be to produce future cows with the greatest possible genetic capability to make a profit. The fulfillment of this goal requires cows that produce high levels of milk of desirable composition and strong healthy cows that can stand the rigors of high production through many lactations without special treatment; that is, cows with a capacity for longevity so that most culling will be voluntary (at the option of the dairyman) and culling will contribute to genetic improvement of the herd. This goal can best be attained through a breeding program that introduces into the herd the best available genetic merit for economically important traits. The attainment of this goal will result in a herd of cows with outstanding genotypes for total economic merit.

7.2 Economic Importance of Predicted Differences and Cow Indexes for Yield

The impact of the breeding program on the economic well-being of a herd should not be underestimated because an economically important relation has been shown between a bull's Predicted Difference (PD) for milk yield and the Income Over Feed Cost (IOF) of his daughters.[1] At a milk price of around $12.00 per hundredweight, a bull's daughters will return approximately $16.00 extra IOF per lactation for each additional 100 lb in the bull's PD for milk yield. Using that relationship, the economic advantage from bulls with different levels of PD's can be calculated for herds of varying size.

Predicted Difference and Cow Index for gross income ($PD_\$$ and $CI_\$$) are important to dairymen because total income for milk depends on the amount of milk, the milk fat percentage, and in some markets the percentage of protein or solids-not-fat. Levels of milk components have decreased slightly due to selection for increased milk yield, caused in large part by the emphasis on fluid volume in most pricing formulas for many years. As greater emphasis on component pricing (sometimes called equity pricing) spreads to more milk markets, the level of milk components will become more important economically to dairymen. Therefore, functions of $PD_\$$ and $CI_\$$ are probably the most economically important values for ranking bulls and cows. $PD_\$$ and $CI_\$$ should be understood by dairymen and the importance of $PD_\$$ percentiles and $CI_\$$ percentiles fully appreciated. The formulas for calculating $PD_\$$ and $CI_\$$ using milk and component percentages are shown in Appendix II-E. USDA-DHIA Sire Summary information for yield traits for the active AI bulls with the highest $PD_\$$ in each breed, based on milk yield, milk fat percent and

Fig. 7.1 USDA-DHIA Sire Summary information on the active AI bull with the highest PD82$ in each breed based on milk, fat and protein in the January 1984 USDA-DHIA Sire Summary run.

	REGISTRATION							*MILK AND FAT*							*PROTEIN AND SNF*					
Breed	*Number*	*Name of Bull*	*NAAB Code*	*PCT ILE*	*No. Hrds*	*No. Dtrs*	*% Cul*	*% RIP*	*Rpt %*	*Predicted Difference-82 Milk*	*%*	*Fat*	*$$*		*PD82 $$*	*PD82 %*	*PD82 Lbs*	*Rpt %*	*No. Hrds*	*No. Dtrs*
Ayrshire	132662	Reidina Dawner Gallant Man	9A32	93	134	367	14	34	92	+827	−.04	+26	+98	PRO	+86	−.07	+17	81	74	149
Guernsey	590165	Welcome Choice Admiral	29G811	96	33	54	5	0	74	+1065	+.01	+50	+154	PRO	+147	−.06	+29	57	19	29
Holstein	1747862	Cor-Vel Enchantment	7H877	99	39	54	9	6	73	+1770	−.03	+58	+213	PRO	+189	−.10	+38	44	13	18
Jersey	623330	Quicksilvers Magic of Ogston	7J121	98	458	1378	8	54	98	+1008	−.03	+44	+140	PRO	+132	−.09	+26	95	270	731
Brown Swiss	170838	Norvic Telstar	14B155	95	156	385	8	30	94	+933	+.08	+49	+144	PRO	+150	+.02	+35	88	100	223
Milking Shorthorn	368707	Korncrest Pacesetter	21M515	90	75	248	6	21	90	+1151	.00	+42	+146	PRO	+147	+.02	+38	63	28	62

	REGISTRATION							*MILK AND FAT*							*PROTEIN AND SNF*					
Breed	*Number*	*Name of Bull*	*NAAB Code*	*PCT ILE*	*No. Hrds*	*No. Dtrs*	*% Cul*	*% RIP*	*Rpt %*	*Predicted Difference-82 Milk*	*%*	*Fat*	*$$*		*PD82 $$*	*PD82 %*	*PD82 Lbs*	*Rpt %*	*No. Hrds*	*No. Dtrs*
a	b	c	d	e	f	g	h	i	j	k	l	m	n	o	p	q	r	s	t	u
Ayshire	132662	Reidina Dawner Gallant Man	9A32	93	134	367	14	34	92	+827	−.04	+26	+98	PRO	+86	−.07	+17	81	74	149

a Breed
b Registration number
c Registration name
d Bull's NAAB code number
e PD$ percentile ranking among active AI bulls
f Number of herds with daughters
g Number of daughters in evaluation
h Percent of 1st lactation daughters that were culled
i Percent of 1st lactations that were records-in-progress
j Repeatability of this summary in percent
k Predicted Difference for milk yield in lb. (PD82)
l Predicted Difference for fat percent (PD82)
m Predicted Difference for fat yield in lb. (PD82)
n Predicted Difference for gross income in dollars based on milk and fat
o Designation of component information, PRO = protein, SNF = solids-not-fat
p Predicted Difference for gross income in dollars based on milk, fat and component
q Predicted Difference for component percent (PD82)
r Predicted Difference for component yield in lb. (PD82)
s Repeatability of component PD82's in percent
t Number of herds with daughters tested for component
u Number of daughters in component evaluation

protein percent in the January 1984 USDA-DHIA Sire Summary run are shown in Figure 7.1. The key to the information is given at the bottom of the figure with an explanation of each item of information.

The Ideal Genotype

The ideal genotype for which to strive varies from herd to herd. Different traits have varying economic importance in different herds depending on the dairyman's objectives, the herd's management regime and the pricing formula for the herd's milk. Breeding towards the ideal genotype for each herd requires the evaluation of many alternatives in the selection program and consideration of how many different traits have important effects on the herd's profitability and satisfy the herdowner's objectives.

7.3 Science—Understanding Genetic Evaluations

All genetic evaluations are expressed as deviations from a genetic base because genetic evaluation procedures estimate only differences among bulls and among cows. The purpose of a genetic base is to provide a reference point so genetic evaluations are understandable and useful. A genetic base is necessary no matter whether genetic evaluations are expressed as pounds, kilograms, dollars, percentiles, rankings, percent of breed average, or any other units. A genetic base can be defined by 4 characteristics; the breed to which it applies, the trait to which it applies, the geographical area to which it applies, and the point in time to which it refers. The first 3 characteristics are usually easy to determine because a separate genetic base is usually established for each trait in each breed for an entire country. The last characteristic is important for correct interpretation of genetic evaluations and is sometimes rather difficult to determine.

Alternative Genetic Bases

Three genetic bases are possible; fixed, moving or stepwise.[2] The 3 bases are depicted graphically in Figures 7.2, 7.3, and 7.4. The PD's of the same two bulls are shown in each figure assuming Bull A was evaluated in 1975 and Bull B in 1984. In all figures a constant genetic progress of 90 lb per year is assumed and shown by the line labeled "genetic progress". Note that Bulls A and B were just average bulls genetically when they were evaluated.

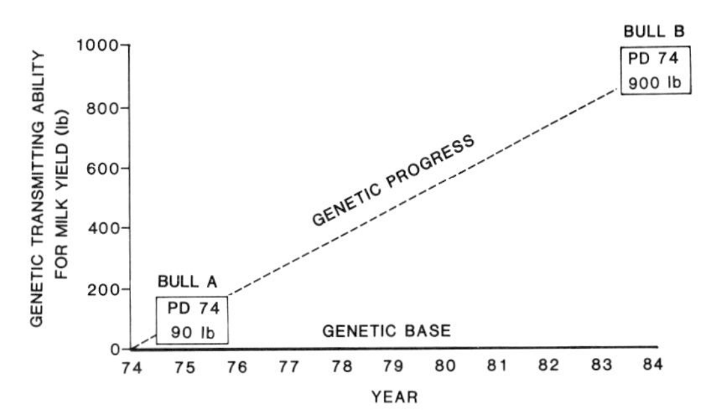

FIG. 7.2 Diagram of fixed genetic base. (Courtesy Agricultural Research Service, USDA)

A fixed genetic base (Figure 7.2) means, as the name implies, continual use of a static base without change. A fixed base eliminates the need to account for changes in the base over time and, in this respect, is different from either a moving or stepwise base. Under a fixed genetic base, PD's and CI's accurately reflect the relative genetic merit of bulls and cows over a long time, no matter when they were evaluated. The word "relative" is important. Bulls and cows always will be ranked correctly under a fixed base, but PD's and CI's of new bulls and cows will become larger as time progresses because of continuing genetic improvement. Therefore, the PD's and CI's reflect bulls' and cows' current contribution to genetic progress less and less because of the continuing divergence of average genetic merit of the population from the fixed base. In Figure 7.2, Bull A has a PD74 of +90 lb (average for 1975) and Bull B has a PD74 of +900 lb (average for 1984). Under a fixed base, it is easy to determine that Bull B is 810 lb superior to Bull A even though they were evaluated at different times but it is not clear just from knowing the PD74s whether Bull B is a breed improver at the time he was evaluated.

A moving genetic base is shown diagrammatically in Figure 7.3. A moving base is updated at least annually. A moving genetic base is most practical in countries with relatively small dairy cattle populations for which all previous genetic evaluations can

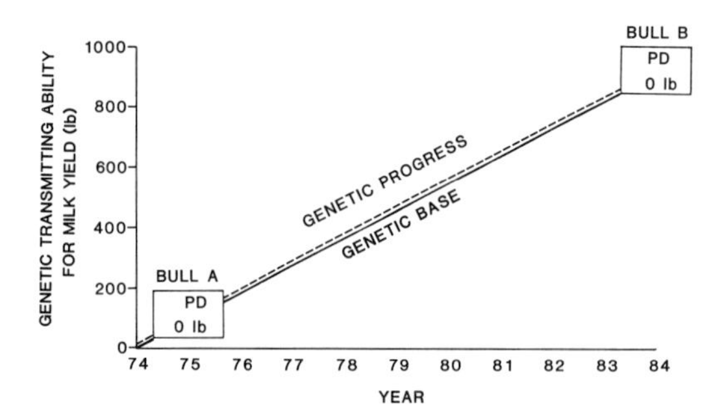

FIG. 7.3 Diagram of moving genetic base. (Courtesy Agricultural Research Service, USDA)

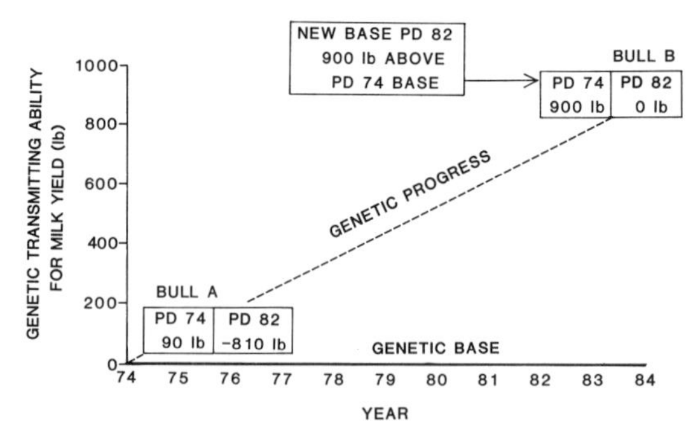

FIG. 7.4 Diagram of stepwise genetic base. (Courtesy Agricultural Research Service, USDA)

be discarded each time new ones are calculated. An advantage of a moving base is that it helps dairymen who do not understand the principles of genetic improvement very well to keep their selection standards current with genetic progress. However, a major disadvantage is that bulls and cows evaluated at different times cannot be compared or ranked correctly because different bases were used for their evaluations. Genetic evaluations expressed to moving genetic bases usually do not have a base time designation on the PD's and CI's as do those under a fixed base. In Figure 7.3, Bulls A and B both have PD's of 0 lb under the moving base because they were both average bulls when they were evaluated. Their PD's would imply that they were of equal genetic merit unless a person knew that Bull B was evaluated much later than Bull A. Even then, determining the superiority of Bull B over Bull A would be difficult especially for a dairyman who might not have detailed figures on all the base changes that had been made over the years.

A stepwise genetic base is shown diagrammatically in Figure 7.4. A stepwise genetic base is changed periodically, usually every 5 to 10 years. The dates and amounts of the base changes are announced ahead of time so everyone using genetic evaluations is informed. The main advantages of a stepwise genetic base are that bulls and cows can be ranked on their genetic merit for long periods of time by accounting for the few base changes that are made, while the magnitude of the PD's and CI's can be adjusted periodically so they reflect more closely the amount of genetic improvement each animal contributes to the current population. Disadvantages of the stepwise genetic base are the extra work required to account for the periodic base changes especially in large computerized files and some loss of the close relation between PD's and CI's and the genetic level of population between base changes. In Figure 7.4, the base designated by PD74 is depicted as changed to a new base designated as PD82. The base change takes place in 1984. The base is raised 900 lb to account for the genetic progress that has occurred since determining the base labeled PD74. Bulls A and B now have new PD's labeled PD82. Each bull's PD82 is exactly 900 lb lower than his PD74 due to the base change. The 2 bulls still can be ranked correctly even if a PD82 is not readily available for one of them because the amount of the base change would be widely publicized and should be available to all dairymen. Stepwise base changes periodically help to reinforce the need for all dairymen to upgrade their selection standards and also provide a periodic educational opportunity for extension and industry leaders.

Genetic Base in the United States

The U.S. dairy industry has adopted a stepwise genetic base for national genetic evaluations. The stepwise genetic base was first established in 1974 with genetic evaluations labeled PD74 and CI74. The genetic base for each trait was changed in January 1984 to a 1982 genetic base labeled PD82 and CI82. The 1982 genetic base for each breed and trait was established by setting to 0 the weighted average PD of sires of first lactation cows calving in 1982. Each bull's contribution to the average was weighted by the number of daughters he sired. Several aspects of this procedure are important. The genetic base is determined from the PD's of sires, not from the production level of cows. Only sires of first lactation cows are included because first lactation cows are relatively unselected. The procedure estimates the amount of genetic progress that has taken place in the cow population rather than just measuring the genetic merit of a group of bulls as is done in some countries. Genetic progress in the cow population is the most important payoff from a genetic improvement program.

Changes in the genetic bases from the 1974 base to the 1982 base are shown in Table 7.1. The effect of the base changes can be represented by the formula: PD82 = PD74 − change in base. The effect of the base changes is to lower the PD's and CI's for most traits. But, notice that some of the base changes are negative so that genetic evaluations for those traits and breeds will be increased slightly.

Genetic Bases in Other Countries

Every country has different genetic bases for each trait in each breed and those bases refer to a variety of time periods. Bases are determined by different methods, so the term "genetic base" has different meanings in addition to the differences caused by the use of transmitting ability in some countries and breeding value in others. A basic formula that can be used to convert genetic evaluations between countries is:

EGM = DGB + CAM, where

EGM represents the Estimated Genetic Merit of the animal in the importing country expressed in the units of the importing country, DGB represents the difference in the Genetic Bases, that is the exporting country's genetic base minus the importing country's genetic base expressed in the units of the importing country, and CAM represents the Converted Animal's Merit that is, the deviation of the animal's genetic merit from the genetic base in the exporting country converted to the units of the importing country.

Use of the correct formula to convert genetic evaluations is important to breeders attempting to acquire superior germ plasm by importation from another country.[3] Incorrect conversions could cause a breeder to import inferior germ plasm in the mistaken belief that it is superior to that available from his own or some other country. Correct conversion of genetic merit can be especially difficult to keep current if a country uses moving genetic bases.

Interpreting Genetic Evaluations

PD82's and CI82's are closer to the average amount of genetic improvement that bulls and cows will contribute to the present population of cows in the U.S. than PD74's and CI74's if they were still used. However, PD82 and CI82 are not estimates of the amount of genetic improvement to be obtained from use of a bull or cow for breeding in a specific herd. The amount of genetic improvement that a bull or cow can be expected to contribute to a specific herd depends on the relation between that animal's PD or CI and the average genetic merit of the cows in the herd. Of course, the cows' genetic merit depends on the PD's and CI's of their sires and dams.

PD's and CI's should be used to rank bulls and cows. No matter what the overall magnitude of PD's and CI's, as determined

Table 7.1 *Changes in the Genetic Bases for Yield Traits from 1974 to 1982.*

Breed	*Milk*	*Fat*	*Fat*	*Protein*	*Protein*	*Solids-not-fat*	*Solids-not-fat*
	(lbs.)	(lbs.)	(%)	(lbs.)	(%)	(lbs.)	(%)
Ayrshire	637	22	−.03	19	−.02	54	−.03
Guernsey	781	30	−.07	25	−.03	68	−.03
Holstein	978	28	−.05	27	−.03	81	−.01
Jersey	993	34	−.16	34	−.04	89	−.04
Brown Swiss	1,094	35	−.07	39	+.01	102	+.03
Milking Shorthorn	992	41	+.04	30	.00	insufficient data	
Red and White	107	3	−.01	2	+.02	insufficient data	

Source: R. L. Powell, and H. D. Norman, AIPL, ARS-USDA.

by the genetic base, dairymen cannot do any better than to use the highest ranking bulls and cows that are available. Various methods for ranking bulls and cows to determine which are the highest for different traits and different economic conditions will be discussed in Section 7.4.

Variation among the Daughters of a Bull

Dairymen sometimes make the mistake of judging a bull's genetic merit for a trait on the basis of 1 or 2 daughters that they see or hear of somewhere, possibly even a couple of daughters in their own herd. Judgment of a bull's genetic merit on so little information can be misleading and cause a dairyman to use an inferior bull or hesitate to use a superior bull. Most quantitative traits including level of production vary considerably among a bull's daughters. Some of the best bulls have some very low-producing daughters and vice versa. This is dramatized in Figure 7.5 which shows a frequency distribution of the yields of daughters of 2 bulls, one of which has a PD of +1100 lb for milk and the other a PD of −1100 lb. The daughters of both bulls range from the 5000- to 6000-lb level to the 23,000- to 24,000-lb level. However, the difference in the average genetic merit transmitted to their daughters is evident in Figure 7.5. Approximately two-thirds of the daughters of the bull with the higher PD are above breed average in milk yield while about two-thirds of the daughters of the bull with the lower PD are below breed average. This variation in level of daughter yield is due to the effects of genetic sampling and environment. Variation of this type can be expected among the daughters of most bulls. This illustrates the difficulty of trying to estimate a bull's transmitting ability accurately from a few daughters.

In addition to the random variation in

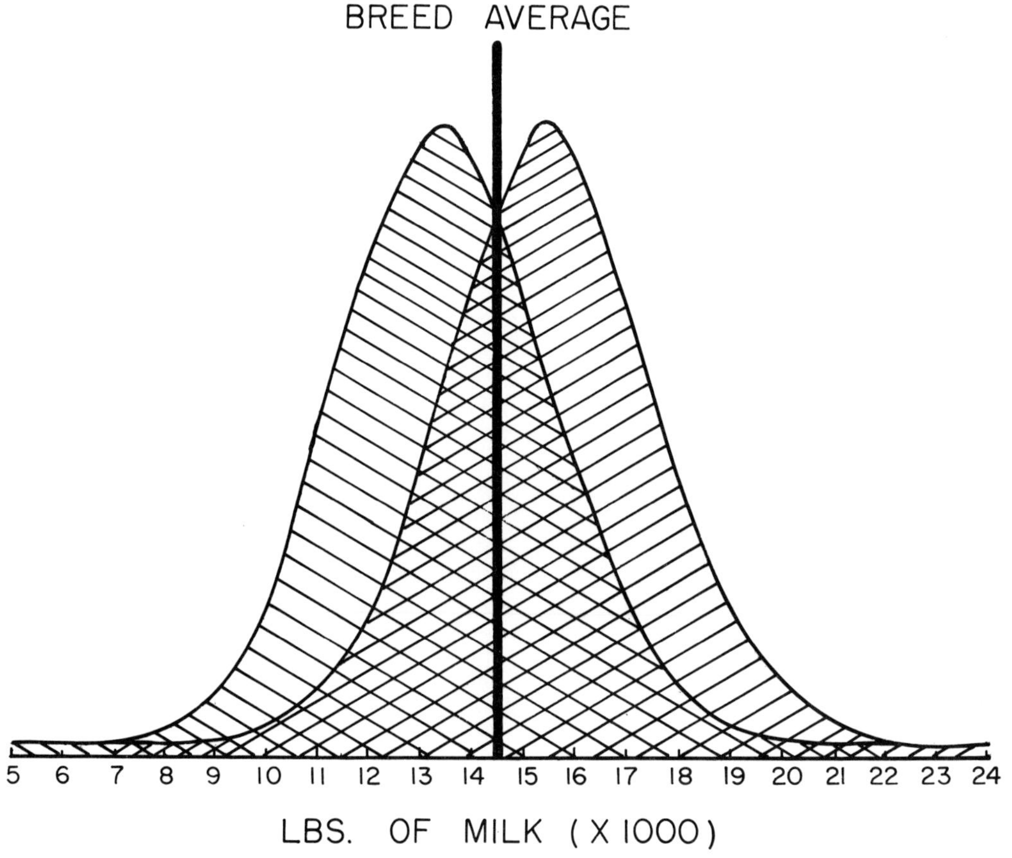

Fig. 7.5 Distribution of the milk yields of daughters of 2 bulls, one with a PD for milk of +1100 lb and the other with a PD for milk of −1100 lb.

milk yield among the daughters of bulls, the daughters of a bull cannot be expected to average the same in milk yield in herds that have different average production levels. In herds above the breed average, a bull's daughters would not be expected to deviate from herdmates by as much as the amount of the PD. In herds below the breed average they would be expected to deviate from herdmates by more than the sire's PD. The amounts by which the daughters of a bull with a PD for milk of +1000 lb would, on the average, be expected to exceed herdmates in milk yield are shown in Table 7.2.

The Role of Repeatability in Selecting Bulls

Repeatability is a measure of reliability of a sire summary as well as a weighting factor in the PD formula. Repeatability should be used to determine how heavily to use each bull in a single herd. It should not be used to decide whether or not to use a bull. Bulls with high repeatabilities can be used heavily in a herd with confidence that their PD's are close to their true transmitting abilities. However, each bull with a low repeatability should be used less heavily in each herd because his PD may vary more from his true transmitting ability. Therefore, groups of low-repeatability bulls can be used with confidence that their average PD is very close to their average true transmitting ability. No "magic" repeatability exists at which PD's suddenly become highly accurate. Accuracy of PD's increases gradually and continuously as repeatability increases.

Repeatability can be converted to a "confidence interval". A confidence interval is a range above and below a PD within which we are confident that a specified percentage of true transmitting abilities lies. The 67% and 80% confidence intervals for PD's at various repeatability levels are shown in Table 7.3. Since confidence intervals are not dependent on PD, the same interval applies to all bulls at a given level of repeatability regardless of PD. For example let us compare the reliability of the PD's of two Holstein bulls, each with a PD82 for milk yield of +1,000 lb, one with a repeatability of 90% and the other 50%. Since both bulls have the same PD82 their estimated transmitting abilities are equal but the one for the bull with the higher repeatability is more reliable. What does "more reliable" mean in terms of confidence intervals? From Table 7.3 we can be confident that 80% of bulls with PD82 of +1,000 lb and repeatabilities of 90% have true transmitting abilities that are within 267 lb above or below 1,000 lb, or between +733 lb and

Table 7.2 *Expected Average Deviation from Herdmates of Daughters of a Bull with a PD for Milk Yield of +1000 lb in Herds with Different Average Levels of Yield*

Herd average milk yield	*Expected deviation of daughters from herdmates*
18,000	+600
17,000	+700
16,000	+800
15,000	+900
14,000 (Breed average)	+1,000
13,000	+1,100
12,000	+1,200
11,000	+1,300
10,000	+1,400

Table 7.3 *67 and 80% Confidence Intervals for Predicted Differences at Given Repeatability Levels*

Repeatability	*67% Confidence Interval*	*80% Confidence Interval*
Holstein, Brown Swiss and Red and White		
(%)	(lbs.)	(lbs.)
99	65	84
90	206	267
80	292	378
70	358	463
60	413	534
50	462	597
40	506	655
30	546	707
20	584	756
Ayrshire, Guernsey, Jersey and Milking Shorthorn		
99	50	64
90	157	202
80	221	286
70	271	351
60	313	405
50	350	453
40	383	496
30	414	535
20	443	572

+1,267 lb; 10% will have true transmitting abilities above +1,267 lb and 10% will be below +733 lb. For bulls with PD82 of +1,000 lb and repeatabilities of 50%, 80% will have true transmitting abilities that are within 597 lb above or below +1,000 lb, or between +403 lb and +1,597 lb; 10% will have true transmitting abilities above +1,597 lb and 10% will be below +403 lb. All confidence intervals in Table 7.3 can be interpreted similarly for other PD's and repeatabilities in all breeds. Note that regardless of the repeatability, that best estimate of a bull's true transmitting ability is his PD82. But, the higher the repeatability, the more confident we are that a bull's PD82 is close to his true transmitting ability. The importance of the relation among PD82, repeatability and confidence intervals when making decisions in breeding programs will be discussed in Section 7.4.

7.4 Art—Managing Your Breeding Program

Breeding a superior herd of cows in an economical manner is an art based on the talents of the breeder, much as painting a masterpiece depends on the talent of the artist. In both cases, the materials needed to produce a masterful product are available to anyone but the results vary tremendously. Certain probabilities of success at genetic improvement are involved in every mating a dairyman makes. Perhaps much of the art of breeding a superior herd consists of the breeder's ability to maximize the probability of producing a genetically superior offspring every time he mates a cow and accomplishing this at a reasonable cost. An element of chance also affects the genetic-economic success of every mating a breeder makes because genetic segregation influences the genotype that results from every mating. Thus, the degree of genetic success from any single mating cannot be guaranteed. Those dairymen who become successful breeders generally do so by applying sound genetic principles consistently and with good judgment.

Establish Breeding Goals—Stick to Them

Herd breeding goals should be established with full consideration given to their economic impact. Once goals are established, they should be adherred to. Breeding goals can vary widely from herd to herd and short term goals may be considerably different from long term goals and receive greatly different emphasis in different herds. But, vacillating from one set of herd breeding goals to another will surely decrease a herd's genetic progress and also result in a loss of potential economic gain to a dairyman.

Determining the goals of his herd's breeding program is a personal decision for each dairyman. Some major factors affecting the selection of breeding goals are: basis of payment for milk (fluid vs components); proportion of income from sale of milk vs breeding stock; interest in registered cattle, type appraisal, the show ring and other breed association programs; participation in bull proving by himself or in cooperation with others, and; influence of extension and agribusiness counselors.

Relative Emphasis on Various Traits

The relative emphasis to be placed on different traits depends on their relative economic importance for fulfilling herd breeding goals, their degree of genetic control (heritability), their genetic relationships (genetic correlations), and the accuracy of information available on them. Traits for which relatively little genetic information is available probably should receive little emphasis because selection for them is likely to be ineffective even though a dairyman may consider them important. Traits for which genetic evaluations are generally available on a national basis include: milk yield, fat yield, fat percent, protein yield, protein percent, solids-not-fat yield, solids-not-fat percent, linear type traits, and dystocia among mates of AI bulls. In almost all herds the yield traits should receive the greatest emphasis because research has showed that yield per cow has the highest relation to herd profitability in most herds. In most parts of the country primary emphasis will be on milk yield under current pricing formulas. The trend in PD_{milk} in sires of first lactation cows since 1960 shown in Figure 7.6 is evidence that large numbers of dairymen have understood the economic importance of selection for higher milk yield. Those farms that produce milk for cheese manufacture

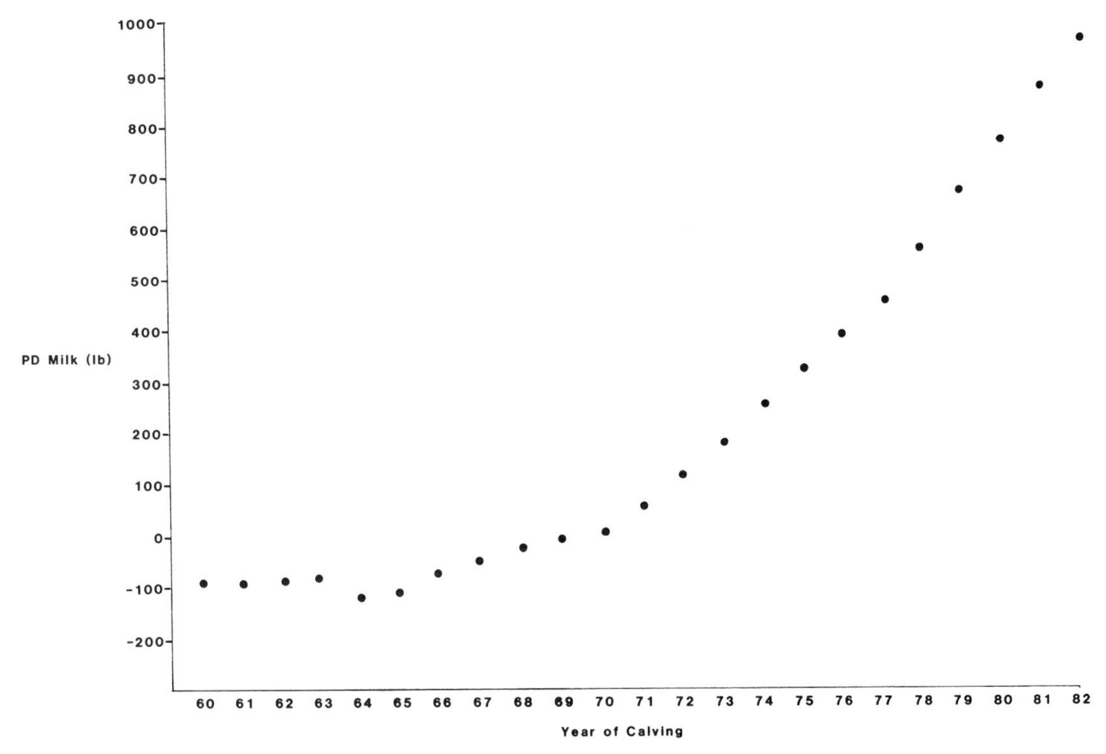

FIG. 7.6 Average PD for milk yield of sires of first lactation cows with records in the DHIA Program. (Courtesy of R. L. Powell, AIPL, ARS-USDA)

should emphasize component yields more than farms that produce for a fluid market. Component percentages should almost never receive the primary selection emphasis in spite of their high heritabilities. The negative genetic correlations of component percentages with yield traits and their lower genetic and phenotypic variances decrease the potential for significant economic progress through direct selection.

Emphasis to be placed on type traits depends on the dairymen's breeding goals. Dairymen expecting to sell a significant amount of breeding stock on a regular basis usually place more emphasis on type than they would if income were exclusively from the sale of milk. Relative progress for yield and type expected from different amounts of emphasis on each are shown in Table 7.4. In most herds and most breeds the optimum ratio of emphasis is about 3:1 in favor of yield over type. Herds that are strongly oriented to type and the show ring may place more emphasis on type but will do so at a sacrifice in genetic progress for yield.

Selection pressure does not need to be exerted for freedom from dystocia. After a group of bulls has been selected for the herd's breeding program, those bulls with desirable evaluations for dystocia can be mated to virgin heifers to minimize dystocia problems.

As was shown in Chapter 6, selection using a selection index is theoretically most efficient and should result in the greatest genetic-economic progress. But, constructing selection indexes is difficult in practice

TABLE 7.4 *Relative Progress for PD$ and PDType from Exerting Varying Amounts of Selection Emphasis on the Top 10% of Jersey Sires.*

Emphasis PD$:PDType	*Progress in PD$*	*Progress in PDType*
0:1	+20	+.69
1:1	+67	+.50
2:1	+80	+.33
3:1	+84	+.23
6:1	+87	+.12
9:1	+87	+.08
1:0	+88	−.01

Source: Norman, H. D., *et al.* J. Dairy Sci. *62:* 1914. 1979.

because varying amounts of information are available on different bulls and cows. Some genetic-economic indexes that may be helpful in assigning appropriate emphasis to traits that are included in a dairyman's breeding goals are discussed in the following section.

Genetic-Economic Indexes

Genetic-economic indexes[4,5] reflecting deviation in gross income per lactation are published as a standard part of the genetic evaluations of all bulls and cows evaluated in the national genetic evaluation program (Section 7.2 and Figure 7.1). The indexes express transmitting abilities as dollar values based on average national price for milk volume and components. Each bull and cow can have up to three indexes depending on the amount of component information available. All animals that are evaluated have PD$ or CI$ based on milk and fat. Those with sufficient protein and/or solids-not-fat information also have PD$ or CI$ that include protein and/or solids-not-fat in addition to milk and fat. A formula for the calculation of PD$ and CI$ using local prices for milk and appropriate components is shown in Appendix II-E. The published indexes based on national pricing values rank bulls and cows reasonably well for most parts of the country. PD$ and CI$ were developed in the early 1970s as selection aids for dairymen who were placing too much selection pressure on milk yield ignoring fat test. Percentile rankings based on PD$ and CI$ were included in genetic evaluations in 1983 for bulls and in 1984 for cows. The percentile rankings were included to emphasize the importance of PD's and CI's as tools to rank bulls and cows and to help users of genetic evaluations to adjust to the genetic base change in January 1984. The PD$ and CI$ values and the percentile rankings are useful to many people as initial screening or ranking criteria for bull and cow selection.

Various indexes have been developed that rank bulls on a dollar net return basis and can be calculated for individual bulls from PD$, semen cost and other information. These indexes account for several economic factors including average number of lactations per daughter, actual milk and fat produced, average feed costs, accumulation of genetic improvement in future generations, and number of units of semen required to produce a daughter that will complete a lactation record. They rank bulls on dollars net return per unit of semen. An index developed in Wisconsin[6] is: $ net return = (.3PD$ + $2.50) − ($ cost per unit of semen). A similar index developed in New York[7] is: $ net return = PD$ − 6($ cost per unit of semen). Both indexes were developed in the mid 1970s and represent some of the first efforts to rank bulls accounting for semen cost, which is an important factor in the economic return received from using different bulls.

Indexes have been developed that rank bulls or cows on PD$ or CI$ and final type score. Research[8] (Table 7.4) has shown that a weighting of about 3:1 for PD$ and final type score for ranking bulls will optimize genetic progress for yield for most dairymen in combination with a reasonable amount of progress for type. These indexes are called Production-Type Indexes in all breeds except Holstein, which refers to its index as a Total Performance Index. Production-type indexes for bulls in each breed are available in breed association publications.

A computerized bull selection system called Net Present Value, abbreviated PV$, was developed in Texas[9] to provide profit-maximizing rankings of individual bulls tailored to each herd's genetic improvement goals. PV$ of each bull represents an optimal weighting of genetic, reproductive and economic information to rank bulls by net profit of their daughters in current dollars. Factors accounted for in ranking bulls by PV$ include; PDs for milk, fat and type, semen price, real interest rate (cost of money), and each herd's first-service conception rate, calving interval and female mortality rate.

PV$ allows dairymen to exercise selection policies with goals of their choosing for milk income and type. Milk income can be the only selection goal or milk income and type can be weighted according to the dairyman's preference. Genetic improvement goals can be based on planning horizons from one generation upwards. Dairymen with large current debt loads would probably choose a one-generation planning horizon to improve their chances of economic survival. Dairymen with small debt loads seeking to maximize long-term gains from their genetic improvement program would use longer term planning horizons.

An example of PV$ ranking of AI bulls from the January 1984 Sire Summary run is shown in Figure 7.7 for the top 10 and bottom 10 Holstein bulls based on a 3:1 ratio of milk to type, 2 generation planning horizon, 50% herd average conception rate, and semen prices reported by AI organizations in January 1984. Note the overlap between the top 10 bulls and the bottom 10 bulls in PD$, PD milk and semen price and the different rankings of the bulls on different traits. In fact, these same bulls might rank quite differently if type were ignored in selection policy, at a different herd conception rate or under a one-generation planning horizon. These bulls would certainly rank differently if their semen prices were changed very much. Dairymen wishing to select bulls on an individual basis should consider carefully the economic consequences of their selection decisions. PV$ ranking provides the information that will maximize profit in each herd's genetic improvement program from selecting bulls individually.

A computerized bull selection procedure called Maxbull was developed in Virginia.[10,11] Maxbull is different from Net Present Value in that it permits dairymen to establish breeding goals for important traits for the average of bulls to be used in the breeding program while also setting minimum and maximum limits as appropriate for individual traits of critical importance. In other words, Maxbull maximizes the overall merit of the group of bulls selected to meet herd breeding goals rather than selecting bulls on how they meet a set of criteria individually. The principle on which Maxbull operates is called Complementary Sire Selection. Complementary Sire Selection does not eliminate a bull that is strong in many traits just because he is weak in one trait, but tends to minimize the impact of his weakness by selecting other bulls for the group that are strong in that trait. This approach tends to maximize the average genetic merit of the group of bulls selected and usually selects a group of bulls that has a desirable average semen price.

Fig. 7.7 Net Present Value Sire Summaries for Holstein bulls from the January 1984 USDA-DHIA Sire Summary run. Based on a two-generation planning horizon, 3:1 milk:type selection policy and a 50% herd conception rate.

PV$ PCTILE	*NAAB Code*	*Bull's Nickname*	*PV$*	*60%CI*	*RPT*	*PD$*	*PDM*	*PD%F*	*R*	*PDT*	*Semen Price ($)*
				TOP 10 BULLS							
99	7H1509	Quick Shot	277	25	50	166	1595	−0.11	36	1.28	18.00
99	3H1012	Forcaster	264	13	86	144	691	0.18	66	1.15	12.00
99	7H0978	Memorial	258	20	69	178	1338	0.03	63	1.27	25.00
98	7H1508	Linc	255	23	58	105	626	0.09	60	1.89	9.00
98	29H4195	Maiz	245	26	45	166	1757	−0.17	44	0.76	18.00
98	14H0474	Del	239	21	66	149	1205	−0.01	60	0.79	14.00
98	7H1502	Camaro	230	27	42	115	465	0.19	41	0.89	6.00
97	9H0626	Saul	229	19	71	105	864	−0.01	73	1.76	12.00
97	21H0568	Victorian	228	23	57	152	1149	0.03	51	1.57	25.00
97	21H0755	Loco	226	27	42	112	711	0.07	53	1.12	8.00
				BOTTOM 10 BULLS							
2	2H0416	Treb	7	14	86	72	163	0.17	51	−1.66	4.00
2	15H0123	Rocket	4	4	99	89	995	−0.11	99	−0.70	20.00
2	1H1378	Bells	−2	24	53	20	416	−0.11	50	−0.36	3.00
1	3H0938	Revenue	−6	17	76	42	713	−0.15	51	−0.89	5.00
1	3H0643	Executive	−48	4	99	90	502	0.09	96	−2.84	7.00
1	23H0206	Tradition	−65	6	97	161	1655	−0.14	94	1.71	80.00
1	29H2960	Sexation	−86	4	99	46	3	0.15	96	1.96	50.00
0	11H1056	Star	−113	4	99	10	128	−0.02	99	0.70	30.00
0	7H0900	Pete	−272	4	99	105	1027	−0.08	99	1.92	100.00
0	29H2851	Valiant	−577	4	99	169	1274	0.03	99	2.14	175.00

The principle of Complementary Sire Selection is explained in more detail with an example in the subsequent section on Planning the Herd's Genetic Improvement Program—First Step.

An example of a group of bulls selected by Maxbull from the January 1984 USDA-DHIA Sire Summary run is shown in Figure 7.8. The goal of this selection run was to obtain a group of bulls with maximum PD$, PD fat % of +.05 and PD type of +1.00 at a price of about $15.00 per unit of semen. Edit limits at the top of Figure 7.8 include minimums of 35% repeatability, 0 lb for PD milk, −.25% for PD fat %, −1.00 points for PD type, $25 for PD$ and a maximum semen price of $30.00. The 10 bulls selected have averages, weighted by units of semen recommended for purchase, of +1,025 lb for PD milk (with a confidence interval of ±258 lb), +1.00 points for PD type (exactly on target), +.06% for PD fat % (slightly above target), and $15.03 per unit of semen, only a few cents above the target price. Interesting comparisons can be made between the top 10 bulls in Figure 7.7 and the 10 bulls selected by Maxbull in Figure 7.8. Seven bulls were selected by both procedures. But the bull (3H1034) recommended for the highest use by Maxbull (36 units of semen out of 150 units) is omitted from the PV$ list because he has no type information. Maxbull includes him without a type proof due to his superiority in yield traits. Note that Maxbull optimizes benefits from the Complementary Sire Selection approach by recommending what proportion of breedings should be to each bull selected in the group. Another option in Maxbull is to maximize the function PD$ − CF$ (confidence interval for PD$) to take into account the risk of using an exceptional bull with low repeatability. When this option is used, the trait to be maximized will usually be lower in the bulls selected but the average repeatability of the group of bulls will be higher.

Some of the genetic-economic indexes described in this section will undoubtedly be refined so that different or additional information may be available in the future. New indexes may be developed. Dairymen are well advised to make use of the best genetic-economic indexes available to them to maximize economic returns from their genetic improvement program.

Sources of Germ Plasm

The most reliable source of superior germ plasm for a herd's breeding program is bulls that have been accurately evaluated for a large number of traits including yield, conformation and calving ease. Basically that means bulls available through AI. AI bulls have been the primary source of genetic improvement in the past and will continue to be for some time in the future. Two categories of bulls are available through AI; those that already have sire summaries and those that are being progeny tested. Bulls' evaluations that are most accurate are those with the highest repeatabilities. But, high repeatability does not necessarily mean genetic superiority. It means that dairymen can be confident that the bulls' evaluations are close to their transmitting abilities. Bulls with high Predicted Differences for several important traits and high repeatabilities usually command a high semen price from AI organizations.

Dairymen should take advantage of the superior germ plasm in the other category of AI bulls, those in progeny test status. Due to the fierce competition among breeders and among AI organizations to produce genetically superior bulls, most bulls in progeny test programs have high pedigree merit, that is high expected transmitting abilities, for economically important traits. Most of the bulls in progeny test status will not have sire summaries but some may have ones with low repeatabilities. But research has shown that for at least a decade progeny test bulls have had higher transmitting abilities than active AI bulls on the average at any given point in time. Therefore, virtually all dairymen should consider using groups of progeny test bulls on some portion of their herd. Dairymen can take advantage of progeny test bulls through the computerized sire selection programs (Maxbull and Net Present Value Sire Summaries) by setting low minimum repeatabilities. Dairymen who participate in AI organization progeny test programs are not necessarily making great sacrifices as they sometimes think. On the average the progeny test bulls are genetically superior for economically important traits and their semen usually can be obtained at highly desirable prices.

Dairymen can take advantage of another

** MAXBULL **

A-FRANK DICKINSON

>>>>>>>>>> BULL EDITING >>>>>>>>>>	BREED	R	PDMILK	PDFAT%	PDTYPE	PD$	PRICE	MISS. PDTYPE	STUD(S)
EDIT LIMITS	3	35	0	-0.25	-1.00	25	30.00	****	1 3 7 9 11 14 15 17 21 24 29 40
SURVIVORS	453	444	411	407	403	385	371	315	371

BULL SIRE LIMITS (%) A- 7H58 100.; B- 40H2025 100.;

MINIMUM # BULLS= 5. MAXIMUM % LOW R BULLS= 15. MAXIMUM % PER LOW R BULL= 5.

MINIMUM AMOUNT STUD(S): 9 = 10% 21 = 10%

SUGGESTED BREEDING PROGRAM FOR MAXIMUM **PD$**

BULL		BS	NO.	%	R	PDM	CFM	PDF%	PDF	PD$	CF$	RANK	R-T	PDT
GOALS			UNITS 150		0			0.05		MAX				1.00
3H1034	BACHELOR		36	24	72	1044	245	0.10	56	163	29	97	0	0.0
3H1012	FORCASTER		23	16	86	691	173	0.18	58	144	21	96	66	1.15
7H1508	LINC		16	11	58	626	300	0.09	38	105	35	83	60	1.89
7H978	MEMORIAL	B	16	11	69	1308	253	0.03	54	178	31	99	63	1.27
9H589	STANDARD	B	15	10	76	637	227	0.09	39	108	27	86	78	1.24
21H568	VICTORIAN	A	14	9	57	1149	304	0.03	46	152	36	97	51	1.57
7H1509	QUICK SHOT	A	14	9	50	1595	327	-0.11	37	166	39	98	36	1.28
29H4195	MAIZ		7	5	45	1757	343	-0.17	31	166	41	98	44	0.76
29H3421	MAPLE		6	4	71	992	249	-0.07	22	101	30	81	65	1.00
21H755	LOCO		1	1	42	711	353	0.07	39	112	42	88	53	1.12
WEIGHTED AVERAGES			148		67	1025	258	0.06	47	146	30	93	44	1.00
PERCENT SUPERIOR														
BONUSES (AVERAGE MINUS GOAL)					67			0.01						0.0
MILK CHANGE FROM RELAXING EACH GOAL ONE UNIT					0			0 /.01%						17

PREDICTED DIFFERENCE – LINEAR TYPE (# = SUPERIOR / = INFERIOR)

BULL		US	RU	FU	TT	HL	FT	ST	PRICE
GOALS		30	15	20	-100	10	0	0	15.00
3H1034	BACHELOR	0.0	0.0	0.0	0.0	0.0	0.0	0.0	10.00
3H1012	FORCASTER	-0.3	-0.1	-1.6/	-0.8	0.1#	-1.7/	-0.4	12.00
7H1508	LINC	0.9#	1.2#	2.0#	0.7	0.4	-0.3	1.3#	9.00
7H978	MEMORIAL	1.3#	0.2	1.4#	1.5#	-1.5/	1.2#	-0.3	25.00
9H589	STANDARD	1.1#	0.4	-0.8	-1.1/	-0.1#	1.1#	0.2	15.00
21H568	VICTORIAN	0.0	0.0	0.0	0.0	0.0	0.0	0.0	25.00
7H1509	QUICK SHOT	0.7	0.6	1.1#	0.6	-1.4/	0.2	1.5#	18.00
29H4195	MAIZ	0.7	-0.1	0.6	-0.2	0.6	-0.1	0.6	18.00
29H3421	MAPLE	-0.1	1.5#	1.6#	1.4#	-0.2#	0.7	-1.1/	14.00
21H755	LOCO	-0.1	0.3	1.1#	0.5	0.4	0.2	-0.3	8.00
WEIGHTED AVERAGES		0.4	0.3	0.2	0.1	-0.2	-0.0	0.2	15.03
PERCENT SUPERIOR		32	15	20	5	9	5	16	
BONUSES (AVERAGE MINUS GOAL)		2	0	0	105	0	5	16	0.0
MILK CHANGE FROM RELAXING EACH GOAL ONE UNIT		0	0	0	0	0	0	0	2

HOLSTEIN STANDARDIZED LINEARIZED TYPE PROOFS ARE CHANGED PERIODICALLY BY HFAA AND MAY DIFFER SLIGHTLY FROM RED BOOK VALUES.

MARCH 1984
UNITS PER BATCH= 1
OPT=0; ITR= 52; CYCLE=2

BENNET G. CASSELL
EXTENSION GENETICIST

MIKE MCGILLIARD & JOHN CLAY
DAIRY SCIENCE - VIRGINIA TECH
BLACKSBURG 24061

Fig. 7.8 Output from Maxbull selection with group average goals of maximum PD$, PD Fat % of +.05 and PD Type of +1.00. (Courtesy of B. G. Cassell, Virginia Tech.)

source of potentially superior germ plasm through participation in bull proving syndicates. Bulls being proven by syndicates are usually of superior estimated transmitting abilities as they must be to be competitive. Participation in syndicates has an added profit potential if some of the bulls are good enough to be used widely in AI. Dairymen participating in syndicates, as well as those in AI organization progeny test programs, should not breed more than a few cows to any one bull. On the average the bulls are superior but gambling that any individual bull is superior through heavy use in one herd is not a wise gamble when the use of a group of those bulls on only a few cows each is almost a sure thing.

The use of herd bulls or pasture bulls should be limited to providing a mechanism for getting cows or heifers pregnant under extraordinarily difficult management conditions. In other words, such bulls should be used only as a last resort where it is impossible to manage an effective genetic improvement program. The unknown genetic merit of those bulls dictates that they should never be used heavily in a herd if an alternative breeding procedure is available.

Embryo transfer and various forms of

FIG. 7.9 The first pair of bulls produced by embryo splitting for progeny testing in artificial insemination; Jackbuilt Duplicate-ETS, 1900136, (top) and Jackbuilt Great Divide-ETS, 1900137, (bottom). (Courtesy of P. D. Miller, American Breeders Service)

genetic engineering have great future potential for genetic improvement of dairy cattle. To date, much of the potential of embryo transfer for genetic improvement of yield has not been realized because of heavy selection emphasis on non-yield traits but that situation is improving. High cost is still a significant obstacle to widespread use of embryo transfer for genetic improvement but as techniques and reliability improve, costs will undoubtedly decrease. Figure 7.9 shows the first two bulls that were produced by embryo splitting and then were progeny tested in AI. Genetically they are identical twins which makes the obvious differences in their color markings particularly interesting.

Planning the Herd's Genetic Improvement Program

Dairymen can maximize the profit from their herd's genetic improvement program by planning it in two steps. The first and by far most important is selecting the group of bulls that will be used in the herd. The second step, that should yield added benefits, is the planning of individual matings between each bull and cow.

First Step. Bulls should be selected for use in the herd with the goal of improving the herd as a whole, not individual cows. This can be done most effectively by selecting a group of bulls according to a set of average minimum and maximum standards for the group rather than selecting individual bulls according to a set of minimum or maximum standards. As mentioned earlier in this chapter, this is called Complementary Sire Selection. It is more successful than individual bull selection because when bulls are evaluated against a set of standards individually each bull is eliminated if he falls outside the acceptable range for any single standard. The best bull in one trait might be eliminated because he is just outside the acceptable range for another trait. However, when a group of bulls is selected against a set of average standards for the group, as in Complementary Sire Selection, bulls just outside the acceptable range for one or even a few traits can be selected for breeding if they are sufficiently superior for other traits as long as the average minimum or maximum standards are met for the group of bulls as a whole. Use of this group selection concept based on average minimum or maximum merit will almost certainly result in selection of a group of bulls that will have higher genetic and economic merit for most traits than will selection of individual bulls one at a time. An example will help to demonstrate this principle. Table 7.5 shows data for three hypothetical bulls for PD milk, PD fat and PD type. Assume that not all bulls can be selected, which, of course, simulates the real-life situation. At first glance it is easy to eliminate bull Y on the basis of low PDs for fat and type and eliminate bull Z on the basis of low PD for milk and select bull X as a reasonable compromise in all three traits. However, the average merits for bulls Y and Z are +1,400 for PD milk, +.05 for PD fat and +.50 for PD type, all higher than bull X's merit for those traits. Therefore, equal use of Bull's Y and Z would lead to greater genetic improvement than just using bull X. This simple example demonstrates how Maxbull helps dairymen maximize genetic-economic improvement. Dairymen should follow closely the proportions suggested by computerized bull selection systems to maximize overall genetic-economic gain from the pool of germ plasm selected.

Second Step. Choosing a bull to mate to each cow so as to accomplish corrective mating can yield an added bonus to the herd's genetic improvement program. Corrective mating as opposed to random mating of bulls and cows within the herd will not result directly in much greater genetic progress. The herd's genetic progress has already been largely determined by the genetic merit of the group of bulls selected for the breeding program. However, corrective mating may, if effective, indirectly contribute to the herd's genetic progress by

TABLE 7.5 *Hypothetical Sire Summaries to Demonstrate Advantages of Selecting Groups of Bulls by Average Standards.*

Bull	PD_{Milk}	PD_{Fat}	PD_{Type}
X	+1,200	0	0
Y	+2,000	−.10	−.50
Z	+800	+.20	+1.50

decreasing involuntary culling. That is, fewer cows may have to be culled for reasons other than low yield thereby raising the average level of yield and also probably the herd's average transmitting ability for yield. The large number of dairymen using planned or corrective mating programs attests to dairymen's confidence in those programs.

Depending on a dairyman's location in the country and his particular management regimen, he may have valid reasons for limiting his source of germ plasm to a single AI organization and may follow that organization's mating recommendations closely. Other factors besides the PD's of bulls legitimately affect a dairyman's choice of source of germ plasm including loyalty to a particular AI organization or inseminator and his ability to breed his own cows successfully. However, dairymen should always keep in mind that the genetic progress in his herd depends much more on the average merit of the group of bulls selected for use in the herd than on corrective matings.

Ideally a dairyman should make new bull selections and semen purchases each time new sire summary information is available, i.e., every 6 months. Purchasing more semen than will be used in the next 6 to 12 months should be done only for investment (speculative) purposes. Dairymen who speculate in semen should be in a financial position to discard their mistakes and write off their financial loss. Many dairymen who keep a semen tank could increase future genetic progress in their herd by culling the genetically inferior semen and replacing it with genetically superior semen. Purchasing semen for investment is similar to purchasing stocks in that you should capitalize on your successes and not hesitate to admit your mistakes. Some of the worst 'deals' dairymen make are with semen suppliers who offer semen from a top bull along with mandatory purchase of semen from mediocre bulls. Ideally you should buy semen only when you can specify the bulls you want in the amounts you want and at prices that will make the investment in your genetic improvement program a profitable one. That statement may be somewhat idealistic in practice but it is a sound principle to follow.

7.5 Summary

The success of a herd's breeding program has an important bearing on the economic success of a dairy enterprise. Breeding goals should be established and adhered to. The most economically important traits should receive the greatest emphasis. Therefore almost all herds should place the greatest emphasis on yield of milk and milk components, with secondary emphasis on type and other non-yield traits. Genetic evaluations should be used to rank bulls and cows. The highest ranking bulls and cows should be used in your breeding program regardless of the level of PD or CI. The magnitude of PD's and CI's is affected by the genetic base. Three genetic bases are possible; a fixed genetic base, a moving genetic base or a stepwise genetic base. The latest PD or CI is the best estimate of an animal's transmitting ability with repeatability indicating how reliable that estimate is. Groups of high PD, low repeatability bulls are genetically superior on the average and can be used for genetic improvement as long as no single bull is used heavily in a herd. Several genetic-economic indexes are available and should be used every 6 months or so by most dairymen to select the group of bulls for use in their herd. Maxbull enables dairymen to benefit from the concept of Complementary Sire Selection and provides excellent guidance for completing the first and most important step in planning a herd's breeding program, that is, selecting the group of bulls that will introduce new germ plasm into the herd. The second step in the breeding program, pairing each cow to a bull, may help to decrease the number of future cows with serious faults in non-yield traits and to decrease involuntary culling, thereby making your culling program a more dynamic tool for genetic improvement.

Review Questions

1. List the primary and secondary objectives of your breeding program if you owned a dairy herd.
2. What are the relative economic benefits of selecting for milk yield versus content of milk fat, protein, or solids-not-fat (SNF) in your region of the country?

3. What are the consequences of frequent changes in goals in a breeding program?
4. How would you explain the terms PD and CI? Give examples with values of your choice.
5. What determines the magnitude of confidence limits on a PD? Give an example using values from Table 7.2.
6. Determine and explain the 80% confidence limits for two bulls with PD_{milk} +1,200 lb and Repeatabilities of 60 and 90%. Which bull is genetically superior? Which bull would you use most heavily in your herd and why?
7. Explain the advantages and disadvantages of fixed, moving and stepwise genetic bases.
8. How were the genetic bases determined for PD82 and CI82?
9. Which genetic base do you favor and why?
10. Calculate PD$ for a bull used heavily in your area using traits important in the milk pricing formula in your area and current prices for fluid and components (see appendix Table II-E).
11. Can you explain all the information in the Sire Summaries shown in Figure 7.1 to a friend from the city who knows little about dairy cattle?
12. Obtain Canadian Sire Summary information on a bull from Canada and convert the genetic evaluations for milk and fat to estimated PD82's using information in Section 7.3.
13. How would you explain Complementary Sire Selection and its advantages over selecting bulls individually?
14. Which genetic-economic index would you use to select the bull semen for use in your herd's genetic improvement program and why do you favor the one you chose?
15. Which of the two steps in the planning of a breeding program is most important and why?

References

1. Dickinson, F. N., and McDaniel, B. T.: Status and potential of artificial insemination for increasing production and income over feed cost for dairymen. J. Dairy Sci. *52:* 1464, 1969.
2. Dickinson, F. N.: Alternative genetic bases for sire summaries and cow indexes. J. Dairy Sci. *63:* 1361, 1980.
3. Dickinson, F. N., and Powell, R. L.: Genetic improvement of yield in dairy cattle. Holstein Science Report, August, 1983, Holstein Association, Brattleboro, Vermont, 1983.
4. Norman, H. D., and Dickinson, F. N.: An economic index for determining the relative value of milk and fat in Predicted Difference of bulls and Cow Index values of cows. Dairy Herd Improvement Letter, ARS-44-223, January, 1971.
5. Norman, H. D.: An economic index for use in selecting bulls evaluated on protein or solids-not-fat. USDA-DHIA Milk Components Sire Summary. U.S. Department of Agriculture Production Research Report No. 178, 1979.
6. Sendelbach, A. O.: New Wisconsin USDA Sire Report has several new features. Dairy Report, Department of Dairy Science, 1975.
7. Everett, R. W.: Income over investment in semen. J. Dairy Sci. *58:* 1717, 1975.
8. Norman, H. D., Cassell, B. G., King, G. J., Powell, R. L., and Wright, E. E.; Sire evaluation for conformation of Jersey cows. J. Dairy Sci. *62:* 1914, 1979.
9. Wilcox, M. L., Shumway, C. R., Blake, R. W., and Tomaszewski, M. A.: Optimal selection of artificial insemination sires to maximize profits. J. Dairy Sci. *67:* in press, 1984.
10. McGilliard, M. L., and Clay, J. S.: Selecting groups of sires by computer to maximize herd breeding goals. J. Dairy Sci. *66:* 647, 1983.
11. McGilliard, M. L., and Clay, J. S.: Breeding programs of dairymen selecting Holstein sires by computer. J. Dairy Sci. *66:* 654, 1983.

Suggested Additional Reading

Dairy Herd Improvement Letters, Agricultural Research Service, USDA.

Literature on dairy cattle breeding and sire summary lists published by state extension service.

Proceedings of National Workshop on Genetic Improvement of Dairy Cattle. Sponsored by Extension Service, USDA. April 1984.

Publications by AI studs serving your area, describing transmitting abilities of bulls available for use.

Publications and articles from breed associations and in agricultural periodicals on genetic improvement of dairy cattle.

PART THREE

Nutritional Principles and Feeding Practices

CHAPTER 8

Anatomy and Physiology of Digestion

8.1 Introduction

The dairy cow, being a ruminant animal, has 4 compartments to its stomach rather than only one, which is common to humans, pigs, rats, and other simple-stomach animals. This gives the cow a decided advantage in digesting and utilizing those parts of plants and other compounds which are practically useless to animals with simple stomachs. Substances such as cellulose, a major constituent of plant tissue, and urea, a nonprotein nitrogen (NPN) compound, are of limited use to nonruminants. However, ruminants, through fermenting and synthesizing actions of microorganisms in their complex stomachs, can efficiently utilize these substances for productive purposes. The dairy cow converts this feed into milk and meat, two highly nutritious and palatable foods for humans. This ability to make use of feed which otherwise would go to waste because it is not directly utilizable by humans is one of the most compelling justifications for the existence of ruminants, especially dairy cattle, in a world that must make use of all possible food sources.

8.2 Anatomy of the Digestive Tract

A simplified diagram of a ruminant digestive tract and the pathway taken by ingested feed is shown in Figure 8.1. Food from the mouth passes through the esophagus and enters the rumen, where it is mixed with ruminal contents and fermented by ruminal microorganisms. Some of the feed is regurgitated for more mastication (chewing the cud) and then is returned to the rumen and reticulum for additional fermenting action. Fatty acids resulting from fermentation of the feed are absorbed into the blood stream from the rumen and reticulum. The remainder of the feed passes through the omasum and abomasum where further digestive action takes place. Finally, it enters the intestines for additional digestion and absorption into the blood stream or excretion as feces. A description of the organs and other structures involved in the digestive process is contained in the following sections.

Mouth, Tongue, and Teeth

The lips of a cow are rather immobile and are not very useful in drawing food into the mouth. The main organ of prehension is the tongue. It is long, strong, mobile, and rough and can readily be curved around forages and other feed, which are drawn between the incisor teeth below and the dental pad above and cut off.

An adult cow has 8 incisor teeth on the lower jaw but none on the upper jaw. The upper jaw contains a dental pad, which is a heavy layer of connective tissue covered with thick, horny epithelium. Cows' jaws

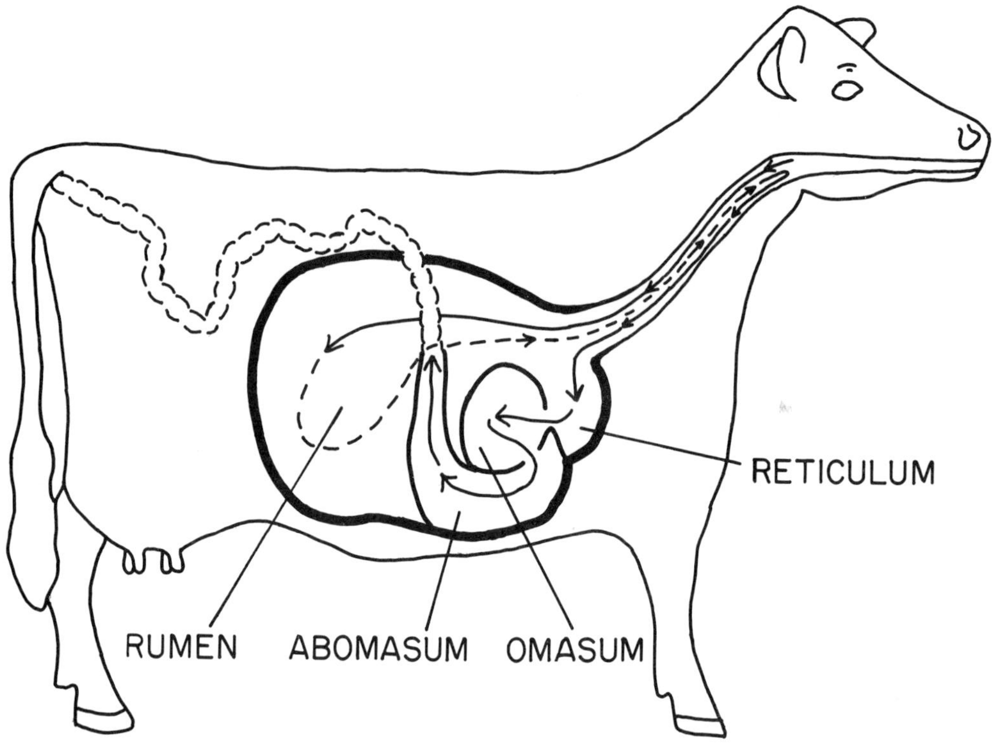

FIG. **8.1** Simplified diagram of the ruminant stomach and pathway of ingested food.

contain no canine teeth above or below but have 6 teeth each in the upper and lower jaws in the premolar and molar categories, which together are called the cheek teeth. The formula for the permanent teeth is $2\ (I\frac{0}{4}C\frac{0}{0}P\frac{3}{3}M\frac{3}{3}) = 32$, where I = incisors, C = canines, P = premolars, and M = molars. For the temporary deciduous teeth, the formula is $2\ (Di\frac{0}{4}Dc\frac{0}{0}Dp\frac{3}{3}) = 20$, where Di = deciduous incisors, Dc = deciduous canines, and Dp = deciduous premolars. Average ages for eruption of teeth in cattle are shown in Table 8.1.

Salivary Glands and Esophagus

Salivary glands located in the oral cavity (mouth) secrete the liquids that together are known as saliva. In cattle, the glands are known as the parotid, submaxillary, inferior molar, ventral sublingual, buccal, palatine, pharyngeal, and labial.[1]

The esophagus is the tube leading from the mouth to the rumen. Its length is about 3.5 feet in mature cattle. Feed and saliva mixed in the mouth pass down the esophagus to the rumen, and ruminal contents also periodically pass back up the esopha-

TABLE **8.1** *Average Time of Eruption of the Teeth in Cattle*

Teeth	*Eruption*
Temporary:	
First incisor (Di 1)	Birth to 2 weeks
Second incisor (Di 2)	
Third incisor (Di 3)	
Fourth incisor (Di 4)	
First cheek tooth (Dp 1)	Birth to few days
Second cheek tooth (Dp 2)	
Third cheek tooth (Dp 3)	
Permanent	
First incisor (I 1)	1½ to 2 years
Second incisor (I 2)	2 to 2½ years
Third incisor (I 3)	3 years
Fourth incisor (I 4)	3½ to 4 years
First cheek tooth (P 1)	2 to 2½ years
Second cheek tooth (P 2)	1½ to 2½ years
Third cheek tooth (P 3)	2½ to 3 years
Fourth cheek tooth (M 1)	5 to 6 months
Fifth cheek tooth (M 2)	1 to 1½ years
Sixth cheek tooth (M 3)	2 to 2½ years

Source: Sisson, S., and Grossman, J. D.: *The Anatomy of Domestic Animals.* 3rd ed. Philadelphia: W. B. Saunders Co., 1938.

gus (regurgitation) for additional mastication before reswallowing.

Rumen

The rumen, or paunch, is divided into 4 areas by muscular bands called pillars. There are a dorsal, a ventral, and 2 posterior sacs. Action of the muscular pillars forces the feed in the rumen to move in a rotary fashion so that it thoroughly mixes with the ruminal fluid. The sacs and pillars are shown in a sagittal section of a rumen in Figure 8.2. Note also the lining of the rumen wall. Many fingerlike projections, the papillae, line the rumen wall, resulting in a greatly increased surface area for absorption of nutrients from the rumen.

Reticulum

The reticulum is only partially separated from the rumen by a low partition, as shown in Figure 8.2. Ruminal and reticular contents intermix freely, and the two compartments frequently are referred to as the reticulo-rumen. The thick walls of the reticulum resemble a honeycomb, a name which has been used in the past for this part of the ruminant stomach. An opening on the right side of the reticulum leads into the omasum, the organ into which food passes after action by microorganisms in the rumen and reticulum. A groove in the reticulum between the esophagus and the omasum, called the esophageal groove, can be seen in Figure 8.3. In the nursing calf, the sides of the groove extend upward by reflex action and form a conduit through which milk passes directly from the esophagus to the omasum before the rumen and reticulum are developed and functioning. Even if ruminal action were normal, it is more efficient for milk to bypass fermentation in the rumen and be digested directly in the abomasum and intestines. Fermenting action by ruminal microorganisms is wasteful of energy and reduces the protein

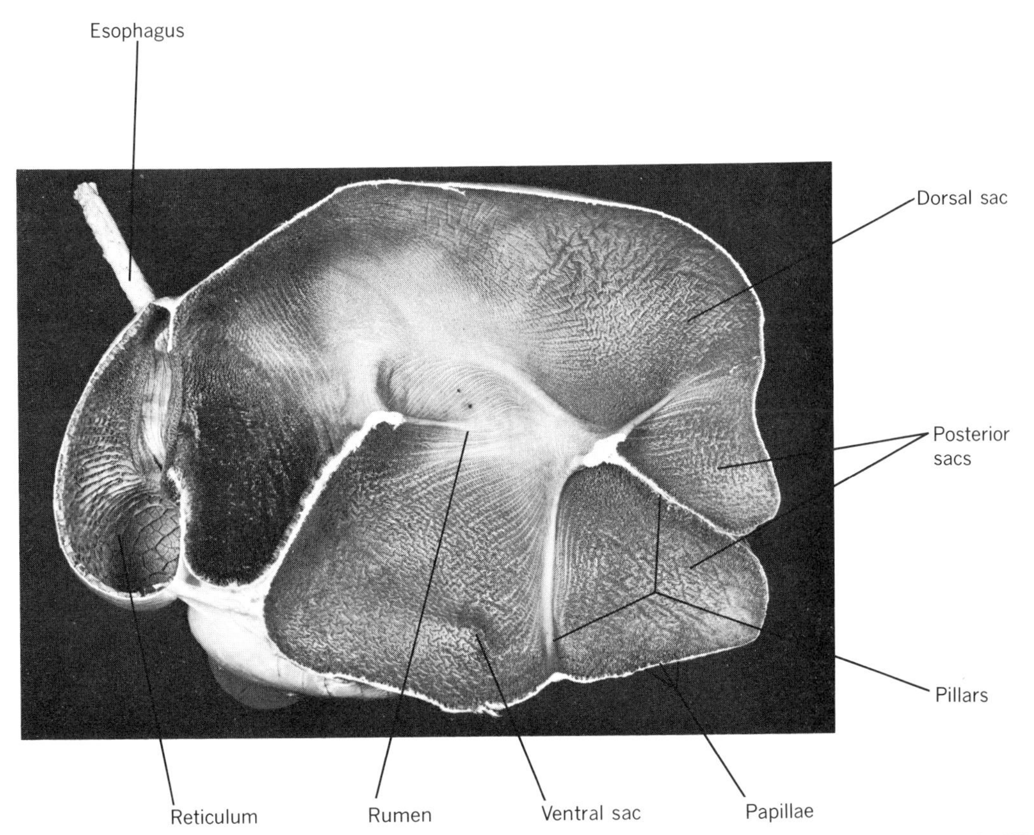

Fig. 8.2 Rumen and reticulum in sagittal section. (Courtesy of N. J. Benevenga, et al., J. Dairy Sci. *52:* 1294, 1969)

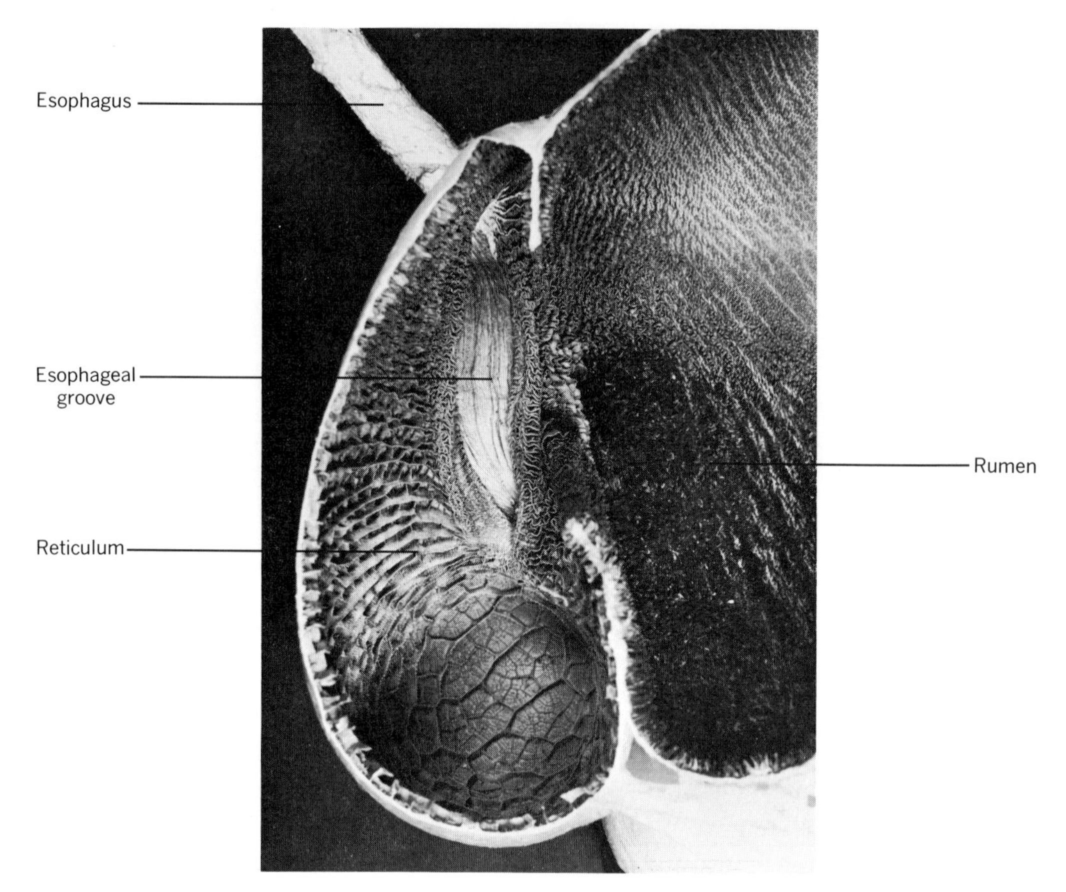

Fig. 8.3 Reticulum showing a close-up view of the esophageal groove. (Courtesy of N. J. Benevenga, et al., J. Dairy Sci. *52:* 1294, 1969)

quality of an easily digested, high-quality food such as milk.

Omasum

After fermenting action in the rumen and reticulum, feed passes through the reticulo-omasal orifice into the omasum, as illustrated diagrammatically in Figure 8.1 and pictured in cross section in Figure 8.4. This organ consists of many layers of muscular leaves, which resulted in its being called the manyplies in older literature.

Abomasum

The abomasum is the true stomach of the cow. It is the only part of the ruminant stomach which contains tissue for secretion of gastric juices. The walls of the abomasum are lined with many folds, greatly increasing the secretory area within the organ, as can be seen in Figure 8.4. This organ makes up 80% or more of the total stomach volume of a newborn calf. The other compartments grow at a much faster rate once the calf starts eating dry feeds, and less than 10% of the stomach capacity of a mature cow is in the abomasum. A comparison of the relative size of the abomasum in a newborn calf and a mature cow is shown in Figure 8.5.

Small Intestine

After digestion in the abomasum, the resulting products pass through the pylorus, illustrated in Figure 8.4, into the duodenum, which is the upper portion of the small intestine. The small intestine, so named because of its diameter rather than its length, is a folded tube about 140 feet long and 2 inches in diameter in mature cattle. Small, finger-like projections, the

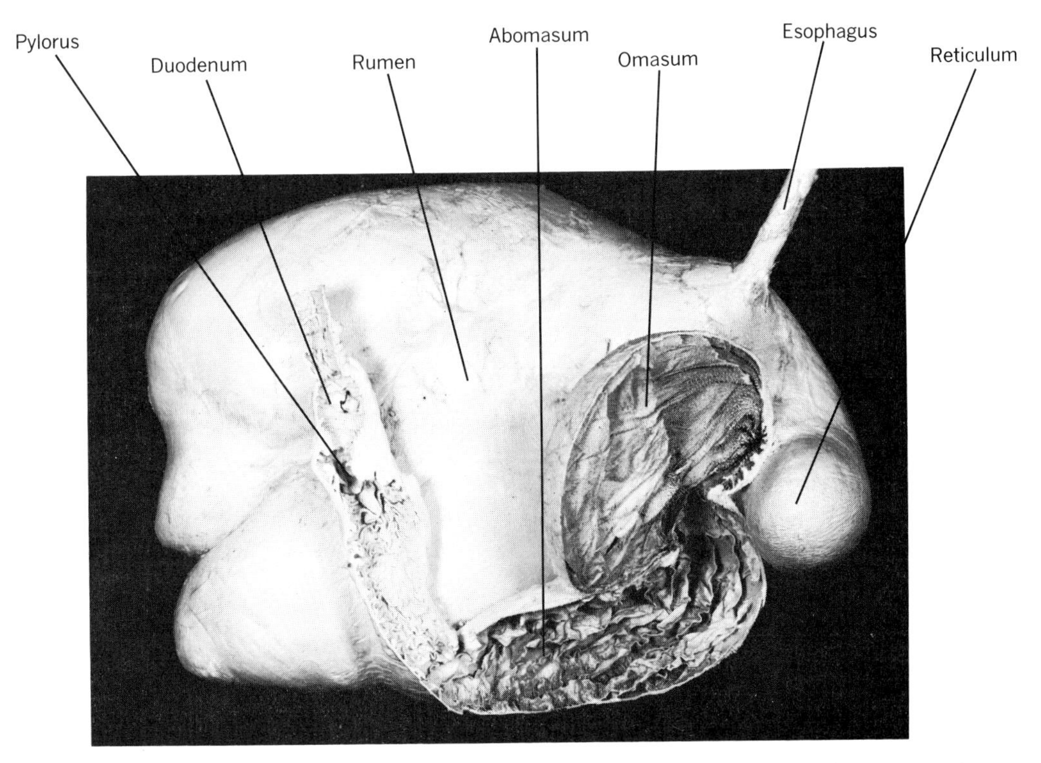

FIG. 8.4 Ruminant stomach from right side showing a cross section of the omasum, abomasum, pylorus, and duodenum. (Courtesy of N. J. Benevenga, et al., J. Dairy Sci. *52:* 1294, 1969)

villi, line the intestinal wall. Action of the villi helps to mix intestinal contents with digestive enzymes and greatly increases the absorptive area of the intestines. The intestinal contents are propelled by peristalsis, which is a movement of alternate waves of constriction and relaxation of the intestinal wall muscles.

Cecum

Although large and important as a site of microbial digestion in some animals such as the rabbit and the horse, the cecum of a cow is small and of little importance. Microbial digestion and synthesis of various compounds take place, but are insignificant compared with these actions in the reticulo-rumen.

Large Intestine

Contents from the small intestine pass into the cecum and large intestine, a tube about 35 feet long varying from 2 to 5 inches in diameter, which ends at the anus.

8.3 Functions of the Digestive Tract

The primary purpose of the digestive tract is to convert plants and other feed consumed by the cow into chemical compounds which can be absorbed into the blood stream for use as nutrients for the tissues of the body. It also provides a means of excreting waste products of tissue metabolism and undigested feed residues. In order to accomplish these functions, many processes are involved, including mastication, salivation, rumination, digestion, and absorption. These processes are the subject of this section.

Mastication

Initial mastication (chewing) of feed by cows is slight, with only enough chewing performed to mix the feed with saliva and form a bolus for swallowing. Complete mastication of regurgitated ruminal contents occurs while cows are resting after eating.

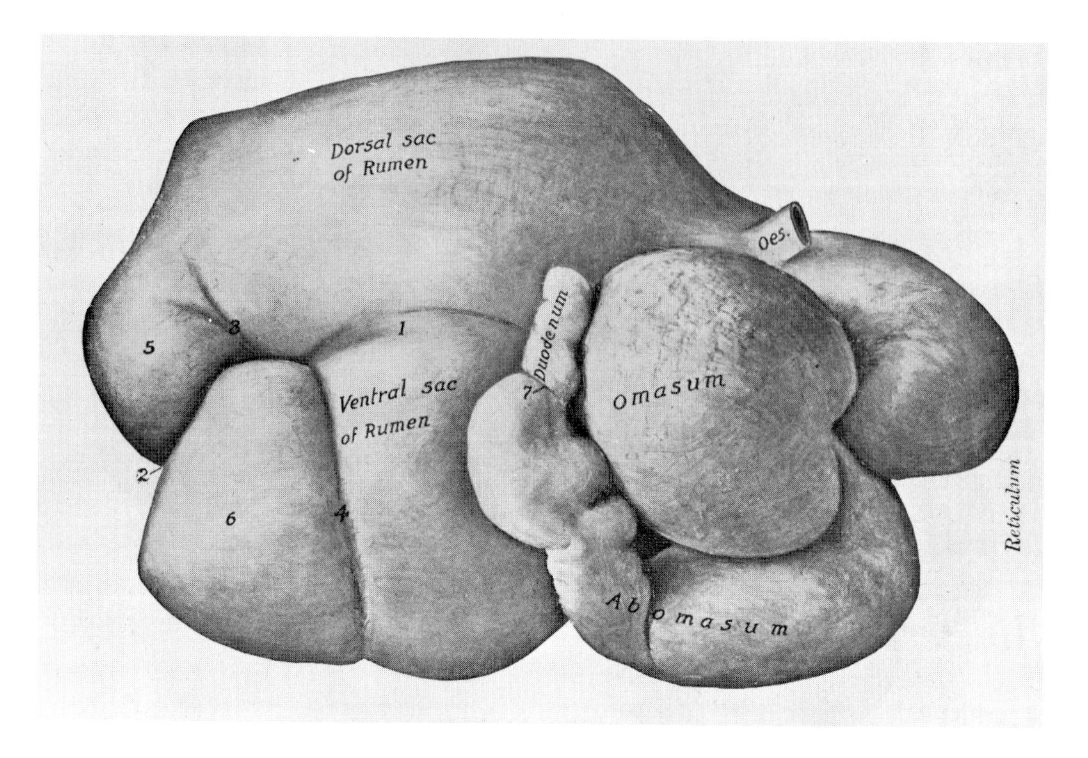

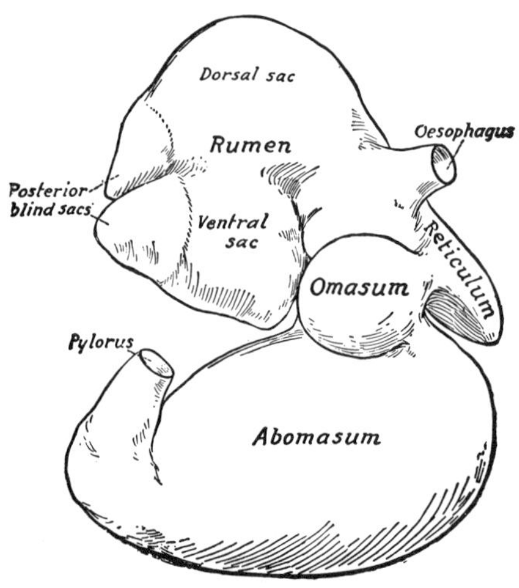

Fig. 8.5 Comparison of the relative size of compartments of the adult ruminant stomach *(top)* and of a new-born calf *(bottom)*. (From Sisson and Grossman. *The Anatomy of Domestic Animals,* Philadelphia: W. B. Saunders Co., 1938)

Mastication is important in that finely ground feed presents a greater surface area than coarse feed for action of the ruminal microorganisms and digestive juices in the abomasum and intestines. The coarse nature of a cow's diet makes mastication of even greater importance than it is for animals fed finely textured diets. It has been estimated that a dairy cow performs about 42,000 jaw movements per day in initial mastication and cud chewing when fed a ration composed of silage, hay, and grain.

Salivation

Tremendous amounts of saliva are secreted by the cow, especially when rations high in dry roughages are fed. Saliva has the dual role of lubricating the feed before it is swallowed and providing buffering action in the rumen with its high content of bicarbonate and phosphate.

Saliva is distinctly alkaline in the cow, having a pH of about 8.2. Organic acids produced by microbial action in the rumen are neutralized by saliva. This keeps the pH of the rumen between 6.5 and 7.5, thereby providing a proper medium for microbial growth and activity. It also tends to suppress frothing action of rumen contents, a condition which can lead to bloat. Bloat is discussed in detail in Section 8.4.

Microbial Digestion in the Reticulo-Rumen

After entering the reticulo-rumen, the feed mixes with ruminal fluid which contains billions of microorganisms of both bacterial and protozoal origin. Bacteria are single-cell plants. Protozoa are single-cell animals which feed on the bacteria and ingested feed. These microorganisms break down the complex carbohydrates, such as cellulose and hemicellulose, by fermenting them to short-chain fatty acids through action of their enzymes. The fatty acids then are absorbed directly from the rumen and reticulum into the blood stream of the cow for use as energy sources or as carbon sources for the synthesis of many important compounds, including milk fat.

Similarly, protein in feed is broken down to peptides, amino acids, ammonia, and amines. The microorganisms use these substances as building blocks for their own cells. Eventually, the microorganisms are passed down the intestinal tract, digested, and used as a protein source by the cow. Therefore, no matter what source of protein you feed the cow, much of it is converted to bacterial and protozoal protein before actual utilization by the cow. It is for this reason that nonprotein nitrogen (NPN) can be utilized as a protein source by ruminants whereas it is practically useless for this purpose by simple-stomach animals which do not have extensive microbial protein synthesis.

Some of the feed which enters the reticulo-rumen resists the initial efforts of the microorganisms to ferment it to usable compounds. Through action of the ribs, diaphragm, rumen, and reticulum, this material is drawn back into the esophagus and formed into a bolus. Negative pressure draws it back to the mouth (Figure 8.1). The liquid it contains is squeezed out in the mouth and immediately reswallowed. The remaining bolus, called a cud, is then rechewed to make it more susceptible to fermenting action by the microorganisms when it returns to the reticulo-rumen. This series of actions, called rumination, is estimated to occupy as much as one-third of a cow's time.

The large storage capacity of the rumen and reticulum, about 50 gallons, and the phenomenon of rumination make it possible for a cow to consume large quantities of feed in a short period of time. She chews it only enough to allow it to pass down the esophagus. Later, at her leisure, she regurgitates any feed which needs additional chewing. This process of rumination may have been very important for survival of the species in the wild state when it was important to eat fast in exposed meadows and then get back quickly to safer surroundings.

End Products of Rumen Fermentation

Carbohydrate Fermentation. Carbohydrates, particularly cellulose and starch, make up the largest percentage of most dairy cattle rations. Both starch and cellulose are composed of chains of glucose (a 6-carbon sugar), but glucose units are linked differently in the 2 compounds. All animals have the enzymes necessary to hydrolyze (break apart) the glucose units of starch, and they can use the resulting glucose as an energy source. Animals do not produce enzymes that hydrolyze the linkages between glucose units of cellulose, but they can be hydrolyzed by cellulase, an enzyme produced by ruminal bacteria and microflora in the cecum. Therefore, ruminants can utilize cellulose and similar compounds as energy sources after fermentation by ruminal microorganisms whereas these substances are practically indigestible to simple-stomach animals such as pigs and humans.

Most of the carbohydrates in the diet are fermented by the ruminal microorganisms to volatile fatty acids (VFA) that contain 2, 3, or 4 carbon atoms. They are acetic acid (2 carbons), propionic acid (3 carbons), and butyric acid (4 carbons). These acids are found in the ionized form in the rumen.

Thus, they frequently are referred to as acetate, propionate, and butyrate. Other fatty acids, such as formic acid (1 carbon) and valeric acid (5 carbons), are produced in small quantities. However these and other miscellaneous fatty acids make up only a small portion of the total fatty acid content of the rumen.

Rations that contain high proportions of forages and other roughage-type feeds favor the production of acetate in the rumen. Ratios among the three major fatty acids when high forage rations are fed are approximately 50 to 65% acetate, 18 to 25% propionate, and 12 to 20% butyrate. Many factors can change these ratios.[2–5] Feeding of a ration high in concentrates, grinding and pelleting of forages, heating and pelleting of concentrates, and high levels of unsaturated fats all tend to reduce the proportion of acetate and increase propionate. Often the percentage of butyrate is changed also. When the percentage of propionate is increased relative to the other fatty acids, a depression in the percentage of milk fat often occurs, accompanied by an increase in body weight as a result of the deposition of fat in the cow.[6] This effect is beneficial for fattening cattle in a feedlot but is a disadvantage for dairymen under present marketing conditions where the value of milk is partially dependent on its milk fat content. Factors other than acetate:propionate ratios also affect fat test, and more research is necessary before a complete understanding of this phenomenon is possible. Various methods of preventing milk fat depression have been tested and are discussed in Chapter 11.

Protein End Products. Proteins that enter the rumen are digested in various ways. Some of the protein escapes fermentation altogether and passes on to the abomasum and intestines where it is digested to peptides and amino acids as in simple-stomach animals. However, much of the dietary protein is broken down by ruminal bacteria to peptides, amino acids, and ammonia. Various types of microorganisms use these compounds for synthesis of their own cell proteins. Some microorganisms can use only peptides or amino acids, others only ammonia. The proportion of various types of microorganisms present varies with the ration being fed. For this reason, a ration should be changed gradually in order to allow the various kinds of microorganisms to adjust their proportions to new feed sources. This adaptation is particularly important for efficient utilization of urea and other NPN compounds. It is estimated that as much as 3 weeks may elapse before sufficient adjustment occurs in ruminal microorganism proportions to make efficient use of urea after it is introduced into the diet.

Carbon chains from proteins that are deaminated can be used as energy sources through fermentation to volatile fatty acids by ruminal microorganisms. Ammonia from deamination is used by ruminal microorganisms for protein synthesis, with any excess ammonia converted to urea in the liver. Some of the urea is recycled to the rumen via the saliva and the remainder is excreted in the urine.

"Protected" protein (protein treated with formalin) passes through the rumen without being broken down by ruminal microorganisms. This protected protein increases the amount of dietary protein that bypasses ruminal fermentation and arrives at the abomasum as intact protein. The protective covering is dissolved in the abomasum by the higher acidity level there, releasing the protein for digestion and absorption as amino acids in the intestines. This technique may have practical application in the future because of the possibility of increasing the amount and quality of protein provided to the cow by a combination of microbial protein synthesis and direct absorption of "protected" proteins. This prospect may be particularly important to high-producing cows in early lactation when the protein requirement is the highest.

Lipid End Products. Much of the lipid (fat) which enters the rumen is hydrogenated by the microorganisms. Some lipids are metabolized and end up as structural lipids of the microorganisms themselves. When disintegration of these microorganisms occurs in the abomasum, this facilitates the subsequent digestion of their structural lipids in the small intestine. Hydrolysis of triglycerides to fatty acids and glycerol also is accomplished in the rumen. The glycerol is fermented principally to propionate but the long-chain fatty acids eventually find their way to the small intestine for digestion.

Although a small portion of fat probably is needed in the diet, excess fat interferes with normal ruminal function. High levels of fats, particularly unsaturated oils, are known to depress fat content of milk and reduce appetite. Two percent of the total ration dry matter in the form of crude fat is adequate for dairy cows, a level that is supplied by all normal rations.

As was the case for "protected" proteins, fats can be encased in proteins treated with formalin, allowing them to pass through the rumen without the usual hydrogenation and hydrolysis by ruminal microorganisms. After the protective covering has been dissolved by the acid in the abomasum, the fats are digested and absorbed in the intestines. This process results in the formation of polyunsaturated fat in the meat and milk products from cattle when "protected" polyunsaturated fats are fed. This technique has been used to protect more saturated fats, such as tallow, resulting in higher milk fat percentage and greater energetic efficiency from dairy cows fed the "protected" fat. Research in progress at several institutions may result in practical and economical application of these techniques to commercial dairy cattle feeding practices in the future.

Other Products of Ruminal Microorganisms. In addition to their effects on carbohydrates, proteins, and lipids, ruminal microorganisms are able to synthesize all of the known B-complex vitamins and vitamin K for use by the cow. Vitamin C is made within the cow's body tissues. Consequently, dietary sources of these vitamins are not necessary. All of the functions of the rumen also take place within the reticulum because the contents of one are in direct contact with the other.

As mentioned previously, microorganisms which have been feeding, growing, and multiplying on the feed the cow has eaten eventually are passed on to the lower tract where they in turn are digested and absorbed from the intestines. The cow has provided food and living quarters for the microorganisms which then have to pay the bill by becoming a source of nutrients for the cow. Without this symbiotic relationship, ruminants would not be able to exist on high-forage rations to which they are so well adapted.

Functions of the Omasum

The reticulo-ruminal contents pass through the omasum on their way to the abomasum and intestines. The function of the omasum is not well understood. It appears to act mainly in the squeezing out and absorption of water and fatty acids from stomach contents passing through it. Action of the leaves and their horny papillae results in some grinding action, and also pumps the contents into the abomasum, which is the true stomach of the cow. From here on, the digestive process is similar to that which takes place in simple-stomach animals.

Digestion in the Abomasum and Intestines

Once the contents reach the abomasum, they are acted upon by gastric juices present there. Hydrochloric acid and the enzymes, pepsin and rennin, are secreted from the walls of the abomasum. Hydrochloric acid activates pepsin which in turn breaks down proteins to peptides, which are short chains of amino acids. The enzyme rennin is particularly important to the young calf because it curdles milk within the abomasum, a process which is necessary for its further digestion and absorption in the intestines. Acidity within the abomasum produced by hydrochloric acid also is important in signaling the sphincter muscle of the pylorus between the abomasum and small intestine to relax allowing abomasal contents to pass into the intestines. Until this acidity level is reached, the contents are held within the abomasum for action by gastric enzymes.

After entering the small intestine, the contents, now known as chyme, undergo reactions with many more enzymes and other substances. Secretions from the pancreas through the pancreatic duct, from the liver via the bile duct, and from glands within the intestines enter the upper portion of the small intestine and mix with the chyme. Pancreatic juice contains the proteolytic enzymes trypsinogen, chymotrypsinogen, and carboxypeptidase, which, when activated in the intestines, convert proteins and partly hydrolyzed proteins into peptides and amino acids. Pancreatic juice also contains a lipase (steapsin), which hydrolyzes fats to fatty acids and glycerol, and an amylase (amylopsin), which hydrolyzes starch and dextrins to maltose, a sugar.

Intestinal glands secrete other enzymes which act on peptides, sugars, and fats to break them down further for subsequent absorption. Additionally bile, which is produced in the liver and stored in the gall bladder, performs several functions in the intestines. It activates pancreatic lipase, assists in emulsification of fats, increases the solubility of fatty acids for ultimate absorption, and acts as a reservoir of alkali for maintenance of optimum pH for reactions in the intestines.

Absorption from the Intestines

Secretion of intestinal juices and most of the digestive reactions take place in the upper portion of the small intestine whereas absorption of the end products of digestion takes place in the lower part. Amino acids and peptides from protein digestion and simple sugars, such as glucose, from carbohydrate digestion are absorbed directly into the blood stream for transport to various tissues in the body. These nutrients then are used for various body functions, such as tissue growth, milk production, and reproduction.

Fat absorption is more complicated. Fatty acids and other lipids combine with bile salts, making them more soluble. These combinations form micelles which penetrate the intestinal lining and enter the lymphatic system. The lymphatics are a series of vessels which drain into the venous blood system anterior to the heart through the thoracic duct. The bile salts eventually are returned to the intestines via the liver, and the fatty acids are recombined into neutral fat with glycerol from the intestinal mucosa. This fat then can be used as an energy source immediately or stored in adipose tissue for future use.

Contents in the large intestine continue to be acted upon by enzymes previously added in the small intestine. Some bacterial digestion also occurs, particularly putrefaction, resulting in the offensive odor of manure. No digestive enzymes are secreted by the large intestine, but absorption, particularly of water, does occur. This results in considerable drying of the contents before excretion. Many breakdown products of metabolic processes within the body are returned to the digestive tract through the walls of the large intestine. These along with undigested feed residues, digestive enzymes, sloughed cells from the digestive tract, and undigested residues of microorganisms are passed out of the body as feces through the rectum and anus.

Rate of Passage

The speed with which feed passes through the digestive tract is largely dependent on the type and amount of feed consumed. Concentrated rations do not require as much rumination and microbial action as coarser diets. Concentrates pass from the reticulo-rumen more rapidly, thereby speeding up rate of passage. Rate of passage is important in obtaining maximal efficiency from a cow because she cannot metabolize additional feed until that which is present moves out. In the case of readily soluble sugars and starches, it is more efficient energetically to bypass ruminal fermentation altogether and digest these carbohydrates to simple sugars in the abomasum for absorption in the intestines. Some carbohydrates follow this route under normal conditions, but the bulk are fermented to volatile fatty acids. Fats and some proteins also may be utilized more efficiently if they are "protected" by formalin treatment, thereby bypassing normal actions of ruminal microorganisms to be digested and absorbed in the abomasum and intestines. Current research may identify practical methods for increasing the proportion of nutrients metabolized in this manner, with resultant increased efficiency of milk production.

There are disadvantages to increased rate of passage also. Incomplete digestion of feed from too rapid a rate of passage increases fecal losses. Additionally, depressed milk fat test, increased incidence of digestive upsets, and displaced abomasums have been reported on diets that increase rate of passage very much beyond the normal rate.

8.4 Disorders of the Digestive Tract

The digestive tract of the cow uniquely equips her to make use of many feeds not utilizable by simple-stomach animals. However, things do go wrong at times leading to ailments and reduced function of the tract. Some of the more common of these are choke, bloat, hardware disease, lactic acidosis, and displaced abomasum. A brief discussion of these and several other ailments follows.

Choke

Occasionally feeds such as beets, potatoes, and hay cubes become lodged in the esophagus of cattle causing them to choke. Manual massage of the throat sometimes will dislodge the material, allowing it to pass on to the rumen. If massage does not work, a stomach tube or rubber hose lubricated with water or oil and pushed down the esophagus usually will free the object. Care must be taken not to damage the lining of the esophagus. It is difficult to heal and may lead to infections and pneumonia. Also the scar tissue may cause a permanent constriction of the esophagus, making it more difficult for food to pass through.

Bloat

Tremendous amounts of gases, principally carbon dioxide and methane, are produced within the rumen. Normally they are expelled by eructation (belching) or through excretory organs. When the gases become trapped within the rumen, a condition known as bloat occurs. Distention of the rumen and paralysis of the rumen walls are the first signs. These are followed by absorption of toxic gases, pressure on other organs, and, if not relieved, eventually death.

Several types of bloat have been reported and various theories as to their causes have been advanced. Excess gas production is not the problem. Rather it is the inability of the animal to get rid of the gas which causes bloat. Lack of scabrous material in the rumen to stimulate eructation seems to be the main cause of pasture bloat. Saliva flow also is reduced under these conditions.

Antifoaming agents, such as oils and poloxalene, have been successful in keeping the esophagus clear of foamy material so that eructation can occur.[7] In extreme cases, a trocar and cannula can be used to punch a hole from the outside into the rumen in order to relieve the pressure. The trocar and cannula should be inserted on the left side of the cow at the center of a triangle formed by the backbone, the hipbone, and the last rib. The trocar is removed but the cannula should stay in place until all gases have dissipated. A knife may be used in emergencies if the above equipment is not available. Treatment of the wound by a veterinarian after removal of the cannula or knife is recommended to prevent subsequent infections or peritonitis.

Treatments such as feeding antibiotics and spraying oil on feed have met with variable success. When cows are turned out onto lush pasture, particularly clover or alfalfa, they should first be fed some coarse hay or have an antifoaming agent included in their feed in order to prevent bloating. For cattle not receiving any feed in addition to lush pasture, blocks containing an antifoaming agent placed in various parts of the pasture have proven to be useful in reducing the incidence of bloat.[8]

Hardware Disease

Heavy foreign materials such as nails, wire, and other hardware that accidentally are swallowed along with feed drop directly into or move forward to the reticulum. These materials may remain in the reticulum for the life of the cow. Although the walls of the reticulum are thick and tough, sharp metal objects sometimes project into or through them. This condition, known as hardware disease, can result in indigestion, poor appetite, lowered milk production, peritonitis, and general discomfort of the cow in the chest region. If the foreign object punctures the heart, which is in close proximity to the reticulum, sudden death occurs.

Metal objects may be removed surgically through the rumen. In most cases, magnets placed in the reticulum through the esophagus attract metal objects and keep them from doing harm within the reticulum. Many dairymen routinely put magnets in their cows to reduce the risk, and most bull studs put magnets in all sires. However, the only sure prevention is removal of metal from the feed before ingestion by the cow.

Lactic Acidosis

If cattle consume large amounts of readily fermentable carbohydrates beyond the fermentative capacity of the microorganisms in the rumen (usually from overconsumption of concentrates by hungry dairy cattle), glucose accumulates in the rumen and can lead to rapid growth of lactic acid-producing bacteria. Lactic acid, a 3-carbon organic acid, is a normal intermediate in the fermentation of carbohydrates to volatile fatty acids. However, when more lactic acid accumulates than can be utilized by the ruminal microorganisms, ruminal acidity increases and many of the normal ruminal protozoa and bacteria are inhibited or killed. Lactobacilli, which are bacteria that tolerate low pH, become predominant in the rumen.

Cattle with lactic acidosis may stop ruminating and go off feed. Some spontaneously recover after a few days, but acute cases lead to death of the animal through hypotension and respiratory failure. Feeding buffers such as sodium bicarbonate, or a drenching with ruminal fluid that contains lactic acid utilizing bacteria (from cattle adapted to high-concentrate rations), reduces the incidence and severity of lactic acidosis when unadapted cattle suddenly consume large amounts of concentrates. However, whenever possible, it is best to make gradual changes in the amounts and types of roughages and concentrates in a ration to allow ruminal microorganism populations to adjust to new feed sources, thus avoiding problems such as the rapid buildup of lactic acid and subsequent lactic acidosis.

Displaced Abomasum

A malady which has been diagnosed with increasing frequency in recent years is displaced abomasum. Normally, the abomasum lies on the right side of the rumen. Sometimes, for reasons that are not entirely known, it slips under the rumen and becomes trapped in that position. The tremendous weight of the ruminal contents tends to reduce passage of food through the abomasum. When this happens, cows go off feed and exhibit many symptoms similar to ketosis, a metabolic disease discussed in Chapter 21. In fact, some secondary ketosis cases result from displaced abomasum because of voluntary decreased feed intake, thereby inducing a ketotic state. However, a well-trained veterinarian can differentiate between displaced abomasum and primary ketosis.

Several treatment methods for displaced abomasum have been used with variable success. Rolling a cow onto her back sometimes frees the abomasum but frequently the problem recurs. A simple surgical procedure has been developed to suture the bottom of the rumen wall to the body wall, thereby preventing the abomasum from slipping between the two. When sutured properly, this usually provides a permanent cure. The operation can be done using local anesthesia and cows thus treated frequently do not miss any milkings. Feed intake and milk production rapidly return to normal if the condition is treated in time.

While the cause of displaced abomasum is not definitely known, several theories have been advanced. Anything which reduces the volume of ruminal contents, thereby allowing a gap between the rumen and body floor, can result in displacement of the abomasum. Cows going off feed for any reason are susceptible. Rations containing high proportions of concentrates and inadequate bulk have been incriminated because of reduced ruminal volume resulting from these diets which allows the rumen to pull away from the body wall. Feeding high levels of corn silage also may be a cause, but cows fed rations with all of their roughage in the form of hay also have been affected. There seems to be a correlation between higher levels of concentrates fed to dairy cows in recent years and increased diagnoses of the malady, but this may be due to improved diagnostic procedures. It is possible that the problem was improperly diagnosed in previous years. A rule of thumb which may help when displaced abomasum is a problem is to provide at least half of the dry matter consumed by a cow in the form of roughage-type feeds, such as hay, silage, or pasture. Improved preventive measures may become available as more is learned about the problem through research and practical experience.

Other Digestive Problems

Various other problems of the digestive tract include indigestion, diarrhea, constipation, and impaction. Cattle get indigestion and go off feed if the composition of the ration is changed rapidly. Also, cows abruptly changed from dry forages to succulent green pasture often develop scours (diarrhea) unless some dry forages are provided in addition to pasture. Many other things can cause digestive upsets, including hardware disease.

Lack of bulk in a dairy ration is a common cause of digestive problems. Highly concentrated rations sometimes cause a cow to cease ruminating, greatly reducing the amount of fluid passing through the omasum. Under these conditions, omasal contents may dry and cake, making passage of feed through the digestive tract difficult. This condition, known as impaction, often occurs in sick cattle which have gone off feed.

It is important that accurate diagnosis and treatment of the above problems be made, or milk production and weight decrease rapidly. Usually a condition which causes loss of weight requires the services of a veterinarian.

8.5 Summary

The digestive tract of a dairy cow, with the 4 compartments of the ruminant stomach, particularly equips her for making more efficient use of fibrous feeds and poor quality plant protein than animals with simple stomachs. The task of the nutritionist and dairyman in formulating rations for cows as compared with simple-stomach animals is greatly simplified by the cow's ability, with the help of ruminal microorganisms, to synthesize all of the amino acids needed for the building of animal protein from plant proteins or from NPN, and to make all of the required B-vitamins and vitamin K. Utilizing feed of less value to humans and other animals by converting it into high-energy, high-protein milk and meat for human consumption makes the dairy cow a valuable asset in a world that needs to utilize all possible food sources. People who claim that we cannot afford the luxury of eating animal products when nutrients can be produced more efficiently and cheaply by plants forget the vast areas of land that are untillable and are covered with grass and other forages. Without ruminants to convert these forages into palatable and nutritious products, milk and meat, this source of energy, protein, minerals, and vitamins would be lost to the human race.

Review Questions

1. What are the four compartments of the ruminant stomach and the main functions of each?
2. Which cow would you expect to spend more time ruminating each day and why: (A) a cow grazing pasture, or (B) a cow being fed a ration of 70% concentrate and 30% hay in a dry lot?
3. Under what conditions would it be better for food to bypass the rumen and reticulum and pass directly from the esophagus into the omasum? How is this accomplished by a calf?
4. Why don't ruminants need B-complex vitamins or vitamin C and K in their feed?
5. Which are the principal fatty acids produced by fermentation of carbohydrate in the rumen, and what are their approximate percentages when normal rations are fed? What changes in the ratios of these fatty acids take place when high concentrate or pelleted rations are fed, and how does this affect the metabolism of the cow?
6. What causes bloat in cows, and what are some methods for preventing it?
7. Why is a magnet sometimes placed in the reticulum of a cow?
8. Why is the abomasum referred to as the true stomach?

References

1. Kay, R. N. B.: The rate of flow and composition of various salivary secretions in sheep and calves. J. Physiol. *150:* 515, 1960.
2. Balch, C. C., Balch, D. A., Bartlett, S., Bartrum, M. P., Johnson, V. M., Rowland, S. J., and Turner, J.: Studies of the secretion of milk of low fat content by cows on diets low in hay and high in concentrates. VI. The effect of the physical and biochemical process of the reticulo rumen. J. Dairy Res. *22:* 270, 1955.
3. Ronning, M.: Effect of varying alfalfa hay-concentrate ratios in a pelleted ration for dairy cows. J. Dairy Sci. *43:* 811, 1960.
4. Adams, H. P., Bohman, V. R., Lesperance, A. L., and Bryant, J. M.: Effects of different lipids in ration of lactating dairy cows on composition of milk. J. Dairy Sci. *52:* 169, 1969.

5. VanSoest, P. J.: Ruminant fat metabolism with particular reference to factors affecting low milk fat and feed efficiency. J. Dairy Sci. *46:* 204, 1963.
6. Ensor, W. L., Shaw, J. C., and Tellechea, H. F.: Special diets for the production of low fat milk and more efficient gains in body weight. J. Dairy Sci. *42:* 189, 1959.
7. Bartley, E. E., Lippke, H., Pfost, H. P., Nijweide, R. J., Jacobson, N. L., and Meyer, R. M.: Bloat in cattle. X. Efficacy of poloxalene in controlling alfalfa bloat in dairy steers and in lactating cows in commercial dairy herds. J. Dairy Sci. *48:* 1657, 1965.
8. Stiles, D. A., Bartley, E. E., Erhart, A. B., Meyer, R. M., and Boren, F. W.: Bloat in cattle. XIII. Efficacy of molasses-salt blocks containing poloxalene in control of alfalfa bloat. J. Dairy Sci. *50:* 1437, 1967.

Suggested Additional Reading

Dukes, H. H.: *The Physiology of Domestic Animals.* 7th ed. Ithaca, N. Y.: Comstock Publishing Associates, Cornell University Press, 1955.

Sisson, S., and Grossman, J. D.: *The Anatomy of Domestic Animals.* 3rd ed. Philadelphia: W. B. Saunders Co., 1938.

CHAPTER 9

Nutrient Requirements

9.1 Introduction

The dairy cow requires five major classes of nutrients—energy, protein, minerals, vitamins, and water. All five are essential for normal health and productive purposes, but some, such as the minerals and vitamins, are needed only in very small amounts. Next to water, the greatest requirement is for energy followed by the need for protein. Without adequate energy, utilization of all other nutrients is impaired. Overemphasis on mineral and vitamin supplementation often is practiced without adequate attention given to energy and protein needs. Because of this, energy and protein are the limiting factors for high milk production in most dairy herds.

9.2 Energy Requirements

The simplest definition of energy is "the ability to do work." Energy is expressed in terms of calories. A calorie is the amount of heat required to increase the temperature of 1 gram of water 1 degree Celsius. A kilocalorie (Kcal) is 1000 calories, and a megacalorie (Mcal) is 1 million calories.

A dairy cow uses energy for a variety of functions in her body. The pathway of energy flow is illustrated in Figure 9.1. A certain amount is used just to maintain her body tissues, which are constantly undergoing the many chemical reactions that sustain life. A growing heifer needs extra energy for the tissue that she is adding to her body during her growth from a calf to a mature cow. A pregnant cow needs additional energy for building the tissues of the fetus developing in her uterus. Finally, a lactating cow requires still more energy to manufacture the milk that is being secreted by her mammary glands each day. A nonpregnant, nonlactating mature cow needs only enough feed each day to provide sufficient energy for maintenance. However, a pregnant, lactating, first-calf heifer would need extra energy for growth, reproduction, and lactation in addition to that needed for maintenance.

When feed is restricted, a dairy cow will use the available energy for maintenance and reproduction at the expense of growth and lactation. Therefore, it is important to supply adequate energy if normal growth, high milk production, and more profits are to be obtained.

To illustrate this point, which is probably the most important single factor in the feeding of dairy cows for high milk production, the comparative energy requirements for 2 nonpregnant, mature, 1400-lb cows producing 40 and 80 lb of milk per day are shown graphically in Figure 9.2. Cow #1 needs 10.1 units of energy per day for maintenance, represented by the bottom segment of the bar in Figure 9.2. In order to produce 40 lb of milk per day, an additional 12.4 energy units are required, represented by the top segment of the bar, for

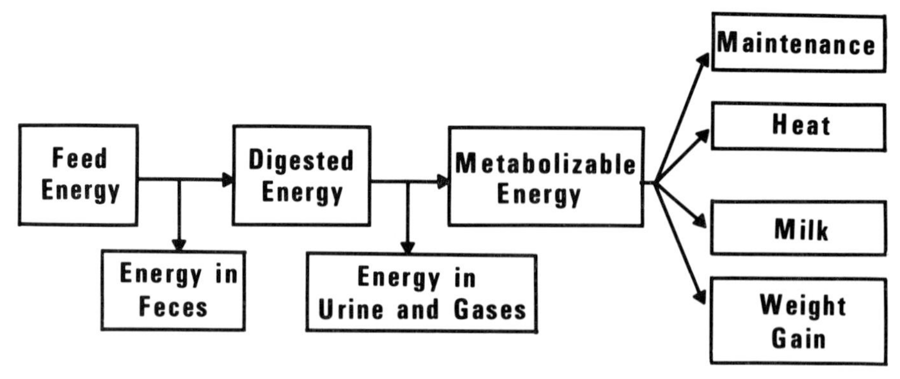

FIG. 9.1 Pathway of energy flow in a lactating cow.

a total of 22.5. Cow #2 requires the same amount of energy for maintenance, 10.1 units per day, because #1 and #2 are the same size. However, rather than only 12.4 energy units for 40 lb of milk, Cow #2 needs 24.8 per day for production of 80 lb of milk, bringing the total to 34.9 per day. If she does not consume 34.9 energy units per day, it is possible for her to continue to produce 80 lb of milk per day only so long as she utilizes energy stores, mainly body fat, to supplement that energy which she consumes. When body fat is depleted, she uses the available energy first for maintenance. Any energy left above the maintenance requirement can be used for milk production. If Cow #2 consumes only 22.5 energy units per day, her milk production soon will drop to 40 lb per day, the same as Cow #1. Drastic drops in milk production after two to three months of lactation often are the result of inadequate energy after body energy stores are depleted. As can be seen in Figure 9.2 an increase in energy intake of approximately 55% is a very profitable investment because it results in 100% more milk from cows with this milk-producing ability. Therefore, it behooves a dairyman to feed cows up to their full potential if he is to obtain maximum economic returns from his herd.

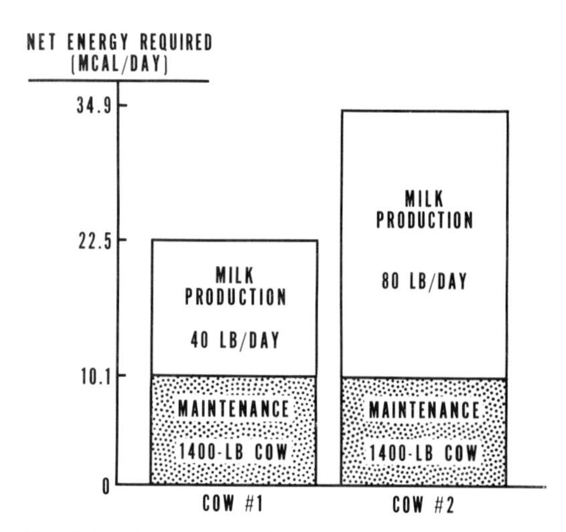

FIG. 9.2 Net energy utilization for maintenance and milk production.

Maintenance

An animal needs energy to maintain its body tissues in energy equilibrium. It must take in sufficient energy to compensate for that lost through basal metabolism and normal activity associated with its environment. Basal metabolism requires a constant amount of energy based upon the metabolic body size of the animal, as described by Kleiber.[1] As body size increases, more energy is needed for its maintenance. Energy expended for activity varies depending upon the temperament of the cow and whether she is grazing pasture or in confined housing. Maintenance requirements in this book apply to dairy cows with minimal activity, such as would be experienced when cows are housed in free-stalls or conventional stanchion barns. To support grazing, allowances for maintenance should be increased by 10% for good pasture and up to as much as 20% for sparse pastures.[2] For cows in large dry lot dairies, allowances should be increased 5% for each mile that they walk to and from the milking parlor. Maintenance energy allowances for cows of various body weights are listed in Appendix Table III-B.

Growth

Young animals which have not reached their mature size require energy above that allowed for maintenance in order to continue growing at a normal rate. A dairy cow usually reaches her mature size at about 5 years of age. Energy allowances for maintenance and growth of calves, heifers, and bulls are listed in Appendix Table III-A. For lactating heifers which have not reached their mature size, add 20% to the mature cow maintenance energy allowance during the first lactation and 10% during the second lactation.

Reproduction

Under ideal conditions, a dairy cow produces a calf once each year, although few dairies actually are able to obtain a calving interval this short. Energy requirements for the developing fetus are very small during the first 6 months of gestation but increase sharply during the last 3 months. The energy allowances for maintenance plus last 2 months' gestation, which are listed in Appendix Table III-B, should be used to cover the increased energy needs of the pregnant cow during this period.

Lactation

Energy for milk production is needed in addition to that required for maintenance, growth, and reproduction. Although other milk components have some effect, energy requirements vary primarily with the amount of milk produced and the fat content of the milk. Energy allowances for production of milk at various fat percentages are listed in Appendix Table III-B.

9.3 Measures Used for Energy Allowances

Energy allowances can be expressed in many different ways. Each system has its advantages and disadvantages. Those in current use include total digestible nutrients (TDN), digestible energy (DE), metabolizable energy (ME), and net energy (NE). Accuracy of the methods in formulating rations generally increases in the order listed. However, difficulty in determining energy values of feeds according to the different systems of measurement also increases in the same order. TDN has been the most extensively used system in the United States in the past. Drawbacks in this measure, as discussed later in this section, and recent breakthroughs in concepts and technology have resulted in the net energy system being adopted as the official standard of the National Research Council (NRC) for ruminant energy allowances.[3] A brief description of each system follows, as all are used to some extent in various parts of the world. Examples in this book will be confined to the use of net energy, however.

Net Energy (NE)

Net energy is the most difficult energy evaluation to make on feeds but is the most accurate for ration formulation. In order to determine net energy, measurements must be made of the energy in the feed, feces, gases, urine, and heat produced by the cow. The formula for determining NE is:

NE = Gross energy − fecal energy − gaseous energy − urinary energy − heat increment

A graphic breakdown of what happens to the energy in a feed is shown in Figure 9.3. The feed which an animal consumes has a specific amount of total, or gross, energy. Not all of this energy is usable by the animal, however. The chart shows where the losses occur. The first and largest loss is the energy contained in the feces. The remainder is called digestible energy. The next losses are gases which escape from the animal and energy lost in the urine, leaving metabolizable energy. The last loss to the animal is the heat given off as a result of the metabolism of nutrients from the ingested

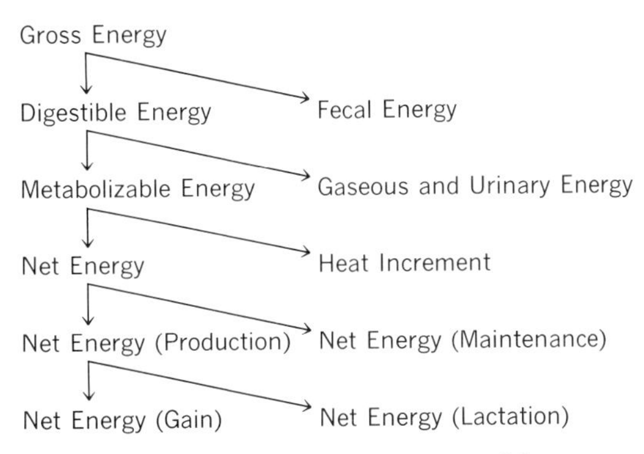

Fig. 9.3 Utilization of energy consumed by a cow.

feed plus the heat from microbial fermentation. It is called heat increment or, in older literature, the specific dynamic action of a feed. The remaining energy is the net energy of the feed. Part of this net energy is used for maintenance of the animal while that which is left over is available for productive purposes, such as growth and milk production.

Approximate percentages of the various energy components for a lactating cow being fed free choice are shown in Figure 9.4. It should be noted that only about 20% of the energy consumed by the cow actually is available for production of milk or body weight gain. Another 20% is used for maintenance whereas about 30% is lost in the feces and 20% as heat increment. Urinary and gaseous energy losses are quite small, making up approximately 5% each on common rations.

Dairy cows use net energy with approximately equal efficiency for maintenance or milk production.[4] Efficiency for growth and fattening is somewhat lower, however. In order to compensate for this, current net energy nomenclature recommended by NRC is as follows:

Net Energy Nomenclature

NE_m = Net energy for maintenance
NE_g = Net energy for gain
NE_l = Net energy for lactating cows

Net energy for maintenance (NE_m) is the fraction of net energy expended to keep the animal in energy equilibrium. In this state the animal's tissues are neither gaining nor losing energy.

Net energy for gain (NE_g) is the net energy required in addition to that needed for body maintenance and is used by growing cattle for body tissue gain.

Net energy for lactating cows (NE_l) is the total net energy needed for maintenance and milk production during lactation and for maintenance plus the last 2 months' gestation for dry, pregnant cows. NE_m and NE_g are used only for growth of youngstock.

Total Digestible Nutrients (TDN)

TDN has been the most extensively used measure for energy allowances in the United States. The formula for calculating TDN is as follows:

$$\% \text{ TDN} = \frac{CP + CF + (EE \times 2.25) + NFE}{\text{feed consumed}} \times 100$$

where CP = digestible crude protein; CF = digestible crude fiber; EE = digestible ether extract; and NFE = digestible nitrogen-free extract.

TDN values approximate digestible energy but the method has several drawbacks. First, it is an empirical formula based upon chemical determinations not related to actual metabolism of the animal. Second, the result is expressed as a percent or in some measure of weight (lb or kg) whereas energy is expressed in calories. Third, TDN takes into consideration only digestive and urinary losses. Urinary losses are approximated by giving protein the same weighting in the TDN formula as crude fiber and nitrogen-free extract, whereas protein actually has about 20% more gross energy per unit weight. TDN ignores gaseous energy and losses due to increased heat production (heat increment). The size of the latter approaches that of digestive losses in ruminants. Fourth, and most important, TDN

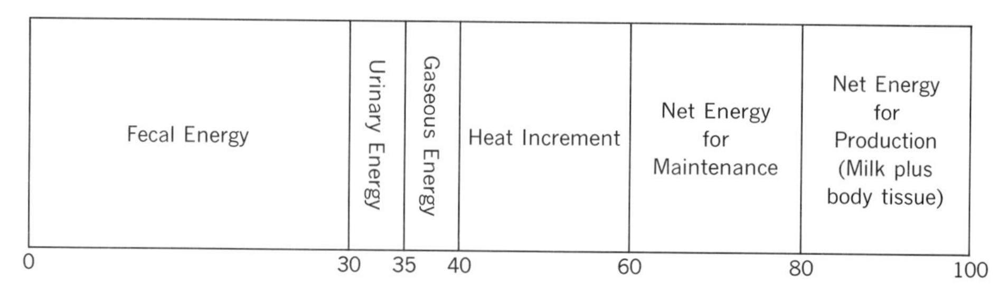

Fig. 9.4 Approximate percentages of gross energy represented by fecal, gaseous, urinary, heat increment, and net energy in a lactating cow. (Courtesy of W. P. Flatt)

overevaluates roughages in relation to concentrates. This factor is not important when comparing ingredients within a class, such as concentrates with concentrates, but is of major importance when comparing roughages with grains and other concentrate ingredients, as is done when formulating a practical and economical ration for a cow.

Because of these limitations, TDN gradually is being replaced in the United States by other energy evaluation systems, particularly net energy. However, voluminous TDN data on many feeds and long-standing tradition insure its continued use by many people for years to come.

Digestible Energy (DE)

Determination of digestible energy is similar to calculation of TDN in that all feed consumed and feces excreted must be weighed, and representative samples of each must be analyzed. However, only dry matter and combustible energy need be determined. The formula for calculating DE is:

$$\text{DE} = \text{energy consumed} - \text{fecal energy}$$

Digestible energy does not have the first 2 drawbacks listed for TDN (i.e., empirical determination and not being expressed in energy terms), but the other criticisms are still valid (i.e., only digestive losses are considered and roughages are overevaluated compared with concentrates). Consequently, it does not offer a significant improvement over TDN as an energy measurement.

Metabolizable Energy (ME)

Determination of metabolizable energy involves the subtraction from digestible energy of gaseous energy and energy lost in urine. It is considerably more tedious and requires more elaborate equipment than determination of digestible energy. Besides collection of feces, all urine must be collected, and measurements of methane excretion must be made. The formula for determining ME is:

$$\text{ME} = \text{energy consumed} - \text{fecal energy} - \text{gaseous energy} - \text{urinary energy}$$

Gaseous and urinary energy represent only a small proportion of the gross energy consumed by ruminants, being about 5 to 7% each in most rations. It is somewhat doubtful if the increase in precision of ME values over DE values justifies the greatly increased time and effort required to determine them. Heat increment is ignored though it is the largest and most variable energy loss, next to fecal losses. Therefore, ME also overevaluates roughages compared with concentrates, as is true for TDN and DE.

A variation of the ME system currently is in use in the United Kingdom. The British system[5] takes into consideration the effect of variation in efficiency of use of ME for various levels of milk production. The amount of ME required is calculated from body size, level of milk production, and partial efficiencies of use of ME of the ration for maintenance and for milk production. Results using this system are very similar to those obtained using the NE system previously described. Application of the British ME system to actual farm situations is more difficult than application of NE because of extra calculations and use of partial efficiencies of ME utilization for maintenance and various levels of milk production.

9.4 Protein Requirements

In order to be utilized by an animal, protein first must be broken down by digestion to the amino acids from which it was made. Proteins from various sources have different combinations of amino acids. The generalized chemical structure of an amino acid is:

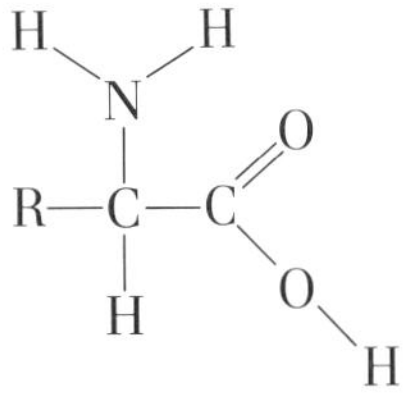

$$\text{R}-\underset{\text{H}}{\overset{\text{NH}_2}{\text{C}}}-\text{C}\overset{\text{O}}{\underset{\text{OH}}{}}$$

where R represents about 20 different carbon chains with varying lengths, configurations, and combinations with other chemical elements. Some amino acids can be made within the body from the other compounds, but others can be made only by plants and bacteria. The latter are called essential amino acids, and they must be present in adequate amounts in the ration of simple-stomach animals. With deficiencies of essential amino acids in the diet, humans, pigs, and chickens, for example, will not grow at a normal rate even though there may be an excess of other amino acids available. Protein quality depends on the proper balance of all essential amino acids and is very important in the feeding of an animal with a simple stomach.

Fortunately, a ruminant animal like the cow has billions of microorganisms in its rumen which are capable of synthesizing proteins for their cells from amino acids and nonprotein nitrogen derived from the cow's diet, as described in Chapter 8. These microbial proteins subsequently are digested and absorbed by the cow, giving her a source of all essential amino acids even though her diet may not have contained all of them in adequate amounts. Therefore, protein quality is not of as much importance to the cow as it is for simple-stomach animals. Cows eating purified diets with all of the nitrogen in the form of urea and ammonium salts not only have survived and maintained themselves, but have produced as much as 10,000 lb of milk in a year and reproduced normally.[6] This level of milk production is not high by today's standards. However, these experiments have demonstrated that ruminants can live without any preformed protein in their ration so long as nitrogen is available to ruminal microorganisms for protein synthesis.

A simplified diagram of the utilization of protein and nonprotein nitrogen by ruminants is shown in Figure 9.5. The nonprotein nitrogen in the diet is converted to ammonia (NH_3) by microorganisms in the rumen. The soluble portion of the true protein and part of the insoluble protein is degraded to NH_3 by ruminal microorganisms. This portion of the true dietary protein is called degradable protein. The remaining insoluble protein is not acted upon by ruminal microorganisms and passes through the rumen without being degraded to NH_3. This undegradable protein also is called "by-pass" or "escape" protein by some.

The NH_3 from the nonprotein nitrogen fraction plus the soluble protein fraction and the degradable insoluble fraction can be used by ruminal microorganisms to synthesize their own cell proteins, as de-

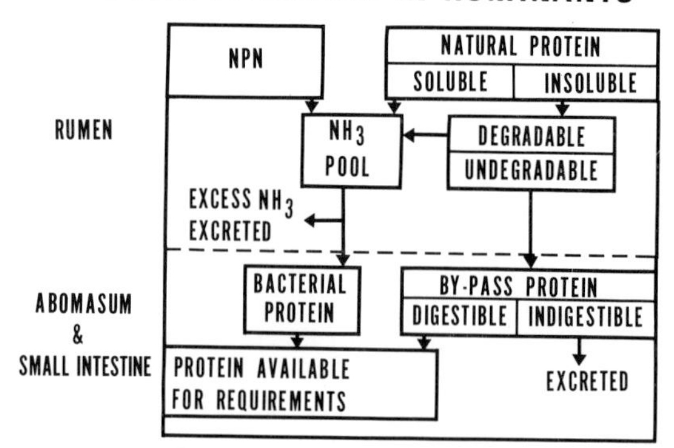

FIG. 9.5 A simplified diagram of protein digestion.

scribed in Chapter 8. Excess NH_3 not utilized by the microorganisms is converted to urea. Part is excreted in the urine but some is transported via the blood and saliva and recycled through the rumen for possible utilization by ruminal microorganisms.

Part of the undegradable "by-pass" protein is digested to amino acids and peptides in the abomasum and small intestine. This digestible fraction and the digestible portion of the microbial protein provide the total protein available to the cow for her metabolic processes. The indigestible portions of the undegradable and microbial protein pass through the digestive tract and are excreted in the manure.

The proportions of the protein that exist as soluble, insoluble, degradable, undegradable, digestible and indigestible protein vary widely among various feedstuffs. Extensive research is being conducted to develop information on the protein characteristics of various feedstuffs and to determine optimum proportions of the various nitrogen fractions in dairy cattle rations.

Maintenance

Protein is needed by the body for building and repair of tissue as part of normal metabolic functions. Besides muscle tissue, such compounds as enzymes and some hormones are proteins which regulate body functions and need to be replaced continuously due to normal degeneration. In order for an animal to remain in protein balance, protein lost from the body in the process of digestion and metabolism of food also must be replaced from a dietary source. The amount lost depends on animal size, type and amount of ration fed, and protein quality.

Protein allowances for maintenance of mature dairy animals of various sizes are listed in Appendix Table III-B.

Growth

In addition to protein for maintenance, immature animals need extra protein for normal growth. Most of the weight gain of young animals is in the form of protein and water in growing tissues and organs. As an animal approaches mature size, weight gain results increasingly from the deposition of fat rather than from increased body protein. Although the protein requirement for growth constantly decreases as an animal approaches maturity, additional allowances above maintenance usually are recommended through a cow's second lactation.

For ease of tabulation and use, allowances of protein for maintenance and growth of youngstock are combined in Appendix Table III-A. Allowances for several rates of growth are listed for each body weight.

Reproduction

As was the case for energy, extra protein is needed by a pregnant cow for the developing fetus, especially during the last 2 months of gestation. It is during this time that the bulk of fetal growth occurs. Much of this tissue growth is in the form of protein, which increases the cow's need for this nutrient. Protein allowances for maintenance plus last 2 months' gestation, which are listed in Appendix Table III-B, should be used to fulfill protein requirements of the pregnant cow during this period.

Lactation

Milk is a rich source of high-quality protein. A deficiency of protein results in lowered milk production and may depress the protein content of milk.[2] When energy is adequate and protein is slightly deficient, cows reduce milk production and tend to become fat. This situation can present a problem when feeding large amounts of a high-energy, low-protein feed such as corn silage unless it is properly supplemented with additional protein or nonprotein nitrogen. Severely protein-deficient cows lose weight rapidly in early lactation and do not regain it normally in late lactation. Excess protein within reasonable ranges is not toxic and can be fed to cows without danger. However, protein supplements usually are expensive and result in higher-cost rations when excess amounts are fed. Higher than normal protein content may be desirable for exceptionally high-producing cows, but very little controlled research is available based upon cows of this caliber.

The amount of milk protein usually varies with the fat content of milk. Therefore, Appendix Table III-B contains protein allowances for milk production at various milk fat percentages.

9.5 Measures Used for Protein Allowances

Crude Protein

Crude protein is calculated by multiplying the nitrogen content of a feed by 6.25. This figure is derived from the fact that most proteins contain about 16% nitrogen. If one divides 100 by 16, the quotient is 6.25. Therefore, if the nitrogen content is multiplied by 6.25, the result will approximate the protein content of a feed.

Not all of the nitrogen in feeds is in the form of protein. Some feeds, particularly green roughages, contain one-third or more of their nitrogen as nonprotein nitrogenous substances such as amides, ammonium salts, amino acids, alkaloids, and other nitrogenous compounds. Because of this nonprotein nitrogen, multiplication of the nitrogen content by 6.25 does not give a value for true protein. Crude protein represents a combination of true protein and nonprotein nitrogen.

As discussed previously, ruminal microorganisms make use of various nitrogen sources for synthesis of microbial proteins which in turn are used by the cow. Consequently, for dairy cows and other ruminant animals, crude protein is as good a measure for protein allowances as is true protein.

Digestible Protein

Digestible protein is the amount of crude protein consumed minus the crude protein excreted in the feces. Calculation of digestible protein is by the following formula:

$$\% \text{ digestible protein} = \frac{\text{crude protein consumed} - \text{fecal crude protein}}{\text{total feed consumed}} \times 100$$

In actuality, the above formula results in apparent digestible protein rather than true digestible protein. Part of the nitrogen in feces is in the form of metabolic fecal nitrogen which is independent of the protein consumed. This arises from metabolic functions in the body such as residues of bile and other digestive juices, epithelial cells from the alimentary tract, and undigested bacterial residues. It is very difficult to separate metabolic fecal nitrogen from truly undigested feed nitrogen, and since both represent losses to the animal, digestible protein has come to mean apparent digestible protein in practical usage.

Digestible protein has been the measure most commonly used in the past for listing protein allowances for dairy cattle. However, digestible protein requirements and feed values no longer are included in the NRC bulletin, "Nutrient Requirements of Dairy Cattle," because recent research has shown that apparent digestibility of feed proteins is directly related to percentage of total crude protein in the ration.[2] Metabolic fecal nitrogen, being independent of the amount of protein consumed, contributes a larger proportion of the total fecal nitrogen excreted when low-protein rations are fed compared with high-protein diets. Thus digestibility coefficients are reduced further for low-protein feeds than for high-protein feeds, giving an unfair advantage to the latter. Therefore, crude protein is a more accurate measure when calculating protein allowances in typical dairy rations composed of both roughages and concentrates. Furthermore, crude protein is much simpler to determine on feedstuffs, thereby making it easier to use under practical conditions.

9.6 Mineral Elements

There are at least 15 mineral elements that are known to be required by dairy cattle. There may be others that are needed in very small traces, but proof of their essentiality is difficult because most feeds contain them in more than adequate amounts. Recommended mineral levels for dairy rations are contained in Appendix Table III-C.

Mineral elements which are known to be required usually are grouped into 2 main categories, according to amounts required by the animal. Major minerals are calcium, phosphorus, sodium, chlorine, magnesium, potassium, and sulfur. The second category, called trace minerals, includes iron, copper, molybdenum, manganese, zinc, cobalt, iodine, and selenium. Other minerals known to be essential for some species, and which may be essential for cattle, include fluorine, chromium, silicon, vanadium, nickel and tin.

Under normal conditions, the minerals which are most likely to be needed by the cow in greater amounts than provided by common rations are calcium, phosphorus, sodium, and chlorine. When legume roughages, such as alfalfa or clover, make up a major part of the ration, extra calcium is not needed because of its high content in these feeds. In areas where soil fertility has been depleted by constant cropping or poor management, other minerals may be deficient in crops grown on such land. Complete dependence on nonlegume roughage, such as all-corn silage feeding, also necessitates supplementation with a greater number of minerals. In general, legumes are much better sources of most mineral elements than nonlegumes. Therefore, when legumes are extensively fed, mineral supplementation usually can be restricted to phosphorus and salt. When legumes are not fed, supplementation with a greater number of minerals is necessary in order to maintain the health and milk-producing ability of a herd of cows. Forage testing, as described in Chapter 10, can be helpful in determining which minerals should be added to the ration.

Major Minerals

Calcium and Phosphorus. By far the most important minerals for the dairy cow are calcium and phosphorus. They are discussed together because of the interactions between them and because the supply of one affects the utilization of the other. The amount of vitamin D in the ration also affects the utilization of both elements, as described later in this chapter.

Calcium and phosphorus are the major constituents of bone and teeth and also are contained in large amounts in milk. Both can be withdrawn from bones during periods of increased need, such as during early lactation when milk production is at its peak and large amounts of calcium and phosphorus are leaving the body in the milk. Continued withdrawal without replacement from dietary sources during periods of lower milk production will result in weakened bones and teeth. Calcium content of milk remains stable even under conditions of severe deficiency, although milk production is decreased. Rations low in calcium also result in reduced growth and development of calves.

Phosphorus is involved in energy metabolism and many other metabolic functions in the body in addition to its function as a major constituent of bone. Consequently, phosphorus deficiency can lead to impaired energy utilization, reduced breeding efficiency, decreased or depraved appetite (chewing of wood, bones, or hair) and stiff joints as well as fragile, easily fractured bones and reduced milk production.

Calcium and phosphorus allowances for maintenance, growth, reproduction, and lactation are contained in Appendix Tables III-A and III-B along with other nutrient allowances.

The ratio of calcium to phosphorus in the total ration is important, as well as the amounts of each. Although many dairy cows receiving alfalfa or other legumes as their sole forage have had 3 to 6 times as much calcium in their rations and have performed well, a ratio of 2 parts calcium to 1 part phosphorus is recommended by some nutritionists. This ratio is the approximate ratio between calcium and phosphorus in bone, which contains the bulk of both elements in the body. Adherence to this ratio in the ration seems to be of greater importance for nonruminant animals than for ruminants. Excesses of either element tend to make the other less digestible because of complexes formed between them in the digestive tract.

Calcium intake affects the incidence of parturient paresis (milk fever), which is caused by a drop in blood serum calcium at, or near, calving. High calcium levels in the dry cow ration aggravate the problem. Feeding a low-calcium diet (less than 0.1 lb per day) before calving followed by normal or high levels after calving has shown promise for preventing milk fever in some experiments. Some limited evidence also has suggested that the calcium to phosphorus ratio in the dry cow ration may be important in preventing the problem, but a low calcium intake during the latter part of the dry period probably is more important.

Sodium and Chlorine. Sodium chloride, or plain table salt, is needed in greater amounts than that provided by most rations. The actual requirement is for sodium because all normal rations provide more than enough chlorine. Lack of appetite, a craving for salt, lusterless eyes, rough hair coat, and a haggard appearance characterize a severe salt deficiency. Loss of weight, a decline in milk production, and sudden death may occur in lactating cows. Continuous shivering and a wavering walk often occur in advanced stages of salt deficiency.

Lactating cows need 20 to 25 g of salt per day for maintenance per 1,000 lb body weight plus 0.8 g per lb of milk produced.[7] Commonly it is provided free choice for dairy animals in the form of loose or block salt. In addition, most concentrate mixtures contain from 0.5 to 1% salt in order to enhance palatability. For complete rations, a level of 0.18% sodium from all sources (equivalent to 0.46% sodium chloride) is adequate for all levels of milk production when energy requirements are fulfilled.[2]

Magnesium. "Grass tetany" is a deficiency disease involving magnesium that occurs in cows grazing some pastures. Recent outbreaks among cattle grazing foothill pastures in California occurred during the rainy and foggy winter months. Supplementation with magnesium and many other feeds have controlled outbreaks, indicating that other factors are involved as well as magnesium.

Grass tetany can be a major problem for lactating cattle grazing lush pastures in cool weather, especially if it has been fertilized heavily with nitrogen or potassium. Under these conditions, it is wise to supplement magnesium up to 0.25% of the ration dry matter. When stored feeds such as hay, silage, and concentrate make up the bulk of the ration, magnesium levels of 0.20% of the dry matter are adequate. Most common rations fulfill this requirement, especially if legume forages comprise a significant portion of the ration. However, magnesium levels in some feeds, including corn and corn silage, vary tremendously. Rations based on these feeds may be deficient. Forage testing, as described in Chapter 10, is very useful in determining when, and how much, magnesium and other minerals should be added to a ration.

Potassium. Potassium is required for normal muscle irritability and is contained in high amounts in milk. However, forages generally are high in potassium. In fact, young lush forages grown in highly fertilized soils in cool weather may be so high in potassium as to interfere with magnesium utilization, thus aggravating a possible "grass tetany" problem. Most common dairy rations contain more than the 0.8% potassium required in the ration dry matter by dairy cattle. Therefore, supplementation with this element normally is not practiced.

Sulfur. Although sulfur is used by animals primarily in organic form as amino acids, ruminal microorganisms are able to use inorganic sulfur to synthesize sulfur-containing amino acids from nonprotein nitrogen. Sulfur supplementation is advisable when nonprotein nitrogen, such as urea, constitutes a large proportion of the crude protein equivalent of a ration because high-protein feeds replaced by urea are the usual sources of sulfur in dairy rations. A nitrogen to sulfur ratio of about 12:1 in the total ration is necessary to maintain maximum feed intake and efficient utilization of urea.

When corn silage with added urea or other nonprotein nitrogen is the major forage in a ration, the need for sulfur supplementation is likely because corn silage often is lower than the 0.20% sulfur requirement estimated for lactating cows. The silage should be tested to determine the proper sulfur supplementation rate. However, addition of inorganic sulfur to rations composed entirely of natural feeds has not proven beneficial, and excess sulfur may adversely affect selenium and copper utilization. Therefore, sulfur supplementation is not recommended unless high levels of nonprotein nitrogen are added to the ration.

Trace Minerals

Iron. Iron is essential to every tissue in the body due to its presence in hemoglobin, the oxygen carrier in blood. It is also a part of several enzymes primarily involved with oxidation reactions in energy metabolism. Thus, although present in very small amounts, it is a very important element in all life processes.

Iron deficiency is not a problem with cattle being fed normal rations because most feeds contain more than the 50 parts per million (ppm) required by dairy cattle in their ration dry matter. Calves up to 3 months of age may need up to 100 ppm. Iron supplementation should be avoided unless actual deficiencies are known. Excess iron combines with phosphorus making it unavailable to the cow and may cause a phosphorus deficiency on an otherwise adequate ration.

Copper. Although not included in hemoglobin, both copper and iron are required for its formation. A deficiency of copper results in decreased absorption of iron, a lower iron content in the body, decreased mobilization from tissues, and a severe anemia. Copper also is a part, or an activator of, several enzymes in the body. Copper is needed in even smaller amounts than iron, but it too is intimately connected with all life processes.

Isolated reports of copper deficiency have come from northern Europe, South Africa, New Zealand, Scotland, and the United States. Some of these problems probably have been complicated by low phosphorus or excess molybdenum. High levels of molybdenum can cause a conditioned copper deficiency, as is discussed in the section of this chapter for that element.

The most common symptom of copper deficiency in cattle is a bleaching of the hair: black hair turns gray, and red hair becomes yellowish.[8] Other symptoms which sometimes occur are anemia, scouring, lowered blood copper content, lameness, swelling of joints, and enzootic ataxia, commonly known as "sway back."

The requirement for cattle is very low. Only 5 to 8 ppm in dry herbage has resulted in healthy animals. A practical minimum requirement suggested by NRC is 10 ppm in the ration dry matter.[2] Most common feeds contain more than this amount. Where copper deficiency or molybdenum toxicity is a problem, supplements containing copper sulfate have been used successfully in counteracting the deficiency. Salt licks with 0.25 to 0.5% copper sulfate or feeding up to 1 g per cow per day of copper sulfate has worked well in the field.

While copper supplementation may be advisable under certain conditions, it should not be practiced indiscriminately. Excess copper is toxic and is a primary cause of oxidized flavor in milk.[9] Therefore, copper should be added to dairy cattle rations only if there is definite evidence indicating copper deficiency or molybdenum toxicity.

Molybdenum. This element is known more for its toxic characteristics, but recently it was discovered to be part of an enzyme found in liver, intestinal tissue, and milk. Its extremely low requirement in animal rations had obscured its essentiality until its biochemical function became known. So far as is known, all common rations provide more than is needed by ruminants.

As mentioned previously in the section on copper, there is an antagonistic relationship between copper and molybdenum. Consumption of forage containing 20 ppm or more of molybdenum results in molybdenosis, which is an induced copper deficiency. Levels as low as 6 ppm of the total ration dry matter may cause some symptoms of molybdenosis, especially if copper levels are low. Although the exact mechanisms are not known, excessive molybdenum intake is known to alter enzyme systems that affect copper utilization. This effect in turn results in the deficiency symptoms described earlier for copper even though levels in the ration may be normal. Under these conditions, supplementation with copper sulfate alleviates the situation. Areas of the United States known to be high in molybdenum are shown in Figure 9.6.

Manganese. A deficiency of manganese results in impaired growth, skeletal abnormalities, reduced fertility, and birth of abnormal calves. Although rations with 20 ppm manganese have been adequate under many conditions, NRC recommends a minimum level of 40 ppm in the ration dry matter to provide a safety factor against the possibility of a borderline deficiency.[2] Corn is very low in manganese but most other feed ingredients, especially forages, contain more than adequate amounts. Except in the case of rations that are predominantly corn, manganese supplementation for cattle probably is not necessary.

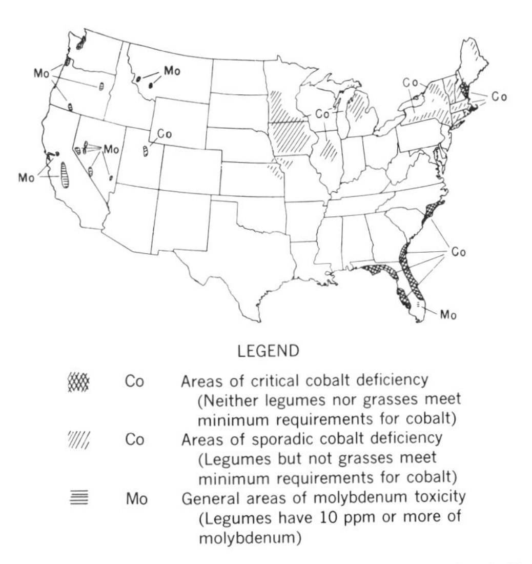

FIG. **9.6** Areas of the U.S. where forages may be deficient in cobalt for ruminants and areas of potential molybdenosis. (Courtesy of J. Kubota, USDA)

Zinc. Zinc is involved in several enzyme systems. Excess calcium in the diet has an antagonistic effect on zinc, probably at the enzyme level. The NRC recommended level of 40 ppm in the ration dry matter provides a safety factor against the possibility of a borderline deficiency.[2] Most common feeds do not contain that amount. However, water that travels through galvanized pipes or is provided in galvanized buckets picks up zinc. Therefore, the combination of zinc provided by feedstuffs and water makes it unlikely that a zinc deficiency would occur under normal conditions. Moderate excesses of zinc are not toxic so zinc supplementation does not produce adverse results if a safety factor is desired.

Cobalt. This element is essential as a part of vitamin B_{12}. Ruminal microorganisms are able to synthesize this vitamin in the presence of adequate cobalt in the diet. On the other hand, most nonruminant animals require preformed vitamin B_{12} because microbial synthesis of this vitamin in their digestive tracts is inadequate. Therefore, cobalt supplementation is of no use to them.

Cobalt-deficient areas occur throughout the world. Areas which have been identified in the United States are shown in Figure 9.6. Deficient areas also have been identified in western Canada and Australia.

Animals with cobalt deficiency are listless, have poor appetites and lose weight, become weak and anemic, and eventually die. A cow suffering from cobalt deficiency and the same cow after cobalt supplementation is shown in Figure 9.7. Early symptoms often are confused with those resulting from phosphorus deficiency. A practical method for differentiating between them is as follows:

If calves under one year of age are healthy and the cows have poor appetites, feed the herd a phosphorus supplement. If the calves are unthrifty and have poor appetites, feed a cobalt supplement. If appetite does not recover within a week, add a phosphorus supplement.[10]

Fig. 9.7 Cow suffering from cobalt deficiency *(top)* and the same cow after receiving cobalt *(bottom)*. (Courtesy of C. F. Huffman, et al., Mich. Agr. Expt. Sta.)

The cobalt requirement for cattle has not been accurately established. The NRC recommended level of 0.1 ppm of the ration dry matter will prevent any deficiencies.[2] Supplements of 30 to 45 g of cobalt sulfate or 20 to 25 g of cobalt carbonate with 100 lb of salt have prevented the problem. Most commercial concentrate mixtures marketed in deficiency areas of the United States now contain cobalt supplements.

Iodine. Much of the small amount of iodine in the body is contained in the thyroid gland as thyroxine and diiodotyrosine. Both amino acids are contained in the protein thyroglobulin, a form of thyroid hormone. Regulation of the metabolic rate of animals is the principal function of the hormone secreted by the thyroid gland.

Lack of iodine in the ration of pregnant cows results in goiter, an enlargement of the thryoid gland, in newborn calves. Necks of the calves are swollen, and they are weak at birth or are born dead.

Iodine-deficient areas have been identified in the Pacific Northwest and the Great Lakes region of the United States. There probably are other borderline areas where occasional deficiencies occur during periods of increased physiological need, such as during gestation or at the onset of puberty. Feeding of iodized salt is recommended in areas of known deficiency. Exact minimum requirements for iodine are not known but inclusion of 0.01% iodine in salt fed as 1% of the concentrate mix is effective in deficient areas. NRC recommends a level of 0.5 ppm iodine in the ration dry matter for lactating cows and cows in late gestation, and 0.25 ppm for other nonlactating dairy cattle.[2] These levels should be increased to 1 ppm for lactating and late gestation cows, and to 0.5 ppm for other cattle when strongly goitrogenic crops such as kale, rape, and turnips make up 25% or more of the ration dry matter. As with most other minerals, supplementation should not greatly exceed recommended levels because iodine is toxic and can cause harm at high levels.

Special advantages sometimes are claimed for organic sources of iodine. For routine supplementation, the cheaper inorganic iodides are equally effective. An organic source has the advantage of slower absorption and less risk of harm from overdosage when massive doses of iodine are given as a therapeutic agent. This advantage of organic iodine is not a factor in the amounts needed to prevent goiter, however.

Addition to dairy rations of iodinated casein, commonly known as thyroprotein, stimulates body processes, such as milk production, but the effect usually is only temporary. Practical application of this principle to dairy cattle feeding is discussed in Chapter 11.

Selenium. Selenium, like molybdenum, was known for its toxic characteristics long before it was discovered to be an essential nutrient. Excess selenium causes a peculiar disease of livestock known as "alkali disease" or "blind staggers." More recently, it was found to be needed in traces to prevent retarded growth, reproductive problems, and white muscle disease, a condition which occurs in calves and lambs in deficiency areas. It is closely associated with vitamin E as both are able to counteract the disease under some conditions.

Deficient and toxic selenium areas are widely scattered throughout the United States. Affected areas are shown in Figure 9.8.

The requirement for cattle is very small, being only about 0.1 ppm of the total ration. Toxicity occurs at 3 to 4 ppm. Supplementation of cattle rations with up to 0.1 ppm selenium in the total ration dry matter is permitted by the United States Food and Drug Administration. When selenium is included in a salt-mineral mixture, maximum allowable level is 20 ppm. This is equivalent to 0.1 ppm if the salt-mineral mixture makes up 0.5% of the total ration dry matter (20 ppm × .005 = 0.1 ppm). Higher levels of selenium supplementation are not allowed because of its toxicity and possible carcinogenic properties.

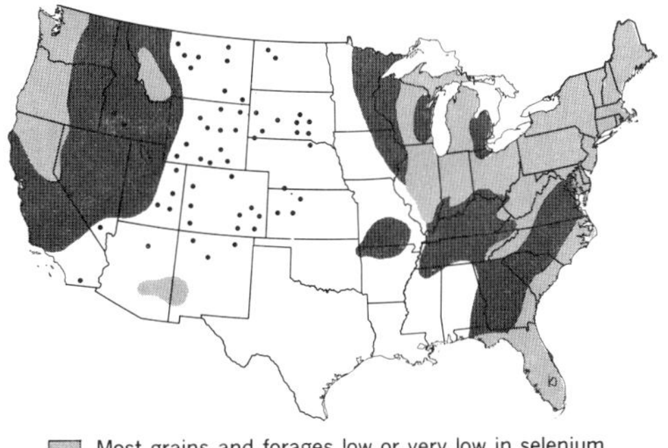

FIG. 9.8 Areas of the U.S. where selenium in plants is excessive or deficient. (Courtesy of J. Kubota, USDA)

Selenium deficiencies are extremely unlikely in dairy cattle which are fed a variety of ingredients in addition to local forage. The requirement is very low, and the chance of all feeds in a dairy ration being produced in deficient areas is very small. Cattle grazing pastures in deficient areas are the most likely to be affected. In this case, providing a salt-selenium mixture free-choice or periodic injection of selenium compounds seems to be the best practical solution at present. Current research with slow dissolving selenium pellets placed in the rumen also shows promise as a possible method of supplementation in the future, but this method has not been approved as yet.

Fluorine. Fluorine makes up about 0.05% of the bones and teeth of mature animals. Although a very small amount in the ration would appear to be essential, slightly higher amounts are toxic. Bones lose their normal color, become thickened and softened, and breaking strength is reduced when excess fluorine is in the diet. Teeth of cattle become mottled, soft and worn down until they become sensitive to cold water and feed consumption is reduced. Eventually growth, reproduction, and lactation are adversely affected due to lowered feed intake.

Beneficial effects of fluorine on the incidence of tooth decay, or dental caries, occur in man when 0.7 to 1 ppm are added to water that normally has lower levels. From this standpoint, fluorine may be considered to be an essential element.

Certain common minerals fed to livestock contain toxic levels of fluorine. Rock phosphate contains about 4% fluorine unless it has been defluorinated, whereas bone meal has only about 0.04%. For cattle, mineral supplements should not contain more than 0.3% fluorine. Phosphorus supplements should contain not more than 0.18% fluorine to avoid toxic effects. Maximum fluorine concentration in the total ration dry matter should not exceed

30 ppm for adult cattle or 20 ppm for immature cattle.

Chromium. Chromium is essential for normal glucose (sugar) metabolism in rats. The same probably is true for other animals, but proof one way or the other is lacking. In any case, required levels are so small that a deficiency would be extremely unlikely on any normal ration.

Others. Suggestions that several other elements, including vanadium, tin, nickel, and silicon, may be useful to animals remain questionable. As with chromium, any requirements are probably more than fulfilled on normal rations because of the low amount needed and/or the ubiquitous nature of these elements.

When and How to Supplement Minerals

Many of the essential minerals have detrimental effects on cattle when consumed in excess, as discussed in previous sections under the individual elements. In addition, interrelationships between minerals are common, and balances are easily upset. Copper, cobalt, iodine, manganese, zinc, selenium, molybdenum, sulfur, iron, fluorine, and salt (sodium chloride) all are poisonous at high levels, and some contribute to the incidence of oxidized flavor in milk. Ratios between calcium and phosphorus and other elements also are important. Therefore, indiscriminate supplementation with mineral mixtures should be avoided.

There is little evidence to indicate that dairy cattle will select the type or amount of minerals that they need if they are provided separately free choice and some evidence that they will not.[11] With the possible exception of salt, it is much better to mix the supplemental minerals with the remainder of the ration. This practice ensures that every cow will get her allowance of minerals each day as she consumes her other feeds.

Except for calcium, phosphorus, and salt, which are needed in greater amounts than are present in most rations, other minerals should not be added to the ration unless it is known that their amounts are deficient. Forage testing, as discussed in Chapter 10, should be used to determine which minerals should be supplemented. This testing is particularly important when non-legume forages, such as corn silage, comprise the bulk of the roughage portion of the ration because mineral content varies widely in non-legume forages. Legume forages, such as alfalfa, generally are good sources of most of the required minerals. However, feeding programs based on legume forages also benefit from forage analyses for energy, protein, and mineral content, and values for molybdenum and copper are particularly important in areas where molybdenosis is a problem. In most cases, adding only the minerals that the forage test indicates as deficient will not only save the dairyman some money on his feed bill, but may also prevent some problems that can occur when minerals are fed in excess. Properly formulated trace mineralized salt available free-choice may be beneficial in some cases. In most cases, however, supplementary trace minerals are more of an insurance factor than a real need of the dairy cow.

9.7 Vitamins

As was the case for protein, ruminal microorganisms are able to synthesize many vitamins for eventual use by the cow, even though some of them may not be contained in adequate amounts in the original ration. All of the B-complex vitamins and vitamin K are synthesized in adequate amounts once the rumen is functioning normally. In addition, vitamin C is synthesized in the body tissues. Consequently, the only vitamins required in the ration of a cow are the fat-soluble vitamins A, D, and E. However, all of the other vitamins except C are needed in the diet of the young calf until ruminal activity is sufficient to fulfill its needs.

Vitamin A

Forage consumed by cattle contains carotene, which is the precursor of physiologically active vitamin A. Under normal conditions, natural feeds provide adequate amounts of this vitamin, or its precursor, which is converted to vitamin A within the body of the cow. When poor quality or limited amounts of forage are fed, vitamin A supplementation may be desirable. Moreover, forage which is stored for a long period loses its carotene through oxidation and, therefore, may require supplementation.

A deficiency of vitamin A causes many problems. Some or all of the following symptoms may occur, depending on the length and severity of the deficiency: (1) night blindness, (2) watery eyes, (3) nasal discharge, (4) coughing, (5) diarrhea, (6) pneumonia, (7) lack of coordination, (8) staggering gait, (9) complete blindness, (10) stratified keratinized epithelium, (11) increased susceptibility to infection, (12) loss of appetite, (13) emaciation, (14) rough hair coat, (15) scaly skin, (16) abortion, and (17) birth of dead, weak, or blind calves. The effect of a vitamin A deficient ration is shown in Figure 9.9.

Requirements of cattle for vitamin A can be met by carotene in feeds, or vitamin A supplements, or a combination of both. For cattle, 1 mg of carotene is considered to be equivalent to 400 I.U. (international units) of vitamin A. Allowances of vitamin A for maintenance, growth, and reproduction are contained in Appendix Tables III-A and III-B. There are no allowances for milk production as levels above those required for normal reproduction do not increase milk yield. Higher vitamin A levels occur in milk when carotene or vitamin A intake is increased, but this may simply be a way for the cow to excrete excess amounts from her system.

Although most common rations provide adequate amounts, many commercial concentrate mixtures contain supplementary vitamin A in a stabilized form. Small excesses of the vitamin are not harmful, and it is the least expensive of all the vitamins. Therefore, this appears to be the best method for providing insurance against a possible deficiency which may occur when limited amounts of forage, or poor quality forage, or forage which has been stored for a long period are fed to dairy cattle. Some recent evidence indicates that high energy rations based primarily on corn silage may benefit from vitamin A supplementation.

A physiological role for carotene per se in addition to its role as a precursor of vitamin A has been indicated by some recent research from Germany. Dairy cattle fed carotene deficient rations had improved fertility when the rations were supplemented with beta carotene, but not when supplemented with vitamin A. The researchers suggested that beta carotene may have a specific physiological role in the function of the ovary. Subsequent research and field trial results have not shown a consistent benefit from beta carotene supplementation, with as many negative results as positive. In any event, one would not expect beneficial results from supplemental beta carotene if cattle are consuming good quality legume forages as a major part of the diet. Clover-grass pastures, greenchop, alfalfa hay and other legume forages are all

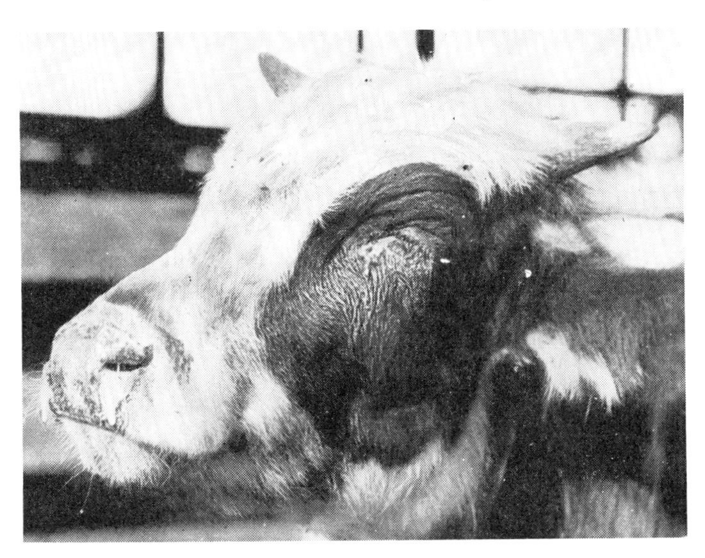

Fig. 9.9 Advanced stage of eye lesion from vitamin A deficiency. (Reprinted with permission from *Animal Nutrition*)

naturally high in beta carotene content. It appears that beneficial effects of beta carotene supplementation are most likely when cattle are fed poor quality forages, or when the forage has lost most of its naturally occurring carotene content through oxidation when stored for a long period of time.

Vitamin D

Cows exposed to sunlight or eating suncured forages do not need supplemental vitamin D. Even green forage, barn-cured hay and silage have some vitamin D activity due to irradiation of dead tissue on stems and leaves of growing plants. Calves housed indoors during northern winters may need vitamin D supplementation due to lack of exposure to sunlight. When exposed to sunlight, the skin synthesizes vitamin D in sufficient amounts for maintenance, growth, reproduction, and lactation.

The most common symptom of vitamin D deficiency is rickets. This condition is characterized by retarded calcification of bones, beading of the ends of the ribs, reduced bone ash, and susceptibility of bones to fracture. In later stages, the forelegs bend forward or sideways, joints become swollen and stiff, the pasterns become straight and the back humped.

As discussed previously in the section on calcium and phosphorus, vitamin D affects the digestibility and utilization of these mineral elements. It is of particular importance in mobilizing calcium from bones during periods of extreme need, such as early lactation. Massive doses (20 million I.U. per day) starting 5 days before expected calving date and continued through the first day after calving, up to a maximum of 7 days, are helpful in controlling milk fever (parturient paresis). However, difficulty in predicting calving dates accurately has reduced the effectiveness of this treatment under practical conditions. Continuous, year-around feeding of about 30,000 I.U. of vitamin D per lb of concentrate reduces the incidence of milk fever in cows with a previous history of the disease.

Feeding or injection of small amounts (1 to 4 mg) of a vitamin D metabolite, 25-hydroxycholecalciferol, has resulted in reduced incidence of milk fever in some trials, but not in others. Recommendations on the use of vitamin D metabolites should await more conclusive results.

Vitamin D requirements for calves are well documented,[2] but information on requirements for cows is scarce. About 6,000 I.U. per cow per day prevents deficiency symptoms. Deficiencies in mature cattle under normal conditions are extremely unlikely because exposure to sunlight provides adequate vitamin D. Many commercial vitamin mixtures contain vitamins A, D, and E in combination, even though vitamin D probably is not needed in most situations.

Vitamin E

White muscle disease in calves can occur from a deficiency of vitamin E as well as a lack of selenium. Supplementation with vitamin E prevents the disease, but the amount needed is dependent on the level of polyunsaturated oils in the diet, as they increase the requirement, and on the level of selenium, which reduces the amount needed.

A practical application of vitamin E supplementation is in the prevention of oxidized flavor in milk. Levels of 400 to 500 I.U. per cow per day have been helpful in reducing the incidence of this flavor defect in milk. The high cost of vitamin E supplements makes this practice very expensive as a routine procedure since only about 1% of the vitamin fed passes into the milk. It would be much more efficient and less costly to add an antioxidant to milk after removal from the cow, but this is prohibited by the United States Food and Drug Administration at the present time.

All green feeds are good sources of vitamin E. Consequently, cows on pasture or receiving greenchop or recently harvested forage have adequate amounts in their rations. Occasionally, cows being fed exclusively dry feeds which have been stored for a long period may need additional vitamin E to prevent the incidence of oxidized flavor in milk. No apparent benefit to growth, reproduction, or lactation in cattle results from vitamin E supplementation of normal rations.

Vitamin K

Vitamin K functions as a stimulant to blood coagulation. Moldy sweet clover contains a high coumarin content which impairs blood coagulation and results in generalized hemorrhage. Vitamin K successfully counteracts this so-called "sweet clover poisoning." Ruminal microorganisms synthesize vitamin K so no dietary source is necessary under normal conditions.

B-vitamins

All of the B-vitamins are synthesized by ruminal microorganisms and are relatively abundant in ordinary feeds. Therefore, routine supplementation of normal rations with B-vitamins is not necessary. However, recent research has shown that supplementation with niacin (nicotinic acid) can be helpful in controlling ketosis in high-producing cows in early lactation. The beneficial effect appears to be associated with the control of fat metabolism and ketone body production and utilization by the ketotic cow. Some experiments also have shown significant increases in milk production with niacin supplementation, particularly with cows that were well-conditioned at calving. Negative results have been obtained when thin cows were fed supplemental niacin. It appears that niacin supplementation should be restricted to well-conditioned or overly fat cows that are particularly prone to ketosis in early lactation.

Before the rumen is functioning normally, young calves have a dietary requirement for thiamine (B_1), riboflavin (B_2), nicotinic acid, pyridoxine (B_6), choline, biotin, pantothenic acid, and vitamin B_{12}. Deficiencies of these vitamins have been produced with calves on experimental diets, but not when they are fed whole milk. Levels of B-complex vitamins in milk replacers, as recommended by NRC, are listed as a footnote to Appendix Table III-C.[2]

Vitamin C

Unlike man and a few other species, all farm animals synthesize vitamin C in their body tissues. Therefore, no dietary source is needed by dairy cattle.

9.8 Water

Although considered last, lack of water will cause death of an animal sooner than the deficiency of any other nutrient. The body can lose practically all of its fat and over half of its protein and still live, but a loss of 10% of its water results in death.

Among its varied functions are regulation of body temperature, digestion, metabolism, excretion, hydrolysis of proteins, fats, and carbohydrates, lubrication of joints, cushion for the nervous system, transport of sound in the ear, and sight in the eye.

Dairy cattle need from 4 to 5 lb of water for each lb of milk produced.[8] They will drink more water if it is available twice a day rather than once a day, and still more if it is available at all times. Cows produce more milk when they have continuous, free access to water. Consequently, fresh water should be available to dairy cattle at all times.

9.9 Summary

Feeding of dairy cattle is greatly simplified by the presence of ruminal microorganisms. Without them, the cow would need a better-quality protein, balanced according to amino acid requirements, and a dietary source of the B-vitamins and vitamin K. Additionally, the cow would not be able to make use, to the same extent, of cellulose from roughage-type feeds, or nonprotein nitrogen. Her ability to make efficient use of these feeds which otherwise would go to waste insures her future usefulness, as competition between humans and farm animals for grains and other concentrates increases. She is able to produce milk and meat, among the most desirable and nutritious foods available to man. And being a ruminant, she can do this utilizing, as a major portion of her ration, roughage feeds not useful to man or other nonruminant animals.

To complete this job with maximum efficiency, the cow needs adequate sources of all 5 types of nutrients: energy, protein, minerals, vitamins, and water. The most common limiting factors in the ration are energy, protein, calcium, phosphorus, sodium, and chlorine. Most rations provide all of the other minerals and vitamins in adequate amounts. Additional supplementation may be indicated in specific places due to area deficiencies or toxicities. Dependence on corn silage as the main roughage source in some areas also increases the likelihood that other minerals and vitamin A should be supplemented.

Review Questions

1. What are the 5 major classes of nutrients required by dairy animals?
2. Which nutrient is the limiting factor for high milk production on most dairy farms?
3. Outline the breakdown of gross feed energy to net energy in a lactating cow, and list the approximate percentages of energy represented by each category.
4. What is the difference between apparent digestible protein and true digestible protein?
5. How is the crude protein content of a feed determined?
6. List the 15 mineral elements known to be required by dairy cattle and one function of each in the body.
7. What 4 major mineral elements are lacking in many dairy cattle rations and need to be supplemented?
8. Give 2 reasons why trace mineral elements should not be indiscriminately added to all dairy rations.
9. Which vitamins are required in the diet of a cow, and why isn't a dietary source of all vitamins necessary?
10. A lack of which nutrient will result in death of an animal the soonest, and why?

References

1. Kleiber, M.: *The Fire of Life.* New York: John Wiley and Sons, Inc., 1961.
2. National Academy of Sciences—National Research Council: Nutrient requirements of dairy cattle. 5th Revised Ed., 1978.
3. National Research Council: Nutritional energetics of domestic animals and glossary of energy terms. 2nd Revised Edition, National Academy Press, 1981.
4. Moe, P. W., Flatt, W. P., and Tyrell, H. F.: The net energy value of feeds for lactation. J. Dairy Sci. *55:* 945, 1972.
5. Technical Committee on the Nutrient Requirements of Farm Livestock: The nutrient requirements of farm livestock. No. 2. Ruminants: Technical Reviews and Summaries. London: Agricultural Research Council, 1965.
6. Virtanen, A. I.: Milk production in cows on protein-free feed. Science *153:* 1603, 1966.
7. Babcock, S. M.: Wisc. Univ. Agr. Expt. Sta. Ann. Rept. 129, 1905.
8. Maynard, L. A. and Loosli, J. K.: *Animal Nutrition.* 6th ed. New York: McGraw-Hill Book Co., Inc., 1969.
9. Dunkley, W. L., Franke, A. A., Robb, J., and Ronning, M.: Influence of dietary copper and ethylenediaminetetra-acetate on copper concentration and oxidative stability of milk. J. Dairy Sci. *51:* 863, 1968.
10. Huffman, C. F., Duncan, C. W., Robinson, C. S., and Lamb, L. W.: Mich. Univ. Agr. Expt. Sta. Tech. Bull. 134, 1933.
11. Coppock, C. E., Everett, R. W., and Belyea, R. L.: Effect of low calcium or low phosphorus diets on free choice consumption of dicalcium phosphate by lactating cows. J. Dairy Sci. *59:* 571, 1976.

Suggested Additional Reading

National Research Council: Nutritional energetics of domestic animals and glossary of energy terms. 2nd Revised Edition, National Academy Press, 1981.

Maynard, L. A., and Loosli, J. K.: *Animal Nutrition.* 6th ed. New York: McGraw-Hill Book Co., Inc., 1969.

National Academy of Sciences—National Research Council: Nutrient requirements of dairy cattle. 5th Revised Ed., 1978.

CHAPTER 10

Feeds for Dairy Cattle

10.1 Introduction

Man depends upon animals and plants for food, clothing, and shelter. Animals, except carnivores, depend upon plants for food. All plants use a process called photosynthesis by which energy from the sun and carbon dioxide (CO_2) from the air are utilized to synthesize plant cells. As this process takes place and plants grow, they give off oxygen as a waste product. The oxygen in turn is used by man and other animals for their metabolic processes, and CO_2 is returned by them to the air as a waste product. By means of this continuous cycle, plants not only provide energy, protein, minerals, and vitamins to the animals which eat them, but also the oxygen which they breathe. Without plants to make use of CO_2 excreted by animals and to give off oxygen during photosynthesis, oxygen in the air eventually would be used up, resulting in the death of all animals.

Classification

Feeds for dairy cattle are divided into 2 main categories: roughages and concentrates. The basis for assignment to the groups is rather arbitrary. Feeds in the roughage category are bulky, fibrous, and relatively low in energy. Examples are hay, pasture, silage, and other forages. Concentrates are so named because they are a more concentrated source of energy or protein and contain less fiber. Corn, barley, and soybean meal are feeds which fit into this category. The 2 main categories are subdivided further based upon the physical form or nutrient content of various feeds. The roughage category includes (A) succulent feeds, such as pasture, soilage (greenchop), and silage; and (B) dry feeds, such as hay, straw, and cottonseed hulls. Major divisions within the concentrate category are grains, by-product feeds, protein supplements, mineral supplements, and other feed additives.

Only the more common feeds fed to dairy cattle are discussed in this chapter. Average nutrient contents of a selected group of roughages, concentrates, and mineral supplements are listed in Appendix Table III-D. A more detailed coverage of this subject is contained in other books which are suggested for additional reading at the end of the chapter.

10.2 Pasture

Pasture has been the basic ingredient in most dairy rations in the past (Figure 10.1). When properly managed, pasture forage is a very nutritious and succulent feed relished by cows. However, as larger numbers of cattle are concentrated on smaller acreages and as milk production per cow in-

FIG. 10.1 Cows grazing on pasture.

creases, dairymen depend less on pasture and more on other feeds. One reason for this is the inability of high-producing cows to consume enough feed to supply their energy requirements when pasture is their main food source. The physical form and volume of pasture fill the rumen to capacity before nutrient needs of high producers are fulfilled.

Other reasons for the decline in popularity of pasture are the variations in pasture quality at different times of the year, and losses due to trampling, urine and manure which greatly reduce pasture yield. Much greater yield of nutrients per acre, particularly energy, is obtained from hay, haylage, silage, and soilage, when it is possible to raise these crops. However, in many areas pasture is the only crop which can be grown profitably due to soil type, topography, amount of rainfall, and other environmental factors. Under these conditions, pasture remains a major source of feed for dairy cows. It is possible for cows grazing pasture to produce well if their ration is supplemented properly with concentrates to supply adequate amounts of energy, protein, and minerals in addition to that provided by pasture. In some areas where pasture is grown very cheaply, this still remains the most economical method of milk production.

Types of Pastures

Pastures can be made up of grasses, such as timothy and orchard grass; legumes, such as alfalfa and clover; or a combination of grasses and legumes. Generally, legumes are higher in protein and minerals than grasses when harvested at the same stage of growth. Bacteria in the root nodules of legumes use nitrogen from the air to build plant protein. Grasses do not have a similar relationship with bacteria and thus usually are lower in protein content, especially at more mature stages of growth.

In most cases, legumes produce greater quantities of feed per year than grasses. When harvested at an early stage of maturity, they also are a good source of energy. The main disadvantage of legume pasture is the tendency for cows to bloat when it is the sole source of roughage. The problem of bloat was discussed in Chapter 8.

A mixture of grasses with legumes greatly decreases the incidence of pasture bloat and lengthens the growing season of pasture. Consequently, most pastures contain both in order to obtain benefits of high protein and high yields from legumes, and bloat prevention plus additional yield due to lengthened growing season from grasses. When fields with high percentages of legumes are pastured, cows must be watched closely and removed from pasture at the first sign of bloat, or other preventive measures must be taken, as described in Chapter 8.

Grazing Patterns

Much of the nutrient value of pasture plants is in the leaves. As plants mature, and yield is increased, more of the plant is present as fibrous stems and less as leaves. A certain amount of maturity is necessary to obtain high yields and provide root stores which increase the longevity of plants. However, plants that are too mature are less nutritious and less palatable. Consequently, a dairyman should have his cows graze a pasture when there is a balance between yield and high nutritive value. For most pasture plants, this occurs when they are between 4 and 6 inches tall.

Grazing cows first select the outer tips of plants because they are the most palatable. If other feed is not available, they progressively graze lower and lower on the stems, which have much less feed value than leaves. If high-producing cows are forced to graze pastures too closely, milk production decreases unless additional concentrates are fed to supplement the pasture. When not supplemented, great variations in milk production occur depending on the quality of pasture being grazed.

Several management schemes are in use to increase the milk production of cows grazing pasture. Rotation of cows in various fields, grazing of fenced strips of pasture for short periods, and selective grazing by high-producing cows before lower producers and dry stock graze are 3 methods which make better use of pasture than continuous grazing.

Stocking Rate

The number of cows per acre (stocking rate) determines the milk production obtained per acre and the level of milk production per cow. Within certain limits higher stocking rates generally increase total milk per acre but decrease milk per cow. Since there are more cows per acre, there is greater utilization of the available forage, but pasture quality is reduced which reduces production per cow. The most economical stocking rate is somewhere between maximum production of milk per acre of pasture grazed and maximum milk production per cow. Excessive stocking rates result in overgrazing which eventually leads to damage of pasture plants and reduces the life span of the plants. Rotational, strip, and selective grazing are helpful in preventing the deleterious effects of overgrazing compared with continuous grazing because pastures have periods between grazings for recovery and growth.

Permanent vs Temporary Pastures

Permanent pastures are made up mainly of perennial plants which continue to produce for years if properly fertilized and managed. Temporary pastures, on the other hand, usually are utilized for only one year or season. Some of the more popular permanent pastures are mixtures of clover and grasses such as Ladino clover or white Dutch clover with orchard grass, blue grass, brome grass, and rye grass. Other legumes such as alfalfa, Salina clover, and trefoil also are used with grasses in areas and on soils to which they are adapted. These pastures have high water requirements and usually are dormant or semidormant during the winter. Growth rate often is depressed in the hot summer months also, especially if there is a lack of moisture. Irrigation is essential in drier areas and is beneficial in most other areas during some months of the year. However, the cost of irrigation may make it impractical in certain areas.

Temporary pastures can be utilized efficiently during periods of dormancy or reduced yield from permanent pastures. Such plants as Sudangrass produce well during hot weather. Millet, oats, and lespedeza also are good as temporary pastures. For winter grazing in the southern States, rye grass, crimson clover, oats, barley, and wheat are used with good results. Relative costs for preparing the land, seeding, and fertilizing temporary pastures are greater than for permanent pastures because the crop lasts only one year. In order to justify this cost, it must be more than compensated for by the extra feed produced during periods when supplemental feeding is necessary.

The most desirable combination of legumes and grasses for pastures varies greatly depending on the area. Current information based upon field trials in every area of the U.S. is available from extension service personnel of the state land-grant universities. They also are the best source of information for local recommendations on use of fertilizers, weed killers, and pesticides.

Importance of Clipping Pastures

Since pasture growth is variable during the growing season, there are times of excess production as well as shortages. When cows are unable to utilize all of the available feed, it is important to clip pastures to keep them from getting too mature and coarse. This excess feed can be baled or ensiled for use during periods of reduced growth. Even if not utilized in this manner, pastures should be clipped periodically in order to reduce coarse growth of less palatable plants. Rapid growth takes place in areas where manure and urine have been deposited, but plants in these areas are unpalatable to cows.

10.3 Soilage Crops

Soilage, commonly referred to as "greenchop," makes use of the same crops as pasture plus others such as corn and sorghums (Figure 10.2). The only difference is that the plants are harvested and brought to the cows rather than being grazed by the cows. Several advantages for this system over pasturing are obvious. Plants are clipped at a constant level thereby making efficient use of all available feed. The amount fed can be adjusted according to need and availability. There is less waste and loss from trampling, manure and urine, and cows do not expend energy walking to, from, and around pastures while grazing. Cost of equipment for harvesting and hauling greenchop and a high labor requirement are the main disadvantages of this system. As with pasture, there are periods when plant growth is not sufficient to fulfill the needs of the herd and other times when there is more than the cows can consume. During periods of excess, hay or silage can be made for feeding during shortages at a later date.

Maturity

There is a tendency for farmers to let greenchop get too mature before cutting during periods of rapid growth. When greenchop is too mature milk production suffers because the cow must eat this poorer-quality forage. Consumption of overmature greenchop has an even greater depressing effect on milk production than overmature pasture because the whole plant is chopped and mixed together. Therefore, cows cannot select the more nutritious leaves and outer parts of stems as they can when grazing pasture. Rather, they must eat a mixture of all parts of the plant as harvested.

Soilage vs Pasture

Whether a pasture or greenchop program is more profitable is determined primarily by the size of the herd. The larger the size, the more likely that greenchopping will pay. Equipment and labor costs for greenchopping are spread over a larger base thereby reducing the overhead cost per cow. Travel time and traffic by cows between milking barn and pastures are major problems in large herds also. Consequently, the trend in large herds is toward elimination of pasture for milking cows and more corral or barn feeding of greenchop, hay, haylage, and silage. Pastures continue to be practical in smaller herds and for

FIG. **10.2** Forage being harvested for soilage (greenchop) feeding.

heifers and dry cows in both large and small herds. The general tendency, however, is a gradual reduction in use of pasture and increased dependence on stored feeds and greenchop as herd sizes and milk production per cow increase.

10.4 Silage

Silage is the product resulting from storage and fermentation of fresh forage under anaerobic (without oxygen) conditions. Bacteria in the forage ferment available carbohydrates to organic acids which cause the ensiled forage to become acidic. When properly made, the pH (a measure of acidity) will be in the range of 3.5 to 4.5. The acids eventually kill the bacteria and preserve the silage in a palatable state so long as air is excluded from the silo. Silage can be stored for years without appreciable change in composition if it is properly put up and sealed, although most silage is fed out within a year after it is made.

Fermentation

Silage fermentation can be divided into 5 phases, as illustrated in Figure 10.3. The first 3 phases take place during the first 3 to 5 days after ensiling and determine the success or failure of making good-quality silage. They merge with each other rather than have definite boundaries.

The first phase starts with the placement of forage in the silo. Plant cells continue to produce heat and carbon dioxide until they cease respiration and die. Heat produced during this phase and carbon dioxide, which reduces the air space and causes anaerobic conditions, are essential for growth of the bacteria which produce organic acids.

During phase 2, acetic acid is the principal acid produced by the bacteria. As the concentration of this acid increases, phase 3 starts with a gradual increase in lactic acid-forming bacteria. Concurrently, there is a decrease in bacteria which form acetic acid because they cannot live at higher lev-

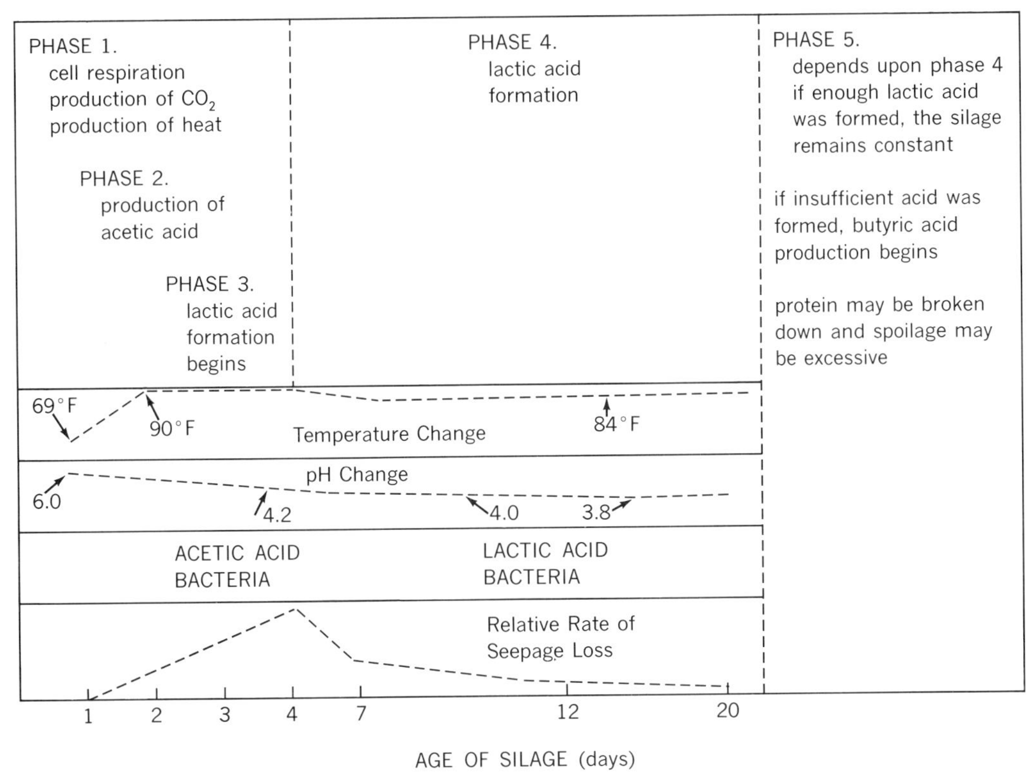

Fig. 10.3 Schematic representation of the 5 phases in silage fermentation. (Reprinted with permission from *Optimum Feeding of Dairy Animals*)

els of acidity. Settling of the forage occurs during the first few days and, if the forage is high in moisture, seepage reaches a peak at the fourth or fifth day.

Lactic acid is the major acid produced during phase 4, which lasts for 15 to 20 days. When acidity reaches the desired level, bacterial action is stopped.

The events in phase 5 depend on the results of the first 4 phases. If there is enough acetic and lactic acid present in the silage to prevent further bacterial action, and the silo is well-packed and sealed to exclude air from the silage pack, no further changes will take place and the silage will be properly preserved. However, if the acid level is too low, undesirable bacteria, such as those that produce butyric acid, may act on the material, resulting in decomposition and foul-smelling silage. Amino acids and protein may be broken down to ammonia and amines, further decreasing the palatability of the silage.

If air gets into the silage pack due to poor packing and sealing, excess heat builds up from oxidation, resulting in invisible silage losses. Furthermore, molds that develop after air entry can use lactic acid formed in phase 4, which lowers the acidity of the silage and allows the undesirable bacterial action described above. This action uses some of the energy in the silage to produce heat, thereby reducing the energy value of the silage when it is fed.

Making Quality Silage

Many factors which affect the fermentation and subsequent quality of silage can be controlled by man. These include moisture content of the forage when ensiled, rapidity of filling the silo, carbohydrates available from added feeds, packing of the material in the silo, and sealing of the silo to reduce exposure to air and other elements. For corn silage, best results are obtained when the dry matter content is between 30 and 36%. If dry matter is below 30%, excessive losses from seepage often occur. If it is above 36% dry matter, packing is more difficult and air may get into the silage mass, resulting in greater oxidation, spoilage, and mold growth.

When forage is ensiled at a desirable moisture content, the silo is filled rapidly, packed well and sealed properly, the chances of making high-quality silage are very good. Unfortunately, great quantities of silage are produced without adherence to proper silage-making methods. As is the case for low-quality forage in all forms, feeding poor-quality silage results in decreased feed intake and, subsequently, lower milk production and reduced profits. Conversely, silage made according to recommended procedures is a very palatable and nutritious forage which will support high milk production when properly supplemented with energy, protein, minerals, and vitamins from concentrates. In fact, many dairymen are feeding all of the roughage in the form of silage with excellent results. Forage testing, as described later in this chapter, is helpful in determining proper supplementation of silage as well as other roughages.

Under some conditions, an all-silage roughage program is the method of choice for obtaining maximum profit from the dairy herd.

Types of Silos

There are two main types of silos, upright and horizontal. Each type has advantages and disadvantages depending upon the crop ensiled, the size of the herd, and the environment in which it is used.

Upright Silos. Upright, or tower, silos are very popular in the eastern and midwestern sections of the United States. They are more commonly used for smaller herds, where required volume of silage is less, than for larger herds. These silos are cylindrical with diameters usually from 16 to 30 ft and heights from 30 to 80 ft. Many types of materials are used to construct them, including wooden staves, concrete staves, poured concrete, tile, brick, and metal. One of the most popular types, the concrete-stave silo, is shown in Figure 10.4.

Losses due to spoilage are reduced in metal or concrete-stave silos lined with glass or other sealing materials to make them airtight. Also, mechanical unloading equipment greatly reduces the labor involved in removing and feeding silage from upright silos. However, the advantages of reduced spoilage and easier unloading and feeding must be weighed against the higher cost of these structures compared with other types of silage storage.

Fig. 10.4 Concrete-stave silos, one of the most popular type of upright silo.

Horizontal Silos. Horizontal silos may be in the form of a stack, a trench, a bunker, or constructed partly above and partly below ground level. Bunker silos (Figure 10.5) are especially popular in the South and West and are increasing in popularity in all areas of the United States. Trench silos have been popular in Europe for many years. Horizontal silos have the advantage of low initial cost and are easily filled and packed with self-unloading wagons and tractors. These silos lend themselves to self-feeding, or the silage can be removed with a scoop on the front of a tractor or by other mechanical equipment. They are particularly popular for large herds because great quantities of silage can be stored at a much lower cost than would be required for upright silos.

Bunker silos usually are made with concrete floors and sides, but wooden sides sometimes are used. Sides must always be well braced because the silage bulk exerts great lateral pressure on the walls.

Trench silos have a lower initial cost than bunkers because the floor and sides usually

Fig. 10.5 A bunker silo, one of the most popular type of horizontal silo.

are made of dirt. However, they are not practical in areas with high water tables and are more difficult to empty than bunker silos.

Stack silos sometimes are used, but it is much more difficult to prevent spoilage losses with this type of silo because of problems in getting adequate packing. Air which gets into the silage mass because of insufficient packing causes spoilage due to heating and poor fermentation. Stacks entirely enclosed in plastic covering (Figure 10.6) are gaining in popularity. Spoilage in plastic enclosed stacks is practically nonexistent when properly packed and sealed.

Sealing of horizontal silos is extremely important because of the large exposed surface area. Tight packing and covering of the silage with a plastic sheet held down by dirt or old tires is a common practice which reduces spoilage losses considerably in bunker, stack, and trench silos. A well-sealed silage stack is shown in Figure 10.7. When these types of silos are filled, packed, and sealed properly, losses are only slightly greater than from upright silos.

Fig. 10.6 Silage stored in large plastic bags.

Several inches of silage must be fed off each day to prevent excess spoilage on the face of horizontal silos once they are opened. The width of the silo should be governed by the number of cows being fed in order to insure that enough silage will be removed from the face of the silo each day. This requirement may make the use of horizontal silos impractical for small herds if not enough silage is fed each day to prevent surface spoilage.

Forages Used for Silage

So many crops are made into silage in various areas of the world that only brief descriptions of the more common ones are contained here.

Corn Silage. By far the most popular silage is made from the corn (maize) plant.

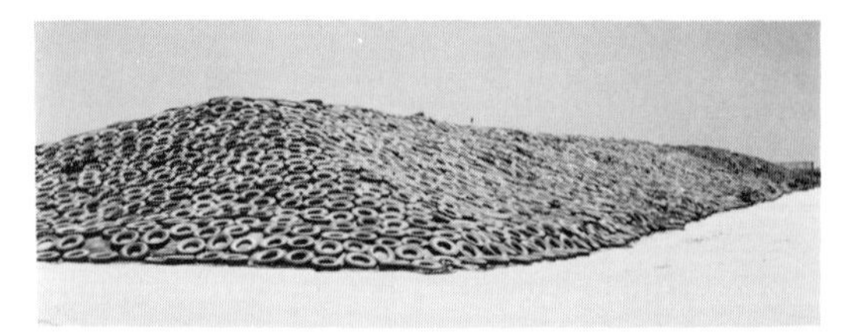

Fig. 10.7 A silage stack covered with black plastic held down by old tires.

Where adapted to the area, this crop is unexcelled as a producer of net energy per acre. Its popularity is enhanced also by the fact that it requires much less labor between the time that it is planted in the spring and harvested in the fall than most other forage crops fed to dairy cattle.

Corn varieties adapted to many different areas are available. As is the case for pasture plants, information on the best varieties and fertilization practices are available from extension service personnel in all areas of the U.S. Similar information is available in other countries. In general, varieties which contain a high ratio of ears to stalk make the best silage for dairy cows. Although other tall-growing varieties with few ears look spectacular in the field before harvest and may produce more tons of green feed per acre, it is the energy per bite which is important to the cow. Therefore, corn silage should be evaluated on the net energy produced per acre rather than by its height or tonnage of green feed per acre.

The main disadvantage of corn silage is its low protein content. Additionally, it is lower than legume forages in most mineral elements, particularly calcium and manganese. When corn silage is fed as the sole forage to cows, particular attention must be given to supplying these nutrients through other feeds or as special supplements.

Addition of urea at the rate of 10 lb per ton of 30% dry matter corn forage as it is ensiled, or equivalent amounts of other NPN products such as anhydrous ammonia, results in a product which contains about 12 to 13% crude protein on a dry basis compared with the 7 to 8% crude protein in the dry matter of regular corn silage. NPN-treated corn silage does not require as much supplementation with high-priced protein feeds as does regular corn silage, thereby reducing feed costs substantially. Furthermore, when corn silage comprises the bulk of the roughage, NPN-treated silage makes it possible to balance the ration with the same concentrate mix even when widely varying amounts of concentrate are fed, according to milk production, within a herd. This balance cannot be achieved when corn silage is the sole, or predominant, roughage without underfeeding protein to the low-producers, or overfeeding protein to the high-producers.

Palatability and intake sometimes are slightly reduced on urea-corn silage, resulting in some depression in milk production.[1] In many cases, reduced feed cost more than compensates for any loss of income due to lowered milk production when urea-corn silage is fed. Recovery and utilization of urea have been best at forage dry-matter levels between 30 and 40%. Addition of urea to corn forage either above or below this range is not recommended.

Ammonia (NH_3) addition to silage has advantages beyond those provided by an equivalent amount of NPN from urea. Some of the NH_3 is converted to bacterial protein during the fermentation process in the silo, resulting in more true protein in the NH_3-treated silage than in urea-treated silage. Furthermore, NH_3 has a preservative effect on the silage when it is exposed to air during feeding, thus reducing the amount of silage that spoils in the feed bunk if too much is fed at one time. A drawback to NH_3 is its objectionable odor and corrosiveness to equipment, requiring special safety precautions when it is added to silage.

When high-protein legume forage, such as alfalfa makes up half or more of the forage dry matter fed, NPN supplementation of corn silage is not warranted. Protein requirements are fulfilled for all except exceptionally high-producing cows when fed a ration containing half of the forage dry matter as corn silage, half as legume, and sufficient amounts of a 15% crude protein concentrate to fulfill energy requirements.

A further discussion of urea and other nonprotein nitrogen supplements is contained in Section 10.10.

Sorghum Silage. Many varieties of sorghum are used for silage and are popular in the South of the U.S. where they grow especially well. Sorghum silage generally has a slightly lower feeding value than corn silage. Unless the sorghum grain is ground, much of it passes through the cow undigested, thereby lowering the energy value of the silage.

Sorghum silage has the same drawbacks as corn silage, being low in protein and calcium. These deficiencies coupled with its lower energy value make sorghum silage of less value for dairy cattle feeding than corn silage. However, it can be useful as a supplementary forage in areas where corn does not grow well.

Oat and Other Small Grain Silages. Silage made from oats or other small grain plants is particularly popular in areas where they can be grown in a double cropping system. Oats planted in the fall are harvested for silage in the early spring. Following the oat harvest the same land is replanted to corn, sorghum, or other summer-growing crops for harvest in the fall. In areas where the growing season is long enough, this type of cropping system results in more feed produced per acre than can any single crop that occupies the land year around, such as alfalfa.

Silage made from small grains is low in protein and calcium. Also, its energy content and palatability are lower than good-quality corn silage. It rarely is fed as the sole forage to dairy cows, but it makes a good supplemental source of feed in conjunction with legumes. A very popular combination in some areas is oats planted with vetch, a legume. Vetch is high in protein and calcium and compensates for the deficiencies of these nutrients in the oat plant, besides increasing total yield per acre. Forage mixes that include oats, ryegrass, wheat, barley, vetch, and other forages also are popular in areas to which they are adapted.

Alfalfa-Grass Silage. Alfalfa and alfalfa-grass mixtures are very popular as silage crops particularly in areas where weather makes haymaking difficult. Field losses due to shattering of leaves are reduced, and weather damage is much less compared with dry hay.

The nutritional value of alfalfa-grass silage is excellent because of its high protein and mineral contents. However, direct-cut silage is high in moisture, and frequently leakage from silos is a problem. Total forage intake usually is less when alfalfa-grass silage is the only roughage fed compared with dry hay, or compared with mixtures of alfalfa-grass silage and other roughages.

Other Silage Crops. Many other crops are ensiled in addition to the main ones discussed in previous sections. Included in the list are pea vines, sugar beet tops, beet pulp, citrus pulp, and potatoes. Although of minor importance compared with corn silage, they often are inexpensive supplemental feed sources in areas where they are available. Ensiling of high-moisture grains is discussed in Section 10.10.

10.5 Haylage (Low-Moisture Silage)

When legumes or grasses are wilted to about 50% moisture before ensiling, a product known as haylage is formed. Haylage usually is stored in upright silos, mostly of the airtight variety, but occasionally it is stored in horizontal silos. The lower moisture content of haylage makes it difficult to pack sufficiently to exclude air from the mass. Therefore, greater spoilage losses occur than from higher-moisture silage unless haylage is stored in an airtight, upright silo.

Lower field losses from leaf shatter and rain damage occur when crops are put up as haylage as compared with dry hay. Furthermore, a cow will consume more dry matter in the form of haylage than silage. These advantages have resulted in great increases in the popularity of haylage in recent years, especially in areas where the climate is not conducive to good haymaking.

Forage stored as haylage retains its nutritive value very well. However, ensiling is not a miracle-producing process and does not improve the value of forage. In order for high-quality haylage or silage to come out of a silo, high-quality forage must be put in.

Haylage has not found much favor in semiarid regions such as the Southwest of the U.S. Irrigation is extensively practiced and rain damage is rare in this area because very little rain falls during most of the growing season. Here, good-quality forage can be produced and stored more cheaply as hay than as haylage. Feeding value of forage when fed as hay or haylage is about the same when proper procedures for each method are followed. Therefore, the main advantage of haylage is the ability to get forage off the field and protected from the elements sooner in areas where rain damage is a factor. Since extensive areas of the U.S. do have considerable rain and high humidity during the growing season, the popularity of haylage probably will continue to grow despite the high cost of the structures needed to store it.

10.6 Hay

Forage which is dried and stored near 85 to 90% dry matter is known as hay. Drying of the forage can be done entirely by the sun while it is in windrows or partly by the sun and completed on wagon dryers or by barn dryers after placement in storage. Storage usually is in the form of bales, but some is stored as chopped hay, long hay, or as wafers. Storage at moisture contents above 20% often results in heat damage and sometimes in spontaneous combustion, a cause of many barn fires. Also, hay becomes moldy if excess moisture is present.

Making Quality Hay

Good-quality hay must contain a high percentage of leaves, as they are much higher in nutritive value than stems. For example, alfalfa leaves contain 2 to 3 times as much protein as alfalfa stems and also are much higher in minerals and vitamins, particularly calcium, phosphorus, and carotene.

To insure that hay is high in leaves, forage must be cut at an immature stage. Yield of dry matter increases with plant maturity, but a greater percentage of the weight is in the form of stems, which contain less protein, minerals and vitamins and more fiber.[2] In addition to reduced nutritive value, hay made from mature forage is less palatable, resulting in less feed intake and lower milk production by cows eating hay of this quality.

Good-quality forage sometimes becomes poor-quality hay due to loss of leaves during the haymaking process. Raking and baling when hay is too dry is the most common cause of excessive leaf shatter. Rain damage and heating of hay put up too wet also reduces the feeding value of hay.

Forages Used for Hay

Forages made into hay fall into two main categories, legumes and nonlegumes.

Legumes

Alfalfa. By far the most important and popular hay crop fed to dairy cattle in the U.S. is alfalfa (Figure 10.8). It combines the virtues of high dry matter yield and high content of protein, minerals (especially calcium), and carotene with palatability unsurpassed by most other forages. Varieties are available which are adaptable to almost any area and climate. Because of its relatively high content of energy, protein, minerals, and vitamins, it commonly is supplemented with very simple concentrate mixes with excellent results.

High-quality alfalfa hay results from cutting at the proper stage of maturity, careful handling to reduce leaf losses and storage under cover to minimize losses due to exposure to the elements. Feeding quality decreases steadily with advancing maturity of the alfalfa plant, but dry matter yield increases. Cows produce the most milk when fed alfalfa harvested in the pre-bud stage. However, continual cutting at this stage reduces the life of the stand and results in increased weediness. In the Western States where alfalfa hay is extensively fed, a compromise which results in high dry matter yields, high-quality hay, and long stand-life is to make the first and last cuttings of the year at the one-tenth bloom stage and the midsummer cuttings at the bud stage. In this area 5 to 10 cuttings per year are possible, rain damage is practically nonexistent, and dry matter yields of 8 tons or more per acre are common with this harvesting procedure.

Other Legumes. Although generally not equal in feeding value to high-quality alfalfa hay, other legumes including the clovers, soybeans, lespedeza, cowpeas, trefoils, and vetch are fed as hay in areas to which they are adapted. Usually they are more difficult to make into high-quality hay because of their growing and drying characteristics, and yields generally are less than alfalfa. For these reasons, they are used more as supplemental feeds than as primary forages in dairy rations.

Nonlegumes. The major nonlegumes are timothy and bromegrass.

Timothy. Timothy is to nonlegume hays what alfalfa is to legumes. However, it is vastly inferior to alfalfa in feeding value due to its low content of protein, minerals, and vitamins. In addition, it is less palatable than alfalfa, resulting in less being consumed. Therefore, energy intake is reduced. Because of these drawbacks, it should be fed in combination with legumes in order to compensate for its deficiencies.

Despite its deficiencies, timothy contin-

FIG. 10.8 Alfalfa hay is one of the most popular forages for dairy cattle.

ues to be popular because of its ability to grow on soil which is too acid or poorly drained for alfalfa. It quickly establishes a good stand and is easily harvested. It can be harvested over a long period of time and cures easily into bright, clean hay, free from dust, which can be handled with minimal waste.

Although yield may be increased somewhat with greater maturity, quality is greatly depressed when timothy is cut later than the early bloom stage. When it is cut at an early stage and properly supplemented with protein and minerals from other sources, cows produce well on rations which include timothy hay.

Bromegrass. Bromegrass is very popular as a companion crop with alfalfa in "mixed hay." It is one of the most palatable grasses and has a long growing season. Due to its high nitrogen requirement, it declines in productivity rapidly when not grown in conjunction with alfalfa unless it is heavily fertilized. Bromegrass retains its feeding value later in bloom than most grasses. However, its nutritive value also is greatly enhanced when fed in combination with alfalfa or other legumes because of their high content of protein and minerals.

Oat and Other Cereal Hays. Cereal hays, of which oat hay is the most prominent, are popular on the Pacific Coast where they make good winter and early spring growth. These hays are low in protein, and their feeding value is greatly increased when fed with legumes. A combination which is very popular is oat and vetch hay. The oat plant, with its strong, upright stems, provides an ideal structure on which the vetch plant can entwine itself. In turn, the vetch, which is a legume, provides the extra protein and minerals lacking in the oat plant, making the combination more palatable and nutritious to the dairy cow.

Early cutting of oats for hay greatly increases its feeding value due to increased energy and protein levels. Even though considerable energy is stored in the kernels at maturity, shattering of the grain during harvesting of mature oats results in energy losses and decreased feeding value compared with early-cut hay.

Other Hays. Other nonlegumes commonly made into hay include Sudangrass, orchard grass, millet, and various prairie grasses as well as many others. They generally are lower in feeding value than legumes because of low protein percentages. Therefore, they seldom make up a large proportion of the forage fed, being used primarily as supplementary feed sources during periods of shortages.

10.7 Other Roughages

Various other roughages occasionally are fed to dairy cattle. Most are by-products such as cottonseed hulls, grain straw, and corn stover. All are low in energy and protein and cannot make up a large percentage of a dairy cattle ration if high milk production is to be maintained. They are included in a ration mainly for their fiber value when other better-quality roughages are not available or cannot be obtained at reasonable prices. In general, they are poor substitutes for high-quality forage and should not be fed to milking cows except under the unique conditions mentioned above. They can be fed to older heifers and dry cows without adverse effects on growth and health if properly supplemented with energy, protein, and minerals.

10.8 Physical Form and Treatment of Roughages

Except for pasture, forages must be handled at least once, and often more times, before being fed. Many methods of handling this coarse type of material are used, including chopping, baling, wafering, and pelleting. Various additives also are used on forages, especially silages, with claims often being made of better preservation and superior feeding value due to these treatments.

Long Hay

Forage that is cut, allowed to dry, and then stacked in shocks or stored in a barn is called long hay. Before modern machinery for chopping and packaging became available, most hay was stored in this form. Loss of leaves and their nutrients and difficulty in handling hay in this form make it practically nonexistent on the larger dairy farms today.

Chopped or Shredded Hay

Dry hay can be chopped into short lengths and stored in this form for subsequent feeding. Popularity of this method also has declined in recent years because of several factors. Chopped hay is bulky, requires great volumes of storage space, is dusty when handled and, unless thoroughly dried, is very susceptible to spontaneous combustion. However, less wastage occurs when hay is chopped or shredded, and it can be easily mixed with other feed ingredients in a complete ration which is not possible with long or baled hay. Also, some dairymen claim better feed intake and higher milk production when hay is shredded and moistened before feeding (Figure 10.9).

Baled Hay

Baling is the most popular form for storing hay. Bales vary in form and size. Small, rectangular bales tied with twine, larger rectangular bales tied with wire (Figure 10.10), round bales (Figure 10.11) and square bales weighing 1,000 pounds or more are all popular in various areas.

Baled hay takes up less storage space and is more easily transported than long or chopped hay due to its compactness and handling ease. However, large losses of nutrients in the field can occur if hay is raked and baled when it is too dry. Losses from leaching also occur when hay is rained on while lying in the windrow before baling. Excessive dryness can be avoided by baling early in the morning while dew is still on the ground, but avoiding rain damage is almost impossible in areas where summer rains are prevalent. This latter problem is responsible for the great increase in popularity of haylage (low-moisture silage) in recent years.

Pelleting

Grinding and pelleting of hay results in a product which is much easier to handle and store than long, chopped, or baled hay. In addition, pelleted hay, pictured in Figure 10.12, is consumed in greater amounts than other forms, resulting in faster gains of cattle, and, in some cases, higher milk production. Unfortunately, ground and pelleted

Fig. 10.9 Baled hay being shredded and moistened before feeding.

hay causes an altered microorganism metabolism resulting in shifts in proportions of fatty acids produced in the rumen, as discussed in Chapter 8. Increased propionate produced on pelleted hay rations is one of the causes of milk fat depression. Therefore, pelleted hay cannot be fed as the sole source of forage to dairy cows if milk with normal fat content is to be obtained.[3]

Fig. 10.10 A hay baler in action *(top)* and baled alfalfa hay stacked for future use *(bottom)*.

Wafering

Wafers, or cubes, shown in Figure 10.13, are the most recently developed hay-packaging form. They have many of the advantages of pellets without the disadvantage of causing severe milk fat depression. Rather than grinding as is necessary for pelleting, hay is chopped into about 1.5 inch lengths before cubing. Consequently, fiber length usually is adequate to maintain normal fat tests although there have been reports of slight fat depression when cubes were the only roughage in the ration.

Cubes that are commercially available have a cross section of 1.25 x 1.25 inches and usually are 2 to 3 inches long. Their density is about 25 lb per cubic foot, which is more than double that of hay bales

Fig. 10.11 A load of round bales.

Fig. **10.12** Alfalfa after being ground and pelleted.

weighing 100 lb. Therefore, less storage space is required and more weight can be carried per unit volume when transporting cubes compared with baled hay. Additionally, wafers, by their small size and compactness, lend themselves to bulk handling by mechanical equipment, greatly reducing the labor requirement for shipment, storage, and feeding.

Dry matter intake of alfalfa cubes by lactating cows averages 20% greater than intakes of baled hay of the same quality. Subsequent increased milk production, reduced waste, ease of feeding, lower transportation costs, and reduced storage space requirements all contribute to their popularity.[4]

Fig. **10.13** Close-up view of alfalfa cubes *(top)*. Cubes feeding from a self-unloading truck *(bottom)*.

Present wafer machines (Figure 10.14) require that hay in the windrow contain about 10% moisture at the time of wafering. As it goes into the machine, water is added to the hay to bring it to 14 to 15% moisture before chopping and compressing it into cubes. Water helps with the binding process but it must be on the outside of the plant rather than in the plant itself in order for present wafer machines to operate properly. Cubes must be allowed to dry and cool overnight on a concrete slab in piles about 3 ft deep before placement in storage in order to minimize heating and possible spontaneous combustion. The requirement for dry hay in the windrow is easily accomplished in semiarid regions, but is practically impossible in rainy and humid areas.

Equipment companies are continuing research and development of wafer machines which will work on higher-moisture hay. Stationary machines also are available which can wafer hay that is field chopped and transported to the machine. These stationary machines can be used to cube hay which previously was baled and also can make mixtures of forage and concentrates into complete-ration cubes.

Silage Additives

Many chemical compounds and feedstuffs are used as silage additives in various areas of the world. In general, feeding

Fig. **10.14** A cuber machine in action.

value of properly made silage is increased only to the extent of the nutrients in the additive. For example, molasses added to silage adds energy and palatability to the resulting mixture due to its high content of sugar. Grains and by-product feeds such as beet pulp added to silage soak up excess moisture from wet silage. Also, they contain readily available carbohydrates which enhance fermentation as well as provide extra nutrients of their own to the mixture. Various acids are used, particularly in Europe, which increase the acidity and help with preservation of silage which would have an acidity too low for optimum fermentation under normal conditions. Addition of NPN to silage was discussed in Section 10.4 under Corn Silage. However, the value of many other chemical additives is unproven, and the appearance and disappearance of many through the years indicate that their value is questionable, at best. As stated in the section on silages, when proper silage-making techniques are followed, good-quality, highly nutritious silage can be made and preserved without additives. Preservatives may have a place in emergency situations when ideal silage-making conditions are not possible, such as having to ensile forage at a moisture content above recommended levels. Formic acid appears to be one of the best products under these conditions but it is expensive and difficult to handle because of its corrosiveness.

10.9 Forage Evaluation

References continually have been made in this chapter to the importance of roughage quality in dairy rations. Quality in this case refers primarily to the two nutrients most often lacking in dairy rations, energy and protein. Immature grasses and legumes are high in both nutrients. As plants grow, increasing amounts of less digestible substances, such as cellulose, hemicellulose, and lignin, are laid down within the plant. These compounds add strength and stability to the plant, but decrease energy and protein levels. Not only are percentages of energy and protein decreased, but substances like lignin depress the digestibility of many nutrients in plants. Also, plants tend to become less palatable with age, due to their more fibrous nature and reduced proportion of leaves. Therefore, when forages are immature, cows consume greater amounts of them, and utilize the forages more efficiently.

Exceptions to the rule of decreased feeding value with age occur with plants which develop considerable quantities of seed with maturity. Examples are corn, barley, and oats. Increased feed value of the kernels more than compensates for decreased value of the stalk as the plant matures. Cereals cut for hay, however, lose many of their kernels if harvested at maturity. Therefore, they should be cut during early flowering or in the soft dough stage. Corn for silage, on the other hand, should be harvested in the medium to hard dough stage with all of the kernels fully dented. Corn at this stage usually will contain 64 to 70% moisture, which is ideal for ensiling.

Various methods for predicting feeding value of forages are in use in the U.S. Typical examples of different methods are described in the following sections. Each method has its advantages and disadvantages, but in each case, the methods have proven valuable in areas to which they are adapted.

Penn State Forage Testing Service

Forage testing is conducted by several state universities in the U.S. The program at Pennsylvania State University[5] is fairly typical. A farmer collects a representative sample of the forage to be tested, encloses it in a special, airtight bag, and sends it to the University's Soil and Forage Testing Laboratory with the appropriate fee. Chemical analyses for crude protein, acid detergent fiber, and moisture are conducted by the laboratory. Upon request, samples also are tested for acid-detergent insoluble nitrogen to determine the amount of heat-damaged protein. From these determined values, the amounts of digestible protein, net energy, and total digestible nutrients are calculated from regression equations developed for various crops from relationships observed in many previous analyses. In addition, quality ratings and hay or grain equivalents are listed.

Formulas for calculating digestible protein (DP), heat damaged protein (HDP), unbound protein (UP), net energy for lactation (NE_1), total digestible nutrients (TDN), estimated net energy (ENE),[6] and crude fiber (CF) for various forages from their crude protein (CP), acid detergent fiber (ADF), and acid detergent insoluble nitrogen (ADIN) contents are shown in Table 10.1.

Analyses for minerals, true protein, non-protein nitrogen, nitrates, and crude fat also are available at an additional cost. Upon completion of the analyses, information on the forage is returned to the farmer on the form shown in Figure 10.15. In addition, he may request a feeding program recommended by university personnel based upon his forage tests and the cost of feeds available to him. Armed with this information, dairy farmers are able to make more efficient use of available feeds and carry on a more scientific and economical feeding program.

TABLE 10.1 *Penn State Forage Testing Service Formulas*[1]

Digestible protein (DP)

1. *Samples without heat damage test*
   DP(%) = .929 CP% − 3.48
2. *Samples with heat damage test*
   a. HDP(%) = ADIN% × 6.25
   b. UP(%) = CP% − HDP%
   c. DP(%) = .68 UP% − .60 HDP% + 1.44

Energy predictions

1. *Legumes*
   NE_1 (Mcal/lb) = 1.044 − .0119 ADF%
   TDN(%) = 89.796 NE_1 + 4.898
2. *Mixed forages*
   NE_1 (Mcal/lb) = 1.0876 − .0127 ADF%
   TDN(%) = 89.796 NE_1 + 4.898
3. *Grasses*
   NE_1 (Mcal/lb) = 1.085 − .0124 ADF%
   TDN(%) = 89.796 NE_1 + 4.898
4. *Corn Silage*
   NE_1 (Mcal/lb) = 1.044 − .0124 ADF%
   TDN(%) = 53.1 NE_1 + 31.4
5. *Sorghum and small grain forages*
   NE_1 (Mcal/lb) = 0.7936 − .00344 ADF%
   TDN(%) = 89.796 NE_1 + 4.898
6. *Complete rations*
   TDN(%) = 93.53 − 1.03 ADF%
   NE_1 (Mcal/lb) = (.0245 TDN% − .12) × .454
7. *Grain-concentrate mixes*
   TDN(%) = 81.41 − 0.6 CF%
   NE_1 (Mcal/lb) = (.0245 TDN% − .12) × .454

Fiber conversions

1. *Forages*
   CF% = 0.75 ADF% + 3.56
2. *Grain-concentrate mixes*
   CF% = 0.8 ADF%
3. *Complete ration*
   CF% = 0.9667 ADF% − 3.8

[1]All formula coefficients on a dry matter basis.

Cornell Date of Cutting Method

A method which works well in the Northeast is based upon date of cutting. The growing season is shorter than in the Southern and Western States, and only a few cuttings per year are possible. Cornell University researchers found digestibility of first-cutting forage to be closely correlated to the date at which it is cut.[7] For New York State, digestibility is reduced approximately 0.5% for each day it is left unharvested after April 30. The formula used is:

$$\text{Digestible dry matter \%} = 85 - 0.48\,x,$$

where x equals the number of days after April 30.

Of course, total forage yield is increased with maturity as digestibility decreases. Therefore, a compromise between maximum digestibility and maximum yield must be determined in order to obtain optimum performance from cows per acre of forage. In New York, feeding trials showed that optimum performance came from hay harvested about June 1. Forage yield is less at this time than for more mature hay, but increased digestibility and palatability of early-cut hay more than compensate for reduced yield. Furthermore, increased yields from subsequent cuttings due to availability of more ground moisture at this time and the longer remaining growing season partially make up for reduced first-cutting yield. However, this method works only on first-cutting hay because the correlation between hay quality and harvest date of subsequent cuttings is low.

Limitations of Forage Testing

Most forage testing in the past has been based upon chemical determination of crude fiber and crude protein. These chemical analyses are good estimators of nutritional value within certain plant species grown under specified environmental conditions. However, accuracy of these estimates is greatly reduced when applied to a plant species grown under varying environ-

H06/19/81	1007	700	CENTRE	MIXED MOSTLY GRASS SILAGE
DATE	LAB NO.	SERIAL NO.	COUNTY	SAMPLE IDENTIFICATION

NAME OF FARMER:

PROC-W/OR W/O M

JOHN DOE
R D # 1
ANYTOWN PA 16801

THE PENNSYLVANIA STATE UNIVERSITY
MERKLE LABORATORY
COLLEGE OF AGRICULTURE
UNIVERSITY PARK, PENNSYLVANIA 16802
TEL. (814)-863-0841

FORAGE/FEEDSTUFF ANALYSIS AND INTERPRETATION

	YOUR SAMPLE ANALYSIS		EXPECTED RANGE * BASED ON AVERAGE FOR THIS TYPE OF SAMPLE DRY MATTER BASIS			QUALITY RATING * (SEE NOTE BELOW) BELOW OPTIMUM / OPTIMUM / ABOVE OPTIMUM
	AS SAMPLED	DRY MATTER BASIS				
MOISTURE (%)	50 (cow symbol)					XXXX
DRY MATTER (%)	50		26	TO	51	XXXXXXXXXX
HAY OR GRAIN EQUIV.	1.8					XX
CRUDE PROTEIN (%)	5.3	10.5	9.0	TO	15.8	X
HEAT DAMAGED PROTEIN (%)	0.8	1.6	1.4	TO	3.3	XXXXXXXX
AVAILABLE PROTEIN (%)	5.0	9.9	6.7	TO	13.3	
DIG. PROTEIN (%), ESTIMATED	3.3	6.5	4.3	TO	9.5	
FAT (%)						
FIBER (%)	20.3	40.6	39.3	TO	49.9	XXXXXXXXXX
TDN (%), ESTIMATED	28 (cow symbol)	56	53	TO	60	XX
ENE (THERMS/CWT.), EST.	24	47	32	TO	44	XX
NE/LACT (MCAL/LB), EST.	0.286	0.572	0.390	TO	0.528	XX
CALCIUM (%)	0.41	0.81	0.36	TO	0.92	XXXXXXXXXX
PHOSPHORUS (%)	0.14	0.27	0.17	TO	0.31	XXXXXXXX
MAGNESIUM (%)	0.10	0.19	0.11	TO	0.23	XXXXXXXX
POTASSIUM (%)	1.07	2.14	1.24	TO	2.36	XXXXXXXXXX
SULFUR (%)						
MANGANESE (PPM)	40	79	18	TO	100	XXXXXXXXXX
COPPER (PPM)	4	7	4	TO	18	XXXX
ZINC (PPM)	15	29	14	TO	40	XXXXXXXX
IRON (PPM)	139	277	47	TO	323	XXXXXXXXXX
ALUMINUM (PPM)	72	144	26	TO	180	
LEAD (PPM)	2	3				
SILICON (PPM)	150	300				
TRUE PROTEIN (%)						
CPE FROM NPN (%)						
NPN (%)						
NITRATE (AS % NO_3)						

NOTE: RATING COMPARES YOUR SAMPLE TO GOAL VALUES, NOT EXPECTED ONES, FOR THIS TYPE OF SAMPLE. IT DOES NOT INDICATE NUTRITIONAL ADEQUACY AS OUTLINED ON THE REVERSE SIDE.

EVERY INDUSTRY
DOLLAR AVE
ANYTOWN, PA 16801

DATE SAMPLE RECEIVED 08/27/81

FOR OFFICE USE ONLY

HYGRO	T	K	M	CO	CT	SA	I	ST
94.74	7	14	3	2				

(cow symbol) - FOR DHI SUPERVISORS - ORDER ON BARN SHEETS IS TDN, MOIST

* - NOT ALWAYS AVAILABLE

REMARKS:

SEE BACK FOR EXPLANATION
FARMER

FIG. 10.15 A sample report form used by the Penn State Soil and Forage Testing Laboratory.

mental conditions or when applied to mixtures of forages.

A method of forage evaluation developed by the USDA is more accurate under a wider range of conditions.[8] Plant tissue is analyzed for its cell wall constituents (CWC) and cellular contents (CC). Digestibility of CC always is very high. However, the digestibility of CWC depends on the amounts of lignin and cellulose present. This lignocellulose is insoluble in acid-detergent, so it is called acid-detergent fiber (ADF). As ADF increases, digestibility of a plant decreases. More accurate prediction of forage quality is possible from ADF content of forages than from other chemical analyses now in use, and it gradually is replacing other methods used in the past.

Near Infrared Reflectance (NIR)

An instrument developed by USDA researchers uses near infrared light reflectance to estimate acid detergent fiber, crude protein, and other chemical constituents and nutritional characteristics of forages and other feedstuffs. When properly calibrated with samples of known chemical composition, NIR analyses are nearly as accurate as actual chemical determinations of forage composition. Rather than the day or more required to run chemical analyses for acid detergent fiber and crude protein, NIR can make the same determinations in a few minutes. The time savings is a tremendous advantage when analyses are needed for buying or selling forage, or for ration formulation. Research and development of NIR technology is continuing. It will have a widespread effect on forage testing when sufficient data are available for calibration of NIR machines for the great variety of forages fed to cattle, and the variable conditions under which the forages are grown.

Heat Damaged Forages

If a forage gets excessively hot during storage, spontaneous combustion can occur. Even if the forage does not reach the temperature for spontaneous combustion, its nutrients can be seriously damaged. This heat damage occurs especially to protein and, to a lesser extent, the energy content of the forage. Proteins that are heat damaged become less degradable by ruminal microorganisms. Some of the undegradable protein may be digested further down the digestive tract in the abomasum and intestines, but severely heat-damaged protein becomes so indigestible that very little is available to the cow.

Heat damage can occur with any type of forage, but is especially a problem with low-moisture silage (haylage).[9] When moisture in the forage is inadequate to dissipate the heat produced by plant respiration and fermentation of the forage, the Maillard reaction takes place. The extent of the heat damage from the Maillard reaction can be estimated by several tests. The 2 tests that have been used the most are (1) pepsin digestion, and (2) acid-detergent insoluble nitrogen (ADIN). The ADIN test is used by the Penn State Forage Testing Service to test for heat damage, as described on page 171. Normally, only about 10% of the forage nitrogen shows up in the ADIN fraction, whereas it may be as high as 90% in severely heat damaged forages. When heat damage is suspected, these tests can be used to estimate the extent of the damage and the amount of protein that actually will be available to the animal.

10.10 Concentrates

In contrast to roughages which are bulky and fibrous, concentrates contain high levels of energy and/or protein and usually are lower in fiber, thus giving them a higher density. Concentrates include the grains and many by-product feeds, including the high-protein supplements. Exceptions to the low fiber rule are some by-product feeds, such as dried beet pulp and dried citrus pulp, which are high in energy, but also contain considerable fiber that is very digestible. These are called bulky concentrates. Nutrient analyses of some concentrates commonly used in dairy rations are shown in Appendix Table III-D.

Grains

The principal grains fed to dairy cattle in the U.S. are corn, barley, oats, milo, and wheat. All are of approximately equal feeding value for dairy cows.[10] Corn, milo, and wheat are lower in fiber and slightly higher in energy than barley and oats. However, variation in energy and protein levels is greater within a type of grain than between them. The cob often is ground with corn kernels, making ground ear corn, a very popular feed in the Corn Belt. Any of the above-listed grains can be used as the major constituent of a dairy concentrate mix with good results. However, palatability usually is increased when no single grain makes up more than 80% of the concentrate mix.

All of the grains are relatively high in energy and phosphorus, and low in protein and calcium, which must be made up from other sources. Legume forages are high in these latter nutrients, making the combination of legume forage and a simple concentrate mix based upon one or more grains an excellent feeding program.

By-Product Feeds

Many by-product feeds are available at times at a reasonable price. These include wheat bran and millrun from the flour industry, hominy feed from corn, beet pulp from sugar beets, citrus pulp from fruit juice processing, distillers' grains from the liquor industry, brewers' grains from breweries, bakery waste, molasses from sugar cane and sugar beets, rice bran and polishings, and dried whey from cheese processors. Usually none of these is used as a primary component of dairy rations, but all can be used efficiently as minor ingredients. Not only do they supply valuable nutrients, but often they are the least expensive source of nutrients, especially in areas close to processing plants producing the by-products.

High-Protein Supplements

Concentrate mixtures high in protein are required when low-protein forage is fed to dairy cattle. Corn silage, other nonlegumes, and poor-quality silage, hay, and pasture of all types are low in protein. By-products of the oilseed industry are relatively high in protein and are used extensively as protein supplements for dairy cows.

The principal high-protein feeds are soybean meal, cottonseed meal, linseed meal, safflower meal, and coconut meal. Protein content of soybean, cottonseed, and linseed meals varies from about 35 to 50%, whereas safflower and coconut meals contain about 20% protein. Energy values of the meals vary also depending on the type of process used to extract oil from the seed. In general, solvent-extraction processes remove a higher percentage of oil than expeller methods, thereby leaving a meal which is lower in fat. Consequently, the solvent-extracted, low-fat meals are lower in energy.

Urea is another substance which commonly is used as a protein supplement in dairy rations. Although urea is a nonprotein nitrogen (NPN) compound, it can be utilized by ruminal microorganisms to synthesize bacterial protein, as discussed in Chapter 8. Feed-grade ureas contain from 42 to 46% nitrogen thereby giving a crude protein equivalent of 262 to 288%.

Urea and other NPN compounds contain no energy, minerals, or vitamins, whereas oil meals are good sources of energy and excellent sources of many of the required minerals. Therefore, these nutrients must be supplied from other sources when NPN replaces oil meal in dairy rations. However, the price per unit of protein equivalent usually is much lower from urea and other NPN components compared with natural protein supplements.

Liquid Supplements

Liquid supplements containing molasses, urea, phosphoric acid, other by-products, minerals, and vitamins are fed in some areas to supplement the regular dairy ration. Usually the liquid is available free-choice in tanks equipped with a wooden wheel which rotates through the sticky liquid as cows lick it (Figure 10.16). Average consumption of the liquid is regulated by varying the amounts of phosphoric acid, urea, or other unpalatable ingredients in the mix. Even when this is done, however, there are large variations in individual intake between cows. Some may consume more than enough to fulfill needs for minerals, vitamins, and protein (from urea), whereas others may not voluntarily consume enough to meet their requirements.

A better method for utilizing liquid supplements is to add them to the forage as it is fed. This works well for automated si-

FIG. **10.16** Liquid supplement being fed with a lick-wheel feeder.

lage-feeding systems or for complete rations fed with mechanized feeding equipment. In these cases, liquid supplement intake is proportional to the forage intake of each cow and will approximate her requirements more closely than when it is fed free-choice.

Cost of nutrients in liquid form usually is higher than the cost of nutrients in dry form. The convenience of handling in liquid form and greater ease of mixing with other ingredients may justify the higher cost in some cases. Any increased cost must be justified by non-nutritional factors, however, because use of nutrients is the same whether they are fed in dry or liquid form.

Other Concentrate Feeds

There are many other concentrate-type feeds fed to dairy cattle in various areas of the U.S. and the world. For a more comprehensive coverage of this subject, the interested student is directed to other books on this subject listed as "Suggested Additional Reading" at the end of the chapter.

Treatment of Concentrates

More efficient utilization of grains is made by dairy cattle if the seed is broken, ground, rolled, or pelleted before feeding. Otherwise, many of the kernels pass through the cow undigested. A cow does not chew her food well before swallowing, and kernels which escape regurgitation and remastication are likely to resist digestion and be excreted whole.

The most common treatment of grains is grinding. A coarse grind is preferable due to the dustiness and reduced palatability which results from fine grinding. Rolled grains also are very palatable, but more elaborate equipment is needed for this process.

Pelleted concentrate mixtures are popular with dairymen using mechanical feeding equipment because of the ease of handling pellets with augers. Pelleting also increases the amount of concentrates that a cow can eat in a short time, which is important when they are milked in parlors and have only a short time for eating. Fat tests of cows eating pelleted concentrate may be depressed about 0.1%, but this reduction is not nearly as marked as was noted earlier for pelleted hay.

Cooking of concentrates may have some beneficial effects for growing beef cattle, but it causes severe milk fat depression in dairy cows. Therefore, it is not recommended for dairy rations.

High-Moisture Grains

Most grain is fed in air-dry form at about 10 to 15% moisture. However, increasing amounts of ensiled, high-moisture grains are being fed, especially by dairy farmers who grow their own grain. Compared with harvesting for storage as dry corn, high-moisture grain is harvested earlier when it contains more moisture, or water is added to bring the moisture content up to about 30% at the time of ensiling. Airtight, upright silos are the principal storage structures used for high-moisture grain, but some is stored in bunkers, pits, or vacuum-stack silos.

Corn is the principal grain handled in this manner, but barley, milo, and other grains also are stored and fed in this form. Although the grains go through a fermentation process in the silo and come out softened and very palatable, they should be ground or rolled to break the seed coat in order to increase their digestibility. Most grain is ensiled whole and crushed as it comes out of the silo, but it can be ground as it goes into the silo. This latter process also is used when high-moisture ground ear corn is ensiled.

Advantages of high-moisture grains are an earlier harvesting date to miss possible bad weather, less field, harvesting, and storage losses, lower storage costs, less rodent damage, and simplified mechanical feeding. Feeding value of high-moisture grains is increased compared with dry grain for fattening beef cattle. However, this is not the case with lactating dairy cows which utilize dry or high-moisture grains with equal efficiency.[11] Therefore, it takes about 20% more high-moisture grain than dry grain to obtain equal production from dairy cows because of the greater amount of water in high-moisture grains.

High-moisture grains treated with chemical preservatives need not be stored in an airtight structure. When treated with propionic acid, or mixtures of propionic with other acids such as acetic, spoilage losses are minimal and the grains remain palatable to cattle.

Mineral Supplements

Although mineral supplements do not contain energy or protein, they are included in the concentrate class because of their high density.

As discussed in Chapter 9, the minerals most commonly lacking in dairy rations are calcium, phosphorus, sodium, and chlorine. Sodium chloride (salt) usually is provided free-choice as well as included in concentrate mixes at levels of 0.5 to 1%. Calcium and phosphorus supplements include bone meal, and mono-, di-, and tricalcium phosphates. Calcium alone can be provided as ground limestone (calcium carbonate) or oystershell flour. Supplements which contain phosphorus but no calcium are mono- and diammonium phosphate, mono- and disodium phosphates, and sodium tripolyphosphate. Any of these can be fed singly free-choice, but palatability and intake of most is low unless they are included in the concentrate or mixed half and half with salt. Most tests have shown bone meal to be the most palatable calcium source and sodium tripolyphosphate the most palatable phosphorus source when fed alone.

Other minerals are required in such small amounts that, when needed, they usually are added as trace mineralized salt or included in purchased concentrates in trace amounts. Many commercial mineral mixtures are available which contain combinations of minerals in proportions designed to supplement various types of dairy rations.

Vitamins

In Chapters 8 and 9 it was pointed out that most vitamins are synthesized by ruminal microorganisms for use by the cow. Dietary sources are needed only for vitamins A, D, and E. Supplies of these vitamins, either singly or in combination, are available commercially either as feed additives or as injectable liquids. Only under very unusual circumstances is their use justified for dairy cows, however, as most common rations supply adequate amounts of all vitamins.

10.11 Summary

Although pasture has been a major feed source for dairy cows in the past, there is a trend towards more year-around dependence on hay and silage now, especially in larger herds. Thus, in addition to pasture, the major roughages fed to dairy cattle in the U.S. are alfalfa hay, corn silage, and silage or haylage made from alfalfa, alfalfa-grass mixtures and cereals.

Corn grain is the most popular concentrate fed in the U.S., but large tonnages of other grains, such as barley, oats, milo, and wheat, also are consumed by dairy cows. In addition, many by-product feeds are utilized by cows, especially in areas near processing plants.

Level of concentrate feeding has increased through the years resulting in great increases in milk production per cow. This high production has made the need for higher-quality forage more critical because of greater nutrient requirements and limited feed capacity in the rumen.

The successful dairyman today must be much more cognizant of the feeding value of various feeds than in the past. Feed costs make up 50 to 65% of the total cost of milk production, so feed is the largest single area where profits can be increased, either by reduced feed costs or by increased efficiency of feed utilization.

Review Questions

1. List several reasons why pasture is becoming less important as a dairy forage in the U.S.
2. What are the 5 stages of silage fermentation, and what factors affect the course of this fermentation?
3. Why is pelleted forage unsuitable as the only source of roughage for dairy cows?
4. Why do various areas of the U.S. rely on forage in different forms? Why isn't one form best for all areas?
5. What does the term "forage quality" mean, and how can it be estimated?
6. If the ruminant stomach is designed primarily for efficient utilization of roughages, why have increasing amounts of concentrates been fed to dairy cows over the years?
7. Besides salt, which minerals are most likely to be needed as supplements when: (a) legume forage is fed free-choice; (b) nonlegume forage is fed free-choice?

References

1. Huber, J. T., Sandy, R. A., Polan, C. E., Bryant, H. T., and Blaser, R. E.: Varying levels of urea for dairy cows fed corn silage as the only forage. J. Dairy Sci. *50:* 1241, 1967.
2. Meyer, J. H., and Jones, L. G.: Controlling alfalfa quality. Calif. Agr. Expt. Sta. Bull. 784, March, 1962.
3. Ronning, M.: Effect of varying alfalfa hay-concentrate ratios in a pelleted ration for dairy cows. J. Dairy Sci. *43:* 811, 1960.
4. Dobie, J. B., and Curley, R. G.: Hay cube storage and feeding. Calif. Agr. Expt. Sta. Circular 550, September, 1969.
5. Adams, R. S.: How to use the Penn State Forage Testing Service. The Pennsylvania State Univ. College of Agr., Extension Service, Univ. Park, Pa., October, 1963.
6. Moore, L. A.: Relationship between TDN and energy values of feeds. J. Dairy Sci. *36:* 93, 1953.
7. Reid, J. T., Kennedy, W. K., Turk, K. L., Slack, S. T., Trimberger, G. W., and Murphy, R. P.: Effect of growth stage, chemical composition, and physical properties upon the nutritive value of forages. J. Dairy Sci. *42:* 567, 1959.
8. Van Soest, P. J.: Development of a comprehensive system of feed analysis and its application to forages. J. Animal Sci. *26:* 119, 1967.
9. Goering, H. K., Gordon, C. H., Hemken, R. W., Waldo, D. R., Van Soest, P. J., and Smith, L. W.: Analytical estimates of nitrogen digestibility in heat damaged forages. J. Dairy Sci. *55:* 1275, 1972.
10. Tommervik, R. S., and Waldern, D. E.: Comparative feeding value of wheat, corn, barley, milo, oats, and a mixed concentrate ration for lactating cows. J. Dairy Sci. *52:* 68, 1969.
11. McCaffree, J. D., and Merrill, W. G.: High-moisture corn for dairy cows in early lactation. J. Dairy Sci. *51:* 553, 1968.

Suggested Additional Reading

Morrison, F. B.: *Feeds and Feeding.* 22nd ed. Ithaca, N.Y.: The Morrison Publishing Co., 1956.

United States-Canadian Tables of Feed Composition, Third Revision. Washington, D.C.: National Academy Press, 1982.

CHAPTER 11

Feeding the Milking Herd

11.1 Introduction

Many types and combinations of feeds for dairy cattle will result in good health and high milk production. No particular feed is essential. Rather, it is the proper balance of nutrients (energy, protein, minerals, and vitamins) offered in a palatable form that distinguishes a good ration from a poor one. The purpose of this chapter is to discuss the development of practical and economical rations for lactating cows which fulfill all of the nutrient requirements set forth in Chapter 9. Rations for dry cows, bulls, and dairy steers will be discussed in Chapter 13 and for youngstock in Chapter 20.

Nutrient requirements of lactating cows vary tremendously during the lactation cycle, as illustrated in Figure 11.1. The example in Figure 11.1 illustrates the milk production, feed dry matter intake, and body weight change for a 1,400-lb cow producing 23,000 pounds of milk during a 10-month lactation period, but larger or smaller cows producing more or less milk would have similar trends in these parameters.

The first 2 months after calving is the most difficult time period for fulfilling nutrient requirements, especially energy and protein. A cow usually reaches peak milk production 6 to 8 weeks after parturition, but her appetite is poor for several weeks. Voluntary feed dry matter intake gradually increases after calving, usually reaching a peak at 12 to 14 weeks. The lag in feed intake behind milk production results in a negative energy balance. A cow in good condition can use her body fat reserves as a source of energy for milk production, resulting in body weight loss, as illustrated in Period 1 of Figure 11.1.

After about 10 weeks, voluntary feed intake usually is adequate to fulfill energy requirements if a nutritionally adequate and palatable ration is fed. Energy balance is attained and the cow can maintain her body weight, as illustrated in Period 2 of Figure 11.1. During Periods 1 and 2, cows should be fed liberal amounts of concentrates in order to minimize the period of body weight loss. Shortening the period of negative energy balance allows the cow to reach her maximum inherited ability for milk production and may also have a beneficial effect on fertility. Conception rates usually are higher when cows are gaining weight compared with cows in negative energy balance.

During Period 3, milk production declines in the later stages of lactation. Dry matter intake remains relatively high resulting in energy intake above her requirements. This allows the cow to go into positive energy balance and regain weight lost

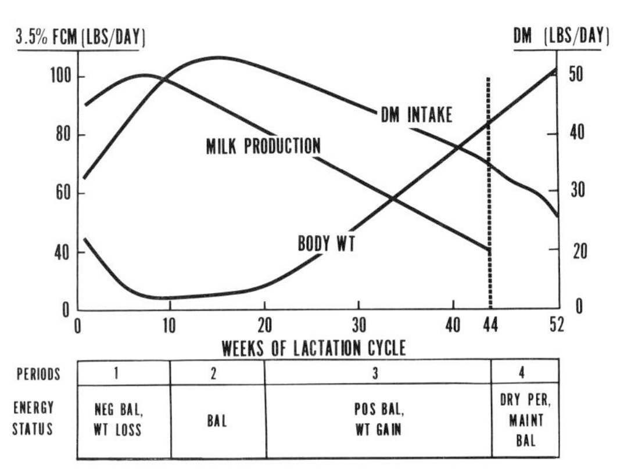

FIG. 11.1 Trends in milk production, feed dry matter intake, and body weight change during the lactation cycle.

in early lactation. Lower energy rations should be fed in the latter part of Period 3 to prevent the cow from becoming overly fat during late lactation.

Period 4 consists of the dry period. Rations for dry cows are discussed in Chapter 13.

The changing milk production, voluntary feed intake, and body weight of lactating cows during the lactation cycle makes formulation of nutritionally adequate dairy rations complex. The following sections illustrate some procedures and methods that can be used to help solve the problem of providing the proper amount of nutrients during the different phases of a cow's lactation cycle.

11.2 Calculating Nutrient Requirements for a Ration

Before we can formulate a ration for a dairy cow, we must know the amounts of each of the nutrients needed to meet the requirements for maintenance of her body and for milk production. Furthermore, lactating heifers need additional nutrients to continue growing and pregnant cows need extra nutrients for growth of the fetus, especially during the last 2 months of gestation.

Energy, protein, calcium, phosphorus, and vitamin A allowances for maintenance, growth, and reproduction for cows of various body weights are listed in Appendix Table III-B. Requirements for these nutrients per pound of milk with various fat percentages are also listed in Appendix Table III-B. Other minerals and vitamins are required, but only in very small amounts. Most rations provide these latter nutrients in sufficient amounts when energy and protein requirements are fulfilled. When needed in greater amounts than provided by a ration, they usually are added as a mineral and vitamin premix or as trace mineralized salt. Therefore, usually only the major nutrients listed in Appendix Table III-B are used to formulate dairy rations.

The following example illustrates the steps in determining amounts of nutrients needed by a 1,400-lb mature cow in a large dry lot dairy producing 60 lb of milk daily with a 3.5% fat test.

Maintenance

In Appendix Table III-B, under the section labeled "Maintenance of Mature Cows," look down column 1 to the row for a cow with a body weight of 1,400 lb. Look across that row to the column labeled NE_1, which shows the maintenance energy requirement to be 10.12 Mcal per day. Continuing across the row, allowances of other nutrients for a 1,400-lb cow are listed. In addition to NE_1, we will use crude protein (CP), calcium (Ca), phosphorus (P), and vitamin A allowances in this example. The values are:

Maintenance of a 1,400-lb cow

NE_1 (Mcal)	CP (lb)	Ca (lb)	P (lb)	Vitamin A (1,000 I.U.)
10.12	1.12	0.048	0.039	48

Activity

The maintenance allowances listed in Appendix Table III-B provide adequate energy for the usual activity of cows housed in conventional stanchion barns, free-stall barns, and many dry lot systems. However, when cows are required to walk unusually long distances to and from the milking parlor, the maintenance allowance should be increased as much as 10% to accommodate extra energy expenditures for activity.

For grazing, allowances for maintenance should be increased 10% for good pasture and 20% for sparse pasture. In this example, the cow is part of a large dry lot dairy where considerable walking is required between the corrals and the milking parlor. Therefore, the activity allowance is estimated to be about 10% of the maintenance energy allowance.

Milk Production

In this example, the cow is producing 60 lb of 3.5% milk each day. Nutrients required for this amount of milk must be added to the preceding allowances for maintenance. Values for milk are found in Appendix Table III-B. Look down column 1 (% Fat) to the row for 3.5%. The allowances in that row are for 1 lb of milk. Each must be multiplied by 60 to obtain the amount needed for 60 lb of milk per day, as shown below.

Nutrient	*Required per lb of milk*		*Lb of milk*	
NE_1	.31	×	60 =	18.60 Mcal per day
CP	.082	×	60 =	4.92 lb per day
Ca	.0026	×	60 =	.156 lb per day
P	.0018	×	60 =	.108 lb per day
Vitamin A	0	×	60 =	0

Now the nutrient allowances calculated for 60 lb of milk are added to those required for maintenance and activity to obtain the total allowances for maintenance and milk production.

If the cow were a lactating heifer, we would add 20% to all maintenance allowances except vitamin A during the first lactation and 10% during the second. Similarly, if she were a dry cow in her last 2 months of gestation, we would use the Maintenance Plus Last 2 Months' Gestation allowances in Appendix Table III-B rather than the allowances for Maintenance of Mature Cows. In this case, however, we will simplify the procedure by assuming that we are working with a mature cow in early or midlactation. Nutrient requirements for this 1,400-lb cow producing 60 lb of milk with 3.5% fat are shown in Table 11.1.

TABLE 11.1 *Nutrient Requirements*

	NE_1 (Mcal)	*CP (lb)*	*Ca (lb)*	*P (lb)*	*Vitamin A (1,000 IU)*
Maintenance	10.12	1.12	.048	.039	48
Activity (10%)	1.01	—	—	—	—
60 lb milk (3.5%)	18.60	4.92	.156	.108	—
Totals	29.73	6.04	.204	.147	48

11.3 Formulating a Ration

The next step is to formulate a ration that has the desired nutrients and will be eaten in adequate amounts to fulfill all of the preceding nutrient requirements. This goal can be accomplished with many different combinations of feeds. We will simplify the procedure by using a limited number of feeds—corn silage, alfalfa hay, corn grain, soybean meal, and dicalcium phosphate.

Voluntary Feed Consumption

Begin with the forages because they usually are the least expensive source of nutrients and the ruminant animal provides its greatest contribution to the world food supply as a converter of roughages to milk and meat for human consumption. The forages could meet all of the requirements if a cow would eat enough of those feeds. However, the capacity of the rumen is limited and cows can eat only enough bulky forages to meet the requirements for maintenance plus about 20 to 30 lb of milk per day, depending on the quality of the forage. Additional nutrients, particularly energy, must be provided by concentrates if higher levels of milk production are to be maintained as illustrated in Figure 9.2.

Maximum daily voluntary intake of forage by lactating cows ranges between about 1.5 and 3 lb of dry matter (DM) per 100 lb (cwt) of body weight. The better the forage quality, the more the cow will consume. Maximum silage DM intake is only about 2.2 to 2.4 lb per cwt, probably because of its physical form and reduced palatability at higher levels of consumption. For the cow in the example, this intake would amount to about 32.2 lb of silage DM (1,400 lb × 2.3 lb/cwt = 32.2 lb silage DM). If the silage has 30% DM, this would be equivalent to 107 lb of silage as fed (32.2 lb of silage DM ÷ 30% DM = 107 lb silage at 30% DM). However, for the example cow producing 60 lb of milk, it would take 143 lb of silage to meet the NE_1 requirement, 247 lb for the protein requirement, 252 lb for calcium, and 245 lb for phosphorus, amounts that greatly exceed the maximum intake possible by a 1,400-lb cow. Therefore, concentrates must be fed in addition to silage in order to balance the ration for the required nutrients. The same situation is generally true for all forages.

In the case of the complete ration (roughage plus concentrate), maximum dry matter intake is closely related to level of milk production as well as body size. Estimated maximum total ration dry matter intakes at various levels of milk production are shown in Table 11.2. Dry matter intakes are listed as percentages of body weights, which is the same as expressing them as lb per cwt. Milk production is listed as pounds of 4% fat-corrected-milk (FCM). Although more research is needed to verify the accuracy of the data in Table 11.2, they can be used as rough guidelines for establishing maximum amounts of dry matter that can be eaten by cows producing various amounts of milk. Estimates of maximum dry matter intake are necessary in many cases because roughages frequently are fed free-choice to dairy cattle making it difficult to determine actual amounts consumed.

Nutrients from Forage

Let us assume that forage supplies are such that we will limit alfalfa hay to 10 lb per cow daily, feed corn silage free-choice to appetite, and feed sufficient amounts of a concentrate mix made up of corn grain, soybean meal, and dicalcium phosphate to fulfill the requirements of the example cow.

The first problem is to determine the approximate amount of silage that the cow is eating free-choice in order to calculate the nutrients provided by the forages. This amount can be approximated using data from Table 11.2 and Appendix Tables III-C and III-D.

Appendix Table III-D contains a listing of the nutrient content of the more common dairy cattle feeds. Nutrient values for the

TABLE **11.2** *Dry Matter Intake Guidelines*

Body wt (lb)	*800*	*1,000*	*1,200*	*1,400*	*1,600*	*1,800*
4% FCM (lb)	*% of body wt*					
20	2.5	2.4	2.3	2.2	2.1	2.0
30	2.8	2.6	2.5	2.4	2.2	2.1
40	3.1	2.8	2.7	2.6	2.4	2.3
50	3.4	3.1	3.0	2.8	2.6	2.5
60	3.6	3.3	3.2	3.0	2.8	2.7
70	3.9	3.6	3.4	3.2	3.0	2.9
80	4.1	3.8	3.6	3.4	3.2	3.1
90	—	3.9	3.7	3.5	3.3	3.2
100	—	4.1	3.9	3.7	3.5	3.4

Source: National Academy of Sciences—National Research Council: Nutrient Requirements of Dairy Cattle (reference No. 4).

example feeds, on a dry matter basis, are as shown at the top of page 182. In Appendix Table III-C under Lactating Cow Rations, the recommended NE_1 concentration in the ration dry matter is 0.73 Mcal/lb for cows between 1,300 and 1,550 lb producing between 46 and 64 lb of milk daily. The maximum percentage of roughage that can be included in the ration and still meet the recommended NE_1 concentration can be calculated from the data in Table 11.3. Assuming that only about one-third of the roughage dry matter comes from alfalfa hay (because it is limited to 10 lb), and about two-thirds comes from corn silage, the NE_1 content of the roughage will be about 0.66 Mcal/lb [(0.33)(0.59 Mcal/lb) + (0.67)(0.694 Mcal/lb) = 0.66 Mcal/lb]. Similarly, if the concentrate mix contains about 70% corn and 30% soybean meal, the NE_1 content would be about 0.90 Mcal/lb [(0.70)(0.921) + (0.30)(0.844) = 0.90]. The percentage of roughage at 0.66 Mcal/lb and concentrate at 0.90 Mcal/lb that will result in a total ration with 0.73 Mcal/lb is calculated as follows, with R = roughage and (1 − R) = concentrate:

$$
\begin{aligned}
0.66R + 0.90\ (1 - R) &= 0.73 \\
0.66R + 0.90 - 0.90R &= 0.73 \\
-0.24R &= -0.17 \\
R &= 0.71 \\
&= 71\%\ \text{roughage}
\end{aligned}
$$

Therefore, the example ration should contain about 71% roughage and 29% concentrate.

Ingredient	Dry matter basis					As fed
	NE_1 (Mcal/lb)	CP (%)	Ca (%)	P (%)	Vit A (1,000 IU/lb)	DM (%)
Alfalfa hay (early bloom)	.590	17.2	1.25	0.23	15.4	90
Corn silage (average)	.694	8.0	0.27	0.20	8.2	30
Corn grain (ground)	.921	10.0	0.03	0.31	0.4	89
Soybean meal (solv-extd 46% protein)	.844	51.8	0.36	0.75	0	89
Dicalcium phosphate	0	0	23.70	18.84	0	96

Roughage Dry Matter Intake

The example cow is producing 60 lb of milk with 3.5% fat. Her daily milk fat production is 2.1 lb ($60 \times 3.5\% = 2.1$). To convert her milk production to 4% fat-corrected-milk, the following equation is used:

$$\begin{aligned} 4\%\ \text{FCM} &= 0.4 \times \text{lb milk} + 15 \times \text{lb fat} \\ &= (0.4)(60) + (15)(2.1) \\ &= 55.5\ \text{lb} \end{aligned}$$

From Table 11.2, a 1,400-lb cow producing 55.5 lb FCM will eat about 2.9% of her body weight (interpolate between the values for 50 and 60 lb) or 40.6 lb DM (1,400-lb cow $\times$ 2.9% = 40.6 lb DM). If 71% of this is roughage, the cow will consume about 28.83 lb of roughage DM ($40.6 \times .71 = 28.83$). The cow is being fed 10 lb of alfalfa hay, which is equivalent to 9 lb of hay DM (10 lb $\times$ 90% DM = 9 lb DM). Next, the free-choice intake of silage can be calculated by subtracting the hay DM consumed from the maximum roughage DM intake calculated previously, which results in 19.83 lb silage DM (28.83 lb roughage DM − 9 lb hay DM = 19.83 lb silage DM). This amount would be equivalent to 66.1 lb corn silage at 30% DM ($19.83 \div 0.30 = 66.1$). The nutrients provided by 10 lb of alfalfa hay and 66.1 lb of corn silage can be calculated as follows:

Concentrate Mix

The alfalfa hay and corn silage provide more than enough vitamin A equivalent for the cow, as shown in the preceding calculations. However, there is a shortage of 10.66 Mcal of NE_1, 2.906 lb crude protein, 0.0380 lb calcium, and 0.0866 lb phosphorus, which must be supplied by a concentrate mix. The biggest deficits are for NE_1 and protein. Corn grain is the best source of NE_1 of the available feeds (0.921 Mcal/lb). Soybean meal is the best source of protein (51.8%).

Simultaneous equations can be used to calculate the amounts of corn grain and soybean meal to feed to fulfill the NE_1 and protein requirements. Any deficits of calcium and phosphorus will be fulfilled subsequently by adding dicalcium phosphate.

Corn grain has an NE_1 value of 0.921 Mcal/lb and 10% crude protein. Values for soybean meal (solv-extd, 46% protein) are 0.844 Mcal/lb and 51.8%. If x = lb of corn DM to feed, and y = lb of soybean meal DM to feed, the calculations are as follows:

Equation (1)

$$0.921x + 0.844y = 10.66\ (NE_1)$$

Equation (2)

$$0.10x + 0.518y = 2.906\ (\text{Protein})$$

Multiply equation (2) by the dividend that

Ingredient	Amount		NE_1 (Mcal)	CP (lb)	Ca (lb)	P (lb)	Vit A equiv. (1,000 IU)
	As Fed (lb)	DM (lb)					
Alfalfa hay (early bloom)	10.0	9.00	5.31	1.548	0.1125	0.0207	139
Corn silage (average)	66.1	19.83	13.76	1.586	0.0535	0.0397	163
Totals from forage	76.1	28.83	19.07	3.134	0.1660	0.0604	302
Amounts required			29.73	6.04	0.2040	0.1470	48
Needed in concentrate mix			10.66	2.906	0.380	0.0866	0

is obtained by dividing the coefficient for x in equation (1) by the coefficient for x in equation (2). In this case, it would be 9.21 ($0.921 \div 0.10 = 9.21$).

Multiplying equation (2) by 9.21, the equations become:

$$(1)\ 0.921x + 0.844y = 10.66$$
$$(2)\ 0.921x + 4.77078y = 26.76426$$

Subtracting equation (2) from equation (1), the result is:

$$-3.92678y = -16.10426$$
$$y = 4.10114$$
$$= \text{lb of soybean meal DM}$$

Substituting this value for y in equation (1), it becomes:

$$(1)\ 0.921x + (0.844)\ (4.10114) = 10.66$$
$$0.921x = 7.19864$$
$$x = 7.816$$

Rounding the values of x and y to the nearest one-hundredth and calculating the nutrients in 7.82 lb corn DM and 4.10 lb of soybean meal DM results in:

	DM (lb)	*NE_1 (Mcal)*	*CP (lb)*	*Ca (lb)*	*P (lb)*
Corn grain	7.82	7.20	.782	0.0023	0.0242
Soybean meal	4.10	3.46	2.124	0.0148	0.0308
	11.92	10.66	2.906	0.0171	0.0550
Needed in concentrate mix		10.66	2.906	0.0380	0.0866
Deficits				0.0209	0.0316

With the addition of 7.82 lb corn DM and 4.10 lb of soybean meal DM, all allowances are fulfilled except calcium and phosphorus. Dicalcium phosphate contains no energy or protein but is high in both calcium and phosphorus (23.7% Ca and 18.84% P on a dry matter basis). Therefore, it can be used to fulfill both the calcium and phosphorus deficits. Dividing the largest remaining deficit of 0.0316 lb P by 0.1884, it takes 0.17 lb of dicalcium phosphate to fulfill the P requirement. This amount of dicalcium phosphate also more than fulfills the calcium requirement.

Dividing the concentrate ingredient DM amounts by their DM percentages gives the amounts on an "as fed" basis.

Corn grain
$$7.82 \text{ lb DM} \div 0.89 = 8.79 \text{ lb}$$
Soybean meal
$$4.10 \text{ lb DM} \div 0.89 = 4.61 \text{ lb}$$
Dicalcium phosphate
$$0.17 \text{ lb DM} \div 0.96 = 0.18 \text{ lb}$$
$$13.58 \text{ lb}$$

Therefore, our concentrate mixture should contain 8.79 lb corn grain, 4.61 lb soybean meal, and 0.18 lb dicalcium phosphate. On a percentage basis, this is 64.7% corn, 34% soybean meal, and 1.3% dicalcium phosphate. This concentrate mix in combination with 10 lb alfalfa hay and 66 lb corn silage fulfills the allowances for all 5 nutrients used as a basis for formulating a ration in the example. In addition, we should provide salt free-choice or add 0.5% salt to the concentrate mix. In areas known to be deficient in certain mineral elements, or where forage tests indicate mineral deficiencies, they should be added to bring mineral concentrations up to the levels recommended for Lactating Cow Rations in Appendix Table III-C.

Our daily ration for a 1,400-lb cow producing 60 lb of 3.5% milk, as shown in Table 11.3, has an excess of calcium and vitamin A, but just meets the NE_1, protein, and phosphorus requirements. However, considerable variation exists in the nutrient content of feeds. We used average values that may be somewhat higher or lower than the nutrient content of the feeds actually being fed. An increase of about 10% in the concentrate allowance, from 13.58 to about 15 lb per day, would be desirable to ensure adequate amounts of all nutrients that are just barely fulfilled by the example ration.

Minimum Fiber Levels

High levels of fibrous feeds (roughages) limit milk production by filling the rumen to capacity before all nutrient needs are met, as pointed out in the previous section on Voluntary Feed Consumption. However, a minimum amount of fiber in the ration is essential for normal ruminal function and production of milk with normal milk fat content. Not only is the level of fiber critical, but also its form. Milk fat percentage is depressed by hay and other forages that are finely ground before feeding, even though the fiber level may be adequate. Many factors are involved, not all of which are completely understood. With present knowledge, a reasonable guideline to prevent milk fat depression is to supply a minimum of 17% crude fiber (CF) in the ration dry matter for lactating cows.[1] This amount is equivalent to about 21% acid-detergent-fiber (ADF), a laboratory test that gradually is replacing crude fiber as a measure of forage quality.

In the example ration, the crude fiber level is 19.3% and the acid detergent fiber level is 24.8%, as shown in Table 11.3. Therefore, the example ration exceeds the minimum fiber levels recommended for lactating cows. If the minimum fiber level has not been met, other higher fiber ingredients should be included in the ration to bring the crude fiber level up to at least 17%.

In addition to minimum fiber levels, roughage DM amounts should be at least 1.35% of the body weight of the cow in order to ensure that a major portion of the fiber in the ration comes from roughages rather than from concentrates because the fiber from concentrates is not as effective in maintaining normal fat test. For a 1,400-lb cow, this amounts to a minimum of 18.9 lb DM per day from roughages that have not been finely ground. The example ration has 28.83 lb roughage DM, which is more than the minimum amount, as shown in Table 11.3.

Economic Considerations

For simplicity in the previous example, we limited ourselves to 5 feed ingredients. Nothing was said about relative prices of available feeds. When a dairyman grows his own forage and grain, he usually will feed them to his own cows because net returns from feeding home-grown crops to cows usually are greater than selling the original crops. In this case, the dairyman probably buys only protein and mineral supplements to mix with homegrown concentrates. Availability and prices of other ingredients are of less interest to him than they would be to a dairyman who buys most of his cattle feed. In this latter case, prices of purchased feeds have a greater effect on the net income of the dairyman.

This adds another dimension to the ration formulation problem. Not only must the ration be palatable to the cows and balanced to provide sufficient nutrients, but it also must be reasonably priced if a profit from the sale of milk is to be realized. This means that the price and nutrient content of many different feed ingredients must be considered rather than the few we used in the previous example. This adds greatly to the complexity and time required to formulate a ration. Fortunately, electronic computers have been programmed to eliminate the drudgery of balancing rations when large numbers of feeds at varying prices are to be considered. These so-called "least-cost" or "maximum profit" rations formulated by computers are discussed in detail in Chapter 12.

Substitution of Feeds in a Ration

At times it is necessary to make changes in a ration due to unavailability of certain feeds. For example, suppose that the dairyman in our previous example has only enough corn silage on hand to feed 50 lb per cow per day rather than the 66 lb that they would eat if it is fed free choice. He must decide whether to increase the amount of alfalfa hay or one of the concentrate ingredients in the ration to make up for the lower amount of silage fed. The most common method for determining which feed to substitute is to calculate the cost per unit of net energy provided by the available feeds and select the one with the lowest cost. The new feed is added to the ration in sufficient quantity to provide the same amount of net energy as was lost by reducing the corn silage fed from 66 lb to 50 lb per day.

The method for calculating relative costs per unit of net energy is shown in Table

TABLE 11.3 *Example Ration Formulated for a 1,400-lb Cow Producing 60 lb Milk (3.5% Fat)*

	Dry matter basis								*As fed basis*			
	Amount fed (lb)	*NE_1 (Mcal)*	*Crude protein (lb)*	*Ca (lb)*	*P (lb)*	*Vit A (1,000 IU)*	*CF (lb)*	*ADF (lb)*	*DM (%)*	*Amount fed (lb)*	*Total ration (%)*	*Concentrate mix (%)*
Alfalfa hay (early bloom)	9.00	5.31	1.548	0.1125	0.0207	139	2.79	3.42	90	10.00	11.2	
Corn silage (average)	19.83	13.76	1.586	0.0535	0.0397	163	4.76	6.15	30	66.10	73.7	
Roughage subtotal	28.83	19.07	3.134	0.1660	0.0604	302	7.55	9.57		76.10	84.9	
Corn grain (ground)	7.82	7.20	0.782	0.0023	0.0242	3	0.16	0.23	89	8.79	9.8	64.7
Soybean meal, 46% (solv-extd)	4.10	3.46	2.124	0.0148	0.0308	0	0.20	0.33	89	4.61	5.1	34.0
Dicalcium phosphate	0.17	0	0	0.0403	0.0320	0	0	0	96	0.18	0.2	1.3
Concentrate subtotal	12.09	10.66	2.906	0.0574	0.0870	3	0.36	0.56		13.58	15.1	100.0
Ration total	40.92	29.73	6.040	0.2234	0.1474	305	7.91	10.13		89.68	100.0	
Composition of DM		0.73[a]	14.8%	0.55%	0.36%	7454[b]	19.3%	24.8%				

[a] $\frac{29.73 \text{ Mcal } NE_1}{40.92 \text{ lb DM fed}} = .73 \text{ Mcal/lb}$

[b] $\frac{305{,}000 \text{ IU vit A}}{40.92 \text{ lb DM fed}} = 7454 \text{ IU/lb}$

11.4. Prices of feeds per lb of dry matter are listed in the second column and Mcal of NE_1 per lb of feed dry matter in the third column. Dividing the Mcal per lb into the price per lb gives the cost per unit of NE_1. For example, alfalfa hay costs $80 per ton of dry matter (4¢ per lb) and contains 0.59 Mcal per lb as seen in Appendix Table III-D. Dividing 4.00 by 0.59 gives 6.78 per Mcal, the figure which appears in column 4 of Table 11.4. Looking down column 4 we see that alfalfa hay provides the cheapest source of net energy of the listed feeds.

To determine the amount of additional alfalfa hay needed to replace 16 lb of corn silage, first calculate the net energy in the silage. Each lb of corn silage dry matter contains 0.694 Mcal (Appendix Table III-D). The example silage has 30% DM, so 16 lb as fed is equal to 4.8 lb DM (16 × 30% DM = 4.8). Therefore, the NE_1 that must be replaced by alfalfa hay is 3.331 Mcal (4.8 lb × 0.694 Mcal/lb = 3.331 Mcal). Dividing the NE_1 of alfalfa hay, 0.590 Mcal/lb DM, into 3.331 Mcal gives 5.65 lb DM. The hay has 90% DM, so it would take 6.3 lb (5.65 ÷ 90% = 6.3) alfalfa hay as fed to replace 16 lb of corn silage on an energy basis.

Similar calculations can be made based upon the protein content of various feeds (columns 5 and 6 of Table 11.4) if the feed to be replaced is high in protein. In some cases it may be necessary to select the least expensive source of energy and protein and use some of both feeds in order to maintain adequate amounts of energy and protein in a ration.

When more than two nutrients are considered, it becomes very difficult to calculate substitutions by hand and still maintain nutrient balances in a ration. Under these conditions, computers become very helpful in calculating least-cost or maximum profit rations, as discussed in Chapter 12.

Variety of Ingredients

Use of a variety of ingredients sometimes increases the acceptability of a ration when less palatable by-product feeds are included, but simple rations can be very palatable and result in high milk production also. Rations made up of alfalfa forage and concentrate mixtures composed mainly of either corn, barley, milo, oats, or wheat are readily accepted by cows and result in a level of milk production equal to that from rations with a greater variety of ingredients. Any of these grains can make up as much as 80% of a concentrate mix without causing palatability problems or affecting feed utilization.[2] Therefore, variety per se is not necessary. However, except where an abundance of home-grown grains is available, use of a variety of ingredients often results in a lower-cost ration.

Roughage: Concentrate Proportions

Before a ration can fulfill the nutrient requirements of a cow, it not only must have the proper nutrient content but it also must be eaten in sufficient amounts. The capacity of the rumen is limited, and a cow can eat only so much feed each day. For the average cow, maximum dry matter intake of common rations is about 3% of body weight. There is considerable variation among cows, however, as shown in Table 11.2. High-producing cows usually consume more feed per unit of body weight

Table 11.4 *Relative Costs per Unit of NE_1 and Crude Protein*

	Dry matter basis				
	Price	NE_1		Crude protein	
	(¢/lb)	(Mcal/lb)	(¢/Mcal)	(%)	(¢/lb)
Alfalfa hay, early bloom	4.0	.590	6.78	17.2	23.26
Beet pulp, dried	7.2	.812	8.87	8.0	90.00
Brewers' grains, dried	6.9	.680	10.15	26.0	26.54
Citrus pulp, dried	6.6	.798	8.27	6.9	95.65
Corn grain, ground	7.5	.921	8.14	10.0	75.00
Oat grain	7.9	.789	10.01	13.6	58.09
Soybean meal, 46% protein	11.1	.844	13.15	51.8	21.43
Urea, 45% N	15.6	.000	∞	281.2	5.55
Wheat bran	7.0	.721	9.71	18.0	38.89

than low producers, a condition which is necessary for intake of sufficient nutrients to maintain high milk production. Although it is true that high producers generally have good appetites, not all good eaters are high producers because some cows use the nutrients to put on body fat rather than produce more milk.

Rumen capacity ceases to be the limiting factor for feed intake when rations with a high proportion of concentrates are fed. Then energy intake becomes the major factor in turning off a cow's desire to eat more feed. This switch in limiting factors for feed consumption occurs when ration digestibility reaches about 68%.[3] Consequently, increasing concentrate levels above about 60% of the ration does not result in greater energy intake because cows eat less forage and total feed consumption is depressed. Additionally, incidences of digestive upsets, fat test depression, and displaced abomasum increase with rations that contain high levels of concentrates. Therefore, there is little to be gained and much to be lost from feeding rations that contain more than about 60% concentrates. When corn silage is the only forage in the ration, concentrates should not comprise more than 50% of the ration dry matter because corn silage contains considerable amounts of grain.

11.4 Allotting Concentrates to Lactating Cows

Cows have different nutrient requirements due to varying body weights and levels of milk production. The procedures for calculating nutrient requirements and formulating rations outlined in Sections 11.2 and 11.3 are very time consuming, especially when many nutrients and large numbers of feeds at varying prices are considered. Consequently, rules of thumb and feeding guides based upon nutrient requirement data are available to simplify the process.

Rations for most dairy cows are based on roughage. This is economically sound, because nutrients from forage usually are cheaper than from concentrates. Roughages usually are fed free-choice, whereas the more expensive concentrates are fed in limited amounts. As mentioned previously, a cow cannot consume enough forage to meet the nutrient requirements for maintenance of her body and high milk production. Part of her ration must be in a more concentrated form to fulfill the energy deficit left by all roughages and the protein deficit left by nonlegume roughages.

Most concentrate mixes contain from 0.71 to 0.78 Mcal NE_l per lb on an air dry basis, which is equivalent to 68 to 75% TDN. In general, concentrate mixes below 70% TDN should be avoided regardless of price, because they usually contain high levels of low-energy feeds, which defeats the purpose of feeding concentrates.

The daily allowance of concentrates that should be fed to a cow depends on her level of milk production and the energy value of the roughages and concentrate mix being fed. If all cows in a herd are fed the same amount of concentrates, the low producers will be overfed and get fat, and the high producers will be underfed and lose weight. The high producers may be able to maintain a high level of milk production for a short time by using their body fat reserves as a source of energy; but when these are depleted, milk production will decrease.

Due to the energy value of the extra fat and solids-not-fat, it takes more energy for a cow to produce a pound of high-fat milk than a pound of low-testing milk. For this reason most methods of feeding concentrates according to production are based on milk fat production rather than pounds of milk.

Concentrate Feeding Guides

The following methods are most frequently used:

Method 1. Divide the monthly milk fat production of the cow by 3, 4, or 5, depending on the milk, concentrate, and forage prices. The divisor most commonly used is 4. When milk prices are high or when concentrate prices are low compared with forage, it may be more profitable to use 3 as the divisor. When the opposite conditions prevail, 5 may be the best to use.

Example A: 80 lb fat ÷ 4 = 20 lb of concentrate to feed daily

Example B: 40 lb fat ÷ 4 = 10 lb of concentrate to feed daily

The method is very simple, but it tends to overfeed low producers and underfeed high producers.

Method 2. Subtract 20, 25, or 30 from the monthly milk fat production depending on the forage quality. Divide the answer by 2.

Example C: 80 lb fat − 30 = 50
50 ÷ 2 = 25 lb of concentrate to feed daily

Example D: 40 lb fat − 30 = 10
10 ÷ 2 = 5 lb of concentrate to feed daily

Using this method, higher producers are fed more and lower producers less than with method 1. Therefore, the amount calculated from method 2 will more nearly approximate the actual requirements of a cow. Subtraction of 30 from the monthly fat production is based on the fact that an average cow can produce about 30 pounds of milk fat per month without any concentrates when fed excellent-quality forage. A dairyman using this method still may want to give a pound or so of concentrates to cows producing less than 30 pounds of fat per month to get them into the barn or parlor and keep them quiet if they are accustomed to receiving some concentrates while being milked.

A dairyman can profitably feed more concentrates to his cows when a high proportion of his milk is sold as Class I (Grade A). Higher levels of concentrates also are warranted when they are relatively inexpensive compared with roughage. The disadvantage of methods 1 and 2 is that the amounts calculated are not related to the price being received for the milk, the cost of the concentrates, nor the cost and quality of the forage. This can be compensated for in part by changing the denominator or the amount subtracted in the equations to reflect varying price and forage-quality conditions. However, because of the great variation in economic conditions in different areas, it would be difficult for each dairyman to figure which factors to use.

Method 3. The preceding methods can be used successfully in small herds where the dairyman knows the production of individual cows and does the milking and feeding himself. In herds of 100 or more cows, it is difficult to remember the production of all cows. Furthermore, in larger herds, cows frequently are milked and fed by hired labor who may not have as much motivation to feed cows strictly according to production as does the dairyman himself. Under these conditions, cows seldom get enough individual attention for method 1 or 2 to be of practical use.

A method that is more practical in larger herds utilizes a color-coding system for marking cows according to their production. Cows within certain production ranges are marked with the same color. Once each month the colors should be checked and changed, if necessary, according to the milk fat production of the cow as determined from the latest production test. Color coding of cows with different levels of production makes it possible for the feeder to know at a glance the amount of feed a particular cow should receive. Colored chicken rings on the neck chain, colored eartags, or colored tape on the neck chain or attached to the tail switch are used with good results (Figure 11.2).

Amounts of concentrates are recommended for cows producing within each range with the amounts based on production at the top of the range. This procedure assures adequate energy intake for the top

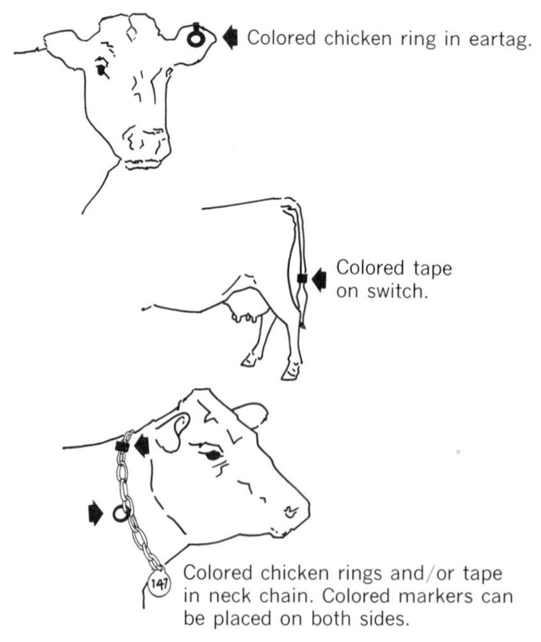

FIG. 11.2 Examples of the use of color coding for allotment of concentrates.

cows in each group to maintain their production. If amounts were calculated for the middle of the range, production of the top cows likely would drop.

On some dairies, mechanical feeders are operated from dials which regulate the amount of concentrates fed to a cow. A color-coding system can be used in this situation as well as for hand-feeding methods. Colored strips are placed on the dial to correspond with the desired feeding level. The feeder turns the switch to the colored strip on the dial which corresponds to the color marker on the cow, and the cow receives the desired amount of concentrates automatically.

Example E: A cow producing 50 lb of milk fat per month is being fed good-quality forage free-choice. To find the amount of concentrates to feed, move down the Milk Fat column in Table 11.5 to the 43–55 lb range. Read across to the third column under "Good" Forage Quality. The cow should be fed 15 lb of concentrates per day (7½ lb per feeding), and should have a yellow marker on her neckchain, ear, or tail, whichever is most visible to the feeder. All cows in the herd with monthly fat production between 43 and 55 lb should have yellow markers and should receive 15 lb of concentrates per day. Other cows could be marked with white, blue, green, red, or black, depending on their level of production. With all cows in a large herd color-coded according to the amount of concentrates that they should receive, it is relatively easy to carry out a concentrate feeding program based on ranges of milk fat production. Some cows will be slightly overfed at the lower end of the production range. However, the level of concentrate feeding will be more nearly correct with this method than when all cows in a herd receive the same amount of concentrates, such as is practiced in some large herds.

TABLE 11.5 *Concentrate Feeding Table for Method 3*

	Forage quality			
	Fair	*Good*	*Excellent*	
Milk fat (lb per month)	*Concentrates* (lb per day)*			*Tag color*
Below 30	5	2	0	Black
30–42	11	9	6	Red
43–55	18	15	13	Yellow
56–68	24	22	19	Green
69–80	30	28	25	Blue
Above 80	**	**	**	White

*Add 2 lb per day for normal growth of first-calf heifers.
**All that the cow can consume safely.

11.5 Protein Level in Concentrate Mixes

There is no single concentrate protein level that is best for all cows in a herd because of differences in body size, level of milk production, type of forage fed, and voluntary forage intake. Since it is impractical to have more than one concentrate mix on most dairy farms, a compromise level must be adopted for the whole herd.

Effect of Forage Protein Content

The protein content required in concentrate mixtures is dependent on the type as well as the amount of forage being fed. Forage is the basis of most dairy rations, and once it is harvested and stored, little can be done to affect its protein content. Legumes are higher in protein content than most nonlegume roughages. The difference between protein required by a cow and the amount supplied by her forage intake must be made up from the concentrates, which usually are allotted according to her production level.

A crude protein (CP) level of 13 to 16% of the dry matter of a complete ration (roughage plus concentrate) satisfies the protein requirement of lactating cows.[4] On an air dry (90% DM) basis, this would be about 12 to 14.5% CP. Most legume forages at 90% DM contain 16% or more CP, so concentrates fed with legumes need not contain more than 12% CP (90% DM basis). However, when the sole source of forage is nonlegume, CP levels as high as 25% may be needed in the concentrate mix to fulfill the requirements of individual cows, depending on the protein content of the forage consumed and level of milk production. When a combination of legume and nonlegume forage is fed, or when corn silage which contains 0.5% urea (or equivalent NPN from other products) is fed, concentrate mixes with 14 to 18% CP are indicated. In order to determine the best concentrate protein level for his herd, a dairyman should know the protein content of his forage. Thus forage testing, as described in Chapter 10, becomes of great importance to the dairyman.

Within reasonable limits, levels of protein higher than needed are neither harmful nor helpful to the cow or her milk production. However, protein usually is one of the more expensive nutrients in the ration, so amounts greatly in excess of requirements should be avoided in order to increase profit potential. Guidelines for dairy concentrate protein levels which have proven useful are summarized in Table 11.6.

Balancing Home-Grown Grains with High-Protein Supplements

Most dairy farmers grow a high proportion of the forage fed to their cows. In some areas, large amounts of grains also are grown on the farm. Grains are excellent energy sources, but are low in protein, calcium, and other minerals. When fed with alfalfa or other legumes, these deficits are balanced by the high content of protein, calcium, and trace minerals in legumes. However, when corn silage or other nonlegumes are the main forages fed, the concentrate mix must contain protein and mineral supplements as well as grains. In this case, the proper combination of grain and supplements must be determined.

Algebraic Method

For example, let us say that we want a concentrate mixture with 20% crude protein (CP). We have available home-grown grain that contains 9% CP and we purchase a high-protein supplement that contains 80% CP equivalent. If we let P equal the proportion of protein supplement and (1 – P) equal the proportion of grain in the mix, the equation is as follows:

$$80\,P + 9(1 - P) = 20$$
$$80\,P + 9 - 9\,P = 20$$
$$71\,P = 11$$
$$P = 0.155$$
$$= 15.5\%\ \text{protein supplement}$$

Substituting the value for P in (1 – P), it becomes

$$1 - P = 1 - 0.155 = 0.845 = 84.5\%\ \text{grain}$$

Therefore, a mixture of 84.5% grain and 15.5% protein supplement would result in a concentrate mix with 20% CP.

Square Method

Another method that can be used to determine proportions of ingredients is called the Square Method. Draw a square as shown in Figure 11.3. In the center, place the CP% desired (20%). At the upper left corner of the square, place the CP% of the grain mixture (9%). At the lower left corner of the square, place the CP% of the protein supplement (80%). Now subtract diagonally, ignoring the sign: 9 – 20 = 11. Place the 11 at the lower right corner of the square, diagonally opposite the 9. Subtract

Table 11.6 *Recommended Levels of Crude Protein, Calcium, and Phosphorus in Concentrate Mixes Fed with Various Roughages*

	90% DM basis			
		Needed in concentrate mix		
Type of roughage	*CP of roughage (%)*	*CP (%)*	*Ca (%)*	*P (%)*
Legume	16 or higher	12	0	0.7
Legume-grass mixture or legume plus corn silage	12–16	14–18	0.5	0.7
Corn silage or other nonlegumes	6–10	18–25	1.0	0.7
Corn silage with 0.5% urea	10–12	16–18	1.0	0.7

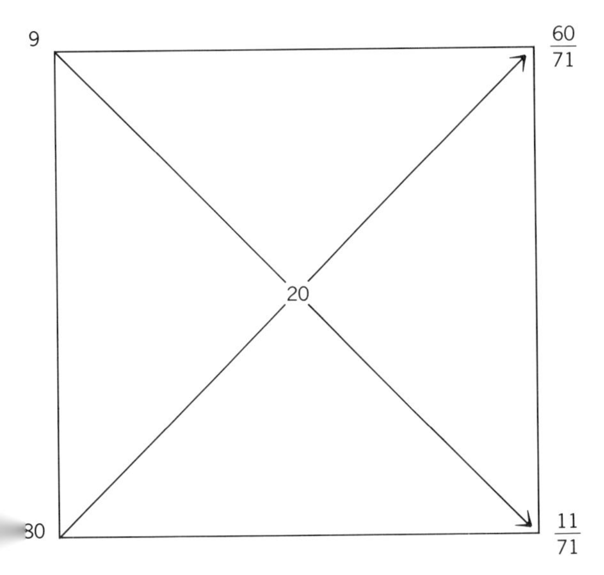

FIG. 11.3 Square Method used to determine feed mixture proportions.

80 − 20 = 60, which goes in the upper right corner, diagonally opposite the 80. Then add the values on the right side of the square, 11 + 60 = 71, and place the sum below each of the numbers on the right corners. A mixture of 60/71 grain and 11/71 supplement will contain 20% CP. To put them on a percentage basis, divide 60 ÷ 71 = .845, and 11 ÷ 71 = .155. Therefore, the mixture should contain 84.5% grain and 15.5% supplement, the same as calculated by the algebraic method.

Use of NPN in Concentrate Mixes

Nonprotein nitrogen (NPN), mostly in the form of urea, is extensively used to increase the protein equivalent of dairy concentrate mixes and corn silage. Utilization of NPN by ruminal bacteria was described in Chapter 8, and the physical characteristics of urea were discussed in Chapter 10.

Urea has received much bad publicity in the past due to its toxic properties and unpalatable nature at high concentrations. However, when fed in proper proportions with other feeds, it can supply part of a cow's crude protein requirement. Corn silage with 0.5% urea is fed successfully in many areas. Urea levels up to 3% of the concentrate mix are fed at times, but more commonly are restricted to 1 to 2% in order to maintain acceptable palatability levels. Feed tags usually contain a minimum crude protein guarantee. If NPN is included in the mix, the tag should show the maximum amount of crude protein equivalent derived from NPN. Dividing the % crude protein from NPN by 2.88 gives the % of urea in the mix because urea with 46% N contains about 288% crude protein equivalent.

Example: A feed tag lists a minimum crude protein content of 16% with a maximum of 4.3% CP in the form of NPN. About how much urea is in the mix? Divide 4.3 by 2.88 (4.3 ÷ 2.88 = 1.5). Therefore, the mix contains approximately 1.5% urea.

Urea contains no energy, minerals, or vitamins. These must be supplied by other feeds when urea replaces part of the protein normally supplied by high-protein supplements, such as soybean meal or cottonseed meal, which are good sources of energy, minerals, and vitamins. Urea adds nothing to a ration already adequate in protein because the excess is excreted in the urine. Urea utilization is best when fed with feeds high in starch, such as corn, barley, or other grains. Molasses frequently is included in mixes with urea and improves the palatability of urea-containing feeds. However, its energy, being mainly in the form of sugars, is metabolized too rapidly for most efficient utilization by ruminal microorganisms for protein synthesis.

Although NPN is unpalatable and toxic at high levels, the feeding of dairy rations which contain NPN probably will continue to increase in the future. The primary reason for this is the much lower cost of protein equivalent from NPN as compared with protein from other sources. When established guidelines are followed, feeds containing NPN can be fed to dairy cows with resulting increased net income due to savings in feed cost and little or no loss in milk production.[5] These guidelines may be modified in the future as additional information is developed through research and practical experience. However, with present knowledge, these recommendations should be followed to prevent toxicity and minimize palatability problems while benefiting from the cost savings possible when urea replaces higher-priced protein supplements in dairy rations. Some of the following guidelines are interdependent, but a ration should meet each of the individual restrictions as well as all combinations of them.

NPN Feeding Guidelines

1. Maximum of 0.92% added NPN in concentrate mix (equivalent to 2% urea).
2. Maximum of 0.23% added NPN in corn silage (equivalent to 0.5% urea).
3. Maximum of ⅓ of total ration protein equivalent from NPN.
4. Maximum of 0.5 lb urea daily, or equivalent from other added NPN, per 1,000 lb body weight.
5. Maximum of 0.5% NPN in total ration dry matter (equivalent to 1.1% urea).
6. Mix thoroughly.
7. Mix with high-energy feeds, preferably grains or corn silage.
8. Convert to urea-containing rations gradually.
9. Check mineral allowances more carefully.
10. Use only if additional protein is needed.

NPN for High-Producing Cows

Some recent research has indicated that NPN may not be utilized as well by high-producing cows as low-producers. Much of the natural protein in the diet is broken down to ammonia (NH_3) and other compounds by ruminal microorganisms and used by them for microbial protein synthesis. When the amount of NH_3 derived from natural protein exceeds the amount that can be used by the microorganisms for protein synthesis, additional NPN in the diet will not be utilized and will be excreted in the urine. High-producing cows need more protein in their rations than low-producers, so they are more likely to exceed the maximum ruminal NH_3 levels that can be efficiently utilized. However, there is no common agreement among researchers as to the level of protein in the diet where this becomes critical, and more research is needed before the problem is completely understood. Current research on protein degradability and "protected" proteins, discussed in Section 8.3, may make it possible to give reliable recommendations on the subject in the future.

11.6 Mineral Levels in Concentrate Mixes

Calcium and Phosphorus

Requirements for the two mineral elements needed in largest amounts, calcium (Ca) and phosphorus (P), were discussed in Chapter 9 and are listed in Appendix Tables III-A and III-B. As was the case for protein, the level of Ca needed in concentrate mixes is dependent on the type and quality of forage being fed. Legumes are high in Ca and often provide adequate amounts of this element when fed free choice to dairy cows. Under these conditions, no additional Ca is needed in the concentrate mix. On the other hand, nonlegumes generally are low in Ca. When they are fed as the sole roughage source, concentrate mixtures should contain about 1% Ca to make up for the deficiency in nonlegumes. Concentrate mixes fed in conjunction with a mixture of 50% legume and 50% nonlegume should contain about 0.5% Ca.

All roughages are low in P. Concentrate mixes fed with either legume or nonlegume forages should contain about 0.7% P. Even higher levels are necessary when poor-quality by-products, such as cotton-seed hulls, are fed as a major roughage source.

A summary of guidelines for Ca and P levels in concentrate mixes is contained in Table 11.6.

When a complete ration (roughage plus concentrate) is considered, minimum levels in the ration dry matter should be 0.6% Ca and 0.4% P. Furthermore, the ratio of Ca to P in the total ration is important as well as the absolute amounts. Ca : P ratios of 1.5 : 1.0 to 2.5 : 1.0 are desirable, but it is difficult to stay within these limits when a high-calcium legume, such as alfalfa, is the major forage in the ration. Above a ratio of 2.5 : 1.0, incidence of milk fever may increase. However, many herds are fed rations with ratios above this range without apparent detrimental effects on health or milk production.

Sodium and Chlorine

Sodium and chlorine usually are provided free choice to dairy cattle as loose or block salt. Also, most commercial concentrate mixes contain 0.5 to 1% salt to increase their palatability. This type of feeding regime provides adequate amounts of both sodium and chlorine under all conditions.

When sodium phosphates are included as a supplemental P source at a rate of 1% or more of a concentrate mix, salt is not needed in the mix. In fact, concentrate mixtures containing a combination of 3% sodium tripolyphosphate and 1% sodium chloride are too salty and less palatable than those with either supplement alone.

Other Minerals

Requirements for other minerals were discussed in Chapter 9. Under most conditions, no additional supplementation of minerals other than the 4 discussed above is necessary, especially if good-quality legumes make up a significant proportion of the roughage being fed. Supplementation with trace minerals, or provision of trace-mineralized salt is justified when corn silage or other nonlegumes are the only forage fed, or when feeds are produced in areas which are iodine or cobalt deficient, or contain excess molybdenum. Indiscriminate supplementation with trace minerals without first testing the available forages for their mineral content should be avoided. Not only does this increase the price of the ration, but some trace minerals can increase the incidence of oxidized flavor in milk when fed in amounts greatly in excess of their requirements.

11.7 Vitamin Supplementation

All B-vitamins and vitamin K are synthesized by ruminal microorganisms, and green or freshly harvested forages are excellent sources of the other vitamins needed by a dairy cow. The content of fat-soluble vitamins A (as carotene), D, and E is reduced by oxidation when forages are stored for long periods. When hay which has been stored for 6 months or more is fed, vitamin A supplementation may be needed. Fortunately, this is the least expensive of the vitamins, and most commercial concentrate mixes contain supplemental vitamin A. Supplementation with large doses of vitamin D for prevention of milk fever, vitamin E for prevention of oxidized flavor in milk, and niacin for prevention of ketosis, was discussed in Chapter 9. Except for these specific cases, vitamin fortification of common dairy rations is not necessary.

11.8 Other Feed Additives

Thyroprotein

Probably the most controversial feed additive for dairy rations is iodinated casein, commonly known as thyroprotein. Promoters of this product claim up to 20% increases in milk production in commercial herds after feeding thyroprotein. Most experiment station research under controlled conditions shows that there is an increase in milk production and fat test when thyroprotein feeding is initiated. However, a compensatory decrease usually occurs in later lactation, resulting in no increase in total production when the entire lactation is considered.[7] The drop in production is particularly sharp when thyroprotein feeding is discontinued in late lactation.

Thyroprotein supplements the action of thyroid hormone in stimulating metabolic processes in the body. Increased milk production and a higher basal metabolic rate resulting from thyroprotein feeding require extra energy, so higher concentrate feeding levels are necessary to prevent excess loss of body flesh when it is fed. Since high-producing cows cannot consume enough feed to maintain energy balance in early lactation, thyroprotein feeding at this time only aggravates the situation. Manufacturers' recommendations currently are that it should not be fed until a cow passes her peak of lactation and can consume enough feed to maintain body flesh. When thyroprotein feeding is started after the peak of lactation, there is a leveling off of the decrease in milk and fat production or, in some cases, an increase to another peak. This is followed by a gradual decline during the rest of the lactation at a rate somewhat greater than normal, resulting in lower production during the latter part of lactation.

In order for thyroprotein feeding to be justified economically, returns from extra milk must exceed costs for the product and for additional feed which must be provided to the cow. Some cows respond with enough extra milk to make it profitable, but others do not. Whether or not milk production response is enough to make it profitable in average herds is debatable. It may be useful to increase production for short periods of time, such as during a base-making period, or for certain cows that tend to get excessively fat in mid-lactation.

Milk production data from cows fed thyroprotein cannot be used in USDA Sire Evaluation Programs because of the bias introduced into their records. Since official Dairy Herd Improvement (DHI) and Dairy Herd Improvement Registry (DHIR) records are used for USDA sire evaluations, herds in which thyroprotein is fed cannot be enrolled in official DHI or DHIR record keeping programs. However, dairymen feeding thyroprotein can receive the benefits of the dairy record keeping programs by enrolling in Owner-Sampler or other unofficial dairy record keeping plans, which were discussed in Chapter 3.

Enzymes

Claims of better utilization of some grains treated with certain enzymes are counterbalanced by other reports that no benefits were obtained. With present knowledge, addition of enzymes to dairy rations is not justified.

Hydroponics

Grains sprouted in chambers under conditions of controlled temperature and humidity provide a source of green feed at times when it is not possible to produce it normally. However, energy value of the hydroponically produced green feed is less than that of the grains used for sprouting.[8] This, plus the high cost of the chambers and low volume of green feed produced, makes the practice of dubious value.

Additives to Help Maintain Normal Milk Fat Percentages

Lactating cows fed rations with high percentages of concentrates (60% or more) frequently produce milk with lower fat tests than milk from cows fed higher-roughage rations. The same situation occurs when extremely low-fiber roughage or roughage which is finely ground and pelleted is fed to dairy cows, as discussed in Chapter 8.

Addition to concentrate mixes of bicarbonates, magnesium oxide, bentonite, or partially delactosed whey helps to maintain fat tests near normal levels when rations with low amounts of roughage, or finely ground roughages, are fed.[9] None of the additives increase fat tests above normal levels. Voluntary intake of concentrates is reduced due to lowered palatability when bicarbonates or magnesium oxide are added, resulting in lower milk production. However, milk fat amounts as well as fat test may be increased even though total milk produced is less.

Partially delactosed whey is palatable but must be fed at about 10% of the total ration to be effective. The high cost of the product and difficulties in blending and pelleting with other feeds are its major drawbacks. In emergency situations, a dairy farmer may be willing to take a loss in milk production or pay more for a ration due to inclusion of these products in the concentrate mix in order to maintain fat tests near normal levels when roughage supplies are low. However, feeding rations with adequate fiber (at least 17% crude fiber or 21% acid-detergent fiber in the ration dry matter) is the best preventive of depressed fat test known at the present time.

11.9 Feeding Methods

It is difficult to manage the concentrate feeding program in large herds to ensure that cows are fed individually according to production. The color-coding system outlined in Table 11.5 can approach this goal in herds where cows are milked in stanchion barns and have ample time to consume their concentrate allowances while being milked. However, recent trends to faster milking in milking parlors make it practically impossible for high-producing cows to consume all of their concentrate allowances while being milked because they spend only a few minutes in the milking parlor each milking. Any concentrates left by a cow when she leaves the parlor are available to the next cow that comes into the stall, whose production may or may not warrant it.

This problem can be partially resolved by feeding additional concentrates outside of the milking parlor. However, when a base amount of concentrates is fed to the whole herd, it results in inefficient feed utilization by the lower producers and increases the incidence of the "fat-cow syndrome" in the herd. This syndrome includes the complex of complications that are common to the fat cow at calving time, such as difficult calving, retained placenta, metritis, milk fever, ketosis, and displaced abomasum. Therefore, overfeeding lower producing cows can cause many problems and may be as unprofitable as underfeeding high producers.

Magnet-Activated Feeders

An automated feeding system designed to allow feeding of more concentrates to high producers outside of the milking parlor relies on a magnetically activated feeder (Figure 11.4). A cow that is to receive the extra feed has a magnet attached to a rope around her neck. Feed is stored in an overhead bin connected to the feeder by an auger. When a cow equipped with a magnet puts her head into the feeder, the magnet makes contact with a metal plate on the front of the feeder that activates an electric motor which dribbles feed into the feeder bowl. Feed is continuously dribbled into the feeder so long as the magnet remains in contact with the front plate. When the cow backs away the circuit is broken, the motor stops, and no more feed drops into the feeder bowl. Cows without magnets on their neck straps cannot operate the motor, so they do not get any feed from the concentrate bin. There is no measure of individual feed intake because the feed is free-choice so long as the cow keeps her head in the feed bowl area. Also, there are problems with boss cows chasing the more timid cows from the feeder, especially if the feeder is not equipped with guard rails to protect the cow while she is eating.

FIG. **11.4** A magnet-activated feeder and feed bin.

Transponder-Activated Feeders

Another system which provides concentrates to specially-equipped cows utilizes a transponder rather than a magnet to activate the feeder. The transponder is similar to the device used to operate an electronic garage door opener. Each cow's transponder is electronically coded according to her milk production. When a cow moves her neck through a loop-formed interrogater antenna built into the feeder, high frequency radio energy flows to the transponder. The transponder's coded memory device begins to charge electronically and drives a signal generator which causes feed to dispense at a rate of one pound per minute. When the transponder is fully charged, the feed delivery system stops. Length of charging time can be varied for each cow. When the cow leaves the feeder, the transponder gradually discharges over a 24-hour period. Therefore, each day the cow is allowed a certain length of time when feed will be delivered to her which she can use up in one feeding or spread out over many feedings during the day. Feed that is available to her at any time is dependent upon the allotted feed she has not eaten during the preceding 24 hours. Transponder feeders are more accurate than magnet feeders in allotting concentrates to cows according to production, but they also are much more expensive.

Electronically-controlled Feeder Doors

Another system uses electronically-controlled feeder doors (Figure 11.5) positioned on a feeding fence or bunk, and a cadmium plated key on the neck strap of the cow. When the cow approaches the door, the key comes close enough to the door to activate the electronic circuit, withdrawing a bolt lock. The cow then can push the door open with her head and reach the feed behind the door. Feed can be available free-choice or limited to certain amounts by the dairyman. Cows without keys on their neck straps cannot open the doors and get to the feed.

Computer-Controlled Feeders

Several companies are marketing computer-controlled concentrate feeders. In each case a mini- or microcomputer controls the feed dispensed to each cow in a herd based on the amounts keyed into the computer by the dairyman. Each cow wears a neck strap with a responder that identifies the cow to the computer when she enters the feeding stall. The computer allocates predetermined portions of her

FIG. 11.5 Electronically-controlled feeder doors.

total daily concentrate allotment and signals the feeder to deliver it slowly to the cow while she is in the feeding stall. When her allotment is used up or she leaves the stall, the feeder stops. Any of her unused allotment is recorded by the computer and is available the next time the cow enters a feeding stall. If a cow does not consume her total allotment, the amount left over is recorded and can be printed out. This can help identify cows off feed sooner than they normally would be noticed.

Complete Rations

Another possibility is to feed all of the concentrates mixed with the roughages as a complete ration outside of the milking parlor (Figures 11.6, 11.7, and 11.8). This practice has many advantages over the conventional method of feeding concentrates in the milking barn and roughages outside. There is less dust, less defecation, and no wasted feed in the parlor. It allows the milkers to be primarily concerned with milking cows. Furthermore, after an initial training period, cows are calmer while being milked when they receive no feed. Better control of the nutritional value of the total diet is also possible. Complete rations can be formulated to fulfill requirements for all nutrients required by the cow. When concentrate mixes are formulated and individually allotted to cows, it is assumed that the cows obtain a certain amount of nutrients from their roughages, which usually are fed free-choice. However, cows vary tremendously in their voluntary forage intake, thus negating much of the value of precise allotment of concentrates based on milk production. When a complete ration is fed, each bite contains a predetermined balance of all essential nutrients. The only limitation on the cow is her ability to consume large amounts of feed and efficiently convert it into milk. Feeding trials in several states have shown this method of feeding to be feasible and, in comparison with more conventional feeding systems, milk production was as good or better from groups fed complete rations.

Importance of Culling. Attention to culling practices is more critical when complete rations are group fed. Low-producing

FIG. **11.6** Feed ingredients stored for on-the-farm mixing of complete rations.

FIG. 11.7 Ingredients of a complete ration being weighed into a mixer truck equipped with load cells.

cows may gain excess weight, making them good candidates for culling. They return more money at slaughter because of their better condition and heavier weight, so the excess feed they consume is not a complete economic loss. Culling cows that do not measure up to the production standard of the herd has merit from a genetic standpoint also because only higher-producing cows and their calves are kept in a herd, resulting in more genetic improvement for milk production.

Labor Savings. Besides being nutritionally sound, the complete ration system has advantages in large herds from a labor-saving standpoint. In one study, a 700-cow dairy in Arizona reduced its feeding labor from 2 men working 10 hours per day when baled hay and silage were fed, to one man for 3 hours per day when the complete ration system was initiated. Additional labor savings were made with the elimination of concentrate feeding in the milking parlor. Considering all of the advantages listed previously, it is likely that feeding of complete rations to dairy cows will continue to increase in popularity in future years.

Grouping Cows According to Production. Feeding complete rations with varying proportions of roughages and concentrates to cows grouped according to milk production results in higher feed efficiency than when all cows in a herd are fed the same ration. In larger herds, a good system is to set up 4 milking cow strings and a fifth string composed of dry cows.

The first group should contain all fresh cows for at least 2 months after calving because cows do not reach their maximum milk production until 6 to 8 weeks after calving. This group should also contain the highest producers in the herd. Selection of the highest producers should be based on fat-corrected-milk (FCM) because it is more closely related to energy requirements than milk alone. The highest energy ration should be fed to this group to allow the fresh cows and highest-producers to reach and maintain their maximum production potential uninhibited by a lack of energy and other nutrients. Cows whose milk production does not warrant this high level of feeding after being fresh 2 months should be moved to a lower producing group.

Fig. 11.8 A complete ration of corn silage, alfalfa cubes, and concentrate ingredients.

The second group should consist of the next group of highest producing cows based on FCM production. Similarly, the next highest group of cows according to FCM would be in the third group, and the fourth group should be composed of the lowest-producers, most of which probably will be near the end of lactation. Dry cows should be handled as a separate fifth group.

Recommended nutrient content of rations for lactating cows of varying sizes at four levels of milk production, and for dry, pregnant cows are shown in Table 11.7. To find the recommended nutrient content for a cow, or group of cows, select the row nearest the average weight of the cows under the column for Cow Wt. Move across that row to the column under Daily Milk Yields, which includes the average amount of milk being produced by the cows. Read down that column for the nutrient concentrations that should be in the ration dry matter. For a group of 1,400-lb cows producing an average of 60 lb of milk daily, the table would be used as follows:

Under Cow Wt., use the 1,300-lb row because it is closest to 1,400. Move across that row to the third column under Daily Milk Yields, which includes milk production levels between 46 and 64 lb. Read down that column for the recommended nutrient concentrations in the ration dry matter, which would be 15% crude protein, 0.73 Mcal/lb NE_l or 71% TDN, 17% crude fiber, 0.54% calcium, 0.38% phosphorus, and 1,450 IU/lb Vitamin A. Concentrations of other minerals and additional nutrients are listed in Appendix Table III-C.

Moving Cows Between Groups. When cows are group-fed complete rations based on the level of production within the group, it is necessary to move cows from high-pro-

TABLE 11.7 *Recommended Nutrient Content of Rations for Dairy Cattle*

Nutrients (concentration in the feed dry matter)	*Cow wt (lb)*	*Daily milk yields (lb)*				
	≤ 900	*<18*	*18–29*	*29–40*	*>40*	
	1100	*<24*	*24–37*	*37–51*	*>51*	*Dry,*
	1300	*<31*	*31–46*	*46–64*	*>64*	*pregnant*
	≥1550	*<40*	*40–57*	*57–78*	*>78*	*cows*
Ration no.		*I*	*II*	*III*	*IV*	*V*
Crude protein (%)		13.0	14.0	15.0	16.0	11.0
Energy						
NE_1 (Mcal/lb)		.64	.69	.73	.78	.61
TDN (%)		63	67	71	75	60
Crude fiber (%)		17	17	17	17	17
Calcium (%)		.43	.48	.54	.60	.37
Phosphorus (%)		.31	.34	.38	.40	.26
Vitamin A (IU/lb)		1,450	1,450	1,450	1,450	1,450

Source: Nutrient Requirements of Dairy Cattle, 5th revised ed., 1978. National Academy of Sciences—National Research Council, Washington, D.C.

ducing groups to lower-producing groups when their production decreases to make room in the high-producing groups for other cows as they freshen. Movement of cows between groups is still a controversial subject, with no general agreement among dairymen relative to the merits of this type of feeding management. However, when certain guidelines are followed, disadvantages are minimized and are greatly overshadowed by the advantages.

Advantages of Grouping Cows by Level of Production

1. Concentrates can be liberally fed to high producers without overfeeding low producers, resulting in better feed efficiency.
2. Better control of both roughage and concentrate feeding is possible, which facilitates adjustment to changes in concentrate-roughage price relationships.
3. All or part of the concentrates can be fed outside of the milking parlor.
4. Fresh cows are together, making it easier to locate those requiring post-calving treatment.
5. Cows to be bred can be limited to 1 or 2 corrals for easier heat detection.

Disadvantages

1. It is difficult to keep all the purebred cows together in herds that include both purebreds and grades.
2. Cows must be moved regularly.
3. Injuries from fighting may occur and milk production of some cows may drop when they are moved, particularly if feeding space is inadequate. This problem tends to subside with time as cows become accustomed to periodic movement between strings.
4. Construction of additional facilities may be required.
5. Remodeling of the milking system may be necessary to handle the large flow of milk in the milk lines when the higher-producing cows are milked together as a string.

Guidelines. A sound management approach is necessary to make the feeding system work when cows are grouped by level of production. Some guidelines[6] that have proven helpful on many dairies using this system are as follows:

1. Divide a herd by production level into 4 strings if possible to reduce milk production variation within strings. More strings are desirable in large herds where corrals and facilities are available. Milk production of cows moved from one string to another is not likely to drop markedly if rations between strings do not drastically differ. Dry cows should be handled in a separate group. Handling first-calf heifers in a separate string also has many advantages because of their smaller size.

2. When fresh cows have recovered from calving and are ready to enter the milking herd, put them in the string fed the most concentrates and leave them there for at least 2 months to allow them to reach their maximum inherited ability for milk production.
3. Regulate the number of cows in a string to fit into regular milking and feeding procedures.
4. Move cows between strings on a regular schedule and move small groups of cows, rather than 1 cow at a time, or large groups all at once. This requires a catch pen or lock-in stanchions. Cows can then be released and moved to a new corral without disrupting the whole string.

A system for moving cows that has worked well is illustrated in Figure 11.9. The example is for a 400-cow milking herd, but the principle can be adapted to any size dairy. The herd is divided into 4 strings of milking cows with 100 cows per corral. Cow movement is based on a 10- to 15-day cycle.

All fresh cows enter Corral 1 a few days after calving. Let us assume that 10 cows (one per day) have entered Corral 1 during the past 10 days. The 10 lowest-producing cows (FCM basis) in Corral 1 that have been fresh more than 60 days are moved to Corral 2. The procedure is repeated in Corrals 2 and 3. In Corral 4, 10 cows are dried up and moved to the dry cow pen. This procedure is repeated every 10 days so that there are never large numbers of cows to move, thus minimizing the commotion and confusion that results when large groups of cows are moved from corral to corral. Cows usually are tested only once each month so the same FCM listing would be used for three 10-day periods.

5. Provide each string with adequate manger space so all cows can eat at the same time. For mangers equipped with stanchions there should be about 110% as many stanchions as cows normally in the string to allow for temporary variations in numbers. This permits recently moved cows to eat without necessarily going through the string's pecking order.
6. Feed each string based on the requirements of that string's higher producers. Milk production of higher producers will decrease if the ration is formulated for the average production of the string.
7. Reformulate the total feeding program

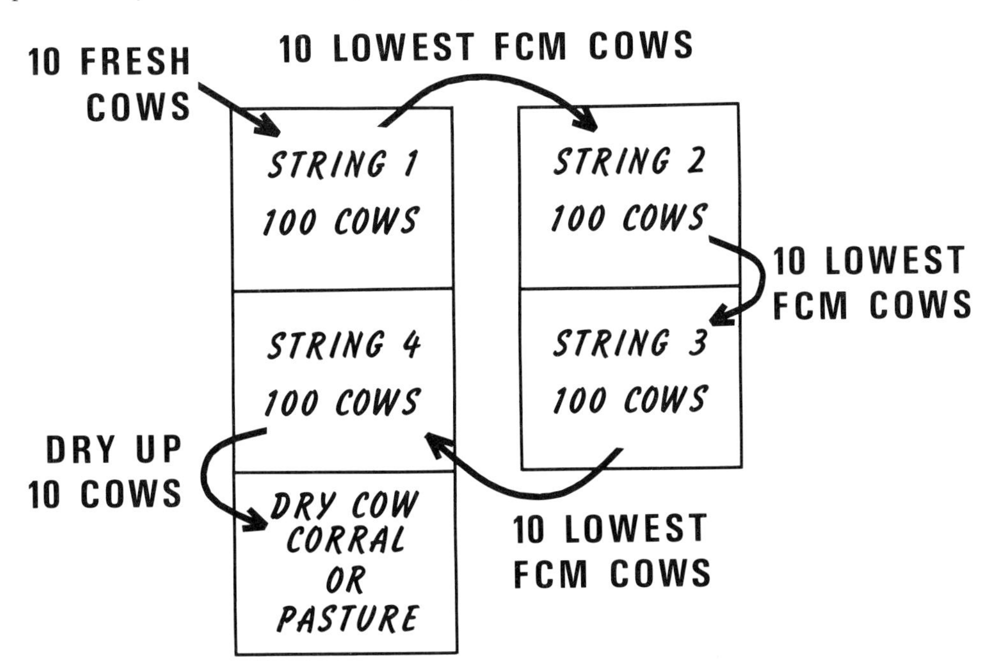

Fig. 11.9 Movement of cows between production strings.

for each string whenever there is a change in the basic ingredients fed, particularly the roughages.

8. Concentrates can be allotted in various ways:
   a. All concentrates can be mixed with the roughages into a complete ration and fed outside of the milking barn.
   b. The same amount of concentrates can be fed to all cows in the milking barn with different amounts group-fed with the roughages outside according to the level of production of the string.
   c. Amounts can be varied both in the milking barn and in the outside mangers.
9. Allot enough extra feed to cows in late lactation for development of unborn calves and for regaining any weight lost in early lactation. Weight gain in late lactation is more efficient than during the dry period.
10. When hay and concentrates are mixed for outside feeding, less feed separation and dustiness occur if the mix is moistened prior to feeding. Moistening is not necessary when silage or other succulent feeds are included in the mixed ration.
11. Provide adequate water facilities for each string.
12. Design the corral and/or free stall housing to provide easy movement of cattle to and from the milking barn.

DHI-EDP Reports. Some computing centers processing DHIA records also provide on request a listing of all cows in the herd ranked by FCM for the previous test day. This list can be used by dairymen to decide which cows should be moved to the lower-producing groups as fresh cows enter the milking herd, thus making it much easier to set up and maintain milk production strings in large herds.

Complete Rations for Small Herds. Dividing cows into groups according to production level and stage of lactation usually is not possible in smaller herds because of limited barn and feeder space. In this case, a single complete ration can be fed to the entire milking herd. Milk production efficiency will be lower than when cows are grouped according to production and fed rations with varying roughage-concentrate ratios. Convenience and labor savings may compensate for this though. Present evidence indicates that a complete ration with about 60% of the ration dry matter as roughage and 40% as concentrates is the most efficient for milk production when fed to all cows in the milking herd during all stages of lactation.

11.10 Other Considerations in Feeding Dairy Cows

Frequency and Regularity of Feeding

Cows are creatures of habit and like to be fed and milked on a regular schedule. Feeding concentrates twice a day while cows are being milked and providing roughage free choice during the remainder of the day is the feeding schedule most frequently followed on dairy farms. Forage may be put out in large amounts just once a day, or small amounts several times a day, depending on the type of roughage fed and the labor situation. Generally, cows will eat more if small amounts are fed more frequently, but the labor requirement may be prohibitive in larger herds. At least twice-a-day feeding of forage usually is desirable to reduce wastage. Spoiled silage in the bunk or hay which previously has been picked over often result from less than twice-a-day feeding, making the remaining forage less palatable. This results in lower forage intake and, subsequently, lower milk production.

Regularity of feeding is important to maintain high milk production. Cows do not take the weekend off from producing milk and must be fed each day on a regular schedule. Feed which goes into the rumen is constantly being digested and metabolized by the cow. If a feeding is missed, milk production is reduced at following milkings. When this happens in mid- or late lactation, cows frequently produce milk at lower levels for the rest of the lactation, even if feeding is increased above previous levels. Once the activity of enzymes and balance of hormones involved in milk synthesis are disrupted, it is often impossible for a cow to regain her previous persistency of milk production until she freshens again after calving.

Palatability and Physical Form of Rations

The importance of dairy rations being palatable as well as being nutritious has been stressed throughout this chapter. No matter how well balanced with all essential nutrients a ration may be, it is of no value if a cow will not eat it. Consumption of large amounts of well-balanced rations is essential for high milk production. When sufficient nutrients are not available from the diet and body fat reserves are depleted, milk production is the first function to suffer, as was discussed in Chapter 9 and illustrated in Figure 9.2.

Many of the factors which make some feeds more palatable than others still are obscure. Variety of ingredients helps in some cases. Inclusion of certain ingredients makes others more acceptable, the most notable case being the ability of molasses to partially mask the bad taste of urea when they are mixed together. Physical form also affects both nutritional value and intake of some ingredients, as discussed in Chapter 10. For example, grinding or rolling is necessary for efficient utilization of grains, but finely ground grain is more dusty and, consequently, less palatable than when it is coarsely ground. Pelleting of both concentrates and roughages increases the consumption rate and total amount eaten by cows, but rumen fermentation and fat test of milk are adversely affected. Wafering of hay retains many of the advantages of pellets without the disadvantages of lowered fat test.

The number of ingredients and proportion of each included in a ration should be determined primarily from their nutrient values and relative prices. However, maximum levels of certain ingredients must not be exceeded if palatability and high intakes by cows are to be maintained. Maximum percentages of ingredients allowable in a ration are variable under different conditions, and any listing of this type is, of necessity, rather arbitrary. This information is essential for computer formulation of rations, however. Otherwise, the computer will formulate a least-cost ration which fulfills the nutrient requirements without regard to its acceptance by cows. Recommended maximum levels of common feeds in dairy rations are listed in Table 12.1 of the next chapter, which deals with computer formulation of dairy rations.

11.11 Summary

Nutrient allowances for cows of all sizes and levels of milk production are listed in Appendix Tables III-A and III-B. Rations which fulfill these allowances can be formulated based upon nutrient values of feeds listed in Appendix Table III-D. Primary concerns in formulating dairy rations are energy, protein, calcium, phosphorus, salt, and vitamin A. When allowances for these nutrients are met, other minerals and vitamins usually are present in sufficient amounts.

Other considerations in feeding dairy cows include palatability of the ration, physical form, fiber levels, protein and mineral contents of concentrates, amount of concentrates fed, relative prices of feeds, voluntary feed intake, and frequency and regularity of feeding. Proper feeding of dairy cows requires that a dairyman have a basic knowledge of nutritional principles to plan an efficient feeding program and the experience and management ability to carry it out.

Review Questions

1. What are the major nutrients that should be considered when formulating rations for dairy cattle?
2. Why is forage the basis of most dairy rations?
3. How important is variety of ingredients in dairy rations and why?
4. What are the factors which influence voluntary feed intake by dairy cows?
5. What effects do levels of fiber have on intake and utilization of dairy rations?
6. What are the factors that determine the amount and protein content of concentrates to feed to dairy cows?
7. What are the advantages and disadvantages of urea in dairy rations?
8. Why would the frequency and regularity of feeding be more important for lactating cows than for non-lactating heifers or dry cows?

References

1. Lofgren, P. A., and Warner, R. G.: Influence of various fiber sources and fractions on milk fat percentage. J. Dairy Sci. *53:* 296, 1970.

2. Tommervik, R. S., and Waldern, D. E.: Comparative feeding value of wheat, corn, barley, milo, oats, and a mixed concentrate ration for lactating cows. J. Dairy Sci. *52:* 68, 1969.
3. Conrad, H. R., Pratt, A. D., and Hibbs, J. W.: Regulation of feed intake in dairy cows. I. Change in importance of physical and physiological factors with increasing digestibility. J. Dairy Sci. *47:* 54, 1964.
4. National Academy of Sciences—National Research Council: Nutrient requirements of dairy cattle. 5th Revised Ed., 1978.
5. Holter, J. B., Colovos, N. F., and Urban, W. E., Jr.: Urea for lactating dairy cattle. IV. Effect of urea versus no urea in the concentrate on production performance in a high-producing herd. J. Dairy Sci. *51:* 1403, 1968.
6. Holtz, E. W., and Bath, D. L.: Managing milk production strings. Leaflet 2889, Division of Agricultural Sciences, Univ. of California, May, 1976.
7. Schmidt, G. H., Warner, R. G., Tyrell, H. F., and Hansel, W.: Effect of thyroprotein feeding on dairy cows. J. Dairy Sci. *54:* 481, 1971.
8. Hight, W. B.: Hydroponically grown grass; its production, laboratory analysis and estimated feed value. 34th Annual Rural Electric Conf. mimeo, Davis, California, January, 1962.
9. Huber, J. T., Emery, R. S., Thomas, J. W., and Yousef, I. M.: Milk fat synthesis on restricted-roughage rations containing whey, sodium bicarbonate and magnesium oxide. J. Dairy Sci. *52:* 54, 1969.

CHAPTER 12

Computer-Formulated Rations

12.1 Introduction

The purpose of this chapter is to acquaint the student with the information required for and available from computer formulation of dairy cattle rations. Programming of computers and actual operation of the machines are not covered as these subjects require whole courses themselves. (Fig. 12.1 shows a typical computer center.)

It is relatively easy to balance a ration when price of the final mixture need not be considered. However, if the ration must be balanced using a combination of ingredients with the lowest possible total cost, the resulting mixture, called a "least-cost" ration, is very difficult to determine by hand. If equal nutrient requirements and ingredient limitations are employed, least-cost rations are neither better nor worse than other rations from a nutritional standpoint. The only difference is the price, which always is lower for the least-cost ration.

Least-cost rations can be calculated by hand using the simplex method,[1] but a tremendous amount of time is required when large numbers of feeds and nutrient requirements are considered. By the time the calculations are completed, many feed prices probably have changed and the formula is no longer least-cost. The benefit from the computer is that it makes possible the determination of the least-cost formula within a few minutes.

A modification of the least-cost computer program results in formulation of rations which maximize income above feed cost. This is a step beyond the least-cost ration concept as it takes into consideration other economic and production factors as well as feed prices. The type of information obtained from these computer programs and its use in dairy cattle feeding programs are discussed in the following sections.

12.2 Linear Programming

The technique employed to calculate least-cost and profit-maximizing rations is called linear programming. A simple definition of linear programming is "the maximizing or minimizing of some function subject to constraints." In the case of dairy rations, it is the minimizing of the cost of a ration or maximizing the income above feed cost. In addition to its use in ration formulation, the technique is extensively used for many other purposes. Oil companies use linear programming to formulate optimum blends of gasolines, airlines and railways use it to establish the most efficient transportation routes and schedules, and agricultural economists determine cropping patterns and acreages for maximum profit from a farm through its use.

It is well known that individual feed ingredients do not always perform as linear

Fig. 12.1 A computer center where "least-cost" and "maximum-profit" rations are formulated.

processes when mixed with other feed ingredients. In addition, many biological functions are curvilinear in nature rather than rectilinear. Although this poses problems in using linear programming, they can be overcome by various techniques which convert curvilinear processes into shorter linear segments which closely approximate the curvilinear function. In any event, the above problems are of no greater consequence for computer-formulated rations than for those formulated by hand. Of greatest importance is the programming of the computer with all of the pertinent information in the form of constraints that would be considered in formulating a ration by hand.

Capabilities of Linear Programming

Least-Cost Formula. The formula of the least-cost ration usually is stated in terms of the percentage of each selected feed ingredient that is included in the final mixture. However, it can be stated as lb per ton, lb per day, or any other convenient method of reporting the output from the computer. The formula is valid only under the specified set of feed prices and ration constraints. A constraint is a specified limitation, either minimum, maximum, or equality, on nutrients, feeds, or other ration characteristics, that must be fulfilled by the formula. As prices change, or as ration constraints change, the least-cost formula also changes.

The number of feed ingredients selected for the ration always is equal to or less than the number of constraints that come into play in formulating the ration. In other words, if the only constraints specified are on net energy and protein, the ration will have a maximum of 2 feed ingredients. One will be the least-cost source of net energy, the other the least-cost source of protein. Under some conditions, it is possible that one feed could be the least-cost source of both energy and protein, in which case only one feed would be selected. On the other hand, if constraints are placed on energy, protein, calcium, phosphorus, fiber, vitamin A, and roughage level, a ration with up to 7 ingredients could be selected.

If more variety of feeds is desired than normally is selected by the computer, minimum levels of certain ingredients can be specified. The quality of the ration usually is not affected by forcing certain feeds into the solution, because the computer balances them with amounts of other ingredients necessary to fulfill all requirements. However, the cost of the ration always is higher when feeds that normally would not be selected are forced into the solution because of the unfavorable relationship between their price and nutrient content.

Price Ranges. In addition to the least-cost formula, the computer calculates a range of prices between which each selected ingredient may vary without changing the least-cost formula. If the price of the selected ingredient drops below the lower range, and all other prices remain the same, more of that ingredient will be used in the least-cost formula. If it goes above the higher range, and other prices remain the same, less of it will be selected. In each case, amounts of other feeds in the formula also change with possible additions or deletions of ingredients.

Opportunity Prices. The computer calculates the prices at which feeds not selected for the ration would start to enter the solution. These "opportunity prices" are valuable for establishing relative dollar values of different feeds and for determining when another run of the computer is necessary to formulate a new least-cost ration as feed prices change.

Shadow Prices. The computer also calculates the costs of the constraints which affect the formula solution. "Shadow prices" are the cost of the last unit of that constraint. Additionally, a range of values is listed between which the shadow prices apply. This information is useful in evaluating the relative contribution of the various constraints to the total cost of the ration.

Limitations of Computer-Formulated Rations

Computers are valuable tools in formulating rations. As is the case with all tools, however, computers can be improperly used.

In the past, computer-formulated least-cost rations have received some criticism because animals fed these rations performed poorly due to improper nutrient specifications and/or ingredient limitations imposed on the computer. An example is the early use of computers to formulate poultry rations. Well-meaning men with expertise in the use of computers, but with little or no knowledge of nutrition, formulated a least-cost poultry ration. The cost was much lower than for other rations being fed at the time, but the chickens would not eat it. An inexpensive ration is of no value if it does not contain the proper balance of nutrients and is not consumed by animals in sufficient quantities to fulfill their nutrient requirements. Therefore, establishment of proper nutrient specifications and limitation of individual ingredients to percentages of a ration which result in a palatable mixture are primary determinants of the success of feeding computer-formulated rations.

Any list of maximum limitations on individual ingredients, or group of ingredients, to ensure palatability of a ration is arbitrary because all possible interactions among ingredients are not known. Maximum ingredient limitations which have worked well under field conditions are listed in Table 12.1.[2]

Some feed limitations in Table 12.1 may be too high or too low for certain conditions. Such factors as pelletability, density, flow rates, and availability dictate different limitations under varying conditions. The limitations in Table 12.1 are based mainly on maximum percentages of ingredients which allow high palatability and efficient feed utilization.

Limitations are listed for concentrate ingredients only and pertain to the maximum percentage of the concentrate dry matter portion of the ration. Restrictions are not assigned to roughages because almost any can be fed as the sole roughage source if properly supplemented with concentrates.

Another limitation of computer formulation is the accuracy of the nutrient data on feedstuffs. Nutrient content of the same type of feed is affected by many factors associated with growing the crop such as soil

Table 12.1 *Recommended Maximum Percentage Composition of Common Feeds for Dairy Concentrate Mixes*

Feed	*Maximum % of concentrate*
Barley	80
Beet pulp, dried	40
Brewers' grains, dried	25
Citrus or orange pulp, dried	25
Coconut meal	50
Corn	80
Cottonseed meal	25
Cottonseed, whole	20
Distillers' corn grains, dried	25
Hominy feed	50
Linseed meal	25
Milo	50
Molasses	10
Oats	80
Rice bran	20
Safflower meal	20
Soybean meal	50
Urea	1.5
Wheat bran or wheat mill run	25
Wheat	50

Source: Bath, D. L., and Bennett, L. F.: Development of a dairy feeding model for maximizing income above feed cost with access by remote computer terminals. J. Dairy Sci. *63:* 1379, 1980.

type, fertilization practices, available moisture, and temperature. In most cases, average values calculated from data from many areas are used as the nutrient content of a particular feed, because actual data are not available on the feed in question. Therefore, the true nutrient content of a computer-formulated ration may be somewhat higher or lower than called for by the nutrient specifications. However, this is no more of a problem for computer-formulated rations than for those calculated by hand if both are based on average "book" values.

To be more accurate, individual feeds should be tested for their nutrient content before formulation of a ration. The tremendous time and expense involved in determining nutrient content, especially energy evaluations, preclude this as a routine practice except for the more easily determined nutrients, such as crude protein, crude fiber, or acid detergent fiber. Until inexpensive and rapid techniques are developed to estimate nutrient content of feeds, average "book" values will continue to be used extensively in ration formulation, whether done by computer or by hand.

Availability and Use of Computer Formulation

Computer formulation of dairy rations is done by commercial computer centers, private nutrition consultants, and, in some states, by the state university. Each offers this service to dairymen for fees ranging from about $5 to $30 per formulation, depending on the number of formulations requested, the number of feeds considered, and the efficiency of the computer program. In some cases, cost of computer formulation is included in the fee charged by private consultants for their advice and services to clients.

Most feed companies and some large dairy farms have made use of this tool in the past. High cost of computer time and lack of knowledge about its possible benefits are the principal reasons that it is not used more extensively by dairymen. Greater use of computer formulation probably will develop as dairy farms become larger, as more efficient and less expensive computer equipment and programs become available, and as more people become better trained in using the tremendous amount of information available from the computer. Availability of on farm computer terminals and microcomputers (Figure 12.2) also will increase the use of computerized ration formulation programs in the future.

The number and frequency of computer formulations needed vary with the type and size of dairy farm. Three to 5 different rations may be used on large dairies where cows are separated into groups according to production, whereas only one or 2 rations may be fed in smaller herds. New formulations are necessary only when feed or milk prices change significantly or when newly harvested feeds become available. In most areas, this occurs only a few times each year. Feed cost savings more than pay for the cost of occasional computer formulations, particularly for larger dairies which buy much of their feed.

FIG. 12.2 A microcomputer being used to formulate a "least-cost" ration.

12.3 Examples of Least-Cost Rations

Concentrate Mixes

Least-cost formulation has been used more for concentrate mixes than for complete rations. This is logical because most dairymen grow their own forage and must feed it regardless of its price and quality. However, some or all of the concentrates usually are purchased by dairymen. Competition is keen among feed companies for the dairyman's dollar, and the company which puts out quality feed at a low cost is most likely to attract and maintain the dairyman's business. Most feed companies formulate least-cost concentrate mixes by computer in order to remain competitive.

Nutrient Specifications. The first step in developing a least-cost concentrate mix formula is to establish nutrient constraints. These will vary depending on the type and quality of forage being fed with the concentrate mix. Energy, protein, calcium, and phosphorus levels recommended for concentrate mixes fed with different types of roughage are discussed in Chapter 11 and summarized in Table 11.6.

Let us assume that a feed company wants to formulate a least-cost concentrate mix that fits the needs of cows consuming 15 lb of alfalfa hay and 45 lb of corn silage daily. On a dry matter basis, this is approximately half legume and half nonlegume forage. A concentrate mix fed with a legume-grass mixture or legume plus corn silage should contain about 16% crude protein, 0.5% calcium, and 0.7% phosphorus as shown in Table 11.6. Additionally, all concentrate mixes should provide about 0.75 Mcal/lb of net energy for lactation (NE_l). Accordingly, the feed company personnel specify the minimum constraints for a concentrate mix shown at the top of Table 12.2. This information is put into the computer along with the nutrient content and prices of all available feeds. The computer then tests all combinations of available feeds and prints out the formula which fulfills or exceeds all of the ration constraints at the lowest cost. An example of a least-cost concentrate mix formulated on the basis of the above specifications is shown in Table 12.2.

Least-Cost Formula. The least-cost mix, as shown in the center of Table 12.2, on an "as fed" basis contains 27.8% barley, 24.9% wheat mill run, 19.7% dried beet pulp, 19.3% whole cottonseed, 5.9% cane molasses, 1.3% urea, and 1.1% dicalcium phosphate. Total cost of the mix is $8.45 per cwt ($169 per ton), which with the available feeds is the lowest-cost mix possible that fulfills the nutrient specifications listed at the top of Table 12.2. Ingredient percentage composition of the mix on a dry matter basis is shown in the second column of figures, the price per cwt of each feed ingredient in column 3, and the lower and upper price ranges in columns 4 and 5.

Taking barley as an example, it is priced at $8.00 per cwt ($160 per ton). At that price, it is included as 27.8% of the mix.

TABLE 12.2 *An Example of a Least-Cost Concentrate Mix*

Nutrient specifications	
Net energy for lactation (NE_l), minimum	0.75 Mcal/lb
Crude protein, minimum	16%
Calcium, minimum	0.5%
Phosphorus, minimum	0.7%

Feeds used in mix	*% of mix*		*Price*	*Price range ($/cwt)*	
	As fed	*DM*	*($/cwt)*	*Lower*	*Upper*
Barley grain, 46#/bu	27.8	27.6	8.00	7.95	8.09
Wheat mill run	24.9	25.0	7.70	2.05	8.44
Beet pulp, dried	19.7	19.9	7.30	7.23	7.34
Cottonseed, whole	19.3	20.0	11.00	3.70	11.63
Molasses, cane	5.9	5.0	5.00	4.66	5.40
Urea, 46% N	1.3	1.3	12.60	0	21.00
Dicalcium phosphate	1.1	1.2	26.00	18.36	30.06
Total cost = $8.45 per cwt					

Estimated analysis of concentrate mix	
Dry matter	89.7%
Net energy for lactation (NE_l)	0.75 Mcal/lb
Total digestible nutrients (TDN)	71.6%
Crude protein	16.0%
Crude protein equivalent from nonprotein nitrogen	3.5%
Crude fat	5.5%
Crude fiber	10.9%
Acid detergent fiber	15.7%
Ash	5.0%
Calcium	0.5%
Phosphorus	0.7%
Magnesium	0.3%
Potassium	0.9%
Sodium	0.2%
Sulfur	0.2%
Iron	148.3 ppm
Cobalt	0.1 ppm
Copper	19.6 ppm
Manganese	42.7 ppm
Zinc	6.0 ppm

However, if the price of barley drops to $7.95 per cwt (column 4), and all other feed prices remain the same, its percentage of the mix will increase. Conversely, if its price increases to $8.09 per cwt (column 5) under the same conditions, its percentage will decrease. Simultaneously, the amounts of other feeds in the mix will vary as the amount of barley changes, so other feeds could be added or deleted.

Estimated Analysis. An estimated analysis of the concentrate mix appears in the lower portion of Table 12.2. All of the minimum specifications listed at the top of the table are met or exceeded. In some cases, it is cheaper to include an excess of a nutrient rather than restrict it to an exact level, which is the reason that minimum or maximum constraints, or both, usually are used rather than exact equalities.

In addition to the nutrients used for minimum specifications, the computer calculates the amounts of any other nutrients for which it has been provided data, any of which could be used as a basis for ration specifications. Usually no constraints are placed on vitamins or trace minerals as it is assumed that they would be added as premix if needed. Their addition would not affect the overall major nutrient content of the ration because they make up such a small percentage of the total ration. If desired, a premix containing minerals, vitamins, and other additives can be specified to be included in the mix at a predetermined percentage. Also, salt can be specified at a certain level or be included as a part of sodium-containing minerals required for the mix.

Opportunity Prices. Feeds that were available but not selected by the computer are listed in Table 12.3 along with their prices at time of formulation, and opportunity prices. Let us take dried citrus pulp as an example. Priced at $7.50 per cwt, it is not a good buy, but it would be selected for the least-cost mix if it were priced at $7.19 per cwt and all other feed prices remained the same. Relative values for all feeds not selected for the least-cost formula are listed by the computer.

Rejection of feed ingredients by the computer does not necessarily mean that they are not good nutrient sources. It simply means that they are priced too high relative to other available feeds that can be used to fulfill the ration specifications.

12.4 Optimum or Maximum-Profit Rations

The main objective in operating a dairy farm is for the dairyman to make a profit. None of the desirable features of farm living is possible for long if the dairy farm is not operating at a profit. Consequently, the modern dairyman must be conscious of all factors that affect his profit. Since feed is the largest single expense item in dairying, it has the greatest potential for affecting profit.

TABLE 12.3 *Opportunity Prices on Feeds Not Used in Mix*

	Price ($/cwt)	
Feeds not used in mix	*At time of formulation*	*At which feed will be included in mix*
Citrus pulp, dried	7.50	7.19
Corn grain	8.60	8.51
Cottonseed meal, 41% sol.	12.00	10.47
Limestone, ground	4.00	2.03
Milo grain	8.30	7.68
Oats grain	8.50	7.55
Soybean meal, 48% sol.	14.00	10.55

Many factors other than feeding affect profit, such as labor costs, land values, taxes, building and equipment costs, and depreciation, but except for labor, they are fixed costs and cannot be significantly changed by individual dairymen. However, a dairyman can control feeding practices and costs to a larger extent. Therefore, maximizing income above feed cost comes close to maximizing profit, and subsequent reference to maximum-profit rations refers to rations which result in maximum income above feed cost.

Differences Between Least-Cost and Maximum-Profit Rations

The primary objective of a least-cost ration is to provide the specific nutrient concentrations at the lowest cost. Milk price and milk production potential are given no consideration. Also, the amount of the ration to feed under different conditions is not specified. Therefore, feeding a least-cost ration does not always result in maximum income above feed cost because nutrient constraints on the ration and amounts fed may not be optimum for the economic and production potential situation of an individual dairy farm.

Maximum-profit ration formulation[3] considers these factors in addition to the factors previously discussed for the least-cost rations. Rations that maximize profit vary in content depending on the price received for milk, producing ability of the cows, and the cost of feed ingredients. Higher energy levels (from more concentrates in the ration) are profitable when milk prices are high relative to the concentrate prices, and even more profitable when fed to high-producing cows compared with medium- or low-producers.

In formulating a ration the computer considers feed costs, nutrient requirements, the price of milk, maintenance requirements for cows of various body weights, production requirements at various levels of milk production and fat tests, maximum voluntary roughage intake as concentrate intake is increased, and minimum fiber and roughage levels to maintain normal milk-fat tests. With this information and data on the nutrient content of the available feeds, the computer tests every feed combination that fulfills the major nutrient requirements of a dairy cow, or group of cows, and selects the combination that results in the highest income above feed cost.

Nutrient constraints specified for the total ration dry matter are from the National Research Council,[4] as summarized in Table 12.4.

Net Energy Requirements. Milk response associated with increasing net energy intake for cows with varying milk producing ability is shown in Figure 12.3. This computer program selects the optimum level of milk production as well as the composition of the ration. For example, in Figure 12.3 as net energy intake increases above the maintenance requirement of the cow, milk production also increases. However, milk production is a curvilinear function. At the higher energy intake levels, the rate of increase in milk production declines because more energy is used for fattening and other metabolic functions. At some point along the milk production curve, the cost of the extra energy and other nutrients exceeds the returns from the extra milk produced. The computer selects the level of milk production at which the value of the next unit of milk equals the cost of the next unit of feed, determining not only the optimum combination of feeds but also the amount to be fed.

Maximum Voluntary Feed Intake. Under some conditions, it might be economical to feed a cow several hundred pounds of feed per day to obtain several hundred pounds of milk; but such a ration could not be consumed by a cow. Consequently, a maximum amount of dry matter, both roughages and concentrates, must be determined in order to prevent the computer from selecting amounts in excess of the intake ability of the cow.

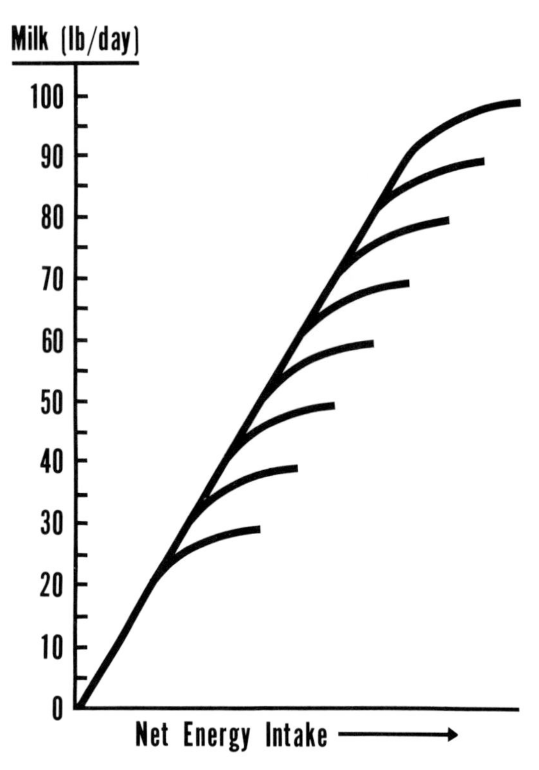

FIG. 12.3 Milk production response with increasing net energy intake.

Maximum voluntary dry matter intake of roughages by cows fed varying amounts of concentrates is shown in Figure 12.4. When no concentrates are fed, a high-producing cow can eat about 3.15% of her body weight in the form of excellent-quality hay. As concentrates are added to the ration, total feed intake goes up, but the amount of hay the cow can eat goes down. The starting point for a good quality hay is about 2.7% of body weight, and for silages and fair-quality hays about 2.25%. Each forage source in the computer program is assigned a starting level so that the computer can calculate the maximum total ration intake that is possible using the available ingredients. A minimum roughage dry matter level of 1.35% of the body weight of the cow is required to ensure that adequate

TABLE 12.4 *Major Nutrient Constraints in Maximum-Profit Ration Program*

	Concentration in total ration dry matter
Crude protein, minimum	13–16%*
Crude fiber, minimum	17%
Calcium, minimum	0.43–0.60%*
Phosphorus, minimum	0.31–0.40%*
Nonprotein nitrogen, maximum	0.5%
Roughage (% of body weight), minimum	1.35%
Net energy (dependent on body size and level of milk production, see Figure 12.3)	

*Requirement varies depending on body size and level of milk production.

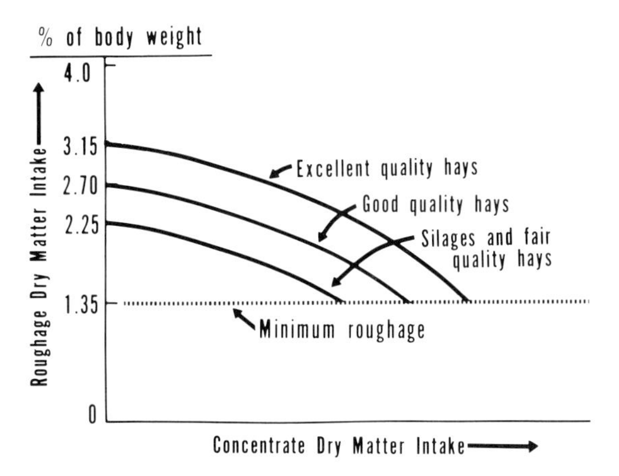

FIG. 12.4 Maximum roughage dry matter intake with increasing concentrate dry matter intake.

coarse fiber is in the ration to maintain normal milk-fat content.

Total dry matter intake recommended by the computer usually varies between 2.5 and 4% of the body weight, depending mainly on level of milk production and forage quality. The higher level of 4% probably could be consumed only by high-producing cows with large appetites. More total feed can be consumed by a cow when excellent-quality forage is fed than when lower-quality forages make up the roughage portion of the ration.

Example Ration. An example of a maximum-profit ration is shown in Table 12.5. The ration was formulated for a group of cows averaging 1,400 lb with medium-production potential. Their average fat test was 3.5%, and the net blend price received for milk (after deduction of transportation costs and assessments) was $12.00 per cwt.

The ration contained 29.55 lb corn silage, 16.52 lb alfalfa hay, 7.49 lb wheat mill run, 7.13 lb barley, 6.35 lb dried beet pulp, 5.8 lb whole cottonseed, 3.06 lb molasses, 0.3 lb urea, and 0.17 lb salt. On an as-fed basis, this amounts to 46.07 lb of roughage and 30.29 lb of concentrates. Corn silage makes up 64.1% and alfalfa hay 35.9% of the roughage on an as-fed basis, as shown in the second column of figures. However, on a dry matter basis, corn silage is 37.4%, and alfalfa hay 62.6% of the roughage (third column of figures). Percentage composition of the concentrate mix, on an as-fed and dry matter basis, is shown in the second and third columns. Prices of feed ingredients selected by the computer are listed in column 4, and the lower and upper price ranges for individual ingredients appear in columns 5 and 6.

Total cost of the ration is $3.72 per cow per day. Milk production on this ration should be about 75 lb per day, which returns $9.00 per cow per day, or $5.28 per cow per day income above feed cost, as shown in the center of Table 12.5. No other combination of feeds and level of production will return more income above feed cost using these milk and feed prices with this set of production functions and nutrient constraints.

Estimated Analyses. Estimated analyses of the concentrate mix, the roughage, and the total ration on a dry matter basis appear in the next section. The following items are calculated for each formulation: dry matter, net energy for lactation (NE_l), TDN, crude protein, crude fat, effective crude fiber, acid detergent fiber, ash, calcium, phosphorus, Ca:P ratio, and NPN.

Following the analysis for the total ration are the minimum and maximum nutrient limits required for the formulation. As can be seen in this case, the total ration dry matter met the minimum requirement of 16% crude protein for this level of production, exceeded the minimum requirement of 17% crude fiber, exceeded the minimum requirement of 0.6% calcium, met the minimum requirement of 0.4% phosphorus, and was below the maximum level of 0.50% NPN allowed for the ration. Estimated mineral analyses of the concentrate mix, the roughage, the total ration, and minimum concentrations recommended by the National Research Council (NRC) appear in the next section of the output form. These minerals are not used as a basis for formulation, but comparison of amounts calculated for the total ration with NRC recommendations gives an indication if any should be included in the ration as a premix.

Feeds Not Used in Ration. The last section of the output form contains a listing of feeds available but not selected for the ration. The fact that they were not selected does not mean that they are not good feed ingredients. It simply means that they were priced higher than they were worth compared with the feeds used in the ration. The names of the feeds appear in the first col-

TABLE 12.5 *Example of a "Maximum-Profit" Ration*

SPECIFICATIONS:

PRODUCTION CURVE MAXIMUM............	=	80 LBS
AVERAGE MILK FAT..................	=	3.5 %
AVERAGE COW WEIGHT.................	=	1400 LBS
BLEND PRICE......................	=	$ 12.00/CWT
NE(L) FOR ACTIVITY................	=	10 % OF MAINTENANCE
FIRST LACTATION HEIFERS IN GROUP...	=	30 %
SECOND LACTATION HEIFERS IN GROUP..	=	20 %

FEEDS USED IN RATION:	LB/DAY AS FED	%ROUGHAGE AS FED	%ROUGHAGE DM	PRICE $/CWT	---RANGE--- LOWER	---RANGE--- UPPER	---CONSTRAINTS---- AS FED -POUNDS- MIN	AS FED -POUNDS- MAX	100% DM -%ROUGH- MIN	100% DM -%ROUGH- MAX
CORN SILAGE,30% DM	29.55	64.1	37.4	1.50	1.39	1.67				
ALFALFA HAY,24% MCF	16.52	35.9	62.6	5.00	4.61	5.39				
TOTAL ROUGHAGE...	46.07	(23.73 LBS DM)								
		%CONCENTRATE AS FED	%CONCENTRATE DM				AS FED -POUNDS- MIN	AS FED -POUNDS- MAX	100% DM -%CONC.- MIN	100% DM -%CONC.- MAX
WHEAT MILL RUN	7.49	24.7	25.0	7.70	4.08	7.96				25.0
BARLEY,46-48#	7.13	23.5	23.5	8.00	7.93	8.09				80.0
BEET PULP,DRIED	6.35	21.0	21.4	7.30	7.06	7.38				40.0
COTTONSEED,WHOLE	5.80	19.1	20.0	11.00	6.32	11.52				20.0
MOLASSES,CANE	3.06	10.1	8.5	5.00	-3.87	5.21				8.5
UREA,46% N	0.30	1.0	1.0	12.60	*****	22.58				1.0
SALT	0.17	0.6	0.6	5.00	-2.10	74.26				
TOTAL CONCENTRATE.	30.29*	(26.97 LBS DM)								

*NOTE: PROVIDE SALT FREE CHOICE IF NOT INCLUDED IN ABOVE RATION AS RECOMMENDED.
PROVIDE OTHER ESSENTIAL MINERALS NOT SUPPLIED IN ADEQUATE AMOUNTS
BY FEEDS IN RATION LISTED ABOVE.

ROUGHAGE:CONCENTRATE RATIO = 47:53 (DM)

	LB/COW	$/COW	PRICE PER CWT	LOWER RANGE	UPPER RANGE
OPTIMUM DAILY MILK PRODUCTION:	75.0	9.00	12.00	11.76	14.70
TOTAL FEED COST		3.72			
TOTAL DAILY INCOME ABOVE FEED COST:		5.28			

ESTIMATED ANALYSIS: (100% DM)	CONCENTRATE	ROUGHAGE	TOTAL RATION	-CONSTRAINTS-- MIN	MAX
DRY MATTER PCT	89.04 %	51.51 %	66.40 %	40.00%	
NE(L)	0.84 MCAL/LB	0.62 MCAL/LB	0.74 MCAL/LB		
TDN	80.21 %	60.61 %	71.04 %		
CRUDE PROTEIN	16.70 %	15.20 %	16.00 %	16.00%	
CRUDE FAT	6.06 %	2.70 %	4.49 %		
EFF. CF	12.21 %	25.25 %	18.31 %	17.00%	
ADF	17.70 %	32.25 %	24.51 %		
ASH	5.48 %	7.96 %	6.64 %		
CALCIUM	0.33 %	0.98 %	0.63 %	0.60%	
PHOSPHORUS	0.55 %	0.23 %	0.40 %	0.40%	
CA:PHOS RATIO	0.60	4.23	1.58	1.50	
NPN	0.46 %	0.00 %	0.24 %		0.50%

Table 12.5 *Example of a "Maximum-Profit" Ration (continued)*

ESTIMATED MINERAL ANALYSIS

MINERAL	ESTIMATED RATION DRY MATTER CONTENT			NRC MINIMUM
	CONCENTRATE	ROUGHAGE	TOTAL RATION	
MAGNESIUM	0.35 %	0.26 %	0.31 %	.20 %
POTASSIUM	1.05 %	2.11 %	1.55 %	.80 %
SODIUM	0.40 %	0.14 %	0.28 %	.18 %
SULFUR	0.23 %	0.21 %	0.22 %	.20 %
IRON	161.04 PPM	395.69 PPM	270.87 PPM	50.00 PPM
COBALT	0.09 PPM	0.08 PPM	0.09 PPM	.10 PPM
COPPER	21.87 PPM	13.33 PPM	17.87 PPM	10.00 PPM
MANGANESE	47.63 PPM	34.00 PPM	41.25 PPM	40.00 PPM
ZINC	7.60 PPM	18.49 PPM	12.70 PPM	40.00 PPM

FEEDS NOT USED IN RATION:	PRICE	
	AT FORMULATION	OPPORTUNITY
CITRUS PULP,DRIED	7.50	7.02
CORN GRAIN,GR OR RLD	8.60	8.54
COTTONSEED MEAL,41 S	12.00	10.91
DICALCIUM PHOSPHATE	26.00	5.28
LIMESTONE,GROUND	4.00	-2.34
MILO,CAL OR MIDWEST	8.30	7.78
OATS,PCS	9.50	7.48
SOYBEAN MEAL,48 SOL	14.00	11.90
COTTONSEED HULLS	4.00	0.95
OAT HAY	4.40	3.28

umn, their prices at the time of formulation are in the second column, and their "opportunity" prices (the prices they would have to drop to before they would be included in the ration) are shown in the third column.

Adaptation to Varying Dairy Farm Conditions

For cows with either higher or lower production potential, the corresponding milk production curves in Figure 12.3 would be used. Also, different sets of factors for cows of different body weights and for milk with different fat tests are employed. Therefore, the program can be run for any size cow, or group of cows, producing any amount of milk with any fat test. This works ideally in large herds where cows are divided into groups according to level of production. A separate ration can be formulated based on the average cow weight, average level of production, and average fat test of each group of cows.

Using the same feed prices and milk price shown in Table 12.5, the computer selected the rations shown in Table 12.6 for a herd divided into 4 production groups.

Rations for the 4 groups varied greatly, particularly the concentrate portion. Concentrate amounts were 8 lb per cow for the low string, 20 lb for the medium string, 30 lb for the medium-high string, and 38 lb for the high string. Although feed costs went up as concentrate amounts were increased, income from increased milk production went up even faster, resulting in much more income above feed cost from the higher-producers. This example illustrates

Table 12.6 *Rations For a Herd Divided into Four Production Groups*

	Production group			
	Low	*Medium*	*Medium-high*	*High*
Milk production, lb	35	55	75	93
Alfalfa hay, lb	16	15	17	14
Corn silage, lb	53	43	30	23
Concentrates, lb	8	20	30	38
Ration cost, $	$2.25	$2.99	$3.72	$4.16
Income minus feed cost, $	$1.95	$3.61	$5.28	$7.00

the need to feed high-producing cows liberal amounts of concentrates in order to obtain maximum profitability from a dairy herd.

Of course, some of the factors in this program may have to be modified as more accurate information becomes available from future research. Of critical importance are more data on maximum dry matter intake as level of concentrate is increased. Additionally, effect of higher energy levels on changes in milk production from cows with varying production potential, and at different stages of lactation, need further study to improve the accuracy of the model. This information is needed also for efficient calculation of a ration by hand, but probably is not the limiting factor because of limited potential for large numbers of calculations. However, maximizing of profit by computer is limited only by the accuracy of the factors in the computer program rather than the ability to test all possible combinations.

12.5 Summary

Feeding dairy cows has moved away from hit-or-miss formulation and allotment of rations to a more scientific approach. Computers are now among the tools that the modern dairyman and his advisors use to formulate and allot rations and to evaluate the performance and efficiency of feed utilization of individual cows. With proper emphasis on nutrient needs of the cow and nutrient values and prices of available feeds, dairymen are able to formulate dairy rations, with the help of computers, that will result in optimum levels of milk production based upon the inherent ability of their cows and prices received for milk.

Review Questions

1. What is the largest single advantage to dairymen of computer-formulated dairy rations?
2. What are the main limitations which must be recognized when feeding computer-formulated rations?
3. In addition to the least-cost ration formula, what information is available from the computer, and how can it be used?
4. Discuss the differences between and advantages and disadvantages of least-cost rations and maximum-profit rations.

References

1. Heady, E. O., and Candler, W.: *Linear Programming Methods,* Ames, Iowa: Iowa State University Press, 1966.
2. Bath, D. L., and Bennett, L. F.: Development of a dairy feeding model for maximizing income above feed cost with access by remote computer terminals. J. Dairy Sci. *63:* 1379, 1980.
3. Dean, G. W., Bath, D. L., and Olayide, S.: Computer program for maximizing income above feed cost from dairy cattle. J. Dairy Sci. *52:* 1008, 1969.

CHAPTER 13

Feeding Dry Cows, Bulls, and Dairy Animals for Veal or Beef

13.1 Introduction

Feeding programs for the milking herd are discussed in Chapter 11 and for calves and replacement heifers in Chapter 20. Feeding the remainder of the herd is the subject of this chapter. This includes dry cows, bulls, and dairy animals raised for veal or beef.

13.2 Feeding Dry Cows

Dairy cows should have a dry period between lactations for about 40 to 60 days. The feeding program during the dry period can be very simple, and it should be based primarily on roughages rather than concentrates. Dry, long-stemmed hay is particularly good for dry cows. The dry hay forces the cow's digestive system to work harder, resulting in stronger muscles in the rumen and less digestive and metabolic problems at, or soon after, calving.

Early Dry Period

The feeding program during the dry period should vary depending on how soon the cow is going to calve. From the time that a cow goes dry until 2 to 3 weeks before parturition, she can be fed only roughages such as hay, pasture, green chop and silage, and mineral supplements. Cows that are very thin when they go dry may need a few pounds of a concentrate mix daily in addition to roughages in order to regain weight lost during lactation. However, it is better to feed thin cows extra concentrates in late lactation to regain body condition, because it is easier to feed concentrates then and cows gain weight more efficiently during lactation than during the dry period.

Salt should be available free choice. A high phosphorus mineral, such as the sodium phosphates, should be available when legume roughage is fed. Trace minerals may be needed under some conditions, as discussed in Chapter 9.

Late Dry Period

For the last 2 to 3 weeks before calving, cows should be fed a concentrate mix in addition to the roughages being fed. This feeding program also applies to heifers nearing their first lactation. The concentrate mix can be the same mix that is fed to the milking cows. With this regimen the dry cows can get reaccustomed to the taste of the concentrate mix that will be fed to them when they return to the milking herd.

Inclusion of some concentrates in the ration during the last 2 to 3 weeks also allows the ruminal microorganisms (bacteria and protozoa) to adjust gradually from an all-roughage ration to the milking cow ration, which will include large amounts of concentrates.

Fulfilling Dry Cow Requirements

Dry cow nutrient requirements for energy, protein, calcium, phosphorus, and vitamin A are listed in Appendix Table III-B. Recommended concentrations in the ration dry matter for these nutrients and other essential minerals and vitamins are listed in Appendix Table III-C.

Many combinations of feeds can fulfill the nutrient requirements of the dry cow. Legume roughages such as alfalfa hay fed free-choice more than fulfill requirements for energy, protein, calcium, and vitamin A. Legumes also are good sources of most trace minerals. Phosphorus and sodium usually are the only nutrients that are lacking. These requirements can be fulfilled by providing salt and a high-phosphorus supplement free-choice. However, when this type of ration is fed, there is a great excess of calcium because legumes such as alfalfa usually have 3 to 6 times as much calcium as phosphorus. When dry cows consume large amounts of calcium above their needs, incidence of milk fever (parturient paresis) increases in those that are susceptible to this metabolic disease, as discussed in Chapter 9.

Non-legume forages such as cereal hays and corn silage are not adequate as the only dry cow feed because they usually are too low in protein and many minerals. In the case of corn silage, another disadvantage is that cows tend to eat too much and get overly fat when they have free-choice access to it. The many problems it entails, commonly referred to as the "fat cow syndrome," include difficult calving, retained placenta, metritis, displaced abomasum, ketosis, and "downer" cows. When corn silage is fed to dry cows, it should be restricted to about 30 lb per day for 1400-lb cows, with proportionately more or less for larger or smaller cows. Feeding the remainder of the forage in the form of dry hay helps to control excessive weight gain and improve rumen muscle tone, thus reducing the problems associated with the "fat cow syndrome."

Although neither legumes nor non-legumes are ideal as the only feed for dry cows, a combination of the 2 results in a very good dry cow ration. Free choice feeding of a mixture of about 30% legumes and 70% non-legumes on a dry matter basis usually fulfills all of the dry cow nutrient requirements except for salt and possibly phosphorus. The ratio between calcium and phosphorus is much narrower than for legumes alone. Providing a phosphorus mineral supplement and trace mineralized salt free-choice would take care of the possible phosphorus deficiency and also provide a safety factor for any trace minerals that may be lacking. During the last few weeks before calving, the above forages should be supplemented with about 5 to 8 lb of a concentrate mix in order to accustom the cows and their ruminal microorganisms to the higher level of concentrates that will be fed following calving. The amount actually fed should be based on body size and condition of the dry cow.

13.3 Feeding Dairy Bulls

Feeding programs for bull calves are identical to those for heifer calves for the first few months of life. However, bulls should be segregated from heifers by about 4 months of age and fed and housed separately.

Nutrient allowances for growing bulls at various weights and ages are shown in Appendix Table III-A. Allowances for maintenance of mature bulls also are contained in the same table.

Young bulls grow faster than heifers. Consequently, they need more energy and other nutrients than heifers of the same age to encourage rapid growth and sexual development. Underfeeding retards the onset of puberty and results in production of poorer-quality semen as well as reduction of growth rate. In addition to liberal concentrate allowances, young bulls should have free-choice access to good-quality hay. At about 10 months of age, free-choice pasture, silage, greenchop, or hay can make up a major portion of a bull's ration. Concentrates should continue to be fed in amounts dependent on the quality of roughage being consumed. Enough concentrates should be fed to encourage rapid growth without excessive fattening. Concentrate mixtures with 12% crude protein are adequate for yearling or mature bulls when fed with good-quality roughages, whether they are legumes or grasses.[1]

A good rule for mature bulls is to feed daily about 1 lb of hay and 0.5 lb concentrate per cwt of body weight. Therefore, a 2,000-lb bull would receive daily about 20 lb of hay and 10 lb of a concentrate mix. These amounts should be adjusted according to the body condition of various bulls because all do not respond in the same way. Excess fatness of mature bulls should be avoided as it reduces their libido and may cause severe stress and strain on their feet and legs.

Excess calcium in bull rations can cause problems also, particularly in older bulls. When legume roughage is fed, the concentrate mix should not contain a calcium supplement. Herd concentrate mixes usually contain added calcium to fulfill needs of the lactating cow, which is losing calcium from her body in her milk. However, bulls do not lose calcium in this way and, in time, excesses may cause vertebrae and other bones to fuse together. Therefore, bulls may need a different concentrate mix than that fed to the milking herd if the herd mix is high in calcium.

13.4 Feeding Dairy Animals for Veal or Beef

About half of the calves born are bulls. Only a very small percentage of these bulls are used as herd sires. Artificial insemination and freezing of semen make it possible for one bull to fertilize thousands of cows each year. With the pressure of genetic improvement in dairy cattle as intense as it is today, only a few of the very best bulls that are born each year are used for breeding purposes.

The remainder of the bull calves can be handled in several ways. Many are sold for slaughter when only a few days old. These are called "deacon" calves or "bob" veal. Others are raised for about 6 to 8 weeks and sold as veal calves. Some are continued on and sold as feeder calves to beef feedlots, or raised as dairy beef by the dairyman himself.

Bulls gain weight faster and more efficiently than steers or heifers. However, bulls are harder to handle as they get older and frequently are discriminated against by buyers. Therefore, most bulls are castrated and raised as steers when fed past the veal stage.

Feeding Veal Calves

Good and choice veal calves should weigh about 200 lb at 6 to 8 weeks and have a well-muscled carcass with a layer of fat over the back. The meat should be light colored indicating that they have not been fed hay or grain.

Whole milk or milk replacer is fed throughout the feeding period. If whole milk is fed, a daily amount equivalent to 10% of the calf's body weight is recommended for the first few weeks. This is increased to a free-choice level near the end of the feeding period. This method of feeding takes about 10 lb milk for each lb of veal produced at 6 to 8 weeks of age. Relative milk and veal prices usually make this an uneconomical practice in most areas of the U.S.

When milk replacer is fed, it takes about 1.3 lb of dry replacer per lb of veal produced. Especially good results have been obtained in Michigan with replacers that contained emulsified animal fat in addition to milk products. The manufacturer's directions should be followed when feeding these products to veal calves.

Feeder Calves

Dairy calves to be sold as feeders or as dairy beef are fed very differently than veal calves. Early weaning and feeding of economical feeds are essential for a profitable operation.

Calves raised to the feeder stage (400 to 725 lb) should be fed up to 5 lb of starter per day and hay free choice until they are 6 months old. Some are sold at this stage. Others continue for up to a year of age on 2 to 3 lb of concentrates per day plus free-choice roughage. Silage can replace hay, but usually only small quantities are eaten until the calf is several months old. Yearling steers should weigh about 725 lb on this program and can go directly into a feedlot for fattening.

Feeding Dairy Beef

Selling dairy animals at a high-standard or low-good grade usually is more profitable than feeding to higher grades. This can be accomplished with several types of feeding programs. In areas where corn silage is the most economical feed source, it can be fed alone to steers between 700 lb and slaughter weight of about 1,000 lb. Daily weight gains on all-silage rations range between 1.7 and 2.0 lb.[2] Similar results are obtained when alfalfa hay replaces part of the corn silage.

Faster gains are obtained when roughages are supplemented with concentrates. Concentrates fed at a rate of 1% of body weight reduce the feeding time necessary to go from 700 to 1,000 lb by 30 to 50 days compared with an all-roughage program. Gains will average from 2.3 to 2.5 lb/day on this program. The most profitable feeding system varies in different areas depending on relative grain, roughage, and beef prices.

An accelerated growth program has proven successful in some areas. Rather than feeding roughage free choice and limited amounts of concentrate, calves are provided practically an all-concentrate ration from weaning to slaughter. Calves have constant access to self-feeders containing the high-concentrate ration. Rations based on barley, dried beet pulp, other by-product feeds, urea, and about 5% alfalfa meal fed free choice result in dairy steers being marketed at 900 to 1,000 lb at 1 year of age. Average daily gains are about 2.8 lb with a feed efficiency of about 6.5 lb of feed per lb of gain. When grain prices are low in relation to beef prices, this method of feeding can be very profitable.

13.5 Summary

Feeding the non-milking members of the dairy herd is much simpler than it is for milking cows. Nutrient requirements are lower because milk production is not a factor. Good-quality roughage can serve as the major source of nutrients, and, in some cases, the only source for dry cows, bulls, and dairy beef. Concentrate feeding can be limited to simple mixes and to amounts which keep young animals growing normally and mature animals in good condition without overfattening. Amounts of concentrates needed vary with the age of the animal and the quality of roughage being fed.

Dairy animals raised for veal or beef can be fed in several different ways, depending on local economic conditions. Vealers usually are fed whole milk or milk replacer for 6 to 8 weeks and then slaughtered. Dairy calves raised for beef can be fed on high-roughage or high-concentrate rations to market weights. Again, choice of feeding program is dependent on economic conditions, particularly relative roughage, concentrate, and beef prices.

Review Questions

1. Why should a dry cow feeding program be based primarily on roughages?
2. What are the important differences between feeding bulls and cows?
3. Contrast the feeding programs for dairy animals being fed for veal or beef production. How would you decide which would be more profitable for a particular dairy farm?

References

1. Branton, C., Bratton, R. W., and Salisbury, G. W.: Total digestible nutrients and protein levels for dairy bulls used in artificial breeding. J. Dairy Sci. *30:* 1003, 1947.
2. Speicher, J. A., and Deans, R. J.: Dairy beef. Farm Science Series, Extension Bull. 485. Michigan State Univ. Coop. Ext. Serv., December, 1964.

PART FOUR

Reproduction and Lactation

CHAPTER 14

General Endocrinology in Dairy Cattle

14.1 Introduction

Maximal milk and meat production requires proper function of all parts of the cow's body. However, the external and internal environment of the dairy cow is in a state of continuous change, and just for survival she must adjust her body functions to meet these environmental challenges. External environmental changes include, for example, season, temperature, humidity, feed, and the temperament of the dairyman. Since man has imposed on the dairy cow requirements for milk production greatly in excess of the amount needed to nourish her calf, she is particularly responsive to changes in external conditions as well as to internal changes associated with estrous cycles, pregnancy, and lactation.

Physiology is the discipline concerned with these problems. Specifically, the science of physiology is the study of function of the body or any of its component parts. The cell is the fundamental building block of the animal's body, and many cells of similar types form a tissue. Organs are made up of several types of tissues, and they are usually organized into systems as outlined in Table 14.1. The digestive (Chapter 8), reproductive (Chapter 15), nervous, and endocrine systems are of primary importance for maximal performance of the dairy cow.

14.2 The Nervous and Endocrine Systems

The nervous and endocrine systems are important because they coordinate body function. In general, the nervous system controls rapid adjustments of the body to changes in the environment, whereas the endocrine system regulates such processes as growth, reproduction, and lactation which require more time.

The nervous system is composed of four parts: brain, spinal cord, peripheral nerves, and autonomic nerves. The basic

TABLE 14.1 *Major Physiological Systems in Cattle*

System	*Primary component*
Skeletal	Bone
Muscular	Muscle
Circulatory	Heart, blood vessels
Digestive	Rumen and other compartments of stomach, intestines
Respiratory	Lungs
Urinary	Kidney, bladder
Reproductive	Ovaries, testes, reproductive tract, udder
Endocrine	Ductless glands
Nervous	Brain, spinal cord, nerves
Sensory	Eye, ear, nose
Integumentary	Skin

cell of the nervous system is the neuron which is made up of a cell body and two or more fibers called axons and dendrites (Figure 14.1). Axons are usually long fibers which conduct impulses away from the cell body. There are often several short dendrites per neuron, and they conduct impulses toward the cell body. The junction between the axon of one neuron and the dendrite of another is the synapse. Many nerve impulses travel over one or several synapses to the spinal cord and then return to some effector organ such as a gland or muscle. They may or may not pass up the spinal cord to the brain. An example of a simple reflex arc that does not involve the brain occurs when a painful stimulus is applied to a cow's teat (Figure 14.2). In this circumstance, the impulse travels the afferent (sensory) nerves of the teat to the spinal cord and returns via efferent (motor) nerves to cause contraction of leg muscles (kick). Premeditated kicking, however, involves the brain.

The brain is composed of four gross structures called the cerebrum, cerebellum, pons, and medulla oblongata (Figure 14.3). The cerebrum is the largest structure and is the center of reasoning, voluntary muscle control, and registration of various sensations. The cerebellum is necessary for coordination of such activities as eating or walking. The pons and medulla oblongata regulate functions such as breathing, swallowing, and rumination.

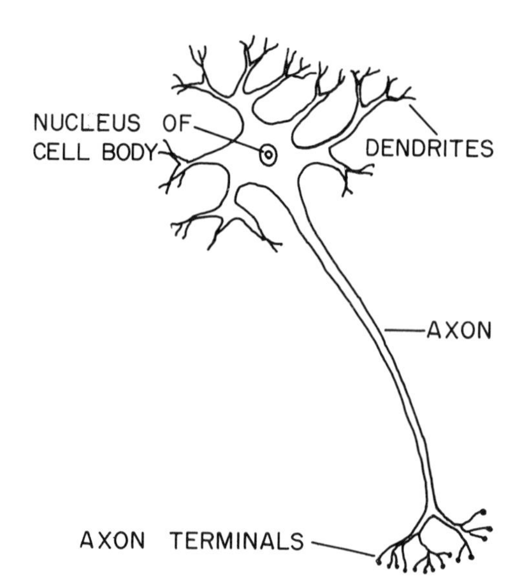

Fig. 14.1 A typical nerve cell (neuron) in cattle.

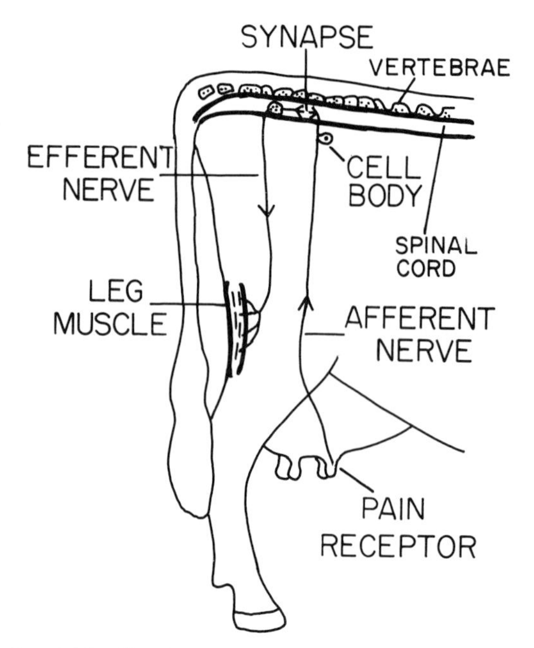

Fig. 14.2 Simple reflex arc involving the kicking reflex in the cow.

The spinal cord is a caudal extension of the medulla oblongata. The spinal cord receives impulses from the periphery and may transmit the message to the brain or relay the message directly back to the periphery via an efferent nerve to an effector organ, such as the leg muscle previously mentioned.

The peripheral nervous system includes all neurons outside the brain and spinal cord. These nerves are routes of communication between the central nervous system and the internal and external environment. Afferent and efferent nerves shown in Figure 14.2 are examples of peripheral nerves.

The autonomic nervous system (which is composed of two antagonistic systems,

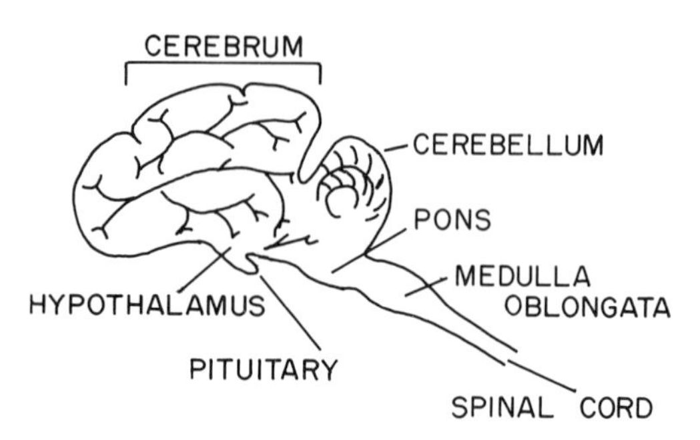

Fig. 14.3 Major components of the brain in cattle.

sympathetic and parasympathetic nerves) is involuntary and acts automatically. These nerves are usually associated with visceral organs such as heart, stomach, and intestines. The sympathetic system secretes the neurohormones epinephrine from the adrenal medulla and norepinephrine from peripheral sympathetic nerves in response to stress. Stress-induced release of epinephrine and norepinephrine inhibits normal milk-ejection (milk let-down) in cows. This topic is covered in greater detail in Chapter 19.

There are many glands scattered throughout the body of the cow. Glands consist of cells that are specialized for secretion or excretion. There are exocrine glands which discharge their secretions via ducts into various body cavities or to the exterior. The cow's udder is a good example of an exocrine gland. In addition, there are several ductless glands which discharge their secretions directly into blood. These ductless glands are called endocrine glands, and their secretions are termed "hormones," which means to stir up or excite.

Each endocrine gland may synthesize one or more hormones. Hormones are chemical messengers transported in blood to target cells where they bind specifically and with high affinity to receptor molecules. Peptide hormones bind to specific protein receptors on cell surface membranes, whereas steroid hormones enter target cells and bind to specific cytoplasmic receptors. The target may be any organ in the body such as the uterus, udder, or even another endocrine gland. Hormones regulate many physiological processes, and two functions of greatest concern to dairymen, reproduction and lactation, are dependent upon several hormones. In many instances the same hormone affects both reproduction and lactation. Thus, a knowledge of endocrinology is basic to understanding how these systems work. Examples of the use of hormones in dairy cattle management include synchronization of estrus and ovulation, stimulation of body weight gain, stimulation of milk production, and treatment of various disorders by veterinarians.

There is functional interlocking between the nervous and endocrine systems.[1] In fact, some neurons synthesize hormones which are called "neurohormones." Oxytocin is an example of a neurohormone which is synthesized in the brain and released into blood at milking to cause ejection of milk. Hormones may act upon the nervous system, and the endocrine organs are influenced by chemical secretions from the nervous system. Moreover, most organs in the body are subject to overlapping control of both systems. Because of this functional interrelationship, the nervous and endocrine systems are often referred to as the neuroendocrine system.

14.3 Location and Function of the Endocrine Glands

Approximate locations of major endocrine glands of cattle are shown in Figure 14.4. With exception of the placenta, endocrine glands are relatively small, usually weighing less than 75 grams in the mature cow.

Endocrine glands most intimately associated with reproduction and lactation in the cow are pituitary, hypothalamus, and ovaries, but other endocrine glands indirectly influence reproduction and lactation. Some functions of hormones are outlined in Table 14.2.

Anterior Pituitary (Adenohypophysis)

The anterior pituitary gland is attached to the base of the brain (Figure 14.4) and rests upon the bony floor of the cranial cavity. The anterior lobe of pituitary receives a rich supply of blood but no direct nerve supply (Figure 14.5). This anatomical relationship requires that control of the anterior pituitary come from substances in blood and not directly from nerves.

Anterior pituitary hormones that affect reproduction specifically are termed gonadotropins because they stimulate gonads (testis and ovary, Figure 14.5). In cattle, the gonadotropins are follicle-stimulating hormone (FSH) and luteinizing hormone (LH).

Other hormones secreted from the anterior pituitary include prolactin, growth hormone (GH), thyroid-stimulating hormone (TSH), and adrenocorticotropic hormone (ACTH).

Major relationships among pituitary hormones are illustrated in Figure 14.5. The

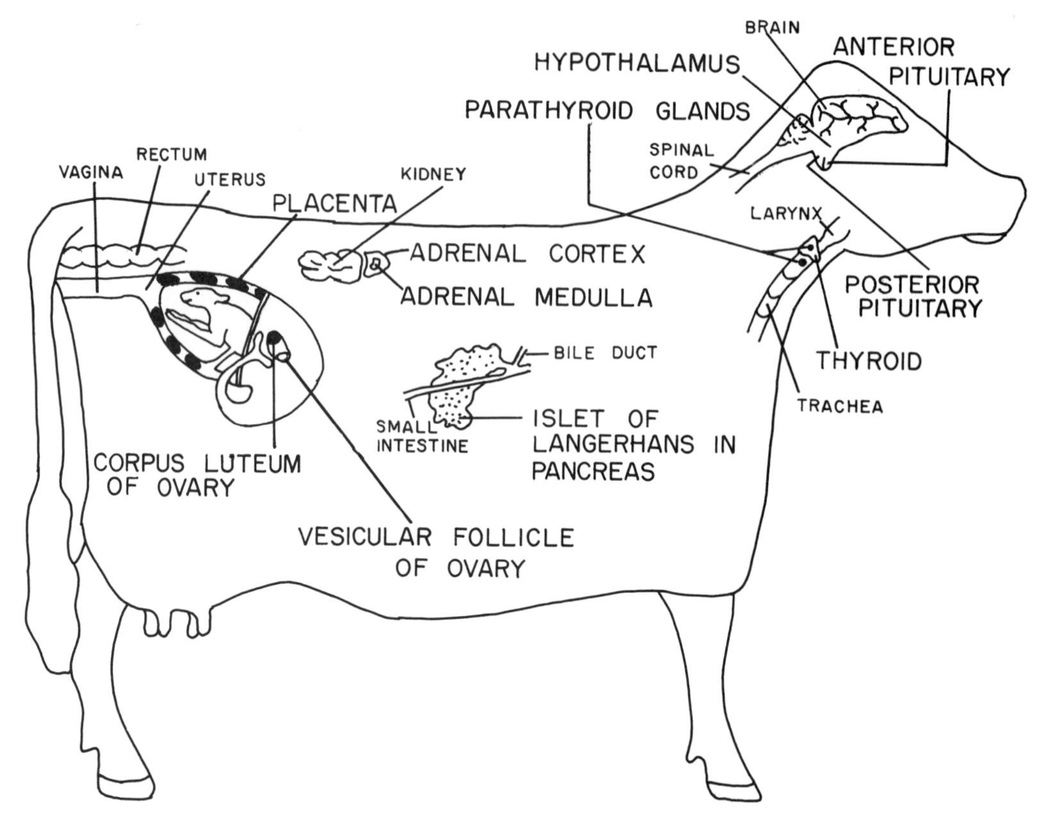

FIG. **14.4** Location of the major endocrine glands of the cow.

primary action of FSH, LH, ACTH, and TSH is to stimulate other endocrine glands (their target organs), to secrete hormones which in turn affect other target organs. However, GH and prolactin stimulate target tissues in cows directly without intervention of other endocrine glands.

Gonadotropins. FSH promotes estrogen secretion from the ovary in the cow whereas LH stimulates progesterone synthesis (Figure 14.5). In the bull, LH stimulates Leydig's cells (also called interstitial cells) of the testes (Figure 15.2) to secrete testosterone. In addition to inducing hormone secretions from gonads, FSH, and LH interact to cause ovarian follicle growth and rupture with release of an egg cell or ovum (ovulation). After ovulation LH regulates formation of the corpus luteum in the ruptured follicle. LH stimulates secretion of progesterone from the corpus luteum.[2]

FSH and LH injections into cattle induce multiple ovulations, and these studies provide a basis for producing many ova (superovulation) from a single cow.[3] Ova may be fertilized and transplanted to other less valuable recipient cows.[4] In general, FSH or pregnant mare's serum gonadotropin (PMS, a gonadotropin with FSH-like activity) is injected into a valuable donor cow twice daily for 4 days during the luteal phase of an estrous cycle. This causes ovarian follicles to grow. On the fourth day of FSH treatment prostaglandin $F_2\alpha$ ($PGF_2\alpha$; see Section 14.6) is injected to terminate the luteal phase of the cycle. Estrus occurs approximately 2 days later. The cow is inseminated every 12 hours from onset of estrus to 12 hours after the end of estrus with semen from a genetically superior sire. Multiple inseminations are used because the interval from first to last ovulation may exceed 12 hours. Meanwhile estrous cycles of several healthy host cattle are synchronized (usually with $PGF_2\alpha$). Five days after insemination of the donor, the oviducts may be exposed surgically and fertilized ova, which are usually in the 8 to 32 cell stage, are collected. Alternatively, embryos may be flushed nonsurgically from the uterus 6 to 8 days after estrus. The fertilized egg may be inserted surgically or

TABLE 14.2 *Hormones Affecting Growth, Reproduction, and Lactation in the Cow*

Endocrine gland	*Hormone secreted*	*Major function*
Anterior pituitary	Follicle-stimulating hormone (FSH)	Ovarian follicle growth; estrogen secretion; spermatogenesis
	Luteinizing hormone (LH)	Progesterone and testosterone secretion
	FSH and LH	Ovulation
	Prolactin	Mammary growth; initiation and maintenance of lactation
	Growth hormone (GH)	Body growth and milk production
	Thyroid-stimulating hormone (TSH)	Stimulates thyroid gland
	Adrenocorticotropic hormone (ACTH)	Stimulates adrenal gland
Posterior pituitary	Oxytocin	Uterine contraction; milk ejection
	Antidiuretic hormone (ADH)	Water balance
Hypothalamus	Releasing hormones (H) or factors (F)	
	Gonadotropin releasing hormone (GnRH)	Stimulates LH and FSH release
	GRF	Stimulates GH release
	Somatostatin	Inhibits GH release
	TRH	Stimulates TSH (also prolactin and GH) release
	CRF	Stimulates ACTH release
	Prolactin inhibiting factor	Inhibits prolactin release
Thyroid	Thyroxine; triiodothyronine	Oxygen consumption, protein synthesis, and milk yield
	Calcitonin	Calcium and phosphorus metabolism
Parathyroid	Parathyroid hormone	Calcium and phosphorus metabolism
Pancreas	Insulin and glucagon	Glucose metabolism
Adrenal cortex	Glucocorticoids	Glucose, protein, and fat metabolism
	Mineralocorticoids	Electrolyte and mineral metabolism
Adrenal medulla	Epinephrine and norepinephrine	Stress response
Ovary	Estradiol	Maturation of female reproductive tract; female sex behavior; mammary duct growth
	Progesterone	Preparation of uterus for implantation; pregnancy maintenance; mammary lobule-alveolar growth
	Inhibin	Inhibits FSH
	Relaxin	Relaxation of cervix and pelvis
Testis	Testosterone	Maturation of male reproductive tract; male sex behavior; spermatogenesis
Placenta	Estrogen	See ovary
	Placental lactogen	Mammary growth

nonsurgically into the lumen of the uterine horn adjacent to the ovary bearing a corpus luteum of a synchronized recipient cow. Pregnancy rates range from 30 to 60% with nonsurgical and 50 to 70% with surgical transfers. Embryos may be split to achieve pregnancy rates in excess of 100%. Treatments frequently yield 6 to 7 normal and 2 to 3 abnormal or unfertilized ova per cow. On average 3 to 4 pregnancies result from one superovulation of one cow. Depending upon the number of fertilized eggs, several recipient cows may be needed for each donor cow. Maintaining a large herd of recipients is the greatest cost in embryo transfer technology. Therefore, methods to store embryos until recipients are in proper stage of the estrous cycle (recipient must be within 1 day of the estrous cycle of that of the donor) permits reducing recipient herd size. Before insemination into a recipient, embryos may be stored in vitro at 37°C for 24 hours in artificial culture medium supplemented with blood serum or bovine serum albumin. Storage of cattle embryos in oviducts of rabbits for as long as 4 days results in subsequent pregnancy rates which are high. Indefinite storage of cattle embryos at −196°C in liquid nitrogen has also been achieved. Although freezing

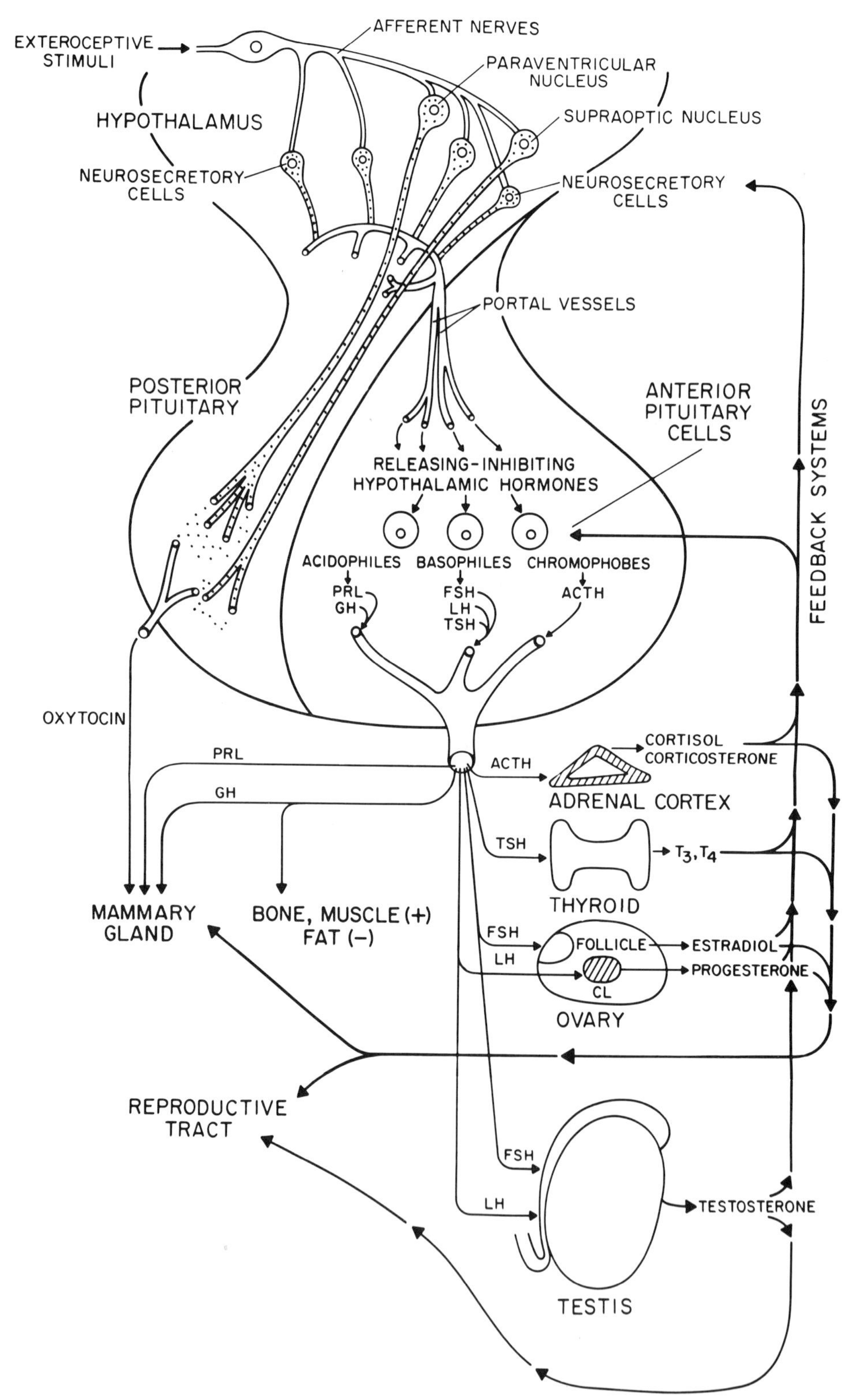

Fig. 14.5 Relationships of the hypothalamus, anterior and posterior pituitary glands and target endocrine glands of major importance to reproduction and mammary function. PRL = prolactin; GH = growth hormone; FSH = follicle stimulating hormone; LH = luteinizing hormone; ACTH = adrenocorticotropin; TSH = thyroid stimulating hormone; T_3 = triiodothyronine; T_4 = thyroxine; CL = corpus luteum.

kills about one third of the embryos, insemination of the remaining two thirds results in pregnancy rates equal to that of non-frozen embryos. At least 2 months should be allowed between superovulations of donor cows.

Superovulation gained in popularity in the 1970s primarily because many breeders of beef cattle desired to introduce "exotic" breeds into the United States, and direct importation was restricted. Initially, several companies performed superovulation. The trend today is for embryo transfers to be performed on the farm. As deep freezing technology improves still more transfers will be performed on the farm. Great variability in ovulation rates of donors is the greatest problem facing the embryo transfer industry. Nevertheless, superovulation may increase reproductive capacity 10 fold and reduce generation intervals if ova are obtained from very young females. Also, it is possible to obtain several full sibs and, with splitting of embryos, identical twins. Estimated costs for on farm embryo transfers are approximately $200 per live calf born.

Although some people are of the opinion that superovulation will lead to rapid improvement of the genetic base of a herd, it should be remembered that genetic worth of female genes will be distributed in a typical bell-shaped curve exactly like that for bulls (Chapter 4). Thus, not all offspring from a superior cow will be superior. Furthermore, the total number of possible ova that may be ovulated from a single cow beginning at 18 months of age is limited to a maximum of approximately 21,000. In contrast, a bull in AI is capable of siring 2,000 offspring per week. Thus, genetic impact of superovulation is relatively insignificant compared with that possible through use of genetically superior sires in AI (Chapter 5). It is possible, though difficult, to determine sex of the embryo before transfer to the host. If females only were selected, genetic impact of embryo transfer would increase. Within a given herd embryo transfer can have a significant influence in development of maternal lines or families.

Other uses of gonadotropins include treatment of cows with cystic ovaries by veterinarians.

Prolactin. This hormone is essential for mammary growth and lactation, and these aspects are examined in Chapter 17.

Growth Hormone. The most obvious function of growth hormone (GH) is to promote growth of most body organs in young animals. For example, surgical removal of the pituitary gland (hypophysectomy) of the calf inhibits further body growth,[5] but injections of GH can restore growth rate. However, GH levels in calves and older animals are not appreciably different, suggesting that function of GH changes as the animal ages because adult animals do not continue to grow.

In addition to promoting growth, GH decreases carcass fat but increases blood glucose. These actions on fat and carbohydrate metabolism may explain why injections of GH stimulate milk production in the cow.[6] This aspect of GH physiology is discussed in Chapter 17. Bovine GH has been synthesized and large quantities of GH should be available for use in cattle in the near future. This aspect is described in greater detail in Section 14.5.

Thyroid-Stimulating Hormone. TSH regulates cells of the thyroid gland. It induces the thyroid to trap iodine from blood; to synthesize increased amounts of thyroxine and triiodothyronine; and to encourage movement of thyroid hormones into blood. These actions elevate blood levels of thyroid hormones. If adequate amounts of iodine are not supplied in the diet of the cow, she may become deficient in thyroid hormone. This lowered thyroid hormone content causes secretion of even more TSH, which induces a conspicuous enlargement of the thyroid gland, or goiter. The condition may persist, especially in calves, in sections of the world where iodine is naturally deficient, but as discussed in Chapter 9, iodized salt prevents goiters.

Adrenocorticotropic Hormone. A cow subjected to stress or excitement releases ACTH from the anterior pituitary. Since high levels of ACTH reduce milk production, all stressful practices on dairy cattle should be avoided. The major function of ACTH is to stimulate the adrenal cortex to secrete glucocorticoid hormones (cortisol and corticosterone). Mineralocorticoids (i.e., aldosterone) are relatively unaffected by ACTH.

Posterior Pituitary (Neurohypophysis)

The posterior pituitary (Figures 14.4 and 14.5) is an outgrowth of the hypothalamus. The hypothalamus is a diffuse area of the brain located immediately above the pituitary. Hormones of the posterior pituitary are oxytocin and antidiuretic hormone (ADH), also called vasopressin. These hormones are synthesized in the hypothalamus where they appear as discrete granules (Figure 14.5). These granules travel the axons of the nerve tracts to the posterior lobe of the pituitary where they are stored for subsequent release into blood. Oxytocin and ADH differ from each other in only 2 of their 8 amino acids. This small difference in chemical structure probably explains why both oxytocin and ADH possess, in varying degrees, biological activity typical of the other.

Oxytocin causes contraction of smooth muscles of the uterus and myo-epithelial cells of the mammary gland. These physiological activities are explored further in chapters concerned with parturition, sperm transport, and milk ejection (milk let-down).

The primary action of ADH is to promote resorption of water in the tubules of the kidney. Thus, ADH reduces urine volume but increases urine concentration. Increased water intake depresses ADH blood levels which results in increased urine output. A secondary effect of ADH is constriction of arterioles and coronary blood vessels which leads to elevation of blood pressure.

Hypothalamus

This structure is located on the lower portion of the brain (Figures 14.3, 14.4, and 14.5) and secretes several low molecular weight neurohormones, or "factors" in addition to oxytocin and ADH. The principal function of these hormones is to regulate secretion of anterior pituitary hormones.[1] Four releasing hormones have been identified: gonadotropin releasing hormone (GnRH; also called LHFSH-RH), growth hormone-releasing factor (GRF), thyroid-stimulating hormone-releasing hormone (TRH), and adrenocorticotropic hormone-releasing factor (CRF). Somatostatin inhibits growth hormone release. A sixth hypothalamic factor, prolactin-inhibiting factor (PIF), inhibits release of prolactin from the anterior pituitary. Most evidence suggests that dopamine, a neurotransmitter, is PIF.

Laboratory synthesis of these small molecular weight releasing hormones has preceded synthesis of the relatively large anterior pituitary hormones. Commercial availability of some hypothalamic hormones has made it possible to regulate some events that anterior pituitary hormones influence. For example, injection of GnRH has been used as treatment for ovarian cystic follicles and can induce ovulation in some dairy cows. However, GnRH will not induce ovulation when a functional corpus luteum is present. In cattle, injection of TRH induces release of TSH, prolactin, and growth hormone and stimulates daily milk yield about 5%.[8] Injections of GRF release growth hormone in cattle, but it is not known if increased growth or milk production ensue.

Thyroid

The thyroid gland consists of 2 lobes on each side of the trachea at the larynx (Figure 14.4). An isthmus of thyroid tissue across the ventral surface of the trachea joins the 2 lobes.

In cattle, the thyroid gland secretes thyroxine and triiodothyronine. These hormones affect nearly every cell. Thyroxine stimulates oxygen utilization by cells, promotes increased use of carbohydrates, and increases protein catabolism. Paradoxically, protein synthesis may be stimulated, and the calorigenic action of thyroxine may be to provide energy for protein synthesis. Thyroxine also enhances oxidation of fats with consequent reduction in body weight. In low doses, thyroxine or its analogues stimulate milk secretion for variable, but limited, periods of time. This practice is discussed in greater detail in Chapters 11 and 17.

Another hormone, calcitonin, is released from the thyroid in response to elevated levels of calcium. Function of thyrocalcitonin is to lower blood calcium and phosphorus by inhibiting resorption of calcium from bone. Overproduction of this hormone may be involved in milk fever (parturient paresis) syndrome in dairy cows.[9]

Parathyroid

Cows usually have 4 parathyroid glands. One is located on the surface of each of the 2 lobes of the thyroid. The remaining pair is adjacent to the thyroid (Figure 14.4).

Parathyroid hormone, secreted from parthyroid glands, acts on bone, gut, and kidneys. Function of this hormone is to interact with metabolites of vitamin D and mobilize calcium from bone, increase calcium absorption from gut, and increase calcium reabsorption from the kidney, all of which lead to elevated blood levels of calcium. Parathyroid hormone increases excretion of phosphorus into urine thereby decreasing blood phosphorus. Although circumstantial evidence implicates the parathyroid in the milk fever syndrome, circulating levels of this hormone are similar in normal and parturient paretic cows. Thus, parathyroid insufficiency is not responsible for the hypocalcemia of parturient paresis.

Pancreas

The pancreas, which is located at the junction of the bile duct and small intestine (Figure 14.4), is an exocrine and an endocrine gland. The exocrine portion secretes digestive enzymes through a duct into the small intestine; the endocrine pancreas consists of small groups of cells, the islets of Langerhans, scattered throughout the gland.

The pancreas elaborates 2 hormones, insulin and glucagon, which are concerned primarily with glucose (energy) metabolism. Cows probably depend less on insulin and glucagon than nonruminants because of the ruminant's ability to use acetate as an energy source. Nonetheless, glucose is of great importance in ruminant metabolism particularly in manufacture of milk. A major function of insulin is to drive glucose from blood into cells of most tissues. Glucagon inhibits this process.

Adrenal

The 2 adrenal glands are located just anterior to the kidneys (Figure 14.4). Each adrenal consists of 2 major parts: an outer shell, the cortex, and an inner core, the medulla. Embryologically the cortex arises from mesoderm in close association with the gonads whereas the medulla is derived from neural tissue. The cortex is devoid of nerve fibers whereas the medulla receives a large nerve supply. The cortical hormones are steroids, whereas hormones of the medulla are amines. Thus, control and function of these 2 parts of the adrenal are different.

Adrenal Cortex. The adrenal cortex of cattle secretes at least 29 different steroid hormones that are usually divided into two broad categories: mineralocorticoids and glucocorticoids. Physiological effects of both groups of hormones are varied and often overlap. Thus, mineralocorticoids act as weak glucocorticoids and vice versa. The primary mineralocorticoid is aldosterone whereas there are 2 major glucocorticoids in cattle, cortisol and corticosterone. Mineralocorticoids are essential for life. Cattle secrete more cortisol than corticosterone.

Mineralocorticoids regulate mineral balance in body fluids by increasing resorption of sodium from the kidney and increasing excretion of potassium in urine. Associated with increased resorption of sodium is an increased retention of chloride and water.

Glucocorticoids increase glucose levels in blood (hyperglycemia) and deposition of glycogen in liver and muscles. Glucocorticoids also increase rate of conversion of tissue protein to glucose, and prolonged use of therapeutic doses leads to extensive depletion of tissue protein. Furthermore, glucocorticoids stimulate breakdown of depot fat to fatty acids which then are converted to glycogen or used directly as an energy source. Both actions spare the use of glucose. The marked hyperglycemic action of glucocorticoids is the basis for their therapeutic use in alleviating hypoglycemia in ketotic cows.[10]

Glucocorticoids also suppress the immune response, delay wound healing, and depress circulating levels of 2 types of white blood cells, the eosinophiles and lymphocytes. Glucocorticoids are anti-inflammatory agents.

Adrenal cortical function is associated with stress. For example, adrenalectomized animals die quickly when subjected to stress, but if given glucocorticoids these animals survive stressful conditions. In intact animals, acute stress causes a marked outpouring of adrenal glucocorticoids, but prolonged stress eventually reduces adrenal secretion to the point where the animal approaches the adrenalectomized state. Such animals are more subject to infections. Depletion of adrenal hormones may play a role in the "shipping fever" syndrome in cattle, but direct evidence is lacking at present.

Progesterone, a hormone normally released into blood in significant quantities from the corpus luteum of the ovary, may be secreted from the adrenal cortex under conditions of high temperature. This condition has been observed during hot summer months in the South of the United States. Because of this syndrome, estrus in cattle is often absent, shortened, or delayed in hot weather. (See Section 14.6 for further discussion of progesterone inhibition of the estrous cycle.)

Adrenal Medulla. Since this part of the adrenal gland is really an extension of the nervous system, the two hormones secreted by the adrenal medulla, epinephrine and norepinephrine, are excellent examples of neurohormones in cows. Norepinephrine is a neurotransmitter released at nerve endings of the sympathetic nervous system. Both hormones are released in response to many types of stress and emergency situations. Epinephrine increases blood glucose, liberates fatty acids from fat depots, and stimulates ACTH release which in turn activates the adrenal cortex to discharge glucocorticoids. Both hormones increase heart rate and blood pressure. Norepinephrine further elevates blood pressure by constricting arterioles of certain tissues whereas epinephrine causes vasodilation of other tissues. Thus, blood can be shifted away from some tissues and toward others in an effort to adjust to emergency situations. Restriction of blood flow may account in part for inhibition of the milk-ejection reflex by epinephrine.

Ovary

Ovaries are paired organs situated at the end of the oviduct of the reproductive tract (Figure 14.4). They are usually found just off the midline below the hips. Ovaries produce hormones and contain the female sex cells (ova). Gross anatomy of the ovary, shown in detail in Figure 15.4, varies according to stage of the estrous cycle. In the mature cow, ovaries contain numerous follicles and a few corpora lutea in various states of development depending upon stage of the estrous cycle or of pregnancy. The ovary produces 4 hormones: estrogen, progesterone, inhibin and relaxin.

Estrogen (female sex hormone) causes growth and maturation of the female reproductive tract, development of secondary sex characteristics and, in concert with other hormones, estrogen regulates the estrous cycle and ovulation, and induces female sexual behavior at heat (estrus). Simultaneous with its effects on the reproductive tract, estrogen causes mammary ductular development.

The theca interna produces androgens which are converted by granulosa cells of follicles to estradiol-17β, the major estrogenic steroid hormone in the cow.

The second major steroid hormone produced from the ovary is progesterone, the hormone of pregnancy. This hormone is secreted primarily by the corpus luteum and usually interacts with estrogen to regulate cyclic activity of the estrous cycle and to stimulate lobule-alveolar growth in the udder. When progesterone dominates, the cow does not undergo estrous cycles. Another major activity of progesterone is to cause thickening of the lining of the uterus in preparation for attachment (implantation) of the fertilized egg. This hormone is essential for maintenance of pregnancy in cattle.[11]

In addition to the two steroid hormones, the ovary also synthesizes 2 other hormones, inhibin and relaxin. Inhibin is a putative protein which is secreted from ovarian follicles and reduces secretion of FSH. Thus, inhibin may be indirectly involved in control of ovarian follicular growth during the estrous cycle. With impending parturition relaxin, a peptide, relaxes the pubic symphysis and cervix and thereby facilitates birth of the calf.

Testis

The 2 testes of the bull, analogous to ovaries in the female, produce male sex cells (spermatozoa) and several steroid hormones called androgens. Testes are found outside the ventral body wall in the scrotum (Figure 15.1). The interstitial, or Leydig, cells of testes which secrete these androgens are located between the seminiferous tubules. Seminiferous tubules produce spermatozoa (Figure 15.2).

Testosterone (male sex hormone) is the major androgen produced. It is responsible for maturation and functional maintenance of the male reproductive tract. Testosterone also stimulates spermatogenic function of the testes. In addition, secondary sex characteristics of the bull are developed under influence of testosterone. Sex drive, or libido, in the bull is partially dependent upon testosterone secretion. Similar to granulosa cells in the cow, Sertoli cells in the testis secrete inhibin which regulates FSH in bulls.

Placenta

The fetal placenta (Figure 14.4) arises from the fetus and becomes attached to the internal surface of the uterus of the cow at structures called caruncles. Although primary function of the placenta is to serve as a point of exchange of nutrients for and waste products from the developing fetus, the placenta also secretes estrogen, plus several other minor steroids and their metabolites.

Placental lactogen, a peptide hormone which chemically and biologically resembles prolactin, has been isolated from pregnant cows.[12] This hormone is probably associated with growth and function of the udder.

14.4 Regulation of Hormone Secretion

Hormones affect many vital functions, and amount of each hormone secreted varies with physiological state and environment of the animal. Since physiological and environmental conditions are changing continuously, hormonal secretion patterns must adjust to maintain normal homeostasis of the cow. There may be more than one mechanism to regulate synthesis of a single hormone and more than one hormone often regulates a given activity. Thus, regulation of hormone synthesis is complex. But this is one of the most active areas of current research in endocrinology of cattle because elucidation of these mechanisms may provide new means to regulate growth, reproduction, or lactation.

The hypothalamus is of primary importance in regulation of anterior pituitary hormones[13] which, as previously mentioned, regulate many other endocrine glands. The hypothalamus acts as a discrete amplifier of stimuli it receives from the internal and external environments. In other words, a given stimulus does not cause secretion of all hypothalamic factors. The close anatomical relationship between the hypothalamus and pituitary is shown in Figure 14.5.

Since a delicate balance of circulating hormones must exist, there are negative and positive feedback control systems that regulate secretion of many anterior pituitary hormones (Figure 14.5). For example, in cows during diestrus (Section 15.7) estradiol inhibits secretion (negative feedback) of LH, whereas during estrus increasing secretion of estradiol stimulates secretion (positive feedback) of LH.[14] Sites for feedback control may be at the hypothalamus or anterior pituitary (Figure 14.5). Since prolactin and GH do not stimulate other endocrine glands, the negative feedback system is absent for these 2 hormones.

Another control mechanism is the "short-loop" feedback system. In this system increasing concentrations of anterior pituitary hormones in capillaries of the anterior pituitary are delivered directly to the hypothalamus to reduce secretion of their respective hypothalamic factor. An exception is prolactin which stimulates secretion of PIF. However, the end result is the same for all anterior pituitary hormones; they are lowered.

In addition, there is evidence that release of hypothalamic neurosecretions negatively modulate their own secretion within the hypothalamus. This is termed the "ultra short-loop" feedback system.

External stimuli influence or regulate production of anterior pituitary hormones through senses such as sight, smell, hearing, and touch. The senses exert their influence through the hypothalamus in a very specific manner (Figure 14.5). For example, the milking stimulus decreases PIF and increases prolactin, but has no known effects on FSH, LH, TSH, or GH in cattle.

To integrate these concepts, an example using adrenal output of cortisol is given. Various external stimuli such as fright, pain, or elevated temperatures stimulate outpouring of corticotropin-releasing factor (CRF) from the hypothalamus which in turn increases anterior pituitary secretion of ACTH. The elevated blood levels of ACTH promote increased production of cortisol from the adrenal cortex. The elevated ACTH and cortisol levels act back on the hypothalamus to depress CRF secretion. Cortisol may also depress pituitary production of ACTH. Eventually as levels of CRF and ACTH decrease, secretion of cortisol decreases. Thus, the brake is released, and once more CRF production is increased; the cycle may be repeated as required. Similar check and balance systems exist for each of the anterior pituitary hormones.

In cattle, injection of GnRH releases FSH and LH, and TRH releases TSH, prolactin, and GH. Additional research is required to explain how these signals are discretely transmitted to the anterior pituitary to permit asynchronous releases of hormones—a process that occurs naturally in cattle. Understanding the physiology of the hypothalamic-pituitary axis and discovery of new hypothalamic hormones may lead to new means of increasing reproductive and lactational efficiency.

Regulation of hormonal secretions from the posterior pituitary or adrenal medulla does not require intervention of releasing or inhibiting factors. In general, hormones of the posterior pituitary and adrenal medulla are released directly via nerve stimulation. The neural pathway from external stimuli usually funnels through the hypothalamus.

Metabolic hormones such as insulin and glucagon have different control mechanisms. They respond to changing concentrations of metabolites. For example, as concentrations of blood glucose rise, insulin is released from the pancreas and lowers glucose levels. When blood glucose falls below a critical level, glucagon is released, and glucose levels are adjusted upward.

Inheritance and secretion of hormones play major roles in regulating reproduction and milk secretion. But the quantitative aspects of genetic regulation of hormone secretion, especially those concerned with reproduction and milk production, have received little attention. This may be a promising area for future research.

How each hormone exerts its action at the target organ is not fully known, although, as described in section 14.2, specific receptors for hormones exist in the various target tissues. Receptor numbers vary from target to target and this determines whether or not a particular tissue is a target for a particular hormone. Hormones regulate receptor numbers and this represents another mechanism to alter biological responses. Thus, studies of receptors may eventually lead to development of additional ways of controlling reproduction as well as increasing milk production.

14.5 Chemical Structure of Hormones

Hormones are often classified according to their chemical structure into one of three groups: proteins (and polypeptides), steroids, and phenols. Major sources of protein hormones in the cow are the pituitary, hypothalamus, parathyroid, pancreas, and placenta. The gonads, adrenal cortex, and placenta secrete steroid hormones, whereas phenolic hormones arise from the adrenal medulla and thyroid gland.

The principal difference between protein and polypeptide hormones is number of amino acids in the molecule. Proteins contain more amino acids than polypeptides. The general formula for an amino acid is

$$H_2N—\overset{\overset{\large R}{|}}{C}—COOH$$

where R represents any series of atoms. To give an idea of complexity of protein hormone molecules, bovine GH and prolactin contain 191 and 198 amino acids, respectively whereas bovine ACTH, oxytocin, and ADH contain only 39, 8, and 8 amino acids, respectively. Several simple peptide hormones (ACTH, oxytocin, ADH, GnRH, TRH, somatostatin, CRF and GRF) have been synthesized in the laboratory and some are available commercially.

Until recently, anterior pituitary protein hormones were available in limited supply only as purified extracts from the pituitary gland. However, by isolating genes for these specific proteins and inserting these genes into host cells, frequently bacteria, large amounts of protein hormones can be synthesized by the host cells. The protein hormone is subsequently isolated and purified from host cells. The protein hormone, though manufactured in bacteria, will be identical to the original mammalian hormone. By growing large volumes of host cells, production of large volumes of protein hormones of interest are theoretically possible. This is generally referred to as recombinant deoxyribonucleic acid (DNA) technology.[15] Injection of recombinantly produced bovine GH into cows increases milk yield.[16] In other pioneering research, rat and human genes for GH have been inserted into fertilized mouse eggs.[17] These genes were incorporated into the genome of some of the mice. These mice subsequently secreted large quantities of rat or mouse GH and grew significantly larger than normal control mice. This represents a potential method whereby secretion of a hormone can be permanently altered. However, insertion of GH genes into the genome of cattle has not been reported.

Steroid hormones are virtually insoluble in water, but they are soluble in organic solvents. Because of their insolubility in water phases, steroid hormones are transported in blood bound to proteins. This protein-bound steroid is physiologically inactive. The steroid becomes functional when liberated from the blood protein at the various target organs.

The general structure for a steroid hormone is:

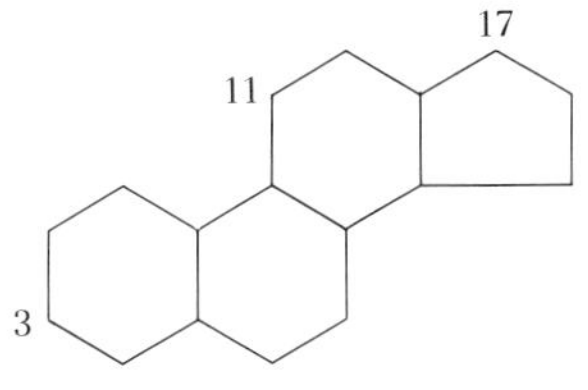

Attachment of O_2, OH, CH_2, or CH_3CH_2 molecules at positions 3, 11, and 17 accounts for differences in biological activity of the various steroid hormones. All major biologically active natural steroids have been synthesized, and some synthetic compounds, not found in nature, possess more biological activity than the native molecule.

The basic structure of the phenolic hormones is:

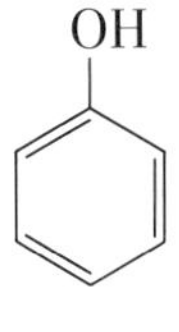

Since these molecules are also simple, they are available in large quantity for therapeutic use in dairy cattle.

14.6 Uses of Hormones in Dairy Cattle Management

Availability of large quantities of low-cost compounds has stimulated use of hormones in various aspects of the cattle industry. Development of biologically active synthetic hormones has been another significant advance in practical application of hormones.

Estrogens

Diethylstilbestrol (DES, stilbestrol) is a synthetic estrogen which increases rate and efficiency of body weight gains in steers by about 10 to 15%.[18] DES is effective when fed or implanted under the skin. It is now illegal to use DES for growth promotion in the United States and many other countries because DES has been implicated in causing cancer in humans.

Stilbestrol may be injected into post-partum cows that have accumulated pus in the uterus (pyometra). It causes cervical dilation and contraction of the uterus which help to expel pus and restore the uterus to a normal condition. Stilbestrol therapy also has been used to expel mummified fetuses.

There are many undesirable side effects in cows from prolonged use of stilbestrol. For example, stilbestrol may induce cystic ovaries, nymphomania, prolapsed vagina, elevated tailheads, and depletion of calcium stores from bone to the point of breakage. Use of this powerful steroid in cows should be done under supervision of a veterinarian.

Progestins

Several orally active progestational compounds have been developed to control ovulation. When these compounds are fed to cows for more than 21 days, final maturation and ovulation of the ovarian follicle are blocked. Upon withdrawal of the drug, maturation of the follicle resumes, and ovulation occurs usually within 3 to 5 days.[19] Thus, the estrous cycle can be regulated (synchronized). The advantage of estrous synchronization is that the breeding period is concentrated into discrete intervals. Thus, labor associated with detection of estrus can be reduced.

Synthetic progestins work only if the animal is healthy, sexually mature, and having normal estrous cycles. However, estrous synchronization with progestins has not been widely accepted because of reduced fertility during the synchronized estrus. A modification of previous methods with progestins offers promise of unimpaired fertility.[20] Animals are injected with 5 mg estradiol benzoate plus 50 mg of progesterone and then implanted intravaginally with silicone rubber containing progesterone. The progesterone implant is removed after 12 days, and most animals come into estrus 2 to 6 days later. A normal calving rate was achieved when the animals were inseminated at the controlled estrus. Recently, synchronization of estrus has been achieved by injecting a synthetic progestin and estrogen and implanting a synthetic progestin in the ear of dairy heifers. After 9 days the implant is removed. Cows show estrus 36 to 60 hours after removal of the implant. Normal fertility is claimed.

An additional effect of the synthetic progestational compounds is their ability to increase rate and efficiency (approximately 12%) of body weight gains in heifers, but not steers. However, exogenous progestins do not affect milk production.

Corticosteroids

The corticosteriods, especially synthetic ones such as dexamethasone, are frequently used by veterinarians as anti-inflammatory agents. They are also prescribed in combination with a diuretic agent to reduce periparturient udder edema.

Prostaglandins

The prostaglandins (PG) are not hormones; rather they are fatty acids synthesized in many tissues. They are mentioned here because of their marked potential to synchronize ovulation and permit insemination without the necessity of detecting behavioral signs of estrus in cattle. Injection of one of the prostaglandins ($PGF_2\alpha$) inhibits progesterone synthesis from the corpus luteum and induces estrus at about 72 hours; ovulation occurs at approximately 95 hours.[21] Since $PGF_2\alpha$ is effective only in cows with functional corpora lutea, cattle exhibiting normal estrous cycles are injected twice with $PGF_2\alpha$ 11 days apart. Most cows will have a functional corpus luteum at time of the second injection, and 90% of cows will be in estrus 48 to 80 hours later. $PGF_2\alpha$ does not affect fertility directly but is an aid in detection of estrus and timing of insemination. Thus, more cows will be inseminated. $PGF_2\alpha$-injected cattle may be inseminated 8 to 12 hours after onset of standing estrus. Another plan is to inseminate cows 80 hours after the second $PGF_2\alpha$ injection without regard to behaviorial symptoms of estrus. Fertility after AI at 80 hours is approximately 15% lower than after AI at estrus induced by $PGF_2\alpha$. Use of $PGF_2\alpha$ reduces time needed for detection of estrus and improves breeding management, labor efficiency and profit through use of genetically superior AI sires. $PGF_2\alpha$ should not be used in repeat service inseminations because it may cause abortion.

Hormone Assays

Techniques are now available which permit detection of many hormones in blood of cows. Routine hormone assays of blood bring about the possibility of relating hormone level with reproductive and lactational performance.

Pregnancy may be diagnosed at a very early stage from assay of progesterone concentration of milk.[22] A milk sample is collected 20 to 22 days after insemination. If progesterone concentration in milk is elevated, the cow may be pregnant or in the luteal phase of an estrous cycle. However, if progesterone concentration is reduced, the cow is not pregnant and the dairyman should watch closely for impending signs of estrus.

14.7 Summary

Endocrine glands play a major role in maintaining normal homeostasis in dairy animals. In addition, they provide the stimulus to regulate the estrous cycle, maintain pregnancy, initiate parturition, promote mammary development, and initiate and maintain lactation. To accomplish all of these events several hormones acting in concert are required. Thus, integration of function is a fundamental characteristic of the endocrine system. Hormones do not operate on only one tissue or organ; rather, several hormones usually interact to affect any single event.

Hormones are used commercially to stimulate body weight gains, to synchronize the estrous cycle, and to stimulate milk production. In addition, they are used extensively by veterinarians to treat many disorders including, for example, cystic ovaries, ketosis, and expulsion of foreign material in the uterus. Hormones are used to produce multiple ovulations from superior cows, to synchronize ovulations within a time-span of a few hours, and to confirm pregnancy. Future uses may be to provide an index of reproductive and lactational performances and to stimulate milk synthesis.

Review Questions

1. What does the endocrine system have in common with the nervous system? What are the differences between the two systems?
2. Which hormones are most intimately associated with reproduction and lactation? Which endocrine glands secrete these hormones?
3. Describe mechanisms that control secretion of an anterior pituitary hormone. How does this mechanism differ from the control mechanism of the posterior pituitary?
4. What functional changes can be expected in the nervous and endocrine systems when a dairy cow receives a painful stimulus?
5. What do the following have in common?
   a. Adrenal medulla and posterior pituitary
   b. Ovary and testis
   c. Growth hormone, epinephrine, cortisol, insulin, and glucagon
6. Why has commercial use of steroid hormones preceded that of protein hormones?

References

1. Greep, R. O., and Astwood, E. B. (Eds.): The pituitary gland and its neuroendocrine control, Part 2. In *Handbook of Physiology, Section 7, Endocrinology,* Vol. IV. Washington, D.C.: American Physiological Society, 1974.
2. Hansel, W.: Pituitary ovarian relationships in the cow. J. Dairy Sci. *53:* 945, 1970.
3. Seidel, G. E.: Superovulation and embryo transfer in cattle. Science *211:* 351, 1981.
4. Sreenan, J. M., Beehan, D., and Mulvehill, P.: Egg transfer in the cow: factors affecting pregnancy and twinning rates following bilateral transfer. J. Reprod. Fert. *44:* 77, 1975.
5. Comline, R. S., and Edwards, A. V.: The growth and development of the calf after hypophysectomy. J. Physiol. *180:* 5 P, 1965.
6. Macklin, L. J.: Effect of growth hormone on milk production and feed utilization in dairy cows. J. Dairy Sci. *56:* 575, 1973.
7. Britt, J. H.: Ovulation and endocrine response after LH-RH in domestic animals. Ann. Biol. Anim. Biochem. Biophys. *15:* 221, 1975.
8. Convey, E. M., Thomas, J. W., Tucker, H. A., and Gill, J. L.: Effect of thyrotropin releasing hormone on yield and composition of bovine milk. J. Dairy Sci. *56:* 484, 1973.
9. Capen, C. C., and Young, D. M.: Thyrocalcitonin: evidence for release in a spontaneous hypocalcemic disorder. Science *157:* 205, 1967.
10. McDonald, L. E.: *Veterinary Endocrinology and Reproduction,* 3rd ed. Philadelphia: Lea & Febiger, 1980.
11. McDonald, L. E., McNutt, S. H., and Nichols, R. E.: On the essentiality of the bovine corpus luteum of pregnancy. Amer. J. Vet. Res. *14:* 539, 1953.
12. Eakle, K. A., Arima, Y., Swanson, P., Grimek, H., and Bremel, R. D.: A 32,000 molecular weight protein from bovine placenta with placental-like activity in radioreceptor assays. Endocrinology *110:* 1758, 1982.
13. Convey, E. M.: Neuroendocrine relationships in farm animals: a review. J. Anim. Sci. *37:* 745, 1973.
14. Hansel, W., and Convey, E. M.: Physiology of the estrous cycle. J. Anim. Sci. *57:* Suppl. 2, 404, 1983.
15. Blattner, F. R.: Biological frontiers. Science, *222:* 719, 1983.
16. Bauman, D. E., de Geeter, M. J., Peel, C. J., Lanza, G. M., Gorewit, R. C., and Hammond, R. W.: Effect of recombinantly derived growth hormone (bGH) on lactational performance of high yielding dairy cows. J. Dairy Sci. *65:* Suppl. 1, 121, 1982.
17. Palmiter, R. D., Norstedt, G., Gelinas, R. E., Hammer, R. E., and Brinster, R. L.: Metallothionein-human GH fusion genes stimulate growth of mice. Science, *222:* 809, 1983.
18. Andrews, F. N., Beeson, W. M., and Johnson, F. D.: The effect of hormones on the growth and fattening of yearling steers. J. Anim. Sci. *9:* 677, 1950.
19. Zimbelman, R. G.: Effects of progestogens on ovarian and pituitary activities in the bovine. J. Reprod. Fert. Suppl. *1:* 9, 1966.
20. Roche, J. F.: Calving rate of cows following insemination after a 12-day treatment with silastic coils impregnated with progesterone. J. Anim. Sci. *43:* 164, 1976.
21. Hafs, H. D., Manns, J. G., and Lamming,

G. E.: Synchronization of oestrus and ovulation in cattle. In *Principles of Cattle Production.* H. Swan and W. H. Broster, Eds. London: Butterworths, 1976.
22. Schiavo. J. J., Matuszczak, R. L., Oltenacu, E. B., and Foote, R. H.: Milk progesterone in postpartum and pregnant cows as a monitor of reproductive status. J. Dairy Sci. *58:* 1713, 1975.

Suggested Additional Reading

McDonald, L. E.: *Veterinary Endocrinology and Reproduction,* 3rd ed. Philadephia: Lea & Febiger, 1980.

Turner, C. D., and Bagnara, J. T.: *General Endocrinology.* 6th ed. Philadelphia: W. B. Saunders Co., 1976.

Goldsworthy, G. J., Robinson, J., and Mordue, W.: *Endocrinology.* New York: John Wiley and Sons, 1981.

CHAPTER 15

Anatomy and Physiology of Reproduction

15.1 Introduction

Production of offspring requires a high degree of synchronization of very complicated processes in both the male and female. Thus, it is not surprising that only about 50% of cows deliver a normal calf from a first-breeding insemination. In fact, sterility and infertility cause economic losses to dairymen which are estimated to cost an average of approximately $140 per cow per year. Sterile animals cannot reproduce, whereas infertile animals reproduce at a rate below normal for the species. Although infertility is a major problem in many herds, enough basic information on reproductive physiology has accrued so that some of these losses can be reduced. Knowledge of normal anatomy and function of reproduction is essential to maximize fertility and to understand and to minimize abnormal reproduction. Thus, more basic concepts of reproduction will be covered in this chapter, whereas more practical aspects will be covered in Chapter 16.

15.2 Male Anatomy

Function of the bull is to produce viable sex cells (spermatozoa) to fertilize the ovum released from the ovary of the cow.[1] The male organs of reproduction are shown in Figure 15.1.

Testis

As described in Chapter 14, the testes produce spermatozoa and hormones, principally testosterone. Each testis is independent of the other, and shortly before birth they migrate from inside the body to the scrotum. The scrotum maintains the temperature of the testes about 4 to 5°C below body temperature. This lower temperature is essential for spermatogenesis and is maintained by contraction or relaxation of muscles in the scrotal wall. Thus, in a cold environment the testes retract in close proximity to the body, whereas in a warm environment the muscles relax.

Failure of the testes to descend from the body into the scrotum, i.e, cryptorchidism, prevents sperm production although testosterone secretion is unaffected. The higher temperature within the body causes this failure in sperm production, and these bulls are sterile. If only one testis descends, sperm production is halved, but fertility after mating is nearly normal. However, the unilateral cryptorchid should be culled because the condition can be inherited.

Sperm are produced by two meiotic di-

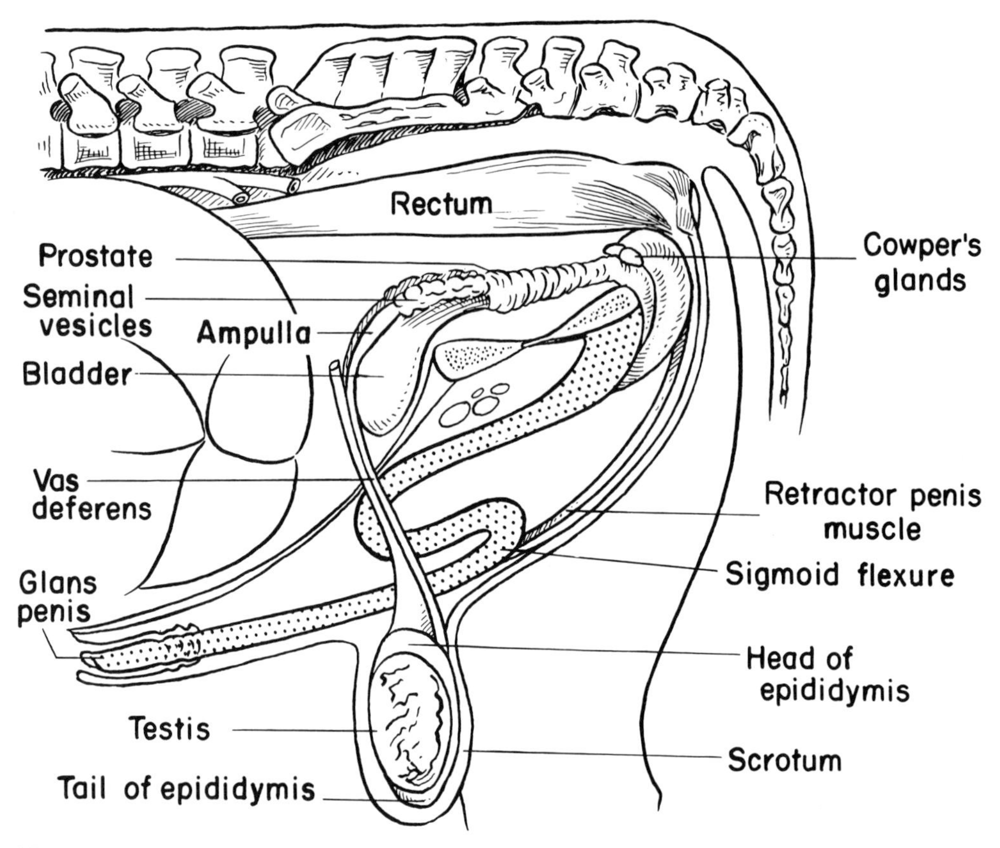

Fig. 15.1 Reproductive organs in the bull. (Courtesy of R. H. Foote; reprinted with permission from *Reproduction in Domestic Animals,* 1969)

visions (Chapter 4) of germ cells which line the coiled seminiferous tubules of the testis (Figure 15.2). Seminiferous tubules are about 3 miles long in the bull and constitute the major portion of the testis. Testes of the mature bull produce on average almost 70×10^9 spermatozoa per week. Total sperm production is correlated with weight of the testes and weight of the bull, thus, larger or older bulls usually produce more sperm than smaller or younger bulls.

After their formation, sperm are transported through the seminiferous tubule to larger tubules, the rete testis. From here the sperm leave at the top of the testis and enter the head of the epididymis.

Other anatomical features of the testes are the interstitial, or Leydig, cells which are located between the seminiferous tubules (Figure 15.2). These cells secrete testosterone which maintains function of the male reproductive tract and produces secondary sex characteristics and influences sex drive. Testosterone secretion, in turn, is regulated by LH from the anterior pituitary, whereas the other anterior pituitary gonadotropin, FSH, stimulates spermatogenic function of the testes.

Epididymis

The epididymis is a single, highly convoluted tube which is more than 100 feet long in the bull. It is composed of head, body, and tail segments which lie close to the testis (Figure 15.1). During passage through the epididymis, sperm mature so that when they reach the tail of the epididymis, they are fertile and ready for ejaculation. The interval from formation of sperm until they arrive in the tail of the epididymis is about 8 weeks. Thus, factors that adversely affect spermatogenesis usually show up 2 months later in the performance of the bull. If a bull suddenly becomes sterile one should

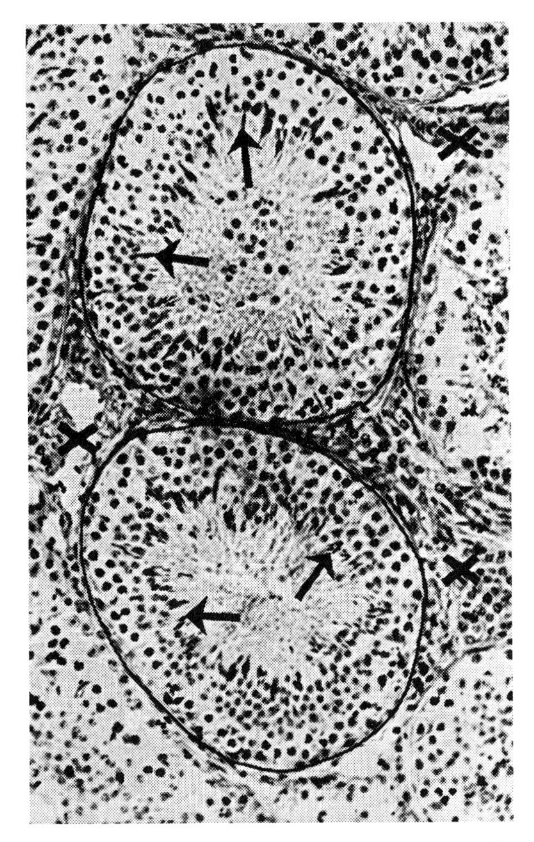

Fig. 15.2 Microscopic cross section of a bull testis. Two seminiferous tubules which contain dividing spermatogenic tissue and sperm heads (*arrows*) are shown. Interstitial cells (X) between tubules secrete testosterone. (Courtesy of H. D. Hafs; reprinted with permission from Hoard's Dairyman, 1964)

attempt to determine what factors changed 2 months earlier.

Vas Deferens (Deferent Duct)

The tail of the epididymis connects to the vas deferens, a tubular duct which enters the pelvic urethra near the bladder (Figure 15.1). Walls of the vas deferens contain longitudinal and circular muscles which contract involuntarily during ejaculation and serve to expel sperm. Each vas deferens enlarges in the pelvic region into a structure called the ampulla. Ampullae contain many glands, and spermatozoa often collect there before ejaculation.

Sperm are not motile in the epididymis or vas deferens, but they become motile immediately after ejaculation. Immobility within the epididymis and vas deferens conserves energy in the sperm cell. Motility of spermatozoa is activated when they are ejaculated into the aerobic environment of the seminal plasma because this medium is rich in fructose, the chief carbohydrate used by sperm for energy.

Urethra and Penis

The two ampullae enter the urethra in the pelvic region of the bull. The urethra receives secretions of the accessory sex glands and serves as the excretory duct for urine. Spermatozoa mix with the seminal plasma of accessory glands in the urethra at ejaculation.

Before the bull can deposit semen in the vagina of the female, the penis must become erect. The penis of the bull, in contrast to most species, contains little erectile tissue, but it is constantly rigid. In the nonerect state the penis of the bull forms an S-shaped curve (Figure 15.1). During erection this curve straightens out and the penis is extended about 12 inches beyond the prepuce, and at the moment of ejaculation it is extended even further.

The prepuce of the bull is located just behind the navel and many long hairs are present in this region. The prepuce is contaminated with many organisms, some of which are important in transmission of certain venereal diseases. This topic is discussed in Chapter 21.

Accessory Sex Glands

Accessory sex glands (Figure 15.1) include the ampullae of the vas deferens, seminal vesicles, prostate gland, and Cowper's glands. Secretions of the accessory sex glands, seminal plasma, of the bull provide an essential medium for transporting sperm from testes to vagina of the cow. This medium is especially rich in those nutrients necessary for life of sperm.

15.3 Ejaculation of Semen

Ejaculation results from muscular contractions of the epididymis and vas deferens which eject the sperm through the pelvic urethra and into the penis. Simultaneously seminal fluids from the accessory glands are expelled by the same route. This mixture of sperm (20%) and seminal plasma (80%) is called semen.

Sperm numbers vary from 5 to 20 billion per ejaculate depending upon the bull and the amount of sexual preparation before ejaculation. But at best only 50% of sperm produced are ejaculated. The unharvested sperm cells may be lost in urine or resorbed in the epididymis. A reduction of these losses would extend use of popular AI bulls whose semen is in short supply.

Frequent use of a bull does not usually impair fertility. For example, mature bulls have been ejaculated daily for as long as 8 months with no decrease in fertility.[2] Moreover, sperm are often abnormal in bulls used infrequently. Many times these abnormal sperm disappear with more frequent ejaculations.

Young bulls begin to produce sperm capable of fertilizing ova at about 9 months of age. Before a bull is purchased he should be checked for cryptorchidism, and his semen should be checked for sperm numbers, abnormal sperm, and for evidence of venereal diseases.

15.4 Female Anatomy

The reproductive system of the cow provides half of the genes for each offspring in the form of an egg cell (ovum). The cow's reproductive system also is a protective environment for nourishment of the developing embryo (Figure 15.3).

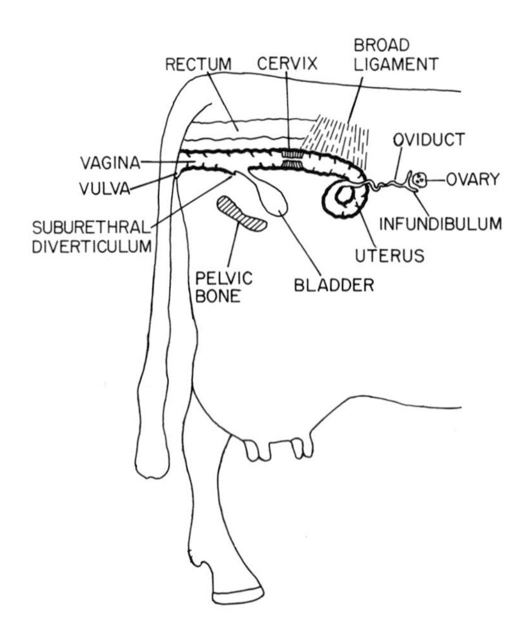

FIG. 15.3 Reproductive organs in the cow.

Ovary

Ovaries of the cow secrete hormones and produce ova. The ovarian structure containing an ovum and surrounding cells is termed a follicle. During the estrous cycle, especially during the last few days before estrus, the follicle enlarges markedly. Usually, only one follicle ruptures and releases its ovum. Cells remaining at the site of ovulation multiply rapidly and form another major structure on the ovary called the corpus luteum. Details of changes in these structures are shown in Figure 15.4 and discussed in Section 15.7.

Oviduct

The ovarian end of the 2 oviducts is enlarged into a funnel-shaped structure, the infundibulum (Figure 15.3) which partially surrounds the ovary especially at ovulation. The released ovum enters the infundibulum and is transported down the oviduct, principally by movement of cilia on cells lining the oviduct. Fertilization of the ovum occurs in the upper half of the oviduct, and the fertilized ovum, or zygote, remains in the oviduct for 3 to 4 days. The junction between the oviduct and the uterus acts like a valve. The valve normally allows passage of sperm up the oviduct only at estrus, and it permits the fertilized egg to enter the uterus only on the 3rd to 4th day after fertilization. This delay in entrance of eggs into the uterus is essential because the uterine environment is not conducive to survival of the developing embryo until 3 to 4 days after estrus.

Uterus

In the cow, the uterus consists of a body and 2 horns (Figure 15.3). The uterus serves as a channel for sperm passage to the oviducts and is the site where the embryo develops and placental attachment occurs. The uterus is a muscular organ capable of enormous expansion to accommodate a growing fetus. The uterus rapidly involutes to normal size shortly after partu-

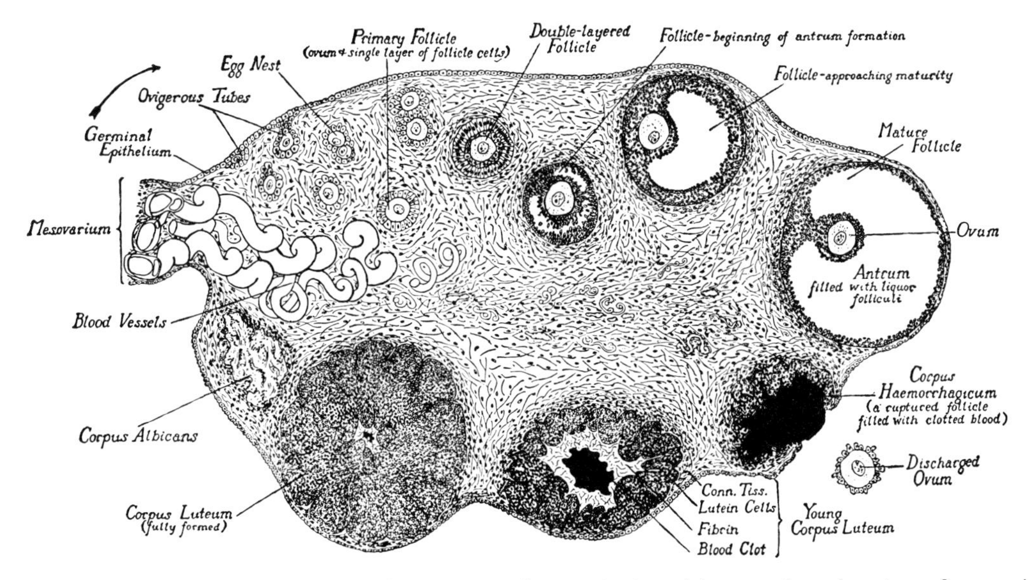

Fig. 15.4 Diagram of an ovary showing the sequence of events in the origin, growth, and rupture of an ovarian (graafian) follicle and the formation and regression of corpora lutea. Follow clockwise around the ovary, starting at the mesovarium. (Reprinted with permission from *Foundations of Embryology,* 1958)

rition. The broad ligament suspends the uterus in the peritoneal cavity (Figure 15.3).

The major portion of the uterine wall, the myometrium, is composed of longitudinal and circular layers of muscle cells. These muscles are responsible for uterine contractions which are essential for birth of the calf. The inside lining of the uterus is called the endometrium. It contains glands that secrete fluids which vary in chemical composition and volume during the estrous cycle. In addition, there are several dozen specialized areas called caruncles, or maternal cotyledons, which are raised slightly above the surrounding surface of the endometrium. During pregnancy the uterine epithelium contacts the fetal membranes at these points to form the placenta.

Cervix

The cervix extends from the uterus to vagina (Figure 15.3). It is composed of thick tight folds of muscle-like tissue which help to protect the uterus from harmful microorganisms which are abundant in the vagina. Muscles around the cervix relax during estrus and at parturition. Cells lining the cervix secrete mucus and are most active during estrus. During pregnancy a plug of cervical mucus seals off the uterus from the vagina.

Vagina

The vagina joins the cervix to the vulva (Figure 15.3) and is the site of semen deposition during natural mating. Although the vagina is lined with cells that secrete a mucous fluid which flushes out bacteria, many low-grade infections persist in the vagina which may result in vaginitis.

Vulva

The vulva is located between the vagina and exterior of the cow (Figure 15.3). It includes the vestibule and suburethral diverticulum, a blind pouch on the floor of the vagina.

15.5 Anatomical Abnormalities

Several surveys have shown that between 8 and 29% of all cows in a general population suffer some type of reproductive anatomical abnormality which could impair fertility. However, one study of clinically normal, but repeat-breeding subfertile heifers revealed that only 10 to 20% of such a selected population of heifers had gross reproductive tract abnormalities which could prevent conception.[3] In other words, reproductive tract abnormalities were no greater in subfertile cows than in the general population. This suggests that anatomical defects are an important, but not a primary nor distinctive, cause of infertility.

Congenital Anomalies

Incomplete development of one or both ovaries, ovarian hypoplasia, occurs in about 13% of all Swedish Highland cattle, but in the United States this condition exists in only 1.9% of the cattle population. Ovarian hypoplasia is inherited, and if it affects both ovaries the animals never show estrus.

During embryonic development there may be arrested development of the oviducts, uterus, or cervix (Figure 15.5). The degree of severity and type of this anomaly vary, but incidence in the general population of cattle is low (1.1 to 3.7%).

Salpingitis and Acquired Anomalies

Inflammation of the oviducts, salpingitis, occurs in about 1.3% of cattle populations. Accumulation of fluid in the oviduct is shown in Figure 15.5. Injury can cause salpingitis as well as adhesions of the ovary, oviducts, and uterus. Such injuries may occur at parturition, when corpora lutea or retained placentas are manually removed or when the reproductive tract is palpated in a rough manner.

Freemartins

About 90% of heifers born twin to a bull, freemartin heifers, are sterile because of underdevelopment or absence of various parts of the reproductive tract; the bull is unaffected. Since the vagina is often closed in the freemartin, a simple test to determine if the condition exists involves insertion of a test tube into the vagina. In the freemartin the tube can be inserted only a short distance, whereas in the normal heifer the tube can be inserted at least 6 inches.

15.6 Puberty

Puberty is the period when the reproductive tract and secondary sex characteristics start to acquire their mature form. Before onset of puberty, the reproductive tract of the heifer grows proportionately to body growth, but beginning at about 6 months of age growth rate of these organs is much greater than body growth.[4] At about 10 months of age the rapid growth phase of

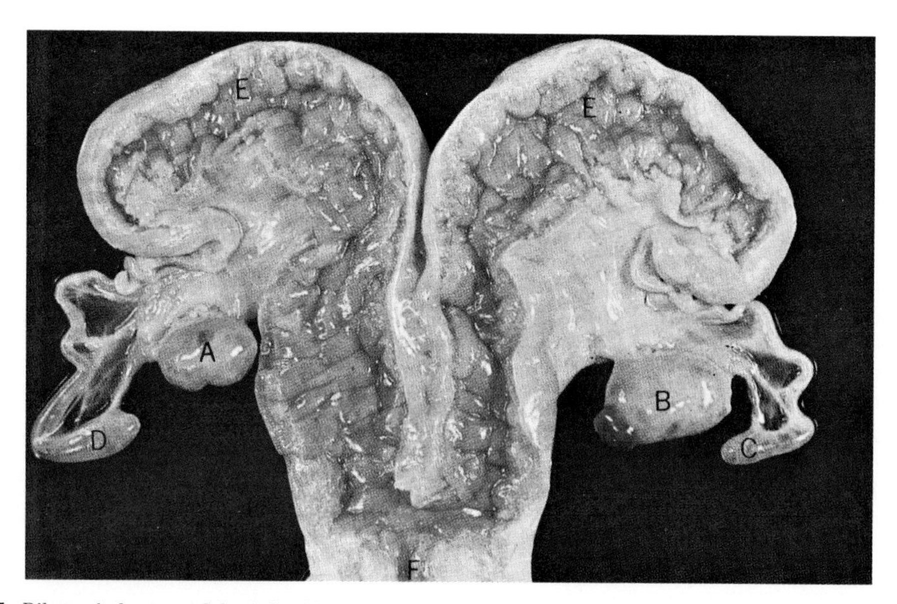

FIG. 15.5 Bilateral absence of the infundibula, which normally surround the ovaries (A, B) of cows. In addition secondary hydrosalpinx (C, D) of right (22 mm) and left (45 mm) terminal segments of the oviducts. The uterine horns (E) and cervix (F) are dissected to show internal surfaces. (Modified and reprinted with permission from Penn. State Univ. Agr. Exp. Sta. Bull. 736)

the reproductive tract ceases, and this signifies the end of puberty.

In addition to rapid growth of the reproductive tract, there are marked changes in the ovary. At birth, ovaries contain all the ova for the animal's lifetime. Vesicular (fluid-filled) follicles appear at about 1 month of age, but estrous cycles, ovulation, and corpus luteum formation do not commence until the heifer is between 5 and 11 months of age. Cattle of larger-size dairy breeds come into puberty at a later age than smaller-size breeds; heifers reach puberty before bulls. As level of nutrition increases, first estrus occurs at earlier ages.[5] Within reasonable limits, average body weight or height at withers at the first estrus is not affected by level of nutrition,[6] which suggests that heifers generally come into estrus at a given skeletal size rather than at a given calendar age.

Onset of puberty in dairy heifers results from interactions among hormones of the hypothalamus, anterior pituitary, and ovaries. The key hormone appears to be LH because secretion of this hormone increases markedly about the time of first estrus, and injections of LH into immature rats will induce puberty at earlier ages. Methods to hasten puberty and allow successful conception and pregnancy with subsequent high milk production are needed to shorten this nonproductive period of approximately 2 years in a cow's lifetime.

First ovulation and formation of the corpus luteum in heifers occur without behavioral signs of estrus more than 70% of the time.[7] Both progesterone and estrogen are required to induce overt signs of estrus. The high incidence of silent heats at first ovulation is probably caused by absence of significant quantities of progesterone since there is no preexisting corpus luteum. The second ovulation in a heifer's life is usually preceded by outward signs of estrus.

15.7 Estrous Cycles and Ovulation

Rhythmic sexual behavior in cows is manifested first during puberty and normally continues at regular intervals throughout the year until conception. Interval from first signs of sexual receptivity at estrus (heat) to start of the next estrus is called the estrous cycle. In dairy heifers average length of the estrous cycle is 20±2 days, whereas in multiparous cows it averages 21±4 days.

Phases of the Estrous Cycle

The cycle consists of four phases: estrus, metestrus, diestrus, and proestrus. The proestrous-estrous period of the cycle is usually referred to as the follicular, or estrogenic, phase, whereas the metestrous-diestrous portion is the luteal, or progestational, phase.

Estrus (heat) is that period when the female accepts mating. Signs of estrus include standing when mounted by other cows, mounting of other cows, clear flow of mucus from the vulva, swollen vulva, restlessness, and sometimes bellowing. Estrus in cattle lasts approximately 18 hours, and ovulation occurs 10 to 14 hours after the end of estrus. Estrogen (estradiol-17β) is the predominate ovarian hormone during this phase of the cycle.

Metestrus is the post-estrus period of 3 to 4 days when estrogen levels decline, and the corpus luteum starts to develop. Progesterone secretion increases gradually during this time. During this phase, about 90% of heifers and 50% of cows show a small bloody discharge from the vagina. This metestrous bleeding is not related to fertility. It means only that the cow was in estrus 2 to 4 days earlier, and if heat was not noted a subsequent heat period should appear within 16 to 19 days.

Diestrus lasts for about 12 to 15 days in the cow and is characterized by predominance of the corpus luteum and secretion of progesterone. Progesterone secretion begins to decline at the end of diestrus, although microscopic appearance of the corpus luteum remains relatively unchanged. Length of diestrus usually determines length of the estrous cycle.

During proestrus, progesterone secretion decreases precipitously as the corpus luteum regresses. The ovarian follicle destined to ovulate commences to grow faster than other follicles and estrogen titers in the blood begin to rise. Proestrus continues for 1 to 3 days until estrus when the cycle is repeated.

Ovarian Changes

Cells lining the follicle multiply during the estrous cycle so that eventually a fluid-filled cavity, or antrum, forms around the ovum (Figure 15.4). The ovum and a layer of surrounding cells, the corona radiata are located on a hillock of cells, the cumulus oophorus, which extend toward the center of the follicle. At this stage the follicle is called a Graafian follicle, and cells lining the follicle are termed granulosa cells. Adjacent to the granulosa cells is the theca interna which is enveloped by another group of cells, the theca externa, which blend into the ovarian stroma.

In the postpuberal heifer, there appear to be at least 2 waves of follicular development during the estrous cycle.[8] One or 2 follicles enlarge to a diameter of about 16 mm by day 6 or 7 of the cycle, but then undergo degeneration (atresia). At this time another group of follicles commences to grow, but usually only one is destined to ovulate.

The follicle which is about to ovulate grows rapidly during estrus and reaches maximal turgidity and a diameter of about 15 to 20 mm a few hours before ovulation. As ovulation approaches, an area of erosion gradually develops in the follicular membrane, intrafollicular pressure decreases, the membrane ruptures, and the ovum is released and enters the infundibulum of the oviduct.

Some hemorrhaging occurs at the ovulation site. The proliferating granulosa cells now form the corpus luteum, the principal source of progesterone. In the cow, growth of the corpus luteum is completed by day 7 of the estrous cycle. In the absence of fertilization, the corpus luteum commences to regress about day 18 and a new vesicular follicle begins to grow rapidly. But if conception occurs, the corpus luteum remains active and continues to secrete progesterone.

Uterine Changes

During the follicular phase of the cycle, the endometrium consists of tall columnar cells, but 2 days post estrus these cells become low and cuboidal in shape. After corpus luteum formation, these cells again become taller reaching a maximum about day 12 of the cycle.

Glandular development in the uterus also varies with the estrous cycle. Uterine glands are rather straight at estrus, but within 2 days these glands develop markedly and begin to secrete a thick fluid termed histotroph, or "uterine milk." These secretions are bacteriostatic and nourish the newly fertilized ovum before it attaches to the uterus at the caruncles. Under the influence of progesterone the uterine glands attain their maximal size by day 12 of the cycle which coincides with maximal endometrial cell height. Regression of uterine glands begins by day 15 if conception does not occur.

Immediately before and during estrus, estrogen causes water retention (edema), increased vascularity, and increased infiltration of leukocytes into uterine stroma and lumen. During the early luteal phase the edema subsides, but the vascularity remains until the later phases of diestrus. This increased blood supply is probably needed to nourish uterine cells for secretion of uterine milk.

Hormonal Interrelationships

Physiological changes in the reproductive tract during the estrous cycle are results of a complex interaction involving the hypothalamus, anterior pituitary, and gonads.[9] Direct measurements of circulating levels of many hormones have been made. At this time control mechanisms are not completely understood, but eventually when all factors are known, it should be possible to regulate precisely the reproductive cycle of the cow.

Ovulation is the most important event of the estrous cycle. Serum concentrations of FSH and LH increase coincidently during proestrus and estrus in cattle,[10] and this surge in secretion of FSH and LH causes ovulation.[11] Shortly after estrus there is a second increase in serum concentrations of FSH which may be important in recruitment of follicles.

FSH causes follicle growth, and FSH plus LH enhances estrogen secretion shortly before estrus (Figure 15.6). This high estrogen concentration probably causes the ovulatory surge of LH.[12] GnRH from the hypothalamus likely participates in the surge of LH at ovulation.

In the cow LH is the principal hormone

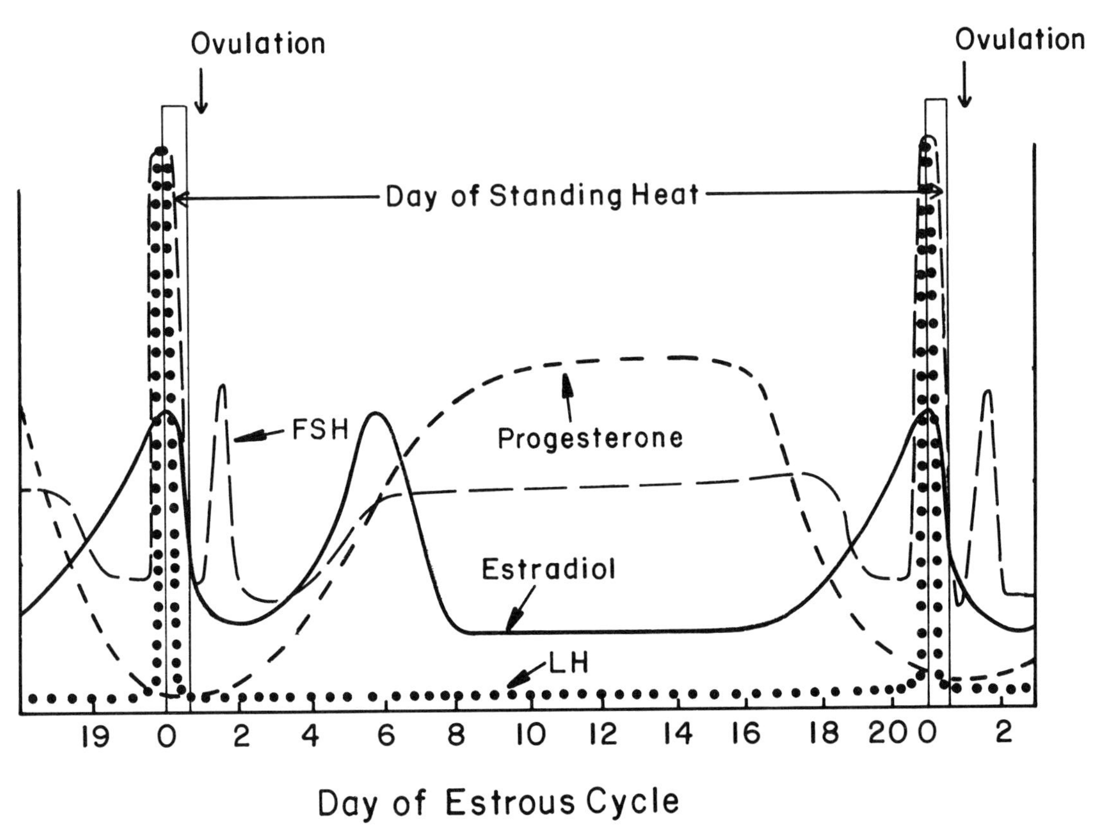

FIG. 15.6 Idealized changes in serum concentrations of LH, FSH, progesterone, and estradiol during the estrous cycle of cattle. Note the synchrony among hormonal changes, behavioral signs of estrus, and ovulation.

that causes corpus luteum growth and progesterone secretion.[13] Plasma levels of progesterone increase to a plateau between days 7 and 16 of the cycle, but between days 16 and 19 levels decline rapidly (Figure 15.6). This decline in progesterone is a most important facet in regulation of the estrous cycle because if progesterone does not decrease, the cow will become acyclic. (See Chapter 16 for a discussion of cystic ovaries.) The reason for regression of the corpus luteum and progesterone secretion at the end of the estrous cycle is not known with certainty. Suggestions range from some luteolytic substance from the uterus, possibly prostaglandin $F_2\alpha$ (see Chapter 14), to elevated secretion of estrogen.

15.8 Insemination—Natural and Artificial

The cow will accept the bull during estrus, but since ovulation does not occur until several hours after the end of estrus the cow should be bred toward the end of the heat period. During natural coitus the bull deposits sperm in the vagina, whereas in artificial insemination (AI) sperm are deposited in the cervix or just inside the body of the uterus. A few cows will show signs of estrus even though they are pregnant. Therefore, cervical deposition of sperm is recommended on repeat services because intrauterine penetration by the AI catheter jeopardizes pregnancy if conception has taken place from a previous service.

Discussion of artificial insemination techniques will be brief because they are often performed by a trained technician who has access to extensive facilities at a bull stud. Several reviews of this subject have been published.[14–16] The major advantage of AI in dairy cattle is that it permits widespread use of superior sires for genetic improvement of the cow population. Other advantages include reduction in risk of venereal diseases and lower annual cost than maintaining a bull of comparable merit.

Many dairymen purchase semen, store it in liquid nitrogen tanks on the farm, and inseminate their own cows. According to one limited study the major reason for shifting to direct service was related to unsatisfactory service of the AI technician.[17] On average, the cost of direct service is about the same as in regular commercial AI, although in herds of less than 80 cows direct service is more expensive. Breeding efficiency is about the same in regular AI as in the self-service program. Dairymen planning to use direct service should complete a course of study on the technique. Also it should be remembered that each insemination takes approximately 20 minutes including preparation, insemination, and record keeping. Direct service is probably of questionable value in herds with fewer than 100 cows because the dairyman will not receive enough practice in the techniques of artificial insemination to assure high conception rates.

Most AI organizations in the United States charge a fee for each dose of semen used, plus a service charge for the AI technician to inseminate the cow. If the cow does not conceive, some organizations do not charge the dairyman with the AI technician service charge at the second or third insemination. Other organizations charge for the technician service at each insemination.

Collection of Semen

Maximal production of sperm of high viability is desirable to achieve optimal use of a given sire. Total sperm output increases as frequency of ejaculation increases, although concentration per ml of semen decreases. Bulls should be stimulated sexually by allowing several false mounts before collection of the ejaculate. This practice increases sperm production. Bulls in routine AI service are usually ejaculated twice daily on 2 days each week.

The most satisfactory method of semen collection involves use of an artificial vagina (Figure 15.7). The rubber liner should be well lubricated, and water temperature inside the device adjusted to about 45°C. The bull is encouraged to mount a dummy or another bull, steer, or cow. The collector directs the penis into the artificial vagina and must coordinate his movements with those of the bull at ejaculation. To prevent thermal shock and maintain high fertility, semen should be maintained at approximately 37°C until processed in the laboratory.

Preservation of Semen

An average ejaculate contains from 5 to 20 billion sperm in 5 to 8 ml of semen. Semen should be processed immediately after ejaculation. Usually gross appearance, volume, sperm concentration, and motility are recorded for each ejaculate. Sperm numbers are measured either by counting under a microscope or by measuring turbidity of the semen in a colorimeter. Degree of motility and morphology of sperm are evaluated under a microscope. Motility should exceed 50%, although it must be recognized that there is a very low correlation between motility and fertility of sperm. The most important criterion for quality of semen is whether or not cows conceive after insemination.

Semen is diluted in an extender such as those given in Appendix Table IV-A or IV-B. The average ejaculate from a normal bull when extended and frozen provides enough sperm to inseminate about 500 to 600 cows. The volume of extended semen deposited in cattle is usually between 0.2 to 1 ml, and this volume should contain at least 5 million motile sperm for liquid semen and 10 to 12 million for frozen semen. Frozen semen from bulls of low fertility should be used at a dilution of 15 million or more motile sperm.

Sperm may be preserved at ambient, at 5°C, at −79°C, or −196°C, temperatures. Lower temperatures preserve fertility of the sperm longer. Use of ambient temperature systems may be of value in places where there is no refrigeration or in cases where there is a very intense, but short breeding season. At temperatures above 40°C, fertility is quickly lost. The 5° storage systems were most prevalent during the early days of artificial insemination, but most methods of semen storage today involve freezing. Specific details for preservation of semen are available in references

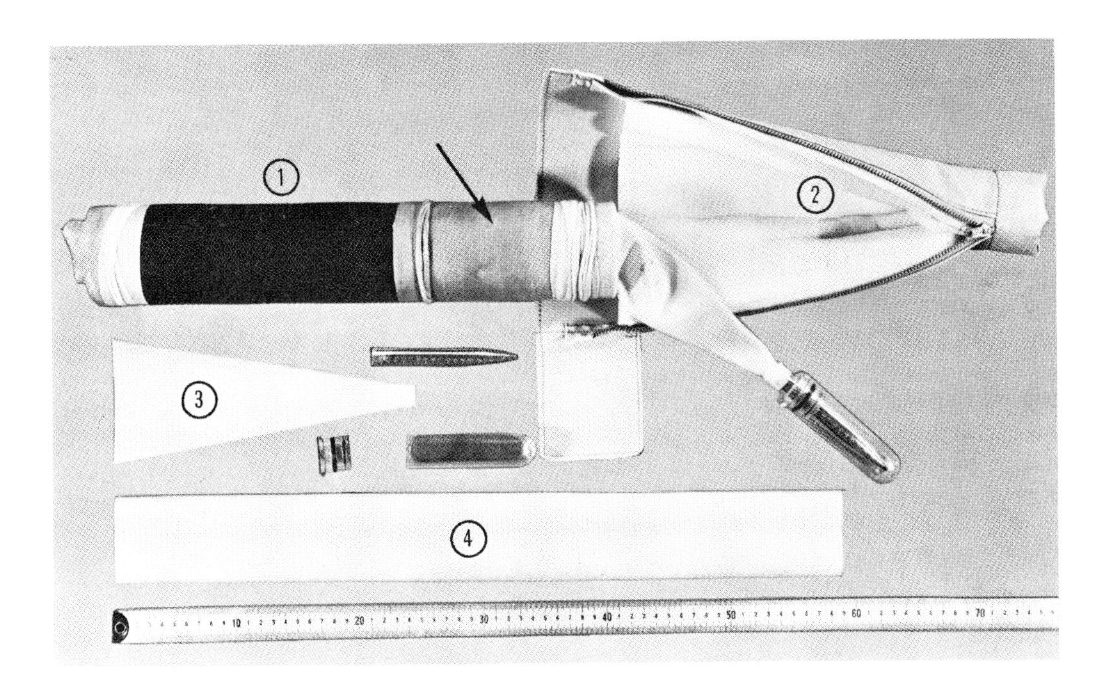

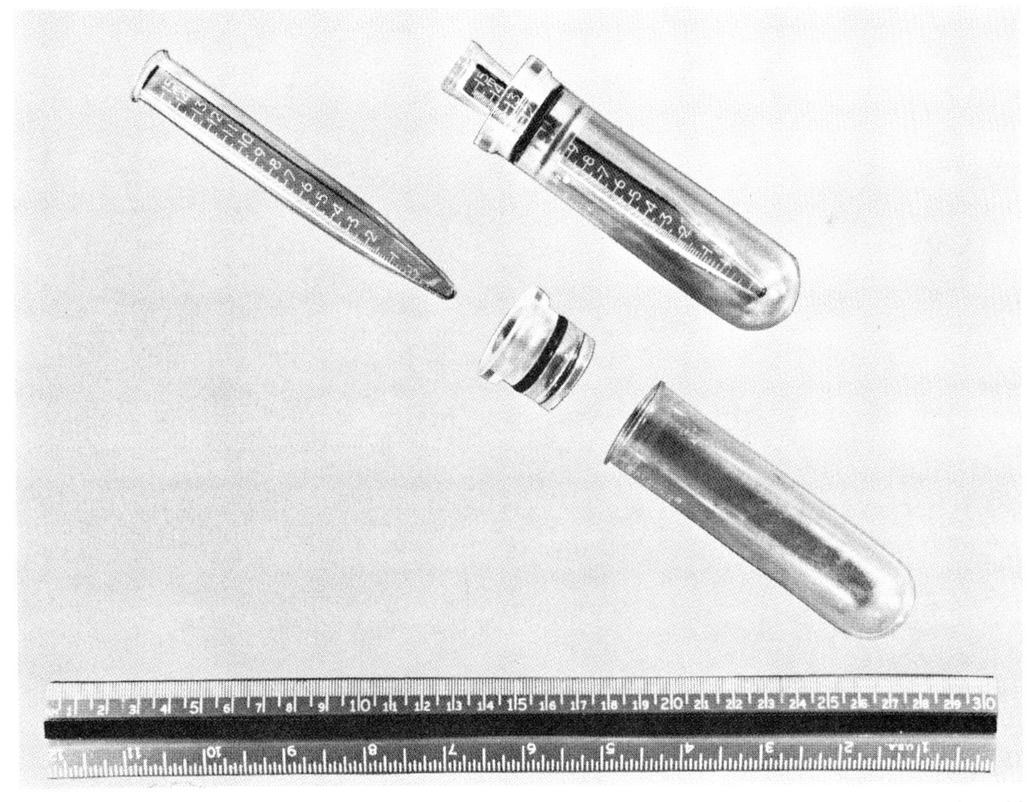

Fig. 15.7 (*Top*) Bull semen collection equipment: 1-assembled artificial vagina; 2-insulated protector for rubber funnel and semen collection tube (made by Hollis Schwartz, Chaska, Minn.); 3-rubber funnel or director cone; 4-rubber inner liner; arrow-latex sleeve, which expands as water is forced through 12 mm holes in outer casing during intromission. (*Below*) Graduated semen collection tube and water bath, consisting of plexiglass stopper and polycarbamate centrifuge tube (stopper developed by American Breeders Service, Inc., DeForest, Wisconsin). (Courtesy of J. O. Almquist; reprinted with permission from *The Artificial Insemination of Farm Animals*, 1968)

14 to 16, and an example method for frozen semen is outlined in Appendix Table IV-C.

Semen Packages

Semen from major AI organizations in the United States usually has high fertility. Use of plastic straws has gained popularity in recent years (Figure 15.8). In comparison with ampules, straws require less storage space and virtually no sperm are lost in the straw whereas 10 to 20% of sperm are lost when frozen in ampules. Furthermore, slightly greater fertility is obtained with semen frozen in straws if: (1) semen is stored in liquid nitrogen at temperatures near −196°C; (2) semen is thawed according to each particular AI organization's recommendations; (3) semen is deposited in the cow within minutes after thawing, and (4) no large temperature variations occur after semen is thawed.

Storage of Semen

Ampules are usually stored on canes whereas straws are stored in goblets or canisters in liquid nitrogen tanks (Figure 15.9). A system for rapid identification of every semen sample in the tank must be devised. An identification number on top of each cane identifies the semen, and a map inside the cover of the tank can locate canes or canisters. To maintain high fertility, do not allow the temperature of frozen semen to rise above −112°C.

Selection of Semen from Liquid Nitrogen Tank

Temperature at the base of the neck of the liquid nitrogen tank is approximately −184°C, in the middle of the neck it is −79°C, whereas at the top of the neck it is near room temperature. Therefore, to select a semen sample and yet maintain temperatures below −112°C:

1. Raise the cane or canister no higher than the frost line (about 1 to 2 inches below the top opening).
2. Only the top ampule should be exposed to the top inch of the liquid nitrogen tank.
3. Pull out the desired straw with long forceps.
4. Return cane or canister into liquid nitrogen as soon as possible.
5. Semen removed from the tank must be either used immediately or discarded.

FIG. 15.8 Plastic straw filled with bull semen. Identification on the straw includes the bull's full name, registration number, code number, breed and a date code. (Courtesy of Fred Dumbroski, Michigan Animal Breeders Cooperative and Select Sires, Inc.)

Thawing of Semen

The best method to thaw bull semen depends upon the package used, the semen extender used, the way in which semen was frozen, and air temperature. Therefore, follow the recommendation of the AI organization that supplied the semen. Some recommend thawing in a water bath ranging from ice water temperature to body temperature. Others recommend thawing in air, in a shirt pocket, or even within the cow. Remove all water from ampules or straws because water kills sperm. Discard all ampules or straws in which leaks are suspected. After thawing, maintain a constant temperature for semen. In cold weather provide a heated area for inseminations. Thaw the semen and place into the breeding gun (straws) or catheter (ampules) near the cow. Insulate the insemination equipment while the thawed semen is carried to the cow. In general, provide conditions so that temperature of semen continues to rise from the time it leaves the liquid nitrogen tank until it is inseminated into the cow. Never chill thawed semen.

Deposition of Semen

To artificially inseminate a cow, one arm covered with a glove is inserted into the rectum. The cervix is grasped carefully through the rectal wall. The breeding catheter (ampules) or gun (straws) is loaded with thawed semen (Figure 15.10) and inserted through the vagina into the cervix. Semen is deposited either in the cervix or

Fig. 15.9 Storage of plastic semen straws in liquid nitrogen tank at −196°C. (Courtesy of R. H. Foote; reprinted with permission from Hafez, E. S. E.: *Reproduction in Farm Animals*, 4th ed. Philadelphia, Lea & Febiger, 1980)

just inside the body of the uterus (Figure 15.11).

15.9 Fertilization and Placentation

Before fertilization sperm must be transported from the site of semen deposition to the ovum in the upper half of the oviduct. Oxytocin is probably released during mating or during artificial insemination to cause contractions of the uterus and oviducts. These contractions propel sperm toward the ovum within a matter of minutes. Sperm motility probably does not contribute significantly to this trip up the reproductive tract until they reach the upper oviduct. About 12 hours after the end of estrus the ovum is released from the ovary and within 6 hours the ovum has traversed the upper half of the oviduct. Fertile life of the ovum after ovulation is no more than 10 hours and sperm in the reproductive tract of the female probably live no more than 24 hours. However, freshly ejaculated sperm will not fertilize the ovum immediately, rather sperm must reside in the uterus and oviducts for about 6 hours before they acquire "capacity" to fertilize ova. Thus, insemination between the middle of estrus to about 6 hours after the end of estrus should provide highest fertility. Sperm deposited before this time die before the ovum is available. Sperm deposited after this time do not spend sufficient time in the reproductive tract to become "capacitated." Furthermore, fertilization of ova older than 10 hours frequently results in abnormal embryos and early embryonic mortality. Fertilization usually occurs within 4 to 6 hours after ovulation or about 16 hours after estrus. These facts emphasize the importance of synchronizing sperm deposition and ovulation. (See Chapter 16 for further discussion.)

Of the millions of sperm deposited in the reproductive tract less than 1000 reach the

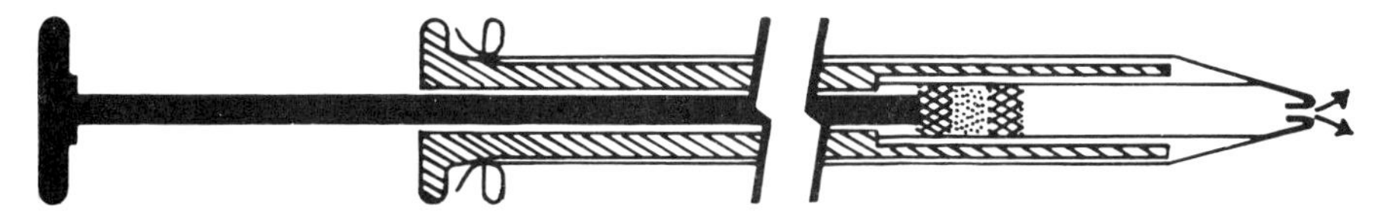

Fig. 15.10 Diagram of breeding gun loaded with plastic straw. (Courtesy of Fred Dumbroski, Michigan Animal Breeders Cooperative and Select Sires, Inc.)

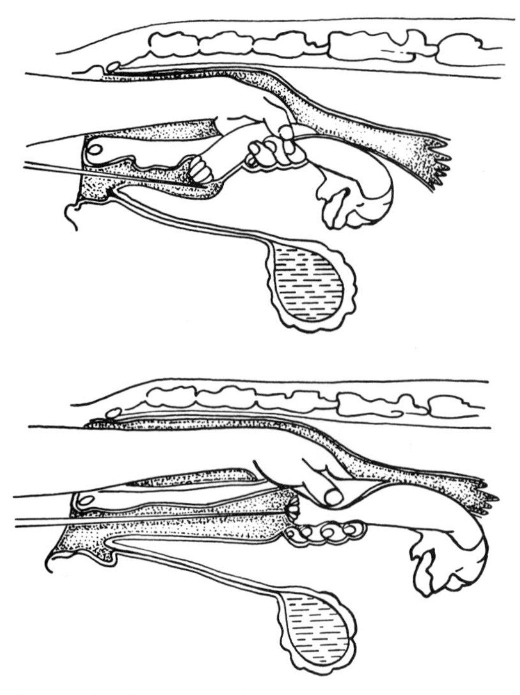

FIG. 15.11 Rectovaginal insemination of cows. (*Top*) Wrong way of holding cervix. If cervix is tipped and/or is not pulled forward it is almost impossible to insert the end of a catheter in the opening of the cervix. (*Bottom*) By using the correct procedure it is relatively easy to deposit semen through the cervix. (Courtesy of R. H. Foote and G. W. Trimberger; reprinted from *Reproduction in Farm Animals,* 1968)

upper oviduct. As shown in Figure 15.12, the sperm cell is much smaller than the ovum. Spermatozoa line up perpendicular to the ovum upon contact. One sperm penetrates the outer membranes of the ovum and at this point all other sperm are excluded from the ovum. Motility of the sperm may assist in penetration of the ovum.

The single sperm cell that penetrates the ovum loses its tail, membranes surrounding the head of the sperm dissolve and nuclei of the 2 sex cells fuse to form a one-celled embryo, or zygote. This is fertilization. It restores number of chromosomes to the diploid state and provides impetus for mitotic cell divisions which successively form the blastula, embryo, fetus, calf, and finally the mature cow or bull.

One-half of sperm carry an X-chromosome and one-half a Y-chromosome. If an X-chromosome-bearing sperm fertilizes the ovum, the resulting calf will be female. Fertilization with a Y-chromosome-bearing sperm produces males. Many attempts to alter the proportion of X- and Y-bearing sperm in semen have been made. To date there is no method of treating sperm which consistently alters the normal sex ratio.[18]

The fertilized egg usually divides once within the first 20 hours, and by the third day these 2 cells divide again to produce a 4-cell structure. By day 5 to 6 the fertilized egg contains 8 to 16 cells (Figure 15.13). These cell divisions occur in the lower oviduct, but the mechanisms which regulate cilia movement and transportation of the fertilized egg down the oviduct are obscure.

Upon entering the uterus the developing ball of embryonic cells gradually forms a fluid-filled cavity and by a series of shifts in cell layers, 3 basic embryonic tissues (ectoderm, mesoderm, and endoderm) are formed. These embryonic tissues differentiate to form specific organs of the developing calf as well as extra-embryonic tissues that form the placenta (Figure 15.14). Within 22 days after fertilization the heart of the calf begins to beat. At 2 months most adult organs are formed, and at 3 months the fetus can be easily recognized as a calf.

During the first 30 days the developing embryo is nourished from fluids in the uterine cavity because it has not yet become attached to uterine caruncles. Extraembryonic membranes grow rapidly after the 15th day and become filled with fluid (Figure 15.14). The embryo is completely enclosed within one membrane, the amnion, which in turn is enclosed by another membrane, the chorion. As the developing membranes become distended with fluid and press against the uterine endometrium, fetal cotyledons develop opposite maternal caruncles, and the two unite firmly 30 to 36 days after fertilization to form the placenta.

Two arteries and a vein grow out from the developing embryo, forming the umbilical cord, pass through the amnion and end as capillaries in the fetal cotyledons on the chorion. Villi (finger-like projections) on the cotyledons grow into the caruncles of the uterus. Arteries of the cow end in capillaries in the maternal caruncles. Blood of the fetus and cow are separated by 5 layers of cells. Nutrients from the cow and waste products from the fetus must diffuse across

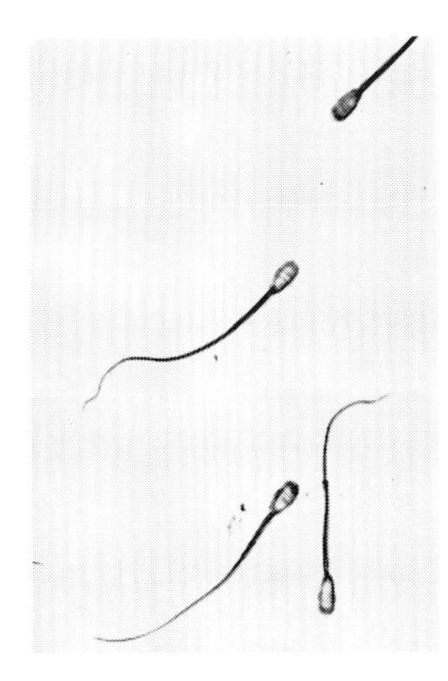
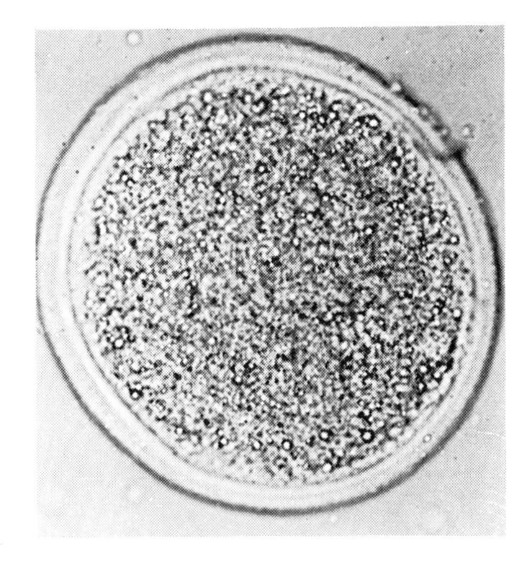

Fig. 15.12 Bull sperm and cow ovum. Each magnified about 300 times. (Courtesy of H. D. Hafs; reprinted with permission from Hoard's Dairyman, 1964)

these cell layers, but not all substances can penetrate the placental barrier, for example, antibodies and vitamin A. Thus, the newborn calf must receive these vital components from colostrum soon after birth.

15.10 Hormonal Maintenance of Pregnancy

Pregnancy is the interval between fertilization and parturition. After fertilization, the corpus luteum does not regress as usually occurs during the normal estrous cycle.[19] Follicle growth on the ovaries and symptoms of estrus are suppressed, although during the first portion of pregnancy some signs of estrus may occur at expected 21-day intervals. The corpus luteum and progesterone secretion are essential throughout gestation for maintenance of pregnancy in the cow, because ovariectomy or removal of the corpus luteum may cause abortion. Thus, it is important, especially when attempting to diagnose pregnancy within 60 days of the last breeding, to palpate the ovary gently to avoid accidental dislodgement of the corpus luteum. However, the source of progesterone shifts as pregnancy advances because ovarian progesterone decreases, but amounts do not change in the jugular vein,

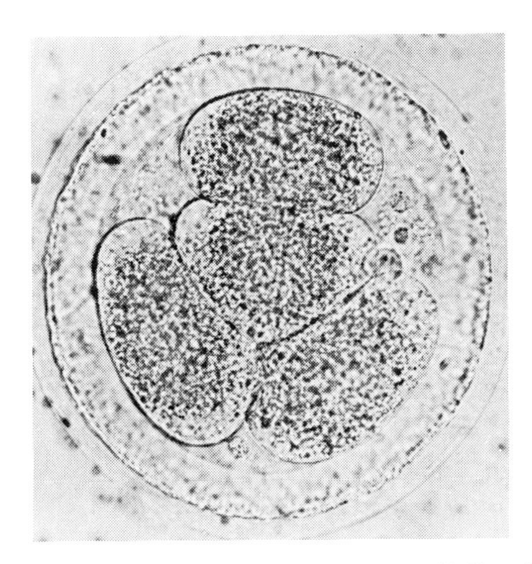
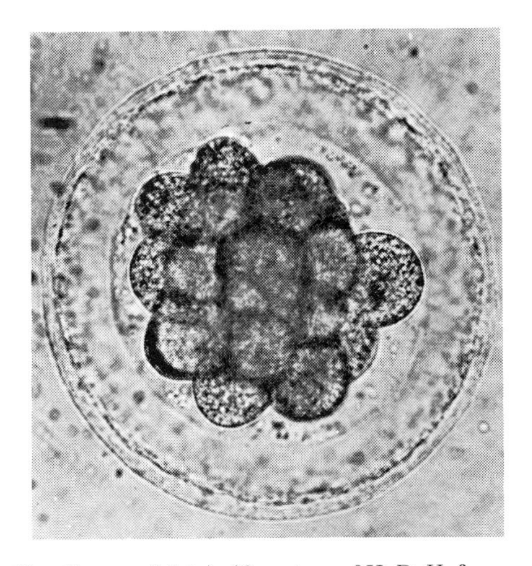

Fig. 15.13 Fertilized cow ova at 4-cell stage (*left*) and at 16-cell stage (*right*). (Courtesy of H. D. Hafs; reprinted with permission from Hoard's Dairyman, 1964)

a measure of peripheral circulation levels. Thus, the corpus luteum of the cow produces progesterone throughout pregnancy, but after 200 days some extra-ovarian source (probably adrenal) of progesterone becomes important.[20]

Estrogen concentrations in blood rise slowly throughout the first 8 months of gestation, but markedly increase during the last month. Estrogen stimulates uterine cell growth and metabolic activity, and if injected in large doses, it causes abortion. However, during early stages of pregnancy progesterone stimulates uterine glandular development and mollifies uterine contractions. Thus, progesterone inhibits the action of estrogen and dominates pregnancy until just before parturition when estrogen becomes predominant.

15.11 Parturition

Length of gestation varies with breed (Table 15.1), sex of calf, age and parity of the cow, and number of calves carried. Bulls are carried about one day longer than heifer calves, and twins are born about one week sooner than single calves.

Signs of Approaching Calving

Some signs of parturition are: relaxation of ligaments around the tailhead and pelvis; swelling of the vulva, which often discharges a mucous secretion; and congestion of the udder. Other nonvisible gross changes include a softening of ligaments of the pelvic girdle (as these ligaments relax, the birth canal enlarges for easier passage

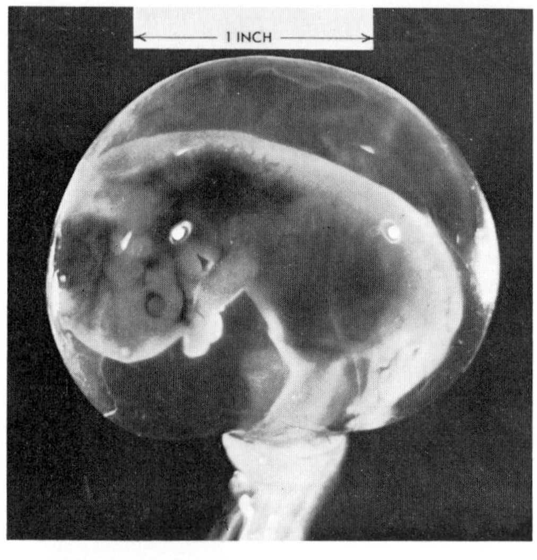

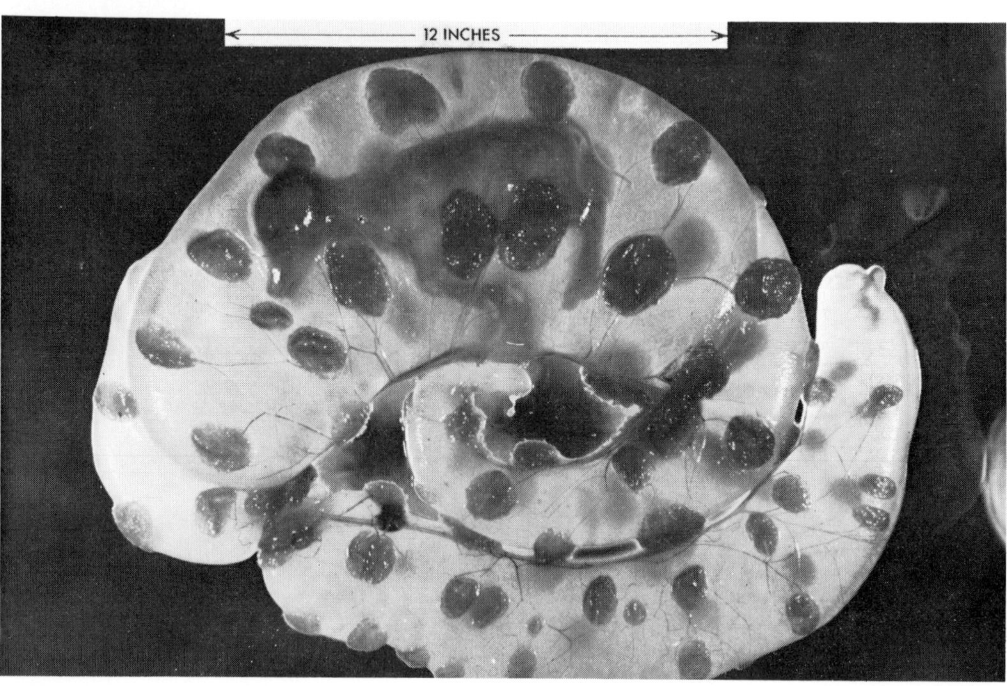

Fig. 15.14 Calf fetus in amnion (*top*). Calf fetus at 4 months of gestation in placental membranes (*bottom*). (Courtesy of H. D. Hafs; reprinted with permission from Hoard's Dairyman, 1964)

Table 15.1 *Effect of Breed of Dairy Cow on Length of Pregnancy*

Breed	*Length of pregnancy*
Brown Swiss	290
Guernsey	284
Dairy Shorthorn	282
Holstein-Friesian	279
Jersey	279
Ayrshire	278

of the calf to the exterior); liquefying of the cervical seal; dilation of the cervix; rotation of the calf to the normal delivery position (Figure 15.15); and then gradual beginning of uterine contractions. Relaxin and estrogen are involved in changes in relaxation of the pelvis and cervix.

Initial mild uterine contractions last about 4 to 20 hours during which time the cow appears increasingly restless. As uterine contractions increase in severity, voluntary abdominal muscle contractions begin and the calf is pushed against the cervix and gradually forced out the birth canal. Fetal membranes break as the forelegs pass the vulva, but cotyledons usually do not separate from caruncles until after parturition. Since the calf is still attached via the umbilical cord to the cotyledons, the calf usually receives an adequate blood supply for the duration of even a lengthy parturition. When the calf leaves the vulva, the umbilical cord breaks, and at this time the calf must commence to breathe. Uterine contractions continue after parturition and serve to expel the fetal membranes. However, about 10% of all cows retain their placenta for 12 hours or longer after delivery of the calf. Supplementation with selenium and vitamin E reduces incidence of retained placenta.

Hormonal Control

Factors that control parturition are not fully understood, but no single factor is responsible for onset of parturition. As the fetus grows, the uterus expands to the point where the myometrium of the uterus becomes more sensitive to factors which initiate contractions.

Progesterone blocks parturition, so this hormone must decrease before parturition. As progesterone decreases, estrogens markedly increase and enhance contractile capacity of the uterus and sensitize the uterus to oxytocin. Indeed, a marked increase in blood concentrations of oxytocin occur just before final expulsion of the calf, and injections of oxytocin induce labor in many species. In addition, ACTH and adrenal glucocorticoids from the fetus rise markedly before parturition, and it has

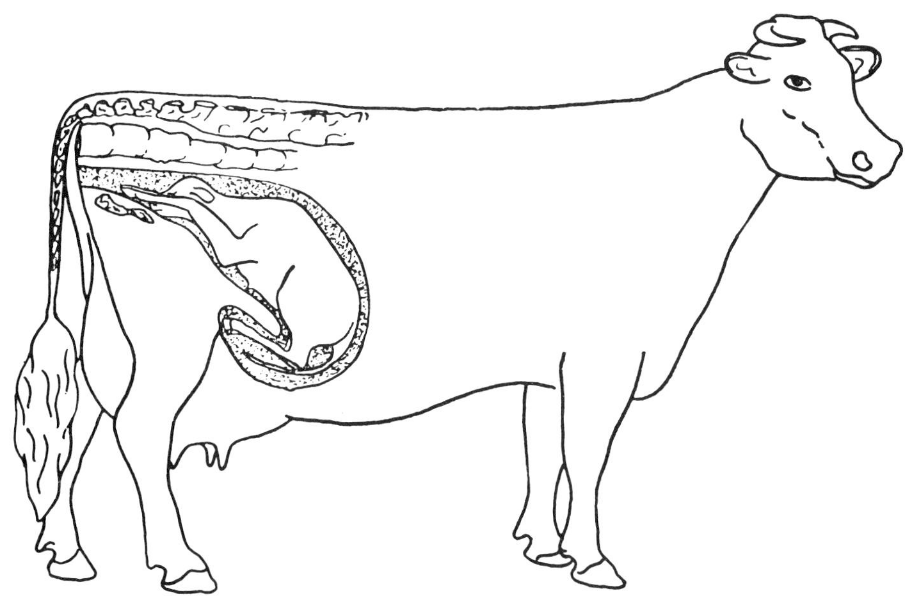

Fig. 15.15 Position of the calf in the uterus after it has been oriented for normal delivery. (Reprinted with permission from *Physiology of Reproduction and Artificial Insemination of Cattle*, 1961)

been speculated that glucocorticoids from the fetal adrenal stimulate prostaglandin $F_{2}\alpha$ synthesis, which initiates the periparturient reduction in maternal progesterone synthesis.[21]

Administration of adrenal glucocorticoids to the cow induces premature parturition, and this technique has been used on a practical basis where it is desirable to have cows calving in discrete intervals.[22] A disadvantage of this technique is greater retention of the placenta. Prostaglandin $F_{2}\alpha$ causes strong contractions of the uterine myometrium and induces abortion and/or parturition.

15.12 The Post-Partum Cow

The term "post-partum" in this section refers to the interval between parturition and subsequent conception.[23] Ideally production of one calf per year per cow results in more calves per lifetime and greatest production of milk per year provided no complications occur around parturition. However, a calving interval of 13 to 15 months is average for most dairy farm operations. An estimated $250 loss per cow per year occurs when calving intervals are extended to 15 months in average herds.

Involution of the Uterus

Time required for the uterus to involute to its pre-breeding size varies from 12 to 56 days. This period is important because fertility is reduced until the uterus involutes. Uterine involution takes longer in multiparous cows and cows that have had complications at parturition. Conception rates at first post-partum breeding from an ovulation on the side of the pregnant uterine horn are lower than conception rates from the non-pregnant side. Overall there is a lower conception rate from matings at the first post-partum estrus than from matings at a later estrus. The cervix usually returns to normal size within 24 to 36 hours post-partum although a retained placenta will delay its closure.

Reestablishment of Estrous Cycles

Interval from parturition to first behavioral estrus in dairy cattle usually ranges from 20 to 72 days. Rapidity of return to estrus varies with milking intensity: cows milked more frequently are delayed in their return to estrus. Furthermore, superior milk producers take about 9 days longer to return to estrus than low milk producers. Ovulation without overt signs of estrus occurs in more than 45% of dairy cows during the early postpartum interval. Conversely, about 10% of dairy cows show estrus without subsequent ovulation.

Ovarian follicular development in early post-partum cows is markedly suppressed. Since gonadotropins are reduced generally during pregnancy, absence of large follicles is not surprising. There is a wide variation in time following parturition before mature follicles commence to grow, but in one study follicular activity started within 10 days of parturition. Re-initiation of follicular growth is associated with increased secretion of LH. As discussed in section 15.7, FSH and LH are responsible for the first post-partum ovulation.

When to Rebreed

Conception must occur within 85 days post-partum to achieve a calving interval of 1 year. The average cow is not inseminated until 90 days post-partum.[24] Thus, considerable reduction in calving interval is possible. The primary limitation is failure of dairymen to detect estrus, not failure of cows to exhibit estrus. Conception rate is only about 25% in the few cows that return to estrus and are bred during the first 20 days post-partum; conception rate averages about 50% between days 40 and 60, and stabilizes at approximately 60% after 60 days. The present recommendation is to have a veterinarian palpate the uterus 20 to 30 days post-partum, and if the reproductive tract is involuted and healthy, the cow should be bred at the first estrus beginning approximately 45 to 50 days post-partum.

15.13 Summary

The male sex cells, spermatozoa, are produced in large quantities from the testes, mature in the epididymis, and at mating are deposited in the vagina of the cow. In the case of artificial insemination, spermatozoa are placed in the cervix or uterus. The cow is receptive to mating only during estrus, although the female sex cell, the ovum, is not released from the follicle until 10 to 14 hours after the end of estrus. In the meantime, sperm are rapidly transported to the oviducts where one sperm cell fuses with the ovum which has been transported from the ovarian follicle to the oviduct. The fertilized ovum descends within 3 to 4 days to the uterus where embryonic growth and fetal development occur.

The gonadotropins (FSH and LH), ovarian steroids (estrogen and progesterone), and a testicular steroid (testosterone) are key hormones that regulate the reproductive process of the cow and bull. Interactions among these hormones are responsible for spermatogenesis, ovogenesis, puberty, the estrous cycle, maintenance of pregnancy, parturition, maintenance of the reproductive tracts, and secondary sex characteristics.

In the average dairy herd only 50% of cows deliver a normal, healthy calf from a single first-breeding insemination, but this rate of efficiency can be improved by following the management practices outlined in Chapter 16.

Review Questions

1. Trace the route of the sperm cell from its point of origin to fertilization of the ovum.
2. Compare hormonal control of spermatogenesis with that of ovulation.
3. Compare in a diagram the changes in reproductive tract development with changes in anterior pituitary and ovarian hormones during the estrous cycle.
4. What are the key hormones needed to maintain pregnancy? What factors are involved in parturition?
5. What is puberty? How could you hasten onset of puberty in dairy heifers?
6. What are the advantages and disadvantages of artificial insemination in comparison with natural service?
7. Why must the breeding of a cow be closely synchronized with ovulation?
8. What factors influence a cow's return to estrus following parturition? Why is this interval of importance to the dairyman?

References

1. Amann, R. P., and Schanbacher, B. D.: Physiology of male reproduction. J. Anim. Sci. *57:* Suppl. 2, 380, 1983.
2. Hafs, H. D., Hoyt, R. S., and Bratton, R. W.: Libido, sperm characteristics, sperm output and fertility of mature dairy bulls ejaculated daily or weekly for thirty-two weeks. J. Dairy Sci. *42:* 626, 1959.
3. Tanabe, T. Y., and Almquist, J. O.: The nature of subfertility in the dairy heifer/III. Gross genital abnormalities. Penn. State Univ. Agr. Exp. Sta. Bull. 736, 1967.
4. Desjardins, C., and Hafs, H. D.: Maturation of bovine female genitalia from birth through puberty. J. Anim. Sci. *28:* 502, 1969.
5. Sorensen, A. M., Hansel, W., Hough, W. H., Armstrong, D. T., McEntee, K., and Bratton, R. W.: Causes and prevention of reproductive failure in cattle. I. Influence of underfeeding and overfeeding on growth and development of Holstein heifers. Cornell Univ. Agr. Exp. Sta. Bull. 936, 1959.
6. Pritchard, D. E., Hafs, H. D., Tucker, H. A., Boyd, L. J., Purchas, R. W., and Huber, J. T.: Growth, mammary, reproductive and pituitary hormone characteristics of Holstein heifers fed extra grain and melengestrol acetate. J. Dairy Sci. *55:* 995, 1972.
7. Morrow, D. A.: Estrous behavior and ovarian activity in prepuberal and postpuberal dairy heifers. J. Dairy Sci. *52:* 224, 1969.
8. Rajakoski, E.: The ovarian follicular system in sexually mature heifers with special reference to seasonal, cyclical and left-right variations. Acta Endocrinol. *34:* Suppl. 52, 1960.
9. Hansel, W., and Convey, E. M.: Physiology of the estrous cycle. J. Anim. Sci. *57,* Suppl. 2: 404, 1983.
10. Rzepkowski, R. A., Ireland, J. J., Fogwell, R. L., Chapin, L. T., and Tucker, H. A.: Serum luteinizing hormone, follicle stimulating hormone and prolactin response to photoperiod during the estrous cycle of Holstein heifers. J. Anim. Sci. *55:* 1125, 1982.
11. Hafs, H. D., Manns, J. G., and Lamming, G. E.: Synchronisation of estrous and ovulation in cattle. In *Principles of Cattle Production.* H. Swan and W. H. Broster, Eds. London: Butterworths, 1976.
12. Hansel, W., Concannon, P. W., and Lukaszewska, J. H.: Corpora lutea of the large domestic animals. Biol. Reprod. *8:* 222, 1973.
13. Hansel, W., and Seifart, K. H.: Maintenance of luteal function in the cow. J. Dairy Sci. *50:* 1948, 1967.
14. Salisbury, G. W., VanDemark, N. L. and Lodge, J. R.: *Physiology of Reproduction and Artificial Insemination of Cattle.* 2nd ed. San Francisco: W. H. Freeman and Co., 1978.
15. Perry, E. J.: *The Artificial Insemination of*

Farm Animals. 4th ed. New Brunswick, N.J.: Rutgers University Press, 1968.

16. Foote, R. H.: Artificial insemination. In *Reproduction in Farm Animals.* 3rd ed. E. S. E. Hafez, Ed. Philadelphia: Lea & Febiger, 1974.
17. Boyd, L. J.: What does direct service cost? Hoard's Dairyman *112:* 1045, 1967.
18. Kiddy, C. A., and Hafs, H. D., eds.: Sex ratio at birth—prospects for control. Amer. Soc. Anim. Sci., 1971.
19. Bazer, F. W., and First, N. L.: Pregnancy and parturition. J. Anim. Sci. *57:* Suppl. 2, 425, 1983.
20. Erb, R. E., Estergreen, V. L., Gomes, W. R., Plotka, E. D., and Frost, O. L.: Progestin levels in corpora lutea and progesterone in ovarian venous and jugular vein blood plasma of the pregnant bovine. J. Dairy Sci. *51:* 401, 1968.
21. Liggins, G. C., Grieves, S. A., Kendall, J. Z., and Knox, B. S.: The physiological roles of progesterone, oestradiol-17β and prostaglandin $F_{2}\alpha$ in the control of bovine parturition. J. Reprod. Fert. Suppl. *16:* 85, 1972.
22. Welch, R. A. S., Newling, P., and Anderson, D.: Induction of parturition in cattle with corticosteroids: an analysis of field trials. N.Z. Vet. J. *21:* 103, 1973.
23. Studies on the postpartum cow. Wis. Res. Bull. 270, 1968.
24. Britt, J. H.: Early postpartum breeding in dairy cows. A review. J. Dairy Sci. *58:* 266, 1975.

Suggested Additional Reading

McDonald, L. E.: *Veterinary Endocrinology and Reproduction,* 3rd ed. Philadelphia: Lea & Febiger, 1980.

Salisbury, G. W., VanDemark, N. L., and Lodge, J. R.: *Physiology of Reproduction and Artificial Insemination of Cattle.* 2nd ed. San Francisco: W. H. Freeman and Co., 1978.

CHAPTER 16

Management Problems Associated with Reproduction

16.1 Introduction

Efficient reproduction in the cow can be attained by careful attention to a number of important details of herd management. The importance of regular calving intervals in the herd is emphasized by the fact that depending upon level of production it costs $2.35 to $4.60 per day in lost income for each day the calving interval is prolonged beyond 365 days.[1]

Although low production is responsible for 50 to 60% of cows culled annually, infertility accounts for another 15 to 20%. Furthermore, some cows culled for low production may have been low producers because of fertility problems.

Causes of reproductive inefficiency include anatomical and genetic defects (see Chapters 4 and 15), and physiological, pathological, and management factors. During the past 30 years, pathological causes have been reduced, but not eliminated entirely, through use of antibiotics, vaccines, artificial insemination (AI) techniques, and herd health testing and culling programs. On the other hand, many authorities believe that, especially in cows with high levels of milk production, physiological causes of infertility are becoming more prevalent or, at least, more clearly recognized. Mistakes in management can be reduced further through educational programs.

16.2 Measures of Reproductive Efficiency

The ultimate goal in reproductive efficiency of dairy cattle would be to deposit one sperm cell in the uterus and obtain a healthy calf 9 months later. However, many cows do not conceive even when millions of sperm are inseminated. Furthermore, some developing embryos are resorbed and some fetuses aborted during pregnancy. Despite the fact that delivery of a normal calf is the result of a single insemination, several unsuccessful inseminations may have preceded it. Therefore, reproductive efficiency is measured in varying degrees of fertility.

Fertilization Rate

Fertilization rate of clinically normal dairy cattle is very high, over 96% when high quality semen is properly used, but this rate decreases to about 77% when semen from "low" fertility bulls is used.[2] Thus the bull has been blamed for most fertilization failures. However, a summary of fertilization rates in repeat breeding

Table 16.1 *Loss in Reproductive Efficiency after a First Service*

Days after insemination	*% Fertility*
1	96–77
30	70
90	58
Calving	50

Source: Summarized from literature.

cows reveals that although 46% of ova are fertilized after insemination, between 16 and 54% of fertilization failures are attributed to the cow.[3] Thus, the bull and cow each contribute to fertilization failure.

In addition to fertilization failure there is a substantial loss in reproductive efficiency between fertilization and parturition (Table 16.1). This represents embryonic and fetal mortality.

Services per Conception

Pregnancy can be determined from a rectal palpation of the uterus of the cow 40 to 60 days post insemination. Based on this information, the well-managed dairy herd averages about 1.3 inseminations (services) per conception (an efficiency rate of 77%). However, after fetal deaths are considered, this is equivalent to about 1.6 services per calf born. The average herd requires about 2 inseminations per calf born.

Nonreturn Rate

Another measure of reproductive efficiency is rate of nonreturn to estrus. Most AI organizations use this method to evaluate fertility of bulls. It is based on the assumption that if the AI technician is not recalled within a reasonable period of time to breed a cow after the initial insemination, the cow is pregnant. The usual period is 60 to 90 days. That is, if a cow is not rebred within 60 to 90 days, the bull stud considers that cow pregnant.

Naturally, many factors can account for a failure to rebreed cows within this period even when the cows did not conceive to a previous insemination (e.g., culling or cystic ovaries). In addition, when a cow returns to estrus some dairymen will breed the cow naturally or switch to another AI organization, and these would be counted as pregnant to the previous service. Abortions after 90 days are ignored in this method. These factors increase the nonreturn rate and give a higher estimate of breeding efficiency than actually exists. On the other side of the coin, about 3.5% of all cows show a false estrus and may be rebred when already pregnant. This lowers the nonreturn rate.

The average 60- to 90-day nonreturn rate is about 70%. However, exceptionally well-managed herds will average 80%, whereas herds with major infertility problems may average only 50% or less. However, the 60- to 90-day nonreturn rate must be reduced substantially to estimate actual calving rate. In general, the nonreturn rate is a relatively good, rapid, inexpensive method to compare fertility of bulls, efficiency of breeding technicians, as well as fertility among herds of cows.

Calving Rate

The number of inseminations needed to produce a live calf is one of the most useful measures of reproductive efficiency. The value for well-managed herds is about 1.6 inseminations per calf born (62% efficiency), whereas for average herds it is about 2.0. The value of 1.6 inseminations per calf born is dependent upon use of semen of high quality. Actual calving rate from a single insemination is about 50%,[4] but between 1951 and 1978 reproductive efficiency has decreased.[5] Breeding nonestrual cows is a major contributor to lowered conception rates.

With succeeding services there is a steady decline in level of fertility (Table 16.2). It is assumed that highly fertile cows

Table 16.2 *Reproductive Efficiency with Succeeding Artificial Inseminations*

Insemination number	*Calving rate (%)*
1	59
2	53
3	48
4	42
5	30
6	31
7+	19

Source: Foote and Hall: J. Dairy Sci. *37:* 673, 1954.

conceive first, and the remaining population of cows now contains a greater proportion of "hard-to-settle" cows which probably explains the decrease in fertility with each succeeding insemination. Percentage of total cows conceiving decreases to very low values, especially after the third service, but on average about 90% of cows conceive within 3 services.

Another measure of a cow's reproductive efficiency is number of calves born in a lifetime. Heifers bred to calve at 30 months of age produce an average of only about 3 calves in their lifetime, whereas if bred to calve at 24 months, lifetime production increases to about 4 calves.

Calving Interval

A calving interval of 13 months for first-calf heifers and 12 months for cows in subsequent lactations maximizes milk production[6] and profits. Higher-producing, more persistent cows, however, may not suffer the production losses observed in average cows when calving interval is extended to 13 or 14 months. Furthermore, some high-producing cows do not return to estrus soon enough after parturition to achieve a yearly calving interval.

Culling for Infertility

Culling cows for infertility can bias all measures of reproductive efficiency. Culling less than 5% of cows per year for infertility should be the goal of every dairyman.

16.3 Physiological Factors Associated with Infertility

Fortunately, most cows do not suffer physiological disturbances which render them infertile. But in some cows physiological problems lower fertility, and these are discussed in the following paragraphs.

Abnormal Estrous Cycles

Cows are normally polyestrus animals, that is, estrous cycles occur regularly in healthy nonpregnant animals throughout the year at approximately 21-day intervals (20-day intervals for heifers). The four most serious problems associated with the estrous cycle are: (1) absence of estrus (anestrus), (2) irregular estrous cycles, (3) silent estrus, and (4) constant estrus (nymphomania).

Anestrus. Underdeveloped ovaries are often associated with anestrus. Genetic defects and malnutrition may cause infantile reproductive organs. However, some heifers have infantile ovaries despite adequate nutrition and no known genetic defects. The most common treatment for infantile ovaries is to administer a gonadotropic hormone, especially one high in FSH. Injection of FSH may produce multiple ovulations, therefore, breeding should be delayed until the second estrus.

Retention of a portion of placenta within the uterus or pyometra (pus in the uterus) usually produces anestrus. In these cases, the corpus luteum may persist and synthesize progesterone which prevents estrus. This may be analogous to maintenance of the corpus luteum during pregnancy. Elimination of the placenta and pus from the uterus is essential for return to estrus.

Sometimes the corpus luteum persists beyond its usual life expectancy for no readily apparent reason, although it may represent an unobserved estrus and ovulation. If the persistent corpus luteum also has a fluid-filled cavity, it is referred to as a cystic corpus luteum (Figure 16.1). Thus, it is one form of the so-called "cystic ovary" condition. One treatment has been to squeeze the corpus luteum off the ovary by rectal palpation. However, this may cause extensive hemorrhage around the ovary and even death of the cow. It may also result in formation of adhesions around the oviduct and infundibulum to such an extent

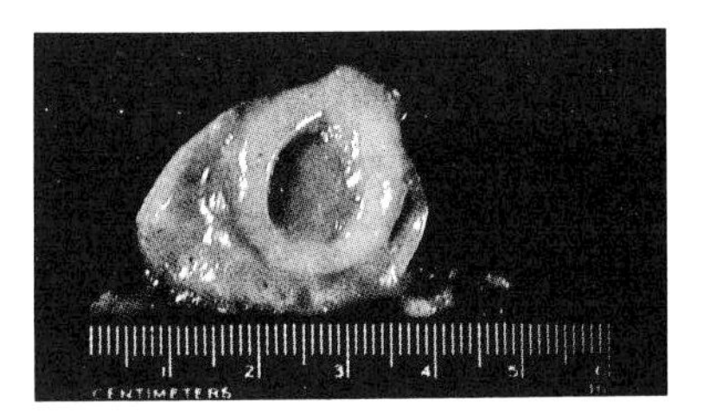

Fig. 16.1 Cystic corpus luteum with fluid-filled central cavity. Protrusion of lutein tissue above surface of ovary indicates occurrence of ovulation. Persistence of the corpus luteum is usually associated with retention of a portion of placenta or pyometra in the uterus. (Reprinted with permission from Veterinary Scope *14:* 1, 1969)

that egg and sperm transport are inhibited. This animal would be unilaterally sterile. Other treatments to initiate the estrous cycle usually involve injection of a gonadotropin high in LH, a small dose of estrogen, or prostaglandin $F_2\alpha$. Since the cystic corpus luteum may be normal and not really persistent, treatment is often not justified.

As discussed in Chapter 15, the interval between calving and first behavioral exhibition of estrus varies from 30 to 72 days. Furthermore, cows milked more frequently than 2 times per day and cows producing large quantities of milk are often delayed in returning to estrus after parturition. In these animals anestrous is of relatively short duration. Very old cows may reach a point where they no longer cycle. Obviously this is a rare condition in cows because most are culled long before they reach reproductive senility.

Irregular Estrous Cycles. Cows normal in every observable respect may undergo estrous cycles of a very short (<18 days) or very long (>24 days) duration. Causes of these irregular cycles in "normal" cows are obscure. The dairyman may have missed the preceding estrus or recorded a "false" estrus. Satisfactory conception rates generally occur in "normal" cows with long cycles, but this is not true for cows with cycles shorter than 18 days. Inflammation of the uterus (metritis) is sometimes associated with irregular estrous cycles. Irregular cycles are an important sign of infertility, and the veterinarian should be informed of these types of estrous cycles to aid in diagnosis of the problem.

Silent Estrus. From 15 to 25% of all ovulations occur without full behavioral manifestation of heat. Since silent heat is accompanied by ovulation, the main problem in these cows is to detect estrus—a management problem. This condition is more prevalent during the early postpartum period and in heifers undergoing puberty. Often the dairyman misses a heat because the heat period is very short, heat occurs at night and is undetected, or heat detection methods are not adequate (Section 16.7). Insemination of the cow at the proper time relative to ovulation during a silent heat usually results in a normal conception.

Constant Estrus. The nymphomaniac, or "chronic buller," shows unusually prolonged periods of estrus. Large fluid-filled (cystic) follicles persist on ovaries of these cows for at least 2 weeks in the absence of a corpus luteum. However, not all cows with cystic follicles exhibit nymphomania. Cystic follicles presumably secrete large amounts of estrogen, but available experimental evidence is not conclusive.[7] An ovarian follicular cyst may reach 10 cm in diameter. Although it undergoes cyclic periods of growth and regression, a follicular cyst does not usually ovulate spontaneously nor does it luteinize.

Incidence of cystic follicles is not directly related to level of milk production, but its heritability is rather high. Thus, certain families of dairy cattle have a high incidence of this condition. Furthermore, cystic ovarian follicles tend to recur in the same cow in subsequent lactations. One treatment for nymphomania is to inject LH. This treatment usually causes luteinization of the follicle wall, the cow returns to a normal estrous cycle, and a normal conception often occurs at the following estrus. Recently, gonadotropin releasing hormone (GnRH) has been observed to be an effective treatment for some dairy cows with follicular cysts.[7] However, many cysts disappear spontaneously.

Abnormal Fertilization, Embryonic and Fetal Mortality

Most anatomical or functional defects in the egg or sperm prevent fertilization, or if fertilization occurs, the developing zygote may be abnormal and resorbed within a few days. Primary causes for abnormal zygotes include aged ova and sperm, immunological reactions involving sperm, and elevated ambient temperatures before breeding (Section 16.6).

As discussed in Chapter 15, life expectancy of the ovum and sperm is limited to only a few hours. But ability of the egg to be fertilized lasts longer than its ability to develop into a normal calf. As the egg ages, various functions are gradually lost. One of the most conspicuous changes is that more than one sperm may enter aged eggs (polyspermy). The resulting zygote contains an excess number of chromosomes and such individuals soon perish in utero. Cows mated naturally are less likely to have aged eggs than cows mated by AI because the period of receptivity to the bull is much shorter than the breeding period used in AI. On the other hand, pretreatment of sperm used in AI with "capacitating factors," such as the enzymes amylase or beta-glucuronidase, is believed to reduce capacitation time required by sperm in the reproductive tract and has increased conception rates.[8] It is suspected that this increase in conception occurs when inseminations are made late in the breeding period, and the eggs are immediately fertilized by the capacitated sperm. Untreated sperm used in AI would have to reside in the reproductive tract for up to 6 hours, and by that time the egg would have degenerated beyond the point of fertilizability. Aging of sperm occurs within the testes and epididymis, but generally there is sufficient loss of sperm into urine or by masturbation to assure continuous production of highly fertile sperm even in infrequently ejaculated bulls. In contrast, maximal survival time of sperm after deposition in the reproductive tract of the cow is about 24 to 30 hours. Sperm survival is longest in the cervix and oviduct and shortest in vagina and uterus. Fertilizing capacity of sperm increases during the first few hours in the cow's reproductive tract, remains constant for several hours, then quickly decreases.

Age of spermatozoa at insemination is an important factor in reproductive efficiency in the AI industry. Fertility of semen increases initially during storage, remains optimum for a period, then decreases (Figure 16.2). These changes in fertility of spermatozoa with age occur at all commonly used storage temperatures. Lowering storage temperature merely delays senescence and the period of optimum fertility.[9] Although cows bred with frozen semen stored almost 20 years have produced normal calves, results in Figure 16.2 support the concept that semen should not be stored longer than 1 to 2 years if reasonable fertility rates and subsequent genetic progress are to be achieved.

As alluded to in several sections of this chapter, there are major individual differences in fertility among bulls (Table 16.3). Thus, dairyman should use bulls that have a high predicted difference for milk production and a high level of fertility. This concept is discussed further in Chapter 7.

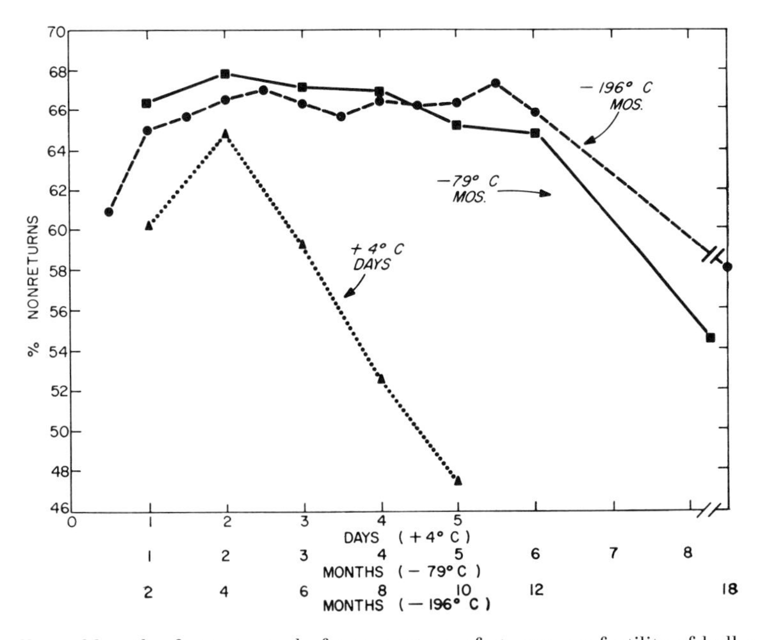

Fig. 16.2 The effect of length of storage and of temperature of storage on fertility of bull spermatozoa. The duration of storage was 5 days at 4C (423,054 inseminations), 1 year at −79 to −88C (174,307 inseminations), and 2 years at −196C (196,448 inseminations). Fertility was assayed 167 to 180 days after insemination. Estimated embryonic mortality varied inversely with fertility level and ranged from about 8 to 16% of the cows inseminated depending upon the source of the semen and the time in the year when collected. (Reprinted with permission from Biol. Reprod. Suppl. *2:* 1, 1970)

TABLE 16.3 *Fertilization Rates and Embryonic Death Rates in Cows Bred to Bulls Classified as High, Medium, and Low Fertility*

Bull fertility	*Recovered normal fertilized ova (%)*	*Nonreturn rate (%-range)*
High	100	70–79
Medium	82	60–69
Low	71	40–59

Source: Nalbandov, A. V.: *Reproductive Physiology.* 2nd ed. San Francisco, W. H. Freeman and Company, 1964.

In some cases, fertilization does not occur because of immunological incompatibility between the uterus (and possibly the ovum) and sperm of certain bulls. Frequency of occurrence of this phenomenon in dairy cattle is not yet known.

As shown in Table 16.1 as many as 46% of fertilized eggs die before parturition. In addition to egg and sperm effects previously discussed, other primary causes of prenatal mortality are endocrine imbalances, hereditary defects, and diseases.

Endocrinological Imbalances. During the first 30 days of life, the embryo and extra-embryonic membranes differentiate very rapidly, and nutrition of these cells is dependent upon secretions from the uterus (uterine milk or histotroph). Thus, the uterine environment, which is dependent upon progesterone secretion, is critical to survival of the embryo. Presence of the embryo inhibits the normal luteolytic mechanism; thus the corpus luteum and progesterone secretion persist throughout gestation. If the embryo does not develop properly during early pregnancy, a luteolytic mechanism may be activated. In this case progesterone secretion declines, estrous cycles resume, and the embryo dies.

Hereditary Defects. Early embryonic death of genetically abnormal individuals is the least expensive method of removing these undesirable animals from a population. Since both sire and dam contribute genes which may cause defects it is especially important for AI centers to watch carefully for undesirable recessive genes in their bulls. Widespread dissemination of undesirable genes must be prevented. The role of genetics in reproductive efficiency has been summarized elsewhere.[10]

The breeding system may contribute to embryonic mortality. Inbreeding and line breeding increase embryonic mortality compared with random mating or outcrossing systems. Breed effects on level of prenatal mortality are not clear, but families of cows have different degrees of fertility, and production of undesirable offspring is more prevalent in some families than in others. When the dairyman suspects that this problem exists, mating to unrelated individuals is recommended.

There is also a sex difference in embryonic mortality. Sex ratio at fertilization favors bulls. In one study, 58% of aborted fetuses were bulls, whereas at normal birth about 51% were bulls. Thus, as gestation advances more bulls than heifers die in utero.

There is greater embryonic and fetal loss among individuals from multiple ovulations than from single ovulations. Since twinning in cattle is heritable, this provides additional evidence that prenatal mortality can be inherited.

Fetal Mummification. Fetal death without either abortion or resorption of the fetus and its membranes usually results in mummification. It usually begins between the 5th and 7th month of pregnancy. Fetal cotyledons separate from maternal caruncles and uterine contents gradually become transformed from a reddish fluid mass to a "dry" mummified fetus which may be carried for several months beyond the normal length of gestation. During this period the cow does not undergo estrous cycles. Palpation of the reproductive tract usually reveals a persistent corpus luteum, absence of an enlarged uterine artery of pregnancy, and no fetal fluids or cotyledons. Estrogen injection is a common method used to induce expulsion of a mummified fetus. After removal of the mummified fetus, the cow usually returns to normal fertility. The syndrome appears to be inherited, but fortunately its incidence is low in dairy cattle. The specific cause is not known, but it may be related to a deficiency in blood supply to the fetus.

Dystocia

Normal presentations of the calf at parturition occur in 95% of all births, and little difficulty is experienced. It is usually better not to assist the cow at calving unless help is obviously needed. In such cases, the dairyman should assist the cow only when presentation of the calf appears normal (Figure 15.15). The calf should be pulled downward rather than straight out from the cow, and this assistance should be applied only when the cow strains.

The most common cause of calving difficulties is an abnormal presentation of the fetus at parturition (Figure 16.3). Abnormalities range from retraction of a foreleg or the head, to a breech, sideways, or upside-down presentation. A veterinarian should be called whenever an abnormal presentation is diagnosed or if the cow has been in labor for as much as 24 hours. Usually the veterinarian will reposition the calf to allow the cow to deliver the calf. If this is not possible, a cesarean section or dismemberment of the calf may be performed to save the cow.

Another cause of dystocia is presentation of a calf which is very large in relation to size of the birth canal. This may be caused by prolonged gestation, fetal monsters, or small cows mated to sires that produce large offspring. In addition, there is a prolonged gestation syndrome which is caused by 2 recessive genes (Chapter 4).

Stillborn calves and dystocia occur more frequently in first-calf heifers than in older cows. For this reason, many dairymen breed first-calf heifers to Angus bulls, which are known to produce smaller calves at birth. Although this practice reduces incidence of dystocia, about one-fourth of the annual calf crop and its accompanying genetic improvement of the herd is lost be-

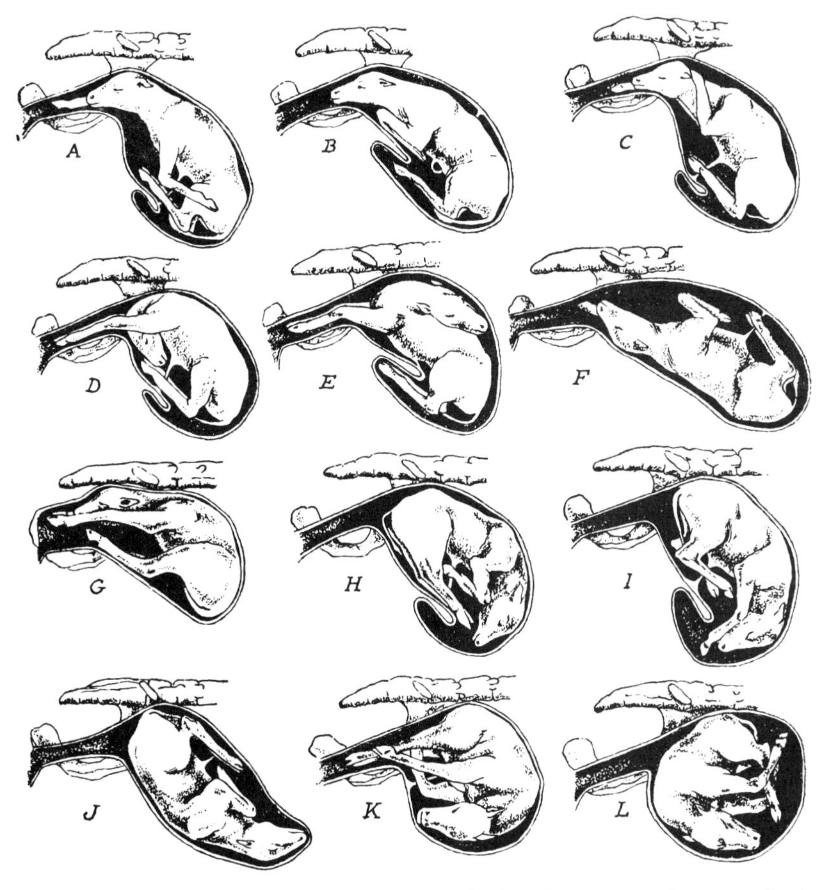

FIG. 16.3 Abnormal presentations of the calf for delivery. A. Anterior presentation—one foreleg retained. B. Anterior presentation—forelegs bent at knee. C. Anterior presentation—forelegs crossed over neck. D. Anterior presentation—downward deviation of head. E. Anterior presentation—upward deviation of head. F. Anterior presentation—with back down. G. Anterior presentation—with hind feet in pelvis. H. Croup and thigh presentation. I. Croup and hock presentation. J. Posterior presentation—the fetus on its back. K. All feet presented. L. Dorsolumbar presentation. (Reprinted with permission from *Physiology of Reproduction and Artificial Insemination of Cattle,* 1961)

cause crossbred Angus-dairy calves are not likely to produce large quantities of milk. A more sensible program is to mate first-calf heifers to bulls of the same breed that are known in the AI centers to sire small offspring. Another practice which minimizes dystocia is to feed dairy heifers adequately before breeding and during their first pregnancy so that they attain large skeletal size before calving. However, as pointed out in Chapter 20, heifers should not be fed excessive amounts of concentrate feeds because high body condition (fat) reduces subsequent milk production.[11]

The vagina and uterus may be partially expelled (prolapsed) following an abnormal parturition. This condition is more prevalent in older cows than in first-calf heifers. A veterinarian can reposition the uterus inside the body cavity if the prolapse is discovered promptly. The prolapse may have been an accident, and such cows will usually have normal fertility afterwards. However, there is a tendency for the syndrome to recur in the same animal. Thus, there is justification for culling such animals.

Age and Size of Cow

Fertility in dairy cows increases up to 4 years of age, remains constant to 6 years, then gradually decreases with advancing age. But age is not a major cause of infertility in dairy cattle because most are culled before they get old.

A more important consideration is size of heifer at first breeding. Heifers can conceive as early as 5 to 7 months of age, depending upon breed. However, as pointed out in the previous section, first-calf heifers have greater difficulty delivering a calf than larger, older cows. Thus, it is important that heifers achieve a sufficient size at first breeding to minimize chances of dystocia.[12] Recommended weights to breed heifers are listed in Table 16.4 for each of the major dairy breeds. Delaying initial breeding of normally fed heifers past 15 months of age reduces total lifetime milk production. Heifers not mated until 3 years of age frequently develop reproductive disorders. Heifers which calve at younger ages reach expected body size at maturity if they are well fed during the first lactation.

Table 16.4 *Recommended Weight of Heifers at First Breeding*

Breed	*Weight (lb)*
Brown Swiss	750
Holstein	750
Ayshire	600
Guernsey	550
Jersey	500

16.4 Pathological Causes of Infertility

This topic is covered in Chapter 21.

16.5 Nutritional Factors Associated with Infertility

Nutrients needed for reproduction are the same as those needed for growth and lactation. Thus, if the dairy cow is fed for adequate growth and maximal milk production, there should be no nutritional limitations on reproduction. However, lactating cows losing body weight have reduced conception rates, delayed first estrus post partum, and greater incidence of silent estrus than cows gaining body weight.[13] Whether it is plane of nutrition per se or milk yield that affects reproductive performance of lactating cows is difficult to determine, because higher producing cows are often losing weight and delay returning to estrus after parturition in comparison with low producing cows. Some important relationships between nutrition and reproduction are outlined below.

Energy and Protein Intake

Probably the most serious nutritional causes of reproductive failures are an insufficient supply of calories (energy) and protein. Severe restriction of intake of these essential components of the ration results in infantile ovaries and delays puberty in heifers. The slightly underfed heifer will have normal fertility but at a slightly older age. Effects of slight to moderate malnutrition on reproduction can be reversed with adequate feeding, but severely stunted heifers may retain their infantile ovaries or, at best, have irregular estrous cycles and low fertility.

Conditions that most commonly produce these nutrient deficiencies in dairy cattle involve pasture feeding. In many parts of the world these conditions are frequently aggravated by deficient rainfall and high temperature. Moreover, some dairymen fail to check on yearling heifers on pasture during summer. If heifers return from pasture in autumn and do not exhibit estrus, the ration should be supplemented with grain for several weeks.

As discussed in Chapter 11, urea is sometimes fed to dairy cattle to replace some of the protein in the feed. Some dairymen and veterinarians are of the opinion that feeding urea adversely affects reproduction. However, there is no experimental evidence to show that feeding urea in recommended amounts significantly affects reproductive efficiency.[14]

Fat dairy heifers may have lower breeding efficiency than heifers in good growing condition although the literature is controversial. For example, in one study of heifers reared on low, medium, or high nutritional levels, percentages of heifers conceiving to first service were 79, 68, and 58, respectively.[15] Other later reports indicate that there are no effects of high nutritional levels on fertility.[16]

Vitamin Deficiencies

Vitamin deficiencies are unlikely to occur in most dairy herds. The only vitamins shown to be involved in reproduction in cattle are A and D. Vitamin A-deficient cows may abort or produce dead or weak calves at full term. Retained placentas are more common in vitamin A-deficient cows. Vitamin D-deficient pregnant cows produce calves with rickets, whereas nonpregnant cows fail to exhibit estrous cycles.

Vitamin E-deficient cows reproduce normally. The beneficial effects upon fertility of feeding wheat-germ oil or sprouted oats (sources of vitamin E) claimed by some have not been substantiated in controlled experiments. As described in section 15.11, supplementation of pregnant cattle with vitamin E and selenium reduces incidence of retained placentas.

Minerals

Phosphorus deficiency is the most common mineral deficiency affecting reproduction in dairy cattle. Symptoms of phosphorus deficit include delayed puberty and, in extreme cases, cessation of estrous cycles. Low phosphorus intake may also be associated with lowered conception rate. Feeding of bone meal, dicalcium phosphate, or mixtures of grain concentrate rations correct phosphorus deficiency.

The trace minerals cobalt, copper, iodine, and manganese are required for general thriftiness of the animal as well as for reproduction. As mentioned previously in Chapter 9, iodine deficiencies may result in goiter in calves. Access to trace-mineralized salt and feeding of most common types of rations, especially legumes, prevent these deficiencies.

Dairymen are sometimes persuaded to feed cows special diets which are claimed to cure many types of ailments, including reproductive disorders. Such feeding programs are not justified. Instead, dairymen should feed cows for highest milk production, and nutritional causes of infertility will usually be eliminated.

16.6 Environmental Factors Affecting Reproduction

Environmental factors such as season, temperature, humidity, and light interact to affect reproduction. Either abnormally high or low temperatures reduce reproductive efficiency. Thus, for most sections of the United States, reproductive efficiency is lowest in summer and winter and highest in spring and autumn.[17] Lowered fertility during winter may be due to shorter daylight hours or to low vitamin content in the forage portion of the ration. Closer confinement of cattle during winter in northern latitudes reduces ease of detection of estrus and contributes to seasonal variation in reproduction.

High temperatures shorten duration and lower behavioral expression of estrus in the cow.[18] Moreover, in hot weather a greater percentage of cows come into heat at night when no one is available to observe them. Thus, missed heats are common in summer. Also, hot climatic conditions often cause anestrus in lactating dairy cows because the adrenal glands secrete large amounts of progesterone.[19] Fertility is reduced if ambient temperature is elevated on the day after insemination, and high temperatures markedly increase embryonic mortality.[20,21] Providing shade and air movement improves fertility in cows subjected to high environmental temperatures.

High temperatures also affect fertility of bulls. Conception rate for semen collected from bulls during hot summer months and then inseminated into cows throughout the year is substantially lower in comparison with semen collected during cool seasons. Furthermore, deleterious effects of aging of frozen semen during storage occur more rapidly in semen collected during summer than in samples collected between November and April.[22]

The most desirable environment for greatest reproductive efficiency is a combination of cool temperatures, low humidity, lengthening hours of daylight, and adequate nutrition.

16.7 Management Programs for Optimal Fertility

The average dairy cow in the United States produces milk for approximately 3.5 years with a total life span of about 6 years. Thus, there is a 20 to 30% turnover in cows in most herds every year. As mentioned earlier, a major cause of this high turnover is reproductive failure. Attention to several management practices outlined in the following paragraphs should enable most herds to achieve a calving rate from first services of more than 60% with not more than 1.6 services per calf born.[14,23]

Visual Detection of Estrus

One of the major management problems confronting dairymen, especially in larger herds, is heat detection. There is no substitute for frequent and systematic observation of the cows. One study has shown that as many as 26% of the cows in estrus could be easily overlooked. Cows should be observed at least 2 times per day, and this should be in addition to regular feeding or milking times. Cows tied in stalls should be turned out twice a day for at least 30 minutes, and during the observation period cows should be moved about. On average this system will detect about 90% of cows actually in estrus. The cardinal sign of estrus is standing while being ridden by other cows. In addition, dairymen should pay particular attention to the cow that is behaving differently than usual as well as to those that are due to be in heat. Red swollen vulvas, a slight vaginal discharge of mucus, or an unexplained depression in milk yield is often the only subtle indication of heat in some cows.

Estrus (Heat) Detection Aids

Use of estrus detectors (Figure 16.4) glued on the rump of cows increases the percent of cows detected in heat.[24] Another system involves marking the tailhead of each animal with a crayon each morning, and when the mark is smudged, the cow probably has been in estrus. Potentially these methods allow continuous heat checking. However, the number of cows falsely detected in heat increases because the aids may be activated if the cow is mounted by another cow in estrus.

Attachment of a chin ball marking device to a cow treated with testosterone or to a bull surgically altered to prevent copulation also aids detection of estrus when such animals are exposed to the herd. Danger associated with dairy bulls and problems associated with surgery are major disadvantages of use of surgically altered bulls.

Other aids for estrus detection being developed include pedometers to measure increased walking in cows in heat, physical and chemical changes in vaginal mucus, and measurement of progesterone in milk. However, the most accurate method of estrus detection remains careful, frequent visual observation of cows by the dairyman.

Average dairy herd size has increased fourfold in the last 25 years, and there is a trend toward housing cattle in groups. When cows are handled in groups, identification of cows in estrus is a major problem. Large numbers painted on haircoats of each side of the animals is perhaps the most reliable system (see Chapter 3).

To manage reproductive performance of cattle in large herds, cows should be grouped as follows: (1) early postpartum cows, cows scheduled to be bred, and problem cows (lactation more than 100 days and not pregnant) (2) pregnant, lactating cows; and (3) pregnant, nonlactating cows.[25] This system is compatible with other grouping systems based on nutritional, herd health, and milking management requirements. As herds become

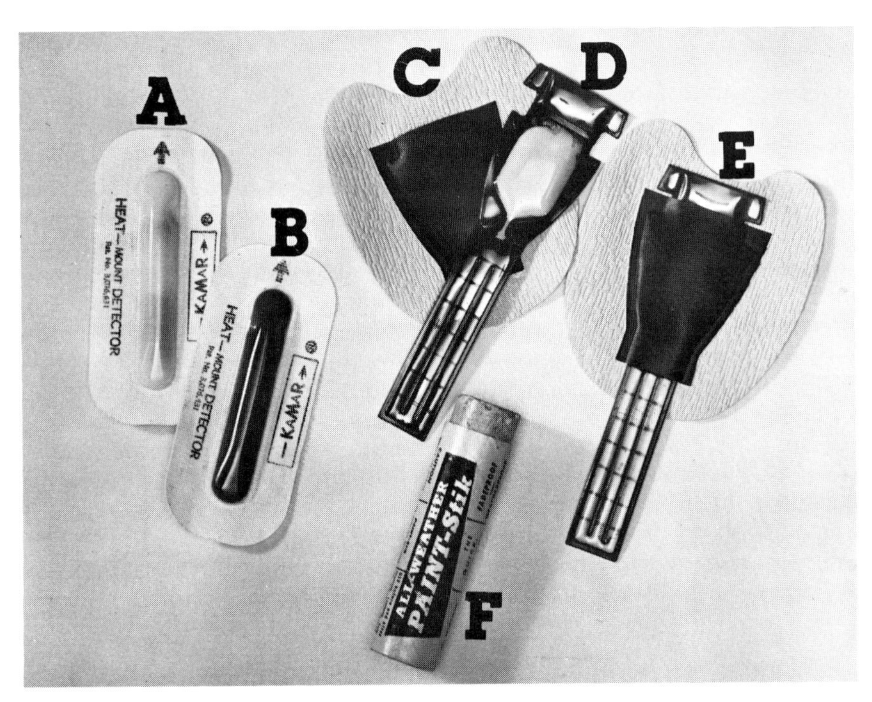

Fig. 16.4 Sample of estrus detection aids. A. Untriggered KaMaR heat mount detector. Heat mount detector is glued to rump of cow. B. Activated (colored) KaMaR heat mount detector. C. Holster for Mate Master heat detector tongue. Holster is glued to rump of cow. D. Mate Master heat detector. Dye has been extruded due to mounting by other cows (approximately 2 cm). E. Mate Master heat detector. Dye has been extruded due to frequent mounting by other cows (approximately 7 cm). F. Marking crayon to be applied to tail head of breeding cows. (Courtesy of J. H. Britt and J. S. Stevenson)

larger, the number of potential management groups increases, and additional grouping strategies may be employed; for example, cows may be grouped according to stage of lactation and this practice may accomplish similar goals to those from grouping by reproductive status.

Other factors associated with lowered detection of estrus and reduced fertility include large herd size, slippery floors, and overcrowding of cattle. These problems have obvious solutions. Natural mating of "problem" breeders with a bull should be used only as a salvage operation to minimize the number of cows culled for infertility.

As described in Section 14.6, use of prostaglandin $F_2\alpha$ is a management technique that minimizes time required to detect estrus in cows.

Time to Breed

Probably the most important management factor associated with infertility in dairy cows is breeding at the wrong time during the estrous period. Fertility is not uniformly distributed throughout estrus. Rather, highest fertility occurs when cows are bred during the final 10 hours of standing heat or within the first 6 hours after the end of standing heat (Figure 16.5). Thus, cows first observed in standing estrus during the morning should be bred in the afternoon of that day, whereas cows first observed in standing heat in the afternoon

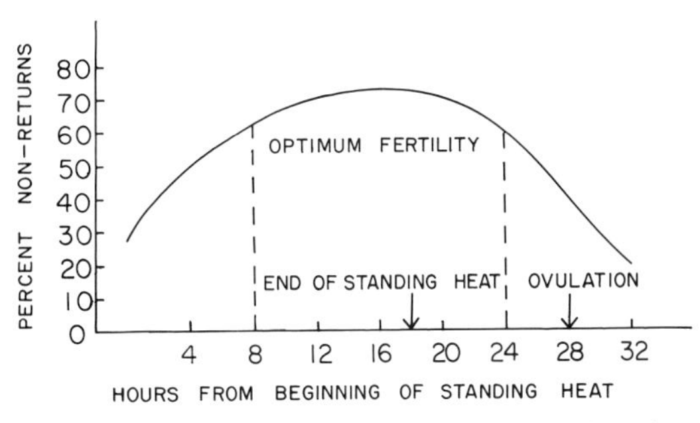

Fig. 16.5 Time to breed cows to achieve highest fertility.

should be bred in the morning of the following day. Some cows, however, do not fit the normal pattern of ovulation, and if bred at the recommended time without conception, it may be wise to breed 2 or 3 hours earlier and/or later than usual. If this new breeding time results in conception, this information should be entered in the breeding record for that cow because she is likely to behave in the same manner in subsequent estrous cycles. Except in cows such as those just described, it does not pay to breed cows more than once in a given estrous period.

As mentioned previously, heifers should be bred at a given skeletal size rather than at some predetermined age. And a cow should not be bred post-partum until a veterinarian determines on the basis of a rectal palpation that she is "ready-to-breed" (Section 15.12).

Care of the Post-partum Cow

Parturition and sudden onset of high milk secretion places great stress on the dairy cow. At this time she is more susceptible to infections such as mastitis as well as metabolic diseases such as milk fever and ketosis.

A clean box stall should be provided for each cow before calving. She should be allowed to stay there for at least a day after calving, but need not be isolated from the main herd for a prolonged period provided she is healthy.

After each use, the box stall should be cleaned, and if disease is suspected, the stall should be disinfected. Uterine defense against disease may be reduced at parturition. Thus, it is important to provide a clean and comfortable box stall to minimize stress at parturition.

Breeding Records

As emphasized in Chapter 3, no business as complex as a dairy enterprise can afford to be without adequate records, and this concept includes the entire reproductive life of the dairy cow. Breeding records for each cow should contain the following: birth date, calfhood diseases and vaccinations, dates in heat, breeding and calving dates. Name of dam and sire should also be recorded. Other health problems or peculiarities about the cow should be recorded because such observations may be the clue needed by the veterinarian to diagnose the cause of infertility. No dairyman should look at his cows without a pencil and notepad in his pocket to record his observations.

The record system should be simple but accurate, and most important it should be used. Successful record keeping systems employ a single sheet or folder for each animal. It is usually best if one person enters notes in the permanent breeding records, although all individuals in contact with the cows should make notes on the "barn" sheets. Use of computer record systems on the farm greatly facilitates reproductive management.

Veterinary Program

Early diagnosis and treatment are essential to rapid recovery of a sick animal. Veterinary service administered at the right time pays dividends, but at the wrong time costs money. The best way to assure optimal health service is to enlist the aid of a qualified veterinarian on a regular basis. In other words, plan to prevent as many health problems as possible through use of a well-planned herd health program (Chapter 21).

The dairyman must provide husbandry, sanitation, and management for the herd, but a veterinarian is essential for providing medical attention. Many specific problems discussed in this chapter require veterinary assistance, and medical advice of the veterinarian as it relates to general dairy herd management also is essential to attain maximal reproductive efficiency and milk production.

Many dairymen and their veterinarians have agreements whereby the veterinarian visits the farm on a regular schedule in addition to emergency calls. During this visit, the veterinarian checks by rectal palpation all cows 30 to 45 days after parturition to determine the degree of uterine involution and state of the ovaries before the animal is rebred. The veterinarian should check cows not exhibiting heat within 40 to 60 days after calving, cows showing aberrant estrous cycles, cows that have been bred 2 or 3 times, as well as virgin heifers ready to breed. Cows inseminated 45 to 60 days previously should be checked to confirm pregnancy; this should be verified again around 90 days after insemination, because about one cow in 10 will lose her calf during this interval.

This regular visit allows the veterinarian to plan, well in advance, group vaccinations, dehorning, and removal of extra teats of calves. Experience indicates that this programmed herd health substantially reduces the number and costs of veterinary emergency calls.

16.8 Summary

Production of one calf per cow per year maximizes milk production in the average dairy cow. Unfortunately this interval is often extended several months, and this delay in conception is costly. Anatomical, physiological, pathological, and management factors affect reproductive efficiency. In all cases of infertility in dairy cattle, the veterinarian is indispensable for prevention, diagnosis, and treatment of the problems. Agreements between the dairyman and the veterinarian whereby the latter checks and treats, if necessary, all calves, heifers, and cows on a regular basis are highly recommended.

The dairyman should provide proper nutrition to heifers for optimal growth and to milking cows for greatest milk production. This practice almost automatically eliminates nutritional causes of infertility. There is no reason why well-grown heifers should not be bred to calve at 24 months of age. Such a practice increases total lifetime milk production. Cows should be observed closely for 30 minutes twice daily for signs of estrus, and if they are to be bred, they should be inseminated sometime between the final 10 hours of standing heat and the first 6 hours after the end of standing heat. Keeping breeding records, providing clean, roomy maternity pens, and use of artificial insemination techniques are all conducive to increased fertility in the dairy herd. If an infertility problem exists in the herd, the dairyman will be well advised to consult with the veterinarian to try to identify the cause and to institute promptly those management practices or specific treatments which minimize infertility problems.

Review Questions

1. What are the most common measures of reproductive efficiency?
2. List factors that may contribute to the decrease in percentage of cows remaining pregnant as gestation advances.
3. Describe briefly the role that nutrition and environment play in affecting reproductive efficiency.
4. What are signs of estrus in cattle?
5. Outline concisely a management program that you believe will minimize infertility in a dairy herd.

References

1. Speicher, J. A., and Meadows, C. E.: Milk production and costs associated with length of calving interval in Holstein cows. J. Dairy Sci. *50:* 975, 1967. (Costs updated to 1984 values.)
2. Bearden, H. J., Hansel, W., and Bratton, R. W.: Fertilization and embryonic mortality rates of bulls with histories of either low or high fertility in artificial breeding. J. Dairy Sci. *39:* 312, 1956.
3. Olds, D.: An objective consideration of dairy herd fertility. J. Amer. Vet. Med. Assoc. *154:* 253, 1969.
4. Spalding, R. W., Everett, R. W., and Foote, R. H.: Fertility in New York artificially inseminated Holstein herds in Dairy Herd Improvement. J. Dairy Sci. *58:* 718, 1975.
5. Foote, R. H.: Reproductive performance and problems in New York dairy herds. Search Agriculture (Animal Science 2), *8:* 1, 1978.
6. Louca, A., and Legates, J. E.: Production losses in dairy cattle due to days open. J. Dairy Sci. *51:* 573, 1968.
7. Kittok, R. J., Britt, J. H., and Convey, E. M.: Endocrine response after GnRH in luteal phase cows and cows with ovarian follicular cysts. J. Anim. Sci. *37:* 985, 1973.
8. Hafs, H. D., Boyd, L. J., Cameron, S., Johnson, W. L., and Hunter, A. G.: Fertility of bull semen with added beta-glucuronidase. J. Dairy Sci. *54:* 420, 1971.
9. Salisbury, G. W., and Hart, R. G.: Gamete aging and its consequences. Biol. Reprod. Suppl. *2:* 1, 1970.
10. Foote, R. H.: Inheritance of fertility—facts, opinions, and speculations. J. Dairy Sci. *53:* 936, 1970.
11. Swanson, E. W.: Optimum growth patterns for dairy cattle. J. Dairy Sci. *50:* 244, 1967.

12. Schultz, L. H.: Relationship of rearing rate of dairy heifers to mature performance. J. Dairy Sci. *52:* 1321, 1969.
13. Spalding, R. W.: Improving dairy cattle reproductive efficiency. Mimeograph of an invitational paper presented at 1976 annual meeting of Amer. Dairy Sci. Assoc., Raleigh, N.C.
14. Boyd, L. J.: Managing dairy cattle for fertility. J. Dairy Sci. *53:* 969, 1970.
15. Reid, J. T.: Effect of energy intake upon reproduction in farm animals. J. Dairy Sci. Suppl. *43:* 103, 1960.
16. Reid, J. T., Loosli, J. K., Trimberger, G. W., Turk, W. L., Asdell, S. A., and Smith, S. E.: Causes and prevention of reproductive failure in dairy cattle: IV. Effect of plane of nutrition during early life on growth, reproduction, production, health, and longevity of Holstein cows. Cornell Univ. Agr. Exp. Sta. Bull. 987, 1964.
17. Tucker, H. A.: Seasonality in cattle. Theriogenology *17:* 53, 1982.
18. Gangwar, P. C., Branton, C., and Evans, D. L.: Reproductive and physiological responses of Holstein heifers to controlled and natural climatic conditions. J. Dairy Sci. *48:* 222, 1965.
19. Wiersma, F., and Stott, G. H.: New concepts in the physiology of heat stress in dairy cattle of interest to engineers. Trans. Amer. Soc. Agric. Eng. *12:* 130, 1969.
20. Ulberg, L. C., and Burfening, P. J.: Embryo death resulting from adverse environment on spermatozoa or ova. J. Anim. Sci. *26:* 571, 1967.
21. Thatcher, W. W.: Effects of season, climate, and temperature on reproduction and lactation. J. Dairy Sci. *47:* 360, 1974.
22. Salisbury, G. W.: Aging phenomena in spermatozoa. III. Effect of season and storage at −79 and −88C on fertility and prenatal losses. J. Dairy Sci. *50:* 1683, 1967.
23. Morrow, D. A.: Diagnosis and prevention of infertility in cattle. J. Dairy Sci. *53:* 961, 1970.
24. Foote, R. H.: Estrus detection and estrus detection aids. J. Dairy Sci. *58:* 248, 1975.
25. Britt, J. H.: Strategies for managing reproduction and controlling health problems in groups of cows. J. Dairy Sci. *60:* 1345, 1977.

Suggested Additional Reading

McDonald, L. E.: *Veterinary Endocrinology and Reproduction.* 3rd ed. Philadelphia: Lea & Febiger, 1980.

Salisbury, G. W., VanDemark, N. L., and Lodge, J. R.: *Physiology of Reproduction and Artificial Insemination of Cattle.* 2nd ed. San Francisco: W. H. Freeman and Co., 1978.

CHAPTER 17

Anatomy and Physiology of the Mammary Gland

17.1 Introduction

Mammary glands are a distinguishing characteristic of all mammals. Mammary glands are modified skin glands, classified as exocrine glands, whose function is to secrete milk for nourishment of young for various periods of postnatal life. These glands grow during pregnancy and commence to secrete milk after parturition. Many hormones that control reproduction also regulate the mammary gland. Thus, mammary development and lactation are integral parts of reproduction.

Since large numbers of mammary secretory cells and a high rate of metabolism are essential for high milk production, the purpose of this chapter is to describe the anatomy of the cow's mammary gland, its normal growth and development during the various phases of reproduction, and endocrine control of these processes.

17.2 External Features of the Mammary Gland

The udder of the cow consists of 4 separate mammary glands. Right and left halves of the udder are clearly separated, whereas front and rear quarters seldom show externally any clear demarcation. When viewed from the side, the floor of the udder should be level, extend anteriorly and be strongly attached to the abdominal body wall. Attachment at the rear should be high and wide and the individual quarters should be symmetrical. These external features contribute to lifetime productivity and constitute important criteria used to judge dairy cattle in the show ring and for breed classification scores.

The cow's udder weighs between 25 and 60 lb or more, exclusive of milk. Udders should be of sufficient size to produce large amounts of milk, but not so large as to weaken attachment to the cow's body. Udder weight is correlated with milk production.[1] Normally rear quarters are larger than forequarters and on average secrete about 60% of the daily milk yield.

Milk from each gland is emptied through a teat. Rear teats are usually shorter than fore teats. Generally cows with long teats require more time to milk than cows with short teats. The most important characteristics of teats for efficient milking are (1) moderate size; (2) proper placement; and (3) enough tension on the sphincter muscle around the teat orifice to allow easy milking while preventing leakage of milk between milkings.

Between 25 and 50% of all cows have extra (supernumerary) teats. They may or may not be directly connected to mammary tissue in the interior of the udder. Extra teats should be removed during calfhood for appearance and to remove a potential route for mastitis-causing organisms to enter the udder.

17.3 Internal Features of the Mammary Gland

The udder is made up of a series of systems which includes the supportive structures; blood, lymph, and nerve supplies; a duct system for storage and conveyance of milk; and secretory units of the epithelial cells, which are arranged into hollow spherical structures called alveoli. Each of these components contributes directly or indirectly to synthesis of milk.

Supportive Structures

Skin. The skin provides little support for the udder, but it protects the interior of the gland from abrasion and bacteria. A fine connective tissue attaches the skin to the udder, and a coarse connective tissue attaches the forequarters to the wall of the abdomen. Excessive udder weight or weakness in this coarse connective tissue leads to a separation between the udder and abdominal wall.

Lateral Suspensory Ligaments. The lateral suspensory ligaments are one of the main supporting structures of the udder (Figure 17.1). These ligaments are chiefly fibrous, nonelastic tissue and they arise from tendons well above and posterior to the udder. The lateral suspensory ligaments extend along both sides of the udder and at intervals send sheets of tissue into the gland to support the interior of the udder. The lateral suspensory ligaments extend to the midline on the floor of the udder where they fuse with the median suspensory ligament.

Median Suspensory Ligament. This ligament is the major supporting structure of the udder. It is composed of elastic tissue that arises from the midline of the abdominal wall and extends between the udder halves joining the lateral suspensory ligaments at the base of the udder (Figure 17.1). Thus, each half of the udder is suspended in a sling of connective tissue. Elasticity of the median suspensory ligament is necessary to allow the udder to increase in size as it fills with milk. Since the lateral ligaments are nonelastic and median ligaments are elastic, the teats of some cows protrude at an oblique angle when the udder is distended with milk. Repeated and excessive stretching of the median suspensory ligament may cause the udder to become pendulous and more susceptible to injury and mastitis. Consideration in selection programs should be given to strong udder attachments because of their negative genetic correlation with milk yields (Chapter 6).

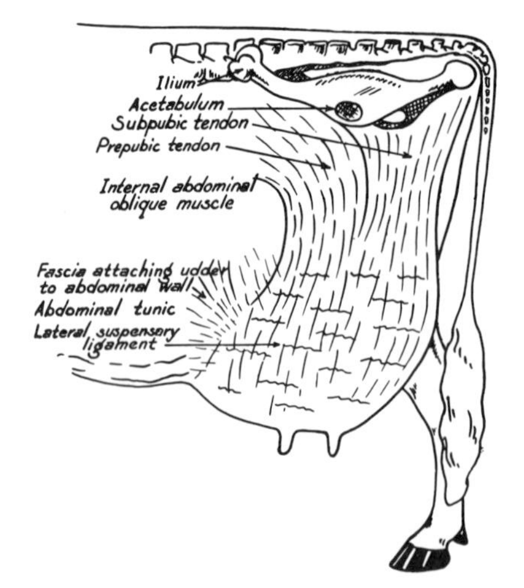

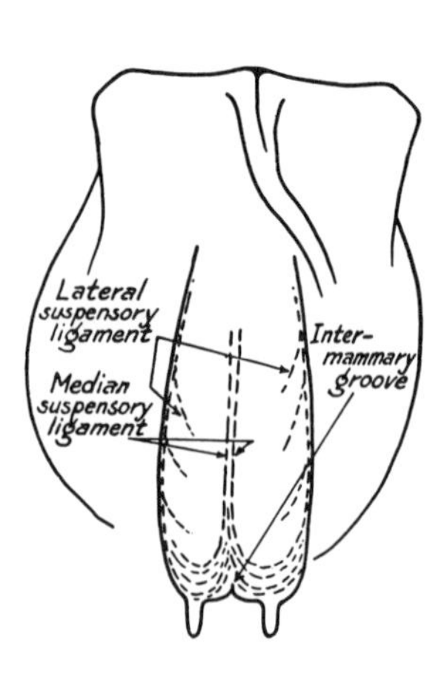

FIG. 17.1 Suspensory apparatus of the udder. (Reprinted with permission from *Physiology of Lactation*, 1959)

Vascular System

Oxygenated blood leaves the heart via the aorta, and through a series of smaller arteries blood is conveyed to the udder from the two external pudendal (pudic) arteries (Figure 17.2). These arteries penetrate the abdominal wall through the inguinal ring, one entering the right and the other entering the left half of the udder. Upon entering the udder, the 2 arteries become the mammary arteries, which divide almost immediately into cranial and caudal branches. These cranial and caudal mammary arteries divide many times to form capillaries, the small blood vessels which deliver blood to cells of the fore and rear quarters of the udder. The perineal arteries (Figure 17.2) are a minor source of blood to the udder.

Venules originate from the capillaries and anastomose with each other to form veins that drain blood from the udder. At the top of the udder, veins converge to form the venous circle (Figure 17.3) where blood can leave the udder by one of 2 routes. One route consists of the 2 external pudendal (pudic) veins which parallel the external pudendal arteries through the inguinal ring and eventually join the vena cava which delivers blood back to the heart. The second route consists of 2 veins (the subcutaneous abdominal or milk veins) which emerge at the anterior edge of the udder. These veins run forward along the ventral abdominal body wall just under the skin. They penetrate the thoracic cavity at the "milk wells" and eventually join the anterior vena cava to enter the heart. The perineal veins join the udder, but because of the valvular arrangement in these vessels, blood is delivered to the udder from the reproductive tract area.

When a cow is in the standing position, most blood returns to the heart via the milk veins. But when a cow lies down and shuts

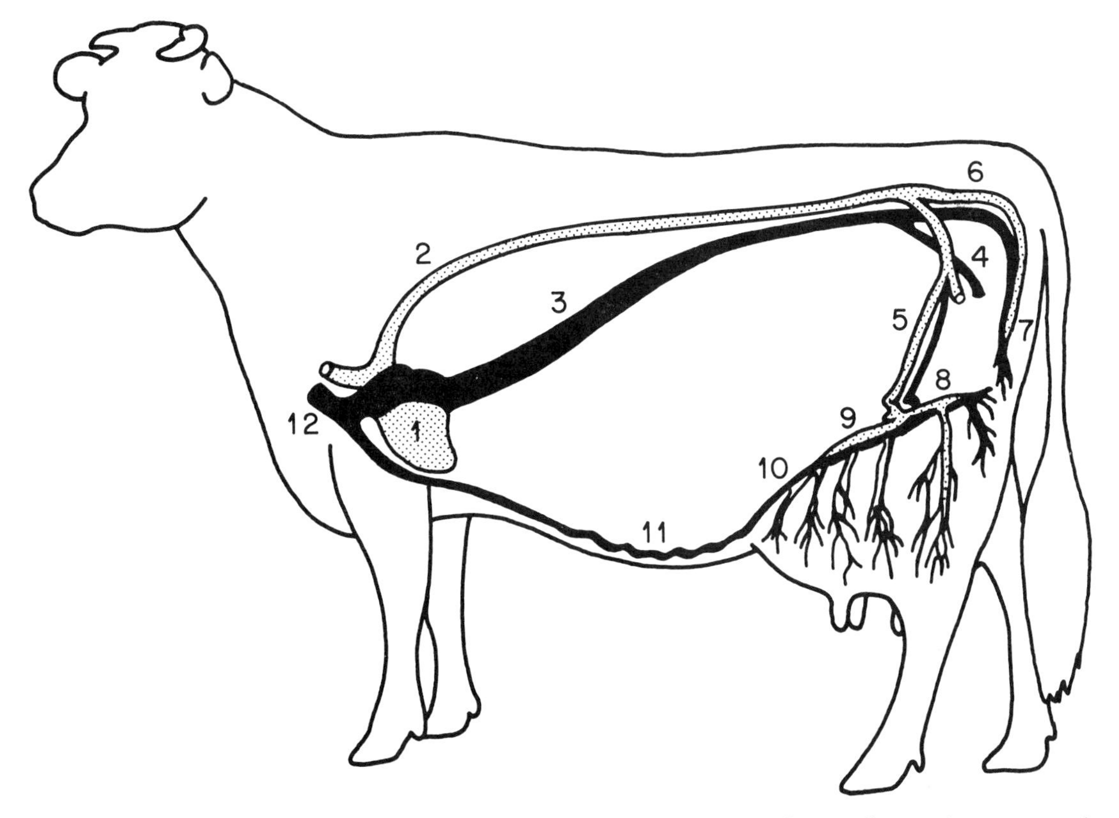

Fig. 17.2 Arterial (stippled) and venous (black) systems of the udder. 1, heart; 2, aorta; 3, posterior vena cava; 4, external iliac artery and vein; 5, external pudendal artery and vein; 6, internal iliac artery and vein; 7, perineal artery and vein; 8, caudal mammary artery; 9, cranial mammary artery; 10, cranial aspect of venous circle; 11, subcutaneous abdominal vein; 12, anterior vena cava.

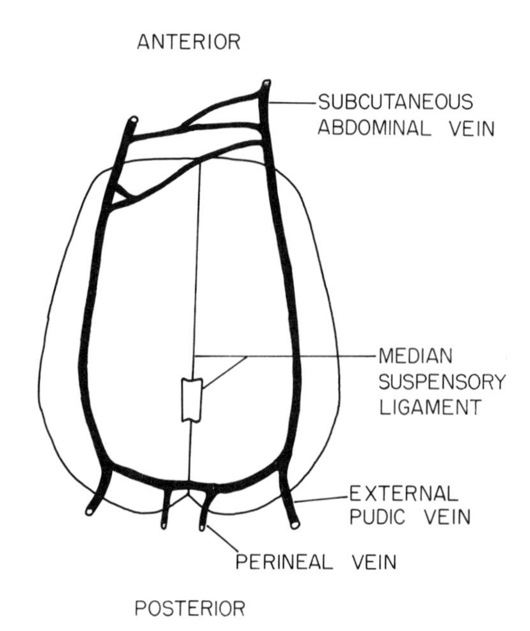

FIG. 17.3 Dorsal view of venous circle at the base of the udder.

off blood flow through the milk veins, milk production is not affected because of the other possible exits for blood.

There is an increased blood flow (averaging about 180%) to the udder in the first few days post-partum.[2] This increased mammary blood flow may be associated with the decrease in uterine blood flow following calving, and it may play an important role in initiating milk secretion by bringing increased amounts of milk precursors and lactogenic hormones to the udder. About 500 volumes of blood flowing through the udder is needed to produce one volume of milk. The rate of blood flow gradually decreases after the peak of lactation parallel to the decrease in milk production. Blood flow rate is an important determinant of milk production in dairy cows.

Lymphatic System

Lymph is a colorless tissue fluid drained from tissue spaces by thin-walled lymph vessels. Lymph originates as a filtrate of blood serum and is similar in composition to blood except that lymph has no red blood cells and about one-half the protein concentration. The major lymph vessels and nodes in the cow are shown in Figure 17.4.

Lymph flows from the udder to the thoracic duct and is eventually discharged into the anterior vena cava. Around the time of parturition, filtration of lymph out of the blood capillaries in the udder may exceed drainage back into blood. This causes an accumulation of fluid in the intercellular tissue spaces which is called edema. Severe edema of the udder occurs in 18 to 28% of cows around parturition, and repetitive occurrence may contribute to pendulous udders and increased development of connective tissue.[3] The swelling is especially noticeable in first-calf heifers and in high-producing cows. More severe udder edema occurs in autumn-winter than in spring-summer, although it is unassociated with feeding practices. Milking before parturition is usually not greatly beneficial in alleviating the swelling. If a cow is milked

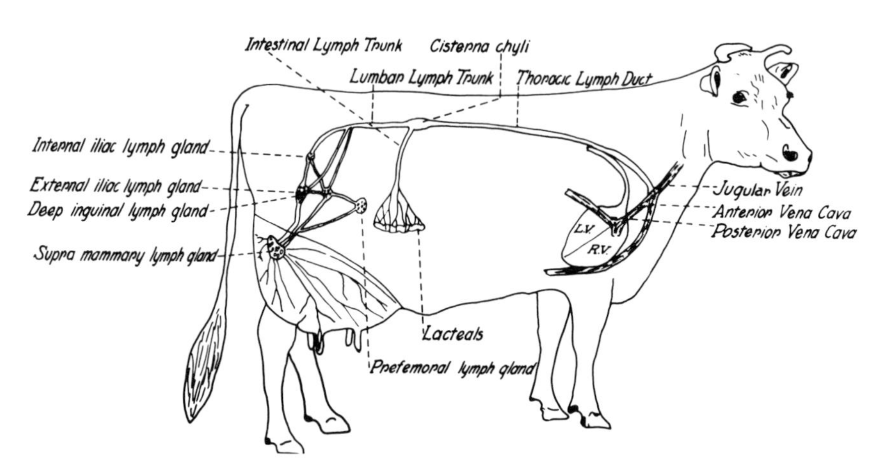

FIG. 17.4 Course of the lymph ducts from the udder to the heart. (Reprinted with permission from *The Mammary Gland,* 1952)

prepartum, colostrum should be frozen for subsequent feeding to the calf because prepartum milking eliminates colostrum after calving. Low salt intakes and use of diuretics, especially if combined with glucocorticoids, are useful in treatment of udder edema.

Lymph nodes of the udder (Figure 17.4) plus other lymph nodes scattered throughout the body are important for disease resistance in cows. Lymph nodes form lymphocytes, a type of white blood cell that is involved in immunity. The nodes also remove bacteria and other foreign material. In response to infections from mastitis, lymph nodes increase their output of lymphocytes into lymph vessels which eventually discharge lymphocytes into the anterior vena cava. Lymphocytes are then transported to the udder to combat infection.

Nervous System

Innervation of the udder consists of afferent sensory and efferent sympathetic nerve fibers (Figure 17.5). The efferent (motor) nerves to the udder automatically regulate blood flow and innervate the smooth muscles which surround the milk-collecting ducts and teat sphincter. Excitation of the cow causes the sympathetic system to discharge the neurohormone epinephrine, which constricts the blood vessels of the udder and reduces milk production. Touch, temperature, and pain receptors are found in the skin of the cow's teat, and washing activates these receptors, thereby initiating impulses that evoke discharge of oxytocin from the posterior pituitary and other hormones from the anterior pituitary. Elimination of the nerve supply to the udder of the cow through surgical section of the various nerves, however, does not significantly affect udder development or milk yields. In fact, complete denervation and transplantation of the udder to a remote site on the body of a goat (Figure 17.6) does not reduce milk yield if the blood supply is restored.[4] Thus, mechanisms other than direct innervation of the udder permit normal secretion of milk.

Mammary Duct System

The milk duct system of the udder consists of a series of drainage channels beginning at the alveoli and ending at the streak canal (Figure 17.7). Bands of connective tis-

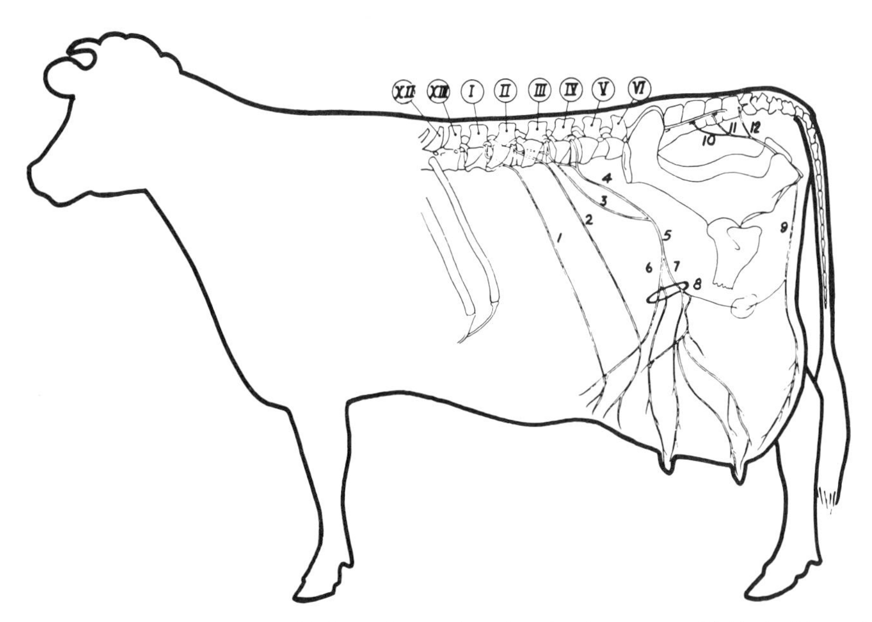

Fig. 17.5 Nerve supply to the udder. (1) Ventral branch of first lumbar nerve. (2) Ventral branch of second lumbar nerve. (3) Ventral component of inguinal nerve. (4) Dorsal component of inguinal nerve. (5) Inguinal nerve. (6) Anterior inguinal nerve. (7) Posterior inguinal nerve. (8) External inguinal ring. (9) Perineal nerve. (10) Second sacral origin of pudic nerve. (11) Third sacral origin of pudic nerve. (12) Fourth sacral origin of pudic nerve. (Reprinted with permission from *Physiology of Lactation,* 1959)

FIG. 17.6 Photograph of goat Hilda with left mammary gland transplanted to the neck. The gland is supplied by the left carotid (exteriorized, visible over black paper) and left jugular vein (visible where it emerges from the gland in front of carotid loop). At the time of the photograph the gland was giving 1,400 ml of milk daily. (Reprinted with permission from Quart. J. Exp. Physiol. *48:* 34, 1963)

sue radiate throughout the mammary glands to support the ductular and secretory tissue. However, excess amounts of connective tissue produce a hard or "meaty" udder which, for its size, does not produce corresponding quantities of milk. A desirable udder should contain maximal amounts of secretory and ductular tissue and minimal amounts of connective tissue.

Teat. The teat is covered with hairless skin which does not contain either sweat or sebaceous glands. At the base of the teat is the streak canal through which milk must be drawn to the exterior. The streak canal is usually 8 to 12 mm long and is lined with cells that form a series of folds that close the streak canal between milkings. These cells produce a lipid-like secretion that is bacteriostatic. Without this secretion, bacteria more easily penetrate the streak canal and cause mastitis. The streak canal is also closed by involuntary circular muscle (teat sphincter muscle). Tightness of the sphincter muscle controls ease and speed of milking and is an inherited trait (Chapter 6). Cows with patent (loose) streak canals milk faster, but they are also more susceptible to udder infections and tend to leak milk between milkings. The teat cistern is located immediately above the streak canal. The wall of the teat receives an extensive blood supply. These blood vessels must not be constantly constricted during milking by the vacuum of the milking machine because this causes blood congestion and irritation to the delicate lining of the teat cistern. As discussed in Chapter 19, the pulsating action of the teat cup liners of mechanical milking machines is designed to minimize blood congestion. The upper walls of the teat may contain a number of small areas of accessory mammary tissue which secrete milk. Irritation of these tissues by improper milking can spread an existing localized infection to the rest of the udder. This occurs most frequently when teat cups crawl up the teat after the cow has been milked dry. Prompt removal of the machine when milking is finished prevents this problem.

Gland Cistern. The teat cistern joins the gland cistern at the base of the udder, and in many cows there is a circular (cricoid) fold of tissue between the 2 cisterns (Figure 17.7). In rare instances this fold may completely separate the 2 cisterns and milk cannot be removed from the gland. Such a condition results in a nonfunctional quarter unless the obstruction is surgically removed. The gland cistern serves as a limited storage space for milk as it trickles down from the secretory tissue. On average the gland cistern holds about a pint of milk, but actual capacity varies markedly among cows. However, size of the gland cistern does not significantly affect milk production.

Mammary Ducts. Branching off the gland cistern are 12 to 50 or more ducts, which in turn branch many times and finally form a terminal ductule which drains each alveolus. There are 2 layers of cells lining the major mammary ducts (Figure 17.8a) but these cells do not secrete milk. The larger ducts function only as a storage space and drainage channel for milk.

Alveoli. The microscopic terminal ductules and alveoli are composed of a single layer of epithelial cells (Figures 17.8b, c

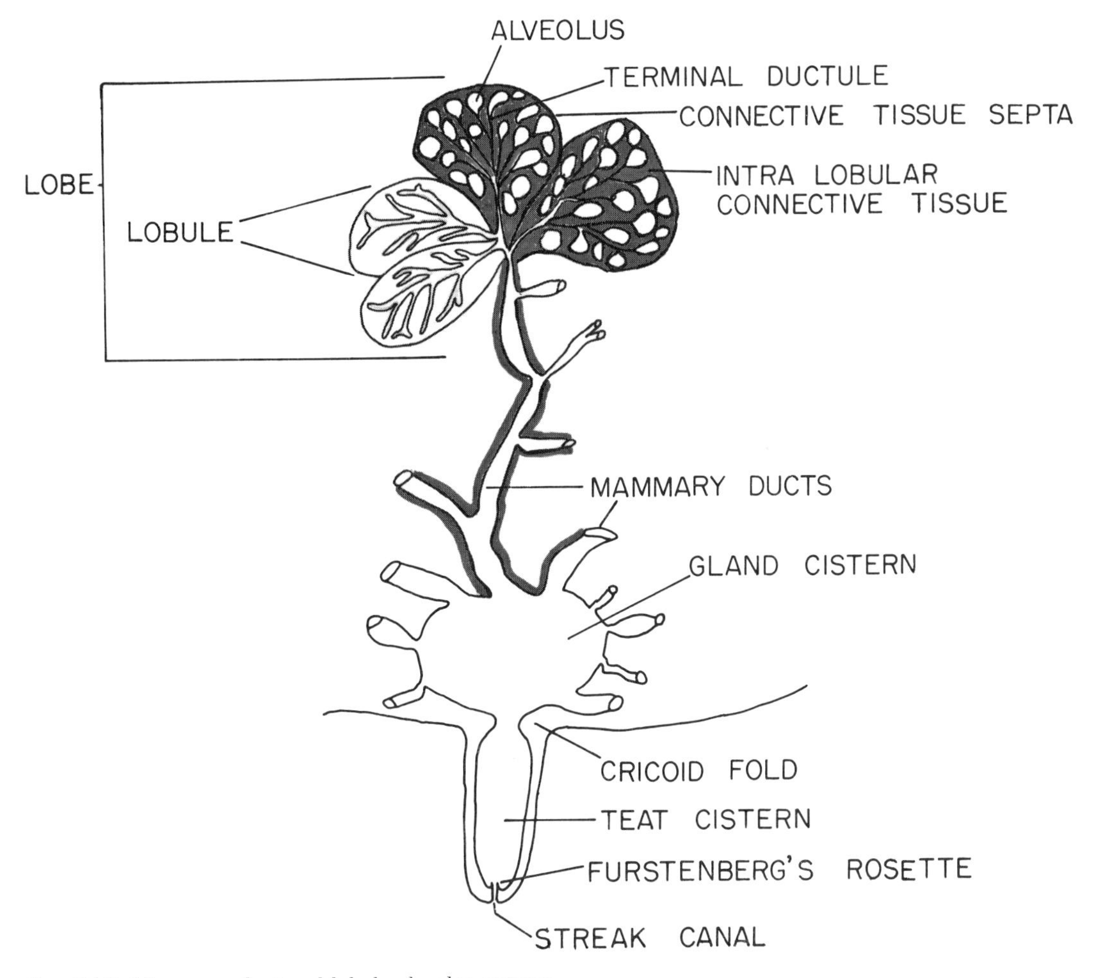

Fig. 17.7 Mammary duct and lobule-alveolar systems.

and 17.9). The function of cells which form these structures is to remove nutrients from the blood, transform them into milk, and discharge the milk into the lumen of each alveolus. The structure of the terminal ductules and alveoli varies with stages of pregnancy, lactation, and mammary involution. In the fully developed condition during lactation, several alveoli are grouped into lobules; several lobules are joined into lobes, which are visible to the naked eye. This pattern of mammary development in the cow is termed lobule-alveolar development. Bands of connective tissue envelop the lobules and lobes (Figure 17.7). Alveoli are attached to delicate connective tissue fibrils which become more obvious microscopically as secretory cells are lost during advancing lactation. Surrounding each alveolus is a capillary network (Figures 17.8b and 17.9) which supplies nutrients and hormones for milk synthesis and removes waste products from the alveolar cells. Also, a network of specialized muscle cells, the myo-epithelial cells, envelopes each alveolus. The myo-epithelium contracts in response to oxytocin forcing milk from the lumen of the alveolus into the ducts and gland and teat cisterns. The milk-ejection reflex is discussed in Chapter 19.

17.4 Normal Mammary Growth and Development

The number of milk-synthesizing cells is a major factor that determines level of milk production.[5] Estimates of the correlation between milk yield and mammary cell numbers range between 0.50 and 0.85. The stages of mammary development discussed in the following paragraphs are divided

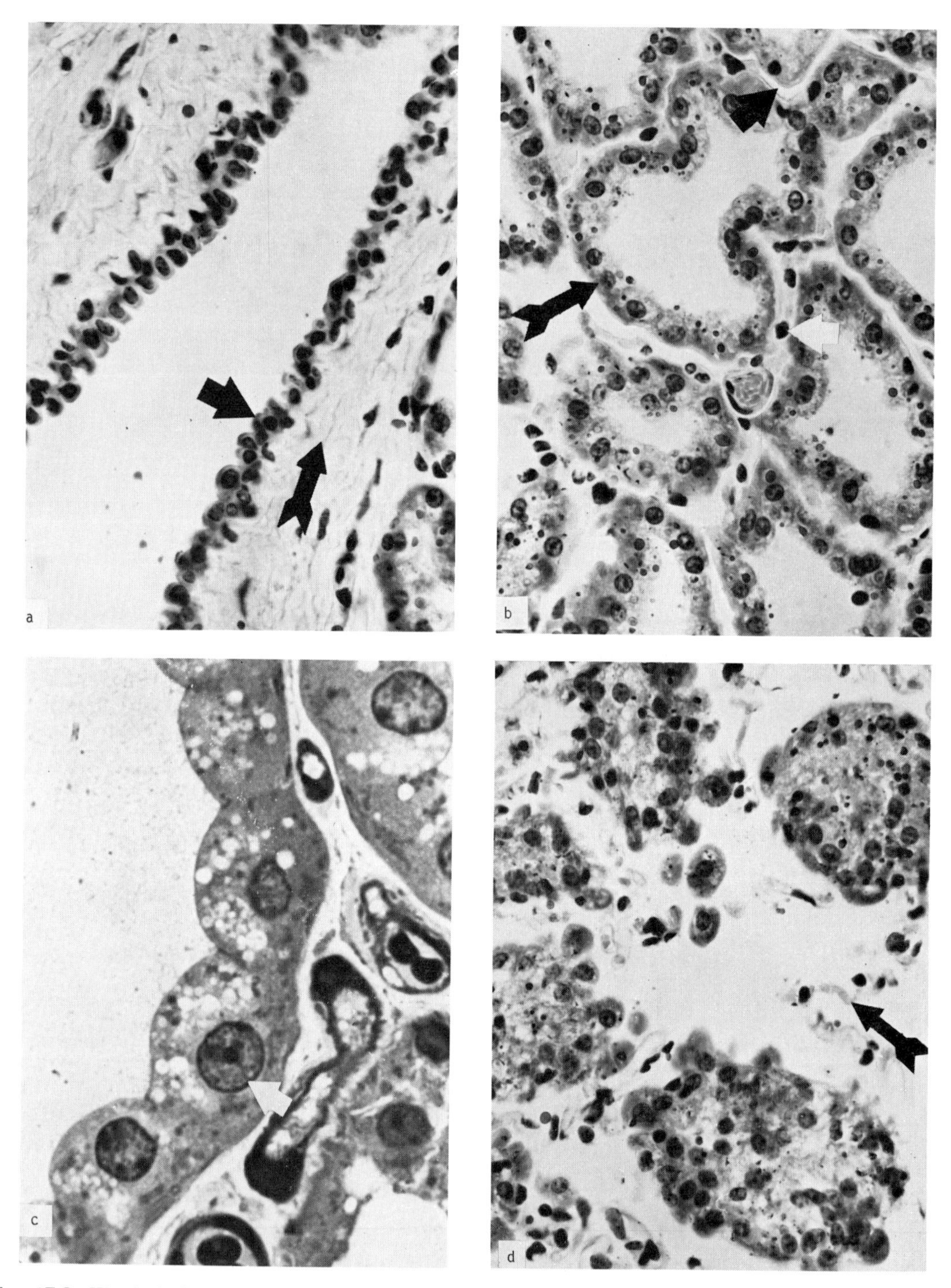

FIG. 17.8 Histological appearance of cow mammary tissue. a. Interlobular duct. 675 ×. Note two cell layers comprising duct wall at short arrow. Large amount of connective tissue (at long arrow) surrounds this duct. b. Fully lactating alveoli. 675 ×. A nucleus of a single epithelial cell, which together with many other epithelial cells make up an alveolus, is at the long black arrow. Note almost complete absence of connective tissue. Capillaries (at short black arrow) and myoepithelial cells (dark nuclei at white arrow) are scattered between individual alveoli. c. Higher magnification of fully lactating alveoli. 2000 ×. Nucleus of individual epithelial cell is at white arrow. Milk fat droplets (appearing as white holes) are located between the nucleus of the cells and the lumen of the alveoli. d. Involuting mammary tissue. 675 ×. Note infiltration of connective tissue (arrow) and absence of alveoli. (Courtesy of R. G. Saacke)

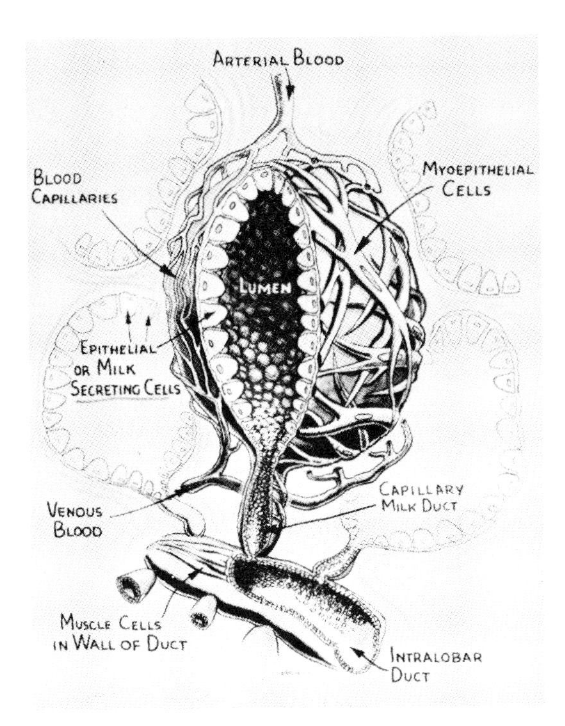

Fig. 17.9 Blood vessels and myo-epithelial cells surrounding an alveolus. (Reprinted with permission from *The Mammary Gland,* 1952)

into various reproductive epochs starting with embryonic development and ending with involution during the dry period.

Embryonic and Fetal Development

The first discernible rudiment of the mammary gland is a thickening of ectodermal cells on the ventral surface of the embryo between the rear legs. This development occurs when the calf is only 1.4 to 1.7 cm long (approximately 30 days after conception). These cells grow and aggregate to form 2 mammary lines on each side of the midline. Distinct areas on this line further differentiate to form 4 structures called mammary buds. Mammary buds are the anlagen of the secretory portion of the mammary gland. Thus, the number of mammary buds determines the number of mammary glands that will develop in the cow. Each bud elongates above and below the surrounding surface of the fetus and upon further cell division gives rise to the primary sprout which is the forerunner of the teat and gland cisterns (Figure 17.10). Several secondary sprouts, which represent future mammary ducts, arise from the

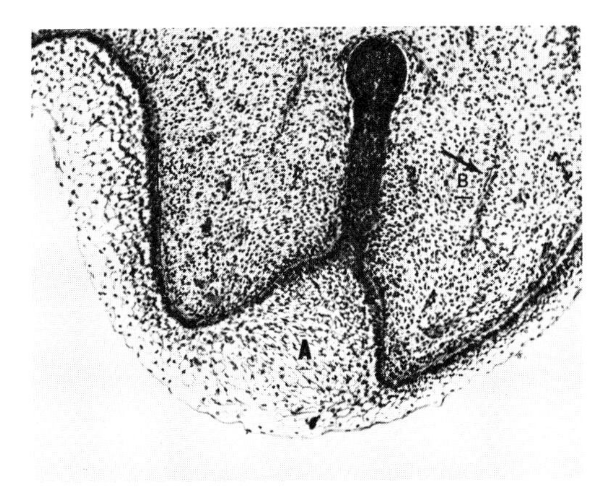

Fig. 17.10 Primary sprout of mammary gland. The primary sprout consists of a solid core of rapidly dividing cells (shown as dark staining cells in the center of this illustration) which invade the embryonic connective tissue (B). Primitive vascular system is under development at arrow. Remnants of the mammary bud (A) have undergone complete cornification. (From Schalm, O. W., Carrol, E. J., and Jain, N. C.: *Bovine Mastitis.* Philadelphia, Lea & Febiger, 1971)

primary sprout. The sprouts become hollow shortly before birth of the calf. The nonglandular portion of the udder (connective tissue and fat pad) is in a rather mature form at birth whereas the glandular portion is very rudimentary. Nonetheless, future milk production is affected by development that occurs before birth. Blind quarters, for example, in first-calf heifers are sometimes the result of aberrant embryonic development of the udder.

Birth to Puberty

Research on the quantitative assessment of mammary development in the cow is far from complete. The rat has been studied more extensively, and as nearly as can be determined, the changes in mammary cell numbers [as measured by deoxyribonucleic acid (DNA) content] in the cow parallel changes observed in the rat during various physiological states. These changes in the rat are shown in Figure 17.11.

Until 3 months of age, the immature mammary duct system of the calf grows into the surrounding fat pad of the udder at a rate which is proportional to increases in body weight. After 3 months, growth of the gland shifts to a rate that is about 3.5 times faster than body growth.[6] This rapid growth continues until the 9th month of

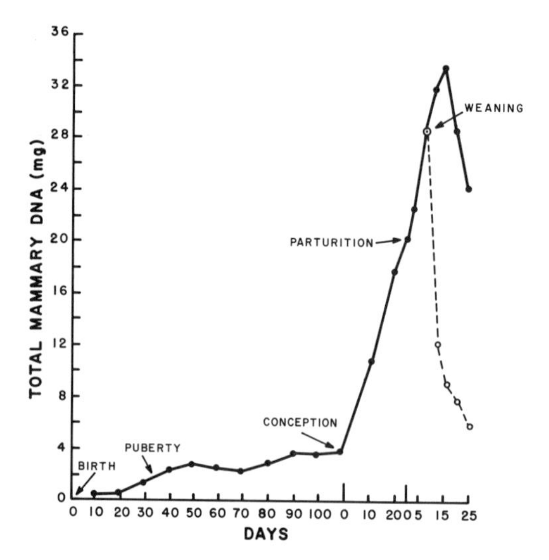

Fig. 17.11 Mammary development in rats from birth through lactation and involution. (Reprinted with permission from J. Dairy Sci. *52:* 721, 1969)

age. Cells of mammary ducts accumulate during the first 3 to 5 estrous cycles after puberty. Most of this growth is initiated during the estrogenic phase of the estrous cycle, but there is a slight decline in cell numbers during the progestational phase.[6] Between 9 months of age and conception, growth and regression of the mammary gland during the estrous cycle reach an equilibrium, and net increases in mammary cell numbers parallel body weight increases. Since a major portion of mammary duct growth before conception is completed by 9 months of age, perhaps dairymen should consider breeding well-grown heifers earlier than the current practice of breeding at 15 to 18 months of age (Chapter 16).

Attempts have been made to predict future milk yields from palpation scores of udder development in calves.[7] Correlation coefficients between milk yield and these subjective evaluations of udder development are positive, but are low. Development of a repeatable method of accurately predicting future milk production would allow early culling of cattle without actual milk production records.

During Pregnancy

Alveoli are not formed in heifers until pregnancy is established. Then, alveoli commence to replace fat tissue throughout the udder. In fact, the extent of the fat pad sets outer limits for mammary gland growth. Mammary growth rates are rapid throughout pregnancy (Figure 17.11), but because of relatively small size of the glands in heifers at the time of conception, udder growth is not conspicuous until after 3 to 4 months of pregnancy. Significant quantities of secretion begin to accumulate in the alveoli between the 7th and 9th month of gestation, and most visible growth of the udder during the last month of gestation is due to accumulation of these secretions.

During Lactation

Mammary cell numbers continue to increase during early lactation. This development probably continues until the peak of lactation. As a result, alveoli are closely packed together during early lactation (Figure 17.8b). Thereafter, the rate of mammary cell loss exceeds the rate of cell division. The result is that the udder contains substantially fewer cells at the end of lactation than at the beginning. Mastitis also causes loss of cells from the udder. Naturally, loss of secretory cells, whether from physiological or pathological causes, lowers milk production. However, it is usually easier to maintain mammary cell numbers than to maintain their rate of metabolism.[5] Maintenance of a mammary cell during lactation does not necessarily indicate that it can continue to synthesize milk at a sustained maximal rate. Nonetheless, maintenance of maximal numbers of mammary cells is conducive to high milk production because if cells are absent, no milk at all can be synthesized.

During Concurrent Lactation and Pregnancy

Since most cows are bred within 40 to 90 days post-partum, a major portion of lactation coexists with pregnancy. Initial stages of pregnancy have relatively little effect on milk production or mammary cell numbers. But as pregnancy advances past 5 months, milk yields and mammary cell numbers decrease in comparison with lactating animals which are not pregnant.

During the Dry Period

Daily milking is usually stopped after the dairy cow has been lactating for about 10 to 12 months (with a range of 6 to 18 months). If the cow is pregnant, this period of nonlactation (dry period) is initiated usually about 60 days before the expected date for parturition. Following cessation of daily milking, the udder in the nonpregnant cow becomes engorged with milk for a few days, but metabolic activity of the cells rapidly declines. Subsequently, there is a marked degeneration and loss of alveolar epithelial cells. Although alveoli are lost, myo-epithelial cells and connective tissue remain. Histologically, the connective tissue and fat cells become more prominent during this period (Figure 17.8d). After complete involution of the udder only the duct system remains. The duct system of the multiparous cow, however, is more extensive than that of the virgin heifer. Although studies in dairy cattle have not been reported, complete involution of the alveoli requires 75 days in the nonpregnant goat.[8]

Cows are normally pregnant during the dry period, and because pregnancy stimulates udder growth, complete involution will not occur in pregnant cows. If the cow is pregnant at least 7 months at the beginning of the dry period, mammary cell numbers do not change appreciably during the dry period.[9] Cows which are not given a normal dry period produce less milk in the subsequent lactation than cows given a rest of 60 days between lactations. Thus, the dry period between lactations is essential for maximal milk production. Absence of a dry period interferes in some manner with the increase in cell numbers that occurs during early stages of the subsequent lactation. This may partially explain the need for a dry period in cattle.

17.5 Hormonal Control of Mammary Development

Significant mammary development does not occur in absence of certain hormones. In general, hormones that stimulate udder growth are the same hormones that regulate reproduction. Thus, most mammary growth occurs only during certain significant events of reproduction; i.e., during puberty, pregnancy, and for a short time following calving (Figure 17.11).

Ovary

Hormones of the ovary stimulate udder development during puberty and pregnancy. Thus, removal of ovaries before puberty abolishes the normal increase in mammary development.

The specific ovarian hormones involved in the udder growth response are estrogen and progesterone.[10] Estrogens stimulate mammary duct growth whereas a combination of estrogen and progesterone is needed to achieve lobule-alveolar development (Figure 17.12).

Anterior Pituitary

Hormones from the anterior pituitary are needed for mammary growth. For example, the mammary gland atrophies following hypophysectomy, and restoration of mammary development occurs by replacing growth hormone and prolactin (Figure 17.12). In the absence of the pituitary, ovarian steroids fail to cause mammary growth. The present concept is that under normal physiological conditions, anterior pituitary hormones synergize with ovarian hormones (estrogen and progesterone) to produce udder development.

Bovine Placental Lactogen

The placenta is a source of estrogen and bovine placental lactogen. The structure of bovine placental lactogen is similar but larger than prolactin and growth hormone. Bovine placental lactogen probably synergizes with anterior pituitary and ovarian hormones to cause mammary development during pregnancy.

Adrenal and Thyroid

Administration of adrenal glucocorticoids and thyroxine enhance mammary development (Figure 17.12). However, these effects probably are related to their general metabolic functions and are not of primary importance in inducing mammary growth.

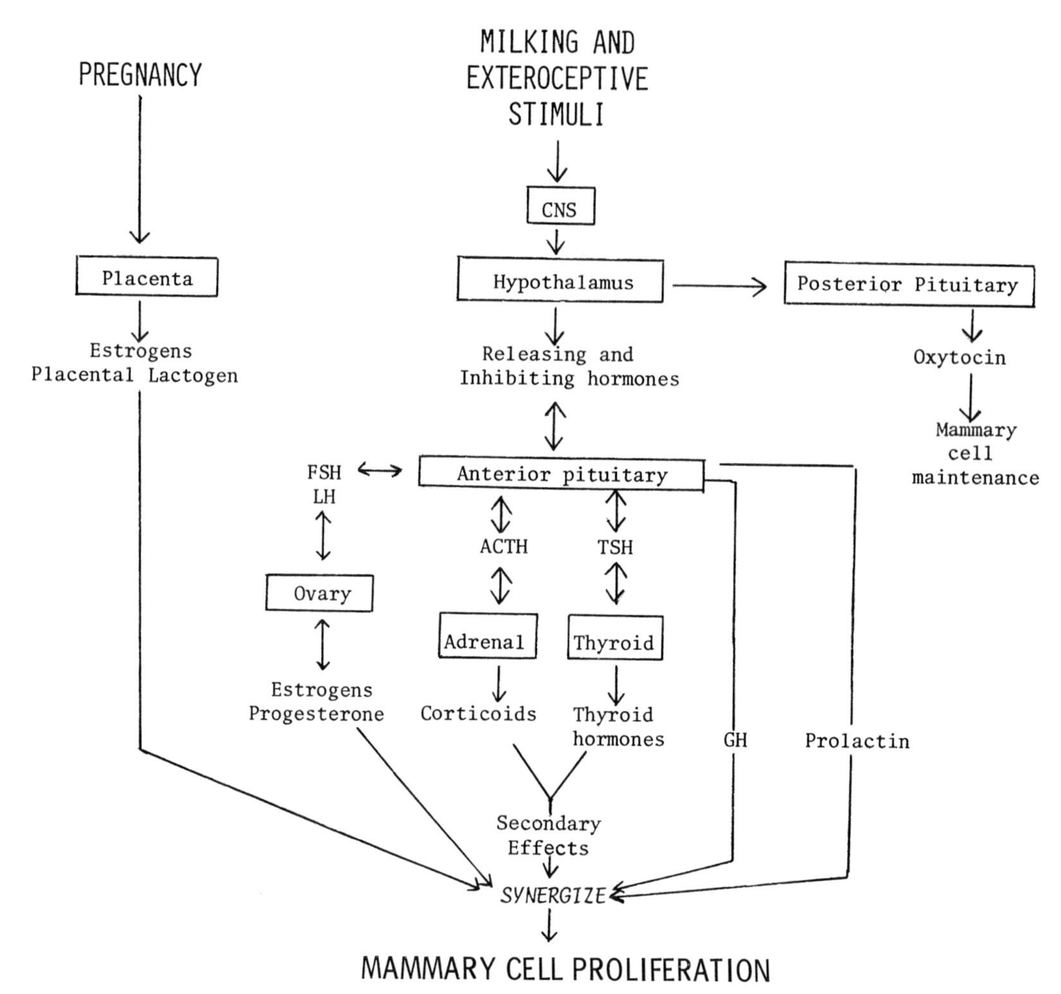

Fig. 17.12 Summary of known factors affecting mammary cell proliferation. CNS, central nervous system; FSH, follicle-stimulating hormone; LH, luteinizing hormone; ACTH, adrenocorticotropic hormone; TSH, thyroid-stimulating hormone; GH, growth hormone. (Modified from J. Dairy Sci. *52:* 721, 1969)

Interaction of Hormones and Nutritional Status

Overfed heifers or, paradoxically, heifers severely restricted in feed intake have markedly reduced development of the udder. Thus, overfed or severely underfed heifers produce significantly less milk than heifers raised on recommended nutrient allowances.

17.6 Hormonal Control of Lactation

Appreciable mammary secretions are produced only after formation of the lobule-alveolar system. Thus, in the pregnant heifer secretion does not commence until the second half of gestation. Many of the enzymes needed for milk synthesis are present within cells of the udder well in advance of calving.[11] At parturition hormones cause a tremendous increase in secretion of milk. Secretions formed prepartum are colostral in nature and not true milk. Subsequent secretion of large amounts of milk is dependent upon preexisting and extensive development of the udder.

Initiation of Lactation

During pregnancy progesterone blocks secretion of α-lactalbumin, a milk protein that is also an enzyme needed for synthesis of lactose, the sugar found in milk. This block is sufficient to inhibit milk synthesis during most of the gestation period of the heifer. Also, high titers of progesterone inhibit initiation of lactation in multiparous cows during the dry period. The progesterone block is ineffective during concurrent lactation and pregnancy; otherwise, lactation would be inhibited as soon as the lactating cow conceived.

Shortly before parturition, progesterone titers fall (which removes the block) and estrogen, ACTH (which stimulates adrenal corticoid secretion), and prolactin levels increase. Administration of adrenal corticoids or estrogen initiates lactation in dairy cows. Although not tested in cows, prolactin will initiate lactation in some but not all species. Use of these hormones to induce lactation in cows is discussed in Section 17.7.

Neural stimuli are not normally required to initiate lactation. However, the milking stimulus (which sends neural impulses to the hypothalamus and pituitary) initiates lactation in heifers during late gestation. This prepartum milking probably causes release of prolactin and ACTH (and adrenal corticoids) which actually initiate lactation. Generally, prepartum milking is not recommended.

Maintenance of Lactation

After parturition in cows, there is a rapid increase in milk yield which reaches a maximum in 2 to 6 weeks. Then milk yield gradually declines. The degree of maintenance of milk production is called persistency. After maximum yield is reached, the decline in milk production each month can be calculated as a percentage of the previous month's production, and this percentage is a measure of persistency. Dairy cows normally lactate 10 to 12 months, but there is a report of one cow that lactated continuously for 5 consecutive years.

The following terminology will be used to describe lactation: *Milk secretion* involves intracellular synthesis of milk and subsequent passage of milk from the cytoplasm into the alveolar lumen. *Milk removal* involves passive withdrawal of milk from teat and gland cisterns and major ducts and active ejection of milk caused by contraction of myo-epithelial cells around the alveolus in response to oxytocin. *Lactation* includes both milk secretion and milk removal. The term *galactopoietic* pertains to the ability of certain factors to enhance an established lactation.

Milking Stimulus. To support intense lactation, the number of secretory cells, their metabolic activity, and an efficient milk ejection reflex must be maintained. The milking stimulus is important for maintenance of mammary structure and lactation. For example, if milking is stopped, milk synthesis stops, and secretory cells of the udder are rapidly lost. In cows, milking causes release of prolactin from the anterior pituitary into blood (Table 17.1). However, the prolactin response to milking probably lasts less than 30 minutes, and significance of milking-induced release of prolactin in relation to basal levels that are continuously present in blood is unknown.

In addition to prolactin, milking causes a discharge of ACTH (and adrenal glucocorticoids) from the anterior pituitary and oxytocin from the posterior pituitary, all of which help to maintain lactation. However, milking stimuli without milk removal will not maintain lactation. The relative importance of milk removal and stimulation of hormone release from the pituitary in the maintenance of lactation is undetermined. In any event, to promote maximal synthesis of milk, cows should be milked out completely at least twice daily (Chapter 19).

Hormonal Factors. The experimental approach used to elucidate the hormonal factors necessary to maintain lactation has

TABLE **17.1** *Serum Levels of Prolactin and Growth Hormone after Normal Milking*

	Prolactin ($m\mu g/ml$)	*Growth hormone ($m\mu g/ml$)*
10 min before wash	45 ± 7	2.1 ± .1
5 min before wash	38 ± 3	1.8 ± .2
5 min after wash[a]	52 ± 16	1.6 ± .1
10 min after wash[b]	119 ± 60	2.1 ± .2
15 min after wash	82 ± 23	1.9 ± .2
20 min after wash	71 ± 20	1.7 ± .1
25 min after wash	55 ± 14	1.7 ± .1
30 min after wash	48 ± 10	1.7 ± .2
35 min after wash	51 ± 8	1.7 ± .1

[a]Milking machine placed on udder.
[b]Milking machine removed from udder.
Source: Tucker, H. A.: J. Anim. Sci. *32* Suppl. I, 137, 1971.

usually involved endocrine gland removal followed by hormonal replacement therapy. Hypophysectomy, adrenalectomy, or thyroidectomy cause a rapid decline in milk production. The following experiments serve to illustrate the hormones needed to develop the gland and to initiate and maintain lactation.[12] Adult rats were hypophysectomized, adrenalectomized, and ovariectomized. Mammary growth was induced with a combination of estrogen, progesterone, prolactin, growth hormone, and adrenal corticoids. Lactation was then initiated with prolactin and adrenal corticoids. Maintenance of lactation required prolactin, growth hormone, and adrenal glucocorticoids.

17.7 Artificial Induction of Lactation

Of all cows culled in dairy herds about 15 to 20% are culled for infertility. Naturally as soon as milk production declines to unprofitable levels, these barren animals are culled from the herd. Much research has been performed in attempts to reinitiate secretion of large quantities of milk in these animals. The problem is twofold: (1) udder growth must be induced and (2) these cells must be stimulated to produce milk.

To grow the udder, most investigators have injected, implanted under the skin, or fed various combinations of estrogen and progesterone. Treatment with these hormones has produced variable amounts of udder growth, and subsequent milk production has ranged from 0 to 80 lb per day. However, the majority of treated cows produce subnormal quantities of milk. Up to the present time, injection of ovarian steroids has not duplicated the mammary growth response of pregnancy.[13]

Provided sufficient numbers of alveolar cells are present, lactation can be induced in dairy cattle within a few days with either adrenal glucocorticoids or high levels of estrogen (Figure 17.13). However, use of these hormones should be restricted to nonpregnant cows because either hormone will cause abortion.

Subcutaneous injection of a total daily dose of 0.1 mg/kg body weight of estradiol-17β and 0.25 mg/kg body weight of progesterone dissolved in 100% ethanol will initiate lactation in about 60 to 70% of barren heifers and cows.[14] Daily dose is divided in half and injected at 12-hour intervals for 7 days. The percent success rate (daily yield greater than 9 kg) increases to 100% if 5 mg of the tranquilizing drug, reserpine, is administered on days 8, 10, 12, and 14.[15] Reserpine increases serum prolactin in these cows. The steroid and prolactin profiles in serum mimic normal changes in these hormones during the periparturient period. Lactation usually commences between days 14 and 21 after initial injections of estradiol-17β and progesterone. Cattle should be restricted in their movement because of intense symptoms of estrus after day 7. The quantity of milk produced in treated cows is usually about 70% of their best previous lactation.

Presently, these treatments are experimental only, and they have not been cleared for commercial use by the Food and Drug Administration of the U.S. government. Until more repeatable responses in milk production are obtained, there is little justification to attempt artificial induction of lactation.

17.8 Hormonal Stimulation of Lactation

Thyroprotein has been cleared by the Federal Food and Drug Administration for commercial use in dairy cows to stimulate lactation. However, there are other hormones that will increase milk production, although their use is restricted currently to experimental purposes.

Thyroprotein

Thyroprotein is a synthetic hormone made by iodination of casein, the major milk protein. Thyroprotein mimics the biological action of thyroxine and triiodothyronine. Thyroprotein is inexpensive and orally active. Thyroprotein fed to dairy cows at the peak of lactation stimulates milk production about 10%, whereas if fed during the declining phase of lactation, average milk production is boosted 15 to 20%.[16] However, the response is extremely

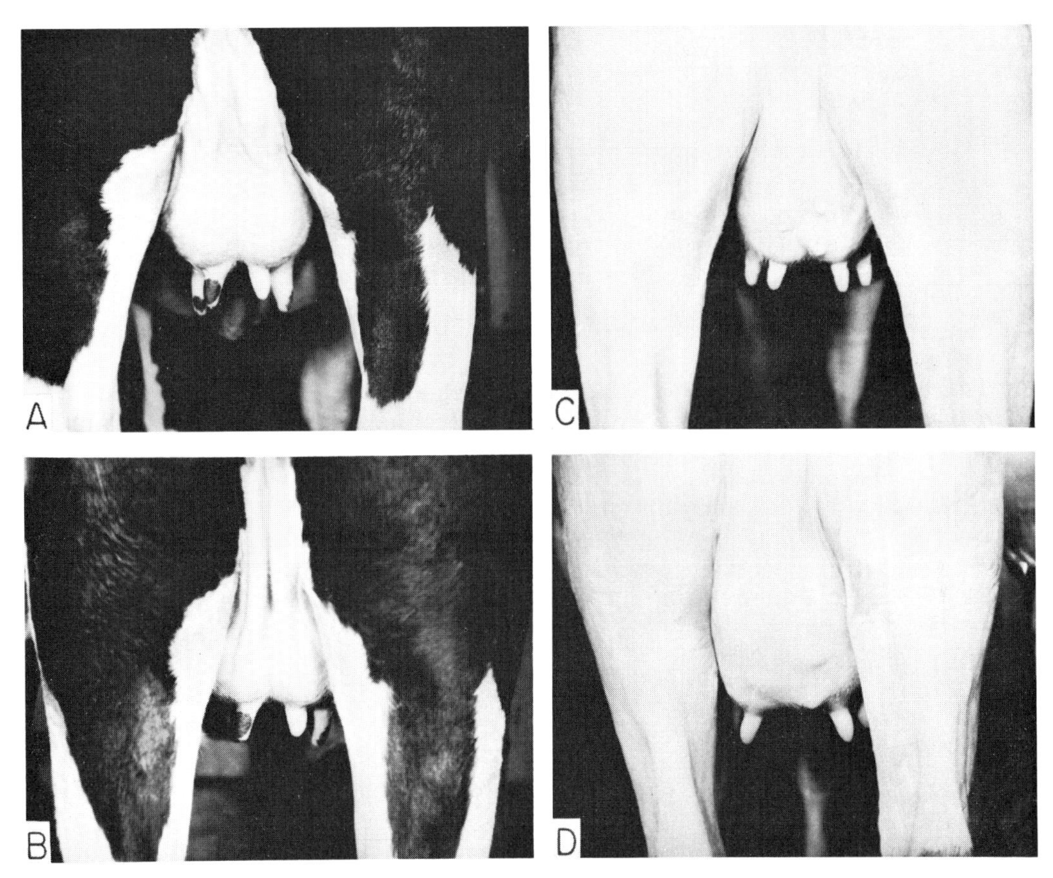

FIG. 17.13 Hormonal induction of lactation. (A) Udder of control heifer at beginning of experiment. (B) Udder of control heifer after 6 days no treatment. (C) Udder of experimental heifer at beginning of experiment. (D) Udder of experimental heifer after 6 days injection with synthetic glucocorticoid (9-fluoroprednisolone acetate). All heifers were pregnant 7½ months at beginning of experiment. (Reprinted with permission from J. Dairy Sci. *48:* 403, 1965)

variable among individual cows. Usually a greater increase in milk production occurs in higher-producing cows and older cows. The maximal increase usually occurs during the first 60 days of thyroprotein feeding. However, beneficial effects disappear within 2 to 4 months. Thereafter, milk production is often below that normally expected. The net result on milk production over an entire lactation suggests that thyroprotein-fed cows do not produce significantly more milk than control cows.

Thyroprotein has no serious effects on health, longevity, or reproductive function. Usually there is some weight loss, and increased heart and respiration rates will be noticed, but these effects are largely dependent upon the dose given.

As recommended in Chapter 11, thyroprotein should not be fed unless energy intake is increased substantially. Thus, if feed intake must be increased and the overall benefit to milk production from feeding thyroprotein is negligible, this practice would not generally be economical. Still, thyroprotein increases milk production at least temporarily. The current practice is to feed thyroprotein when milk production declines below 50 to 60 lb of milk daily. Thyroprotein should be given only to cows gaining in body weight, and it should be discontinued if milk production does not increase within 2 weeks.[17] A marked decline in milk production to below normal levels occurs whenever thyroprotein is abruptly removed from the ration. A better practice is to remove the hormone from the ration gradually. Even with this precaution, reduced milk yields may occur. Thyroprotein should be withdrawn about 2 weeks before the cow is dried off.

Growth Hormone

Experiments in the 1930s showed that crude extracts of the anterior pituitary are galactopoietic in ruminants. Growth hormone content of these extracts probably caused most of this stimulation. Because growth hormone is not commercially available, only short-term lactation trials have been conducted, but about 18% increases in milk production occur within a few days after injecting growth hormone only 3 times per week into dairy cows.[18] Furthermore, feed intake per kg milk is reduced 29%. In 6-day experiments bovine growth hormone, synthesized by recombinant DNA procedures, stimulates milk production 13% and feed efficiency 15% in high-producing cows.[19] Whether or not growth hormone will have long-term beneficial effects on milk yields and feed efficiency has yet to be reported.

Adrenal Glucocorticoids

Many published reports suggest that injection of ACTH or adrenal glucocorticoids depresses lactation in cows. Although there are reports to the contrary, a study shows that feeding a low dose of a synthetic glucocorticoid stimulates milk yield 14 to 18%.[20]

Calcitonin and Parathyroid Hormone

Calcitonin normally suppresses serum concentrations of calcium and phosphorus especially during lactation. In contrast, parathyroid hormone interacts with metabolites of vitamin D to increase serum calcium and stimulate milk yields. In view of the large loss of calcium into milk, calcitonin, and parathyroid hormone are essential for maintenance of lactation.

Ovarian Steroids

Feeding of progestins used in synchronization of the estrous cycle has no significant effect on milk production in dairy cows. Within a low and narrow dosage range, estrogen and the synthetic estrogen, stilbestrol, are galactopoietic in dairy cows. More commonly, however, estrogens in dosages usually employed in veterinary medicine inhibit lactation in cows. A combination of estrogen and progesterone is more inhibitory than estrogen alone. Thus, we are left with a paradox whereby estrogen can initiate lactation and then subsequently reduce milk yields.

Photoperiod

Exposure of growing Holstein heifers to 16 hours of fluorescent light and 8 hours of darkness daily between November and March in Michigan increases serum prolactin fourfold and increases rate of body weight gain approximately 12% in comparison with heifers receiving day lengths characteristic of Michigan's fall and winter.[21] The 16-hour light period stimulates milk production 6 to 10% during fall and winter in lactating dairy cows.

17.9 Summary

The basic milk-secreting units of the udder are epithelial cells which are arranged into hollow balls, called alveoli. Milk is secreted into the cavity of the alveolus. Each alveolus is drained by a tubular duct which joins other ducts to form larger and larger channels as they approach the gland cistern and teats. The fully lactating udder receives an extensive blood supply which delivers the nutrients and hormones needed by the alveoli for milk synthesis. Blood can be returned to the heart by 2 different routes, and extracellular tissue fluids are removed through the lymphatic system. The udder also receives a nerve supply which is especially concentrated in the teat. Connective tissues hold the udder together in close proximity to the body wall.

The calf is born with a very rudimentary mammary system. The glands are stimulated to grow in response to hormones associated with puberty, pregnancy, and the milking stimulus. Approximately 11% of mammary development occurs before pregnancy, 41% occurs during pregnancy, and the remainder takes place during early lactation. Numbers of secretory cells in the udder are important determinants of the level of milk production. Thus, as cell losses occur with advancing lactation, milk production gradually declines. Maintenance of high rates of metabolism in mammary cells and an efficient milk-ejection reflex are also essential for high milk production.

The hormones needed for mammary growth are estrogen, progesterone, prolactin, growth hormone, and adrenal glucocorticoids. To initiate lactation progesterone must be absent, and prolactin and glucocorticoids must be present in high amounts. Maintenance of lactation requires prolactin, growth hormone, thyroxine, and glucocorticoids. Administration of thyroid hormones, growth hormone, glucocorticoids, and 16-hour daily light periods may increase milk production in cows. Determination of exact hormonal requirements for lactation in dairy cows and commercial production of synthetic hormones may lead to new practical methods for stimulating milk yields in the future.

Review Questions

1. What is the importance of each of the following to milk production?
   a. Mammary cell numbers
   b. External pudic arteries
   c. Subcutaneous abdominal mammary veins
   d. Supramammary lymph nodes
   e. Nerves in the teat
   f. Pendulous udders
2. Diagram the changes in mammary development between birth and onset of a second parturition. Which hormones are involved in this process?
3. Discuss the factors regulating the induction of milk synthesis at parturition. What physiological factors regulate persistency?
4. Which hormones stimulate lactation? Which hormones inhibit lactation?
5. Disregarding the economical, legal, or practical significance, what physiological factors would you alter to obtain the most milk from a herd of dairy cows? State reasons why each practice was chosen.

References

1. Matthews, C. A., Swett, W. W., and Fohrman, M. H.: Weight and capacity of the dairy cow udder in relation to producing ability, age, and stage of lactation. USDA Tech. Bull, 989, 1949.
2. Reynolds, M.: Relationship of mammary circulation and oxygen consumption to lactogenesis in *Lactogenesis: The Initiation of Milk Secretion at Parturition.* M. Reynolds and S. J. Folley, eds. Philadelphia: University of Pennsylvania Press, 1969.
3. Hays, R. L., and Albright, J. L.: Udder edema: Its incidence and severity as affected by certain management practices. Illinois Res. *8:* 6, 1966.
4. Linzell, J. L.: Some effects of denervating and transplanting mammary glands. Quart. J. Exp. Physiol. *48:* 34, 1963.
5. Tucker, H. A.: Factors affecting mammary gland cell numbers. J. Dairy Sci. *52:* 721, 1969.
6. Sinha, Y. N., and Tucker, H. A.: Mammary development and pituitary prolactin content of heifers from birth through puberty and during the estrous cycle. J. Dairy Sci. *52:* 507, 1969.
7. Swett, W. W., Book, J. H., Matthews, C. A., and Fohrman, M. H.: Evaluation of mammary gland development in Holstein and Jersey calves as a measure of potential producing capacity. USDA Tech. Bull. 1111, 1955.
8. Turner, C. W., and Reineke, E. P.: A study of the involution of the mammary gland of the goat. Mo. Agr. Exp. Sta. Res. Bull. 235, 1936.
9. Swanson, E. W., Pardue, F. E., and Longmire, D. B.: Effect of gestation and dry period on deoxyribonucleic acid and alveolar characteristics of bovine mammary gland. J. Dairy Sci. *50:* 1288, 1967.
10. Tucker, H. A.: Physiological control of mammary growth, lactogenesis, and lactation. J. Dairy Sci. *64:* 1403, 1981.
11. Baldwin, R. L.: Enzymic and metabolic changes in mammary tissue at lactogenesis in *Lactogenesis: The Initiation of Milk Secretion at Parturition.* M. Reynolds and S. J. Folley, eds. Philadelphia: University of Pennsylvania Press, 1969.
12. Lyons, W. R.: Hormonal synergism in mammary growth. Proc. Roy. Soc. (Biol) *149:* 303, 1958.
13. Sud, S. C., Tucker, H. A., and Meites, J.: Estrogen-progesterone requirements for udder development in ovariectomized heifers. J. Dairy Sci. *51:* 210, 1968.
14. Smith, K. L., and Schanbacher, F. L.: Hormone induced lactation in the bovine. I. Lactational performance following injections of 17β-estradiol and progesterone. J. Dairy Sci. *56:* 738, 1973.
15. Collier, R. J., Bauman, D. E., and Hays, R. L.: Effect of reserpine on milk production and serum prolactin of cows hormonally induced into lactation. J. Dairy Sci. *60:* 896, 1977.
16. Blaxter, K. L., Reineke, E. P., Crampton, E. W., and Petersen, W. E.: The role of thy-

roidal materials and of synthetic goitrogens in animal production and an appraisal of their practical use. J. Anim. Sci. *8:* 307, 1949.
17. Thomas, J. W.: What about thyroprotein for dairy cattle? Hoard's Dairyman *114:* 1237, 1969.
18. Machlin, L. J.: Effect of growth hormone on milk production and feed utilization in dairy cows. J. Dairy Sci. *56:* 575, 1973.
19. Bauman, D. E., de Geeter, M. J., Peel, C. J., Lanza, G. M., Gorewit, R. C., and Hammond, R. W.: Effect of recombinantly derived growth hormone (bGH) on lactational performance of high yielding dairy cows. J. Dairy Sci. *65:* Suppl. 1, 121, 1982.
20. Swanson, L. V., and Lind, R. E.: Lactational response of dairy cows to oral administration of a synthetic glucocorticoid. J. Dairy Sci. *59:* 614, 1976.
21. Peters, R. R., Chapin, L. T., Leining, K. B., and Tucker, H. A.: Supplemental lighting stimulates growth and lactation in cattle. Science *199:* 911, 1978.

Suggested Additional Reading

Kon, S. K., and Cowie, A. T., eds.: *Milk: The Mammary Gland and Its Secretion.* Vol. I. New York: Academic Press, 1961.

Larson, B. L., ed.: *Lactation.* Ames: Iowa State University Press, 1985.

Larson, B. L., and Smith, V. R., eds.: *Lactation: A Comprehensive Treatise.* Vol. I. New York: Academic Press, 1974.

CHAPTER 18

Biosynthesis of Milk

18.1 Introduction

Milk is secreted primarily in the interval between milkings. The principal components of milk are: water; fat; solids-not-fat (SNF), which are made up of proteins, lactose, and minerals; vitamins and several types of cells, i.e., bacteria, leukocytes, and mammary secretory cells. The rate at which the mammary cell removes nutrients from blood, transforms them into components of milk, and discharges them into the alveolar lumen is a major physiological factor in controlling the level of milk production. Important points to be discussed in this chapter will be: (1) to analyze how individual components of cow's milk are synthesized and then blended into milk; and (2) to evaluate some factors that influence these processes.

18.2 Cytology of the Mammary Secretory Cell

A diagram of a typical epithelial secretory cell of a cow's udder is shown in Figure 18.1. The mammary cell is a highly organized factory whose rate of metabolism is very high.[1] To produce milk, the udder utilizes about 80% of the total available glucose, acetate, and amino acids in blood. As a result of the udder's tremendous demand for nutrients, the cow's body loses considerable weight during intense lactation (Figure 18.2). Thus, the amusing suggestion has been made that the body of the dairy cow is merely an appendage of the lactating udder.[2]

Nucleus

Function of the mammary cell nucleus (Figure 18.1) is to transmit genetic information contained in the genes for synthesis of milk proteins and certain enzymes. This is in contrast to the function of sperm and ovum nuclei which transmit genetic information for the entire animal. A large molecule in the chromosome of the nucleus, deoxyribonucleic acid (DNA), contains the genetic information and transcribes this information onto another molecule, messenger ribonucleic acid (mRNA).

Endoplasmic Reticulum

This organelle consists of a system of canals located in the basal two-thirds of the cytoplasm of the mammary cell (Figure 18.1). mRNA moves from the nucleus to endoplasmic reticulum and specifies the order in which amino acids are linked together to make proteins of milk and enzymes in the mammary cell. Surfaces of some of the canals of the endoplasmic reticulum are studded with RNA-protein particles called ribosomes. Ribosomes are the

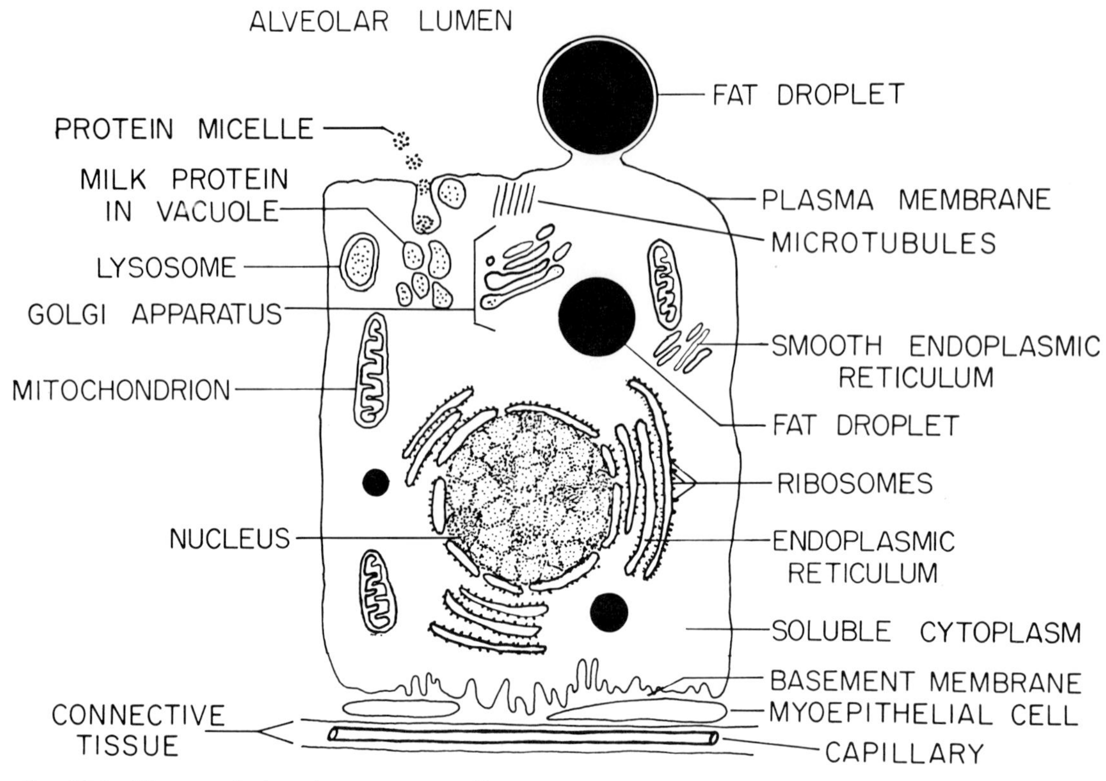

Fig. 18.1 Diagram of a lactating mammary cell as seen through the electron microscope.

sites of protein synthesis; that is, peptide bonds between adjacent amino acids are formed on ribosomes. Other parts of the endoplasmic reticulum are devoid of these particles and serve simply as channels for intracellular transfer of materials. These smooth canals connect to another organelle, the Golgi apparatus (Figure 18.1). In addition to synthesis of proteins, synthesis of milk fat triglycerides is associated with endoplasmic reticulum.

Golgi Apparatus

The Golgi apparatus serves as the site where proteins are packaged. For example, Ca and P are added to casein molecules and the particles (micelles) of casein are formed within the Golgi apparatus. Lactose synthesis also occurs in the Golgi apparatus. Secretory vacuoles (Figure 18.1) containing milk proteins, lactose and water bud off from the Golgi and rise to the apex of the cell where membranes of the secretory vacuoles fuse with the plasma membrane. Thus, membranes from the secretory vesicles replenish plasma membrane which is lost with secretion of fat droplets.[3] The secretory contents of the Golgi are discharged into the lumen of the alveolus by reverse pinocytosis.

Mitochondria

The mitochondria (Figure 18.1) are numerous in metabolically active tissues. Thus, mammary cells from lactating cows contain many mitochondria, whereas few are found in mammary cells of nonlactating cows. The mitochondria are often called the "powerhouses of the cell" because they generate the energy required for synthesis of milk fat, lactose, and protein.

Lysosomes

These membrane-bound particles (Figure 18.1) contain degradative enzymes which, if released, cause breakdown and death of the cell. One of the mechanisms whereby hormones maintain mammary cells during lactation is thought to be stabilization of the lysosomal membrane which

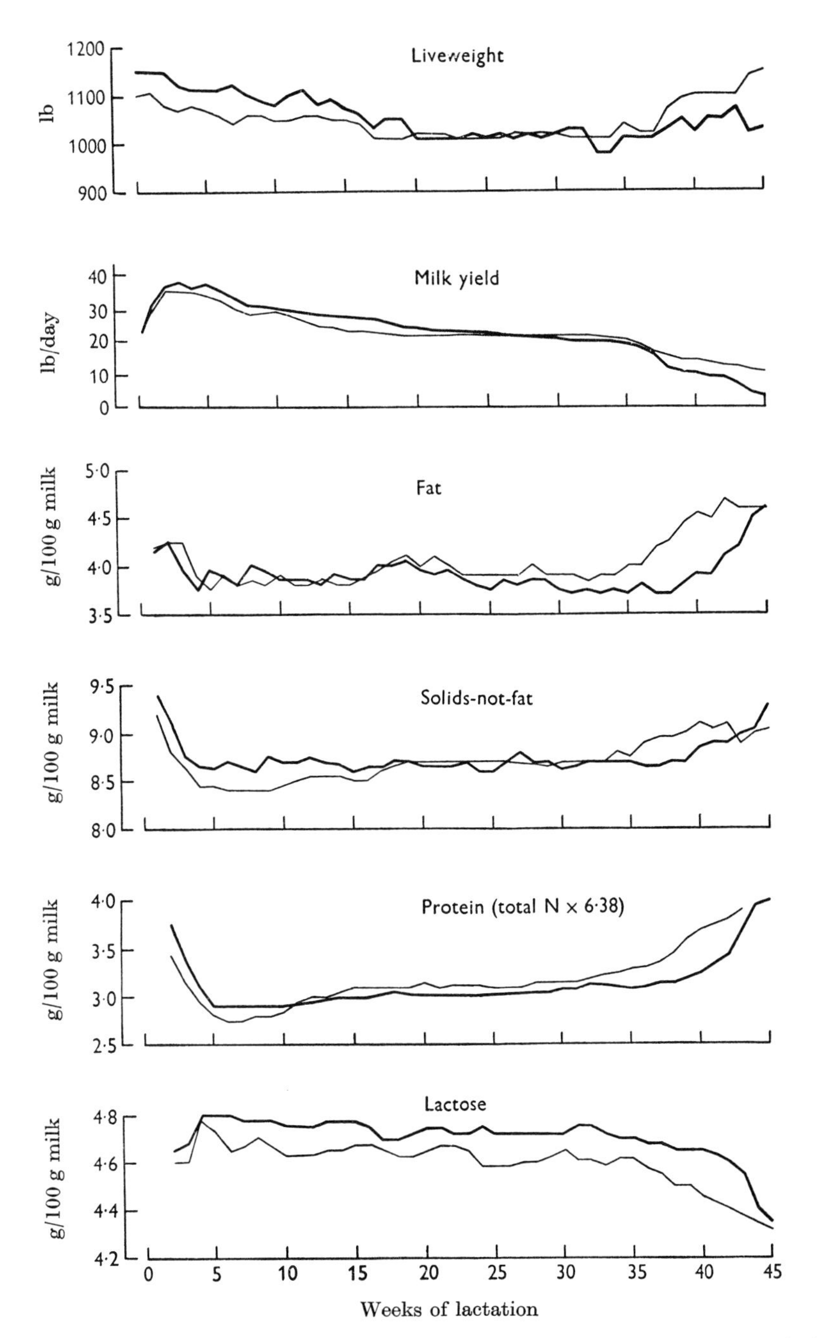

FIG. 18.2 Variations throughout the 1st lactation in live-weight, milk yield, fat, SNF, lactose (anhydrous), and protein (total N × 6.38) contents of milk. The 2 lines on each graph represent different nutritional treatments on 2 groups of cows. (Reprinted with permission from J. Dairy Res. *32:* 45, 1965)

prevents leakage of its enzymes into cytoplasm. When a cell dies, these enzymes are released and aid in digestion and removal of the cell from the body. Lysosomes are particularly active during involution of mammary tissue such as that which occurs in the early dry period or during mastitis.

Cellular Membranes

Membranes envelop all organelles, and one membrane, the plasma membrane, sets the outer limits of the entire mammary cell (Figure 18.1). Membranes exert considerable specificity as to the passage of chemicals into the various compartments of the cell. For example, some nutrients from the capillaries enter the cell through the plasma membrane with ease and, in fact, they may be concentrated many times. Other nutrients in blood are excluded. While milk is isotonic with blood, individual constituents in milk and blood are not in equilibrium. For example, milk contains 9 times more fat, 90 times more sugar, 5 times more potassium, 10 times more phosphorus, 13 times more calcium, 1/7 as much sodium, and 1/2 as much protein as blood.

Microtubules

Microtubules (Figure 18.1) are essential for cell division, shape the mammary cells, and aid in movement of secretory vacuoles to the apex of the cell.

Cytoplasm

The cytoplasm is a fluid matrix which constitutes a large portion of the mammary cell. Much of the material in this fraction, which includes enzymes, nutrients, and macromolecular products, is soluble. Anaerobic breakdown of glucose, which is necessary before it can be oxidized in the mitochondria to provide energy, synthesis of fatty acids and activation of amino acids for synthesis of proteins occur in the soluble cytoplasm.

Discharge of Milk Into the Alveolar Lumen

Discharge of milk into the lumen of the alveolus occurs without exposure of the interior of the cell.[1] Individual components of milk are kept separate inside the mammary cell. Thus, milk is not truly formed until it reaches the alveolar lumen where the individual components mix together. Lipid droplets form in the lower portion of the cell, and as they increase in size, they gradually move toward the alveolar lumen. The cell membrane envelops the fat droplet as the fat droplet pushes out of the cell. Then the fat droplet is pinched off at the outer membrane of the cell surface and becomes free within the alveolus. Milk proteins, in contrast, are packaged in the mammary cell as discrete granules within vacuoles, and they are discharged into the lumen of the alveoli without acquiring a covering of cell membrane. Lactose is contained in the secretory vacuoles and released into the alveolar lumen simultaneously with proteins. A portion of water in milk is discharged via the vacuoles. The mechanisms whereby the remaining chemical components of milk enter the lumen of the alveoli are not known.

18.3 Biosynthesis of Milk Proteins

Most proteins in the diet of humans are composed of essential and nonessential amino acids. Essential amino acids must be supplied in the diet, whereas nonessential amino acids are synthesized in the body from essential amino acids or from carbohydrates. Milk proteins contain more essential amino acids than any other natural food. Thus, it is primarily the protein content of milk that allows one to say that "milk is nature's most nearly perfect food."

Precursors

Primary proteins in milk are α-casein, β-casein, κ-casein, γ-casein, α-lactalbumin, and β-lactoglobulin.[4] These proteins comprise more than 90% of the total proteins in milk and are found only in milk, nowhere else in nature. All are synthesized within the mammary secretory cell from a common pool of free amino acids (Table 18.1). Lactating mammary cell uptake from blood of some essential amino acids exceeds output of these amino acids in milk. The excess amino acids are used as an energy source and to synthesize nonessential amino acids. In milk, casein becomes aggregated into spherical structures called micelles. Although the primary function of casein is to provide amino acids for the calf, other functions have been noted. For example, κ-casein stabilizes casein micelles; otherwise curds would form in milk. β-lactoglobulin provides the characteristic cooked flavor of heated milk. Heat denaturation of β-lactoglobulin prevents curd formation which is necessary in the manufacture of cottage cheese.

Immune globulins and blood serum albumin enter the mammary cell from blood and appear unchanged in milk (Table 18.1).

TABLE 18.1 *Blood Precursors of Milk Constituents*

Milk constituents	*Precursor in blood*
Protein	
α-casein	Free amino acids
β-casein	Free amino acids
κ-casein	Free amino acids
γ-casein	Free amino acids
α-lactalbumin	Free amino acids
β-lactoglobulin	Free amino acids
Immune globulins	Immune globulins
Milk serum albumin	Blood serum albumin
Carbohydrate	
Lactose	Glucose
Fat	
Long-chain fatty acids	Long-chain fatty acids
Short-chain fatty acids	Acetate and β-hydroxybutyrate
Vitamins	Vitamins
Minerals	Minerals
Water	Water

Synthesis of these proteins from amino acids within the mammary cell is not required.

Biochemical Reactions

The total number of milk proteins is relatively small, and a given protein is always composed of the same number of amino acids arranged in the same sequence. Furthermore, each individual cow always produces the same milk proteins, but these may be different from the proteins secreted by other cows in the herd. A few rare milk proteins are found only in certain cows, cow families, or breeds.[5]

Synthesis of milk proteins with their specific amino acid sequences is a rigidly controlled process, and it is the gene, or DNA, which directs protein synthesis. To accomplish protein synthesis, the genetic message of DNA in the nucleus is transcribed onto mRNA which moves to the ribosome. There mRNA translates the message which specifies the amino acid sequence of milk proteins.

Protein synthesis requires energy, and energy comes from breakdown of adenosine triphosphate (ATP) to adenosine monophosphate (AMP). In ruminants ATP is generated from oxidation of carbohydrates, primarily glucose, from acetate, and from fats. Thus, optimal milk protein synthesis cannot occur unless adequate energy is supplied in the diet.

The sequence of events required to synthesize a milk protein is given in Figure 18.3. Initially, there is activation of amino acids in cytoplasm of the secretory mammary cell by enzymes and ATP (Step 1 of Figure 18.3). The activated amino acids link to another kind of RNA, termed transfer RNA or tRNA (Step 2 of Figure 18.3). Each of the 18 common amino acids found in milk protein has its own activating enzyme and tRNA. Amino acid-tRNA complexes move from the cytoplasm to the ribosomes which contain the genetic message in the form of mRNA. A third type of RNA called ribosomal or rRNA joins the amino acid-tRNA with the mRNA (Step 3 of Figure 18.3). Thus, individual amino acids are bound, one at a time, to a growing chain of amino acids to form milk proteins in ribosomes of the mammary secretory cell. The chain of amino acids emerges from a tunnel in the ribosome and enters the lumen of the canals of the endoplasmic reticulum. The chain of amino acids is cleaved to a shorter length, characteristic of milk proteins, as the chain crosses the membrane of the endoplasmic reticulum. The proteins move through the lumen of the endoplasmic reticulum to the Golgi apparatus and secretory vacuoles which discharge their contents as described in Section 18.2.

OVERALL REACTIONS OF PROTEIN SYNTHESIS

$$1)\ AA + ATP \longrightarrow AMP \sim AA + PP$$

$$2)\ AMP \sim AA + tRNA \longrightarrow AA \sim tRNA + AMP$$

$$3)\ (AA \sim tRNA)_n \xrightarrow[\text{mRNA}]{\text{ribosomes, rRNA}} AA_1 - AA_2 - AA_3 - AA_4 \ldots\ldots AA_n$$

FIG. 18.3 Overall steps in the synthesis of proteins. Steps 1 and 2 are catalyzed by a single enzyme (amino acyl synthetases) and a tRNA; in both cases they are specific for each amino acid. Probably several enzymes and factors are involved in Step 3. (Reprinted with permission from J. Dairy Sci. *52:* 737, 1969)

18.4 Carbohydrate Metabolism

The principal carbohydrate in blood of cows is glucose. As described in Chapter 8, most dietary carbohydrates are fermented to volatile fatty acids in the rumen of the dairy cow. One of these fatty acids, propionic, is converted to glucose in the liver. Another important source of blood glucose in ruminants comes from breakdown of proteins (gluconeogenesis) in tissues peripheral to the udder. However, glucose levels in blood of ruminants are only about one-half those found in nonruminant animals. Mammary uptake of glucose is a major limiting factor to maximal milk secretion in dairy cows.[6]

LACTOSE

GALACTOSE GLUCOSE

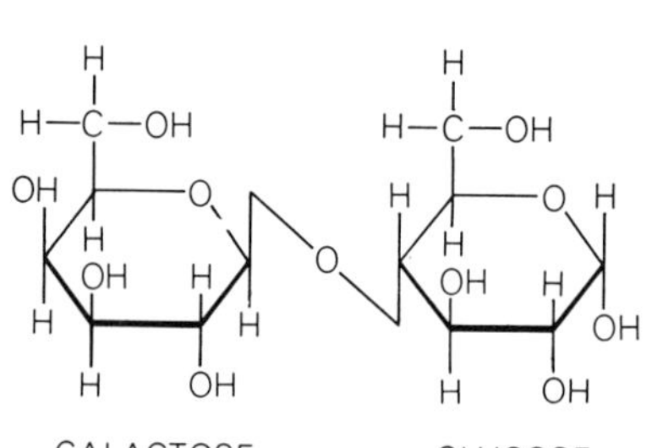

FIG. 18.5 Configuration of the lactose molecule.

Utilization of Glucose

Blood glucose in the cow is used by the mammary cells in a variety of ways, each critical to synthesis of milk (Figure 18.4). For example, glucose (1) is used to synthesize the major sugar in milk, lactose; (2) is a primary source of energy (ATP); (3) can be used to synthesize glycerol of milk triglycerides; and (4) is used in synthesis of RNA. Without glucose milk synthesis could continue only a few minutes.

Biosynthesis of Lactose

The primary sugar of milk is a disaccharide, lactose, which is composed of a molecule of glucose and a molecule of galactose (Figure 18.5). Lactose is responsible for the slightly sweet taste of milk. Lactose also promotes growth of certain bacteria which form lactic acid in the small intestine of the calf. And lactic acid is believed to promote absorption of Ca and P for formation of bone in the young calf.

Glucose is the only precursor of lactose (Table 18.1). Two molecules of glucose must enter the mammary cell for each molecule of lactose formed. One of the glucose units is converted to a form of galactose. The condensation of the second glucose molecule with galactose occurs in the Golgi apparatus and is catalyzed by an enzyme, lactose synthetase. This enzyme is composed of 2 subunits. One of these subunits is α-lactalbumin,[7] a major protein component of milk (Table 18.1). Thus, α-lactalbumin functions as an enzyme and as an export protein for nourishment of the calf. Chapter 17 contains additional information on hormonal control of this protein at initiation of lactation.

18.5 Biosynthesis of Milk Fats

Fats in milk from cows are characterized as mixed triglycerides (Figure 18.6) with a rather high proportion (approximately 50%) of short-chain fatty acids (C_4—C_{16}). The other one-half of milk fat is made up of long-chain fatty acids (C_{18}—C_{20}). Another characteristic of cows' milk is a high proportion of saturated fatty acids. Fat content of milk assumes added importance because it is a major factor in pricing milk (Chapter 2).

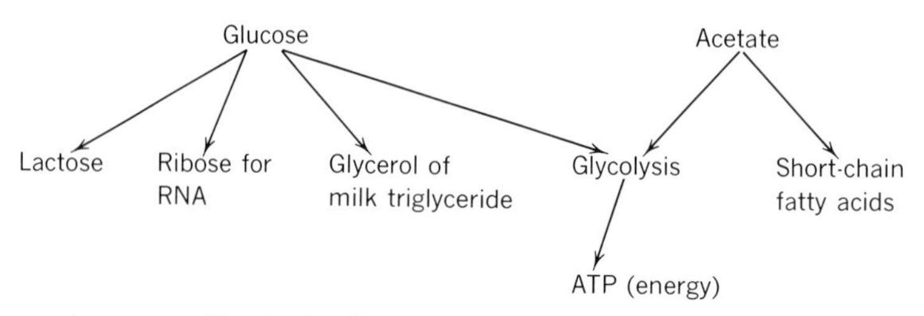

FIG. 18.4 Glucose and acetate utilization by the mammary cell.

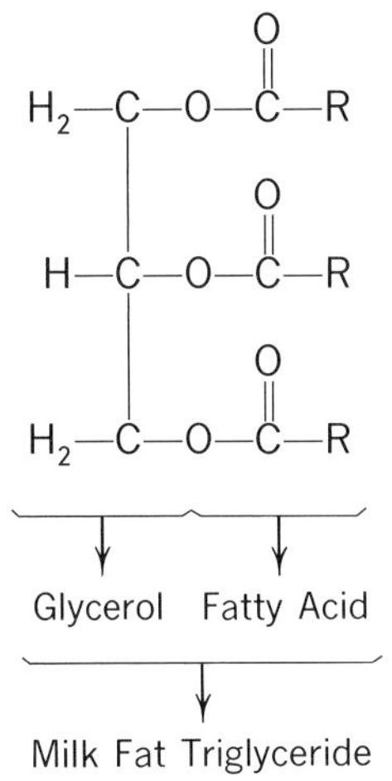

$H_2—C—O—C(=O)—R$

$H—C—O—C(=O)—R$

$H_2—C—O—C(=O)—R$

Glycerol Fatty Acid

Milk Fat Triglyceride

Fig. 18.6 Typical milk fat triglyceride. Representative fatty acids include:
butyric (C_3H_7COOH)
caproic ($C_5H_{11}COOH$)
palmitic ($C_{15}H_{31}COOH$)
oleic ($C_{17}H_{33}COOH$)
stearic ($C_{17}H_{35}COOH$)

Precursors of Long-Chain Fatty Acids

Fatty acids in the cow's ration provide directly about one-half of the fatty acids found in milk. These fatty acids are almost exclusively of the long-chain variety (Table 18.1). Most fatty acids of plants in the cow's diet are long-chain acids and are unsaturated (i.e., they contain a high proportion of double bonds between carbon atoms). Many unsaturated dietary fatty acids become hydrogenated (saturated) in the rumen (Chapter 8). This alteration accounts for the high proportion of saturated fatty acids in cows' milk.

After passing through the rumen, long-chain fatty acids are absorbed from the small intestine into the lymph system (lacteals, Figure 17.4), become bound to a protein, move into blood, and are absorbed from blood by the mammary secretory cell.[8] The type of diet affects chain length of fat secreted into milk, and to a lesser extent the degree of saturation (because the rumen is not 100% efficient in saturating all fatty acids in feed). Additional details on the effects of diet and fiber length on milk fat percentage are described in Chapter 10.

The highly saturated fatty acid content of cows' milk has led some doctors to recommend reducing intake of milk fats in the human diet. Some have theorized that saturated animal fats, as compared with unsaturated plant fats, contribute to a greater circulating level of cholesterol, which in turn forms fatty deposits in arteries of people (atherosclerosis). Evidence at present suggests that total caloric intake relative to expenditure of energy plus a host of other factors are of as much importance in predisposing a person to atherosclerosis as is the type of fat present in the diet.

Precursors of Short-Chain Fatty Acids

Short-chain fatty acids, which comprise the 50% of milk fat not derived directly from fatty acids of the diet are synthesized in the mammary secretory cell from acetate and a ketone body, β-hydroxybutyrate (Table 18.1).[9,10] Acetate is a 2-carbon unit whereas β-hydroxybutyrate is a 4-carbon molecule, and both molecules originate from fermentation of plant carbohydrates to volatile fatty acids in the rumen (Chapter 8). Short-chain fatty acids are very odorous, and they account in large measure for the characteristic smell and flavor of many cheeses.

Short-chain fatty acids are synthesized by stepwise addition of a 2-carbon derivative of acetate, acetyl-coenzyme A (CoA). Initially, CO_2 combines with acetyl-CoA to form a 3-carbon intermediate, malonyl-CoA. Then, an additional molecule of acetyl-CoA joins with malonyl CoA, a molecule of CO_2 is lost, and the 4-carbon fatty acid, butyrl-CoA is created. By successive repetition of this process short-chain fatty acids of various lengths are formed. The mammary secretory cell is also capable of synthesizing short chain fatty acids by converting β-hydroxybutyrate to butyrate which after addition of CoA forms butyrl-CoA, the same intermediate described in utilization of acetate. A secondary pathway exists for utilization of β-hydroxybutyrate in which this 4-carbon fatty acid is cleaved to 2-carbon units and utilized as acetate.

More acetate than β-hydroxybutyrate is used for milk fat synthesis. Also, as shown in Figure 18.4, acetate can provide energy for the mammary cell. Because of its large contribution to synthesis of milk, production of acetate in the rumen of the dairy cow is essential for optimal milk production.

18.6 Vitamins, Minerals, and Water

The mammary secretory cell cannot synthesize vitamins or minerals; therefore, all vitamins and minerals in milk are supplied from blood (Table 18.1).

Calcium, phosphorus, potassium, chloride, sodium, and magnesium are the principal minerals in milk. Although minerals in milk are derived from blood, it is not conclusively known whether they are absorbed in proportion to their concentration in blood or if there are mechanisms which allow selective uptake. There is evidence that the epithelial cell can discharge minerals back into blood as well as into the milk which would suggest some type of active transport mechanism.

There is usually a constant percentage of lactose, sodium, and potassium in milk. These constituents plus chlorides maintain osmotic equilibrium in milk.[11] There is an inverse relationship between lactose concentration and potassium-sodium (and chloride) concentrations in milk. A similar relationship exists between lactose and potassium alone.

Water in milk is derived partially from the potassium-rich intracellular fluids of the alveolar cell and partially by movement from the blood into the cell to maintain osmotic equilibrium as a result of lactose, protein and fat synthesis. Since milk is in osmotic equilibrium with blood, and lactose accounts for almost 50% of osmotic pressure of milk, an increase in lactose concentration causes an influx of water and a decrease in sodium and chloride content of milk. This process profoundly affects milk production, primarily because water makes up such a large proportion (87%) of milk.

Cows with mastitis or near the end of lactation almost invariably have a depressed milk yield with lower amounts of lactose and potassium and elevated levels of sodium and chloride. This accounts for the salty taste of milk from cows during advanced lactation.

18.7 Factors Affecting Composition and Yield of Milk

Some constituents in milk are almost always found in the same proportions whereas others vary markedly. A major factor that alters milk composition is total amount of milk produced at a given milking. Thus, many factors alter milk composition,[12] but the mechanism involved may be indirect with the direct cause of the compositional change being the total amount of milk produced. In addition, many changes in milk composition from milking to milking cannot be ascribed to a definitive cause. For example, milk fat percentage can vary as much as 30% from unknown causes.

Many elements within the cow and her external environment affect milk composition and production. As discussed in the following section, dairymen can alter many of these factors to achieve greater milk production and increased profits.

Genetics and Nutrition

Genetics and nutrition markedly affect milk composition and yield and these aspects are described in detail in Chapters 1, 6, 7, 11 and 13.

Stage of Lactation and Persistency

Secretion produced by the udder immediately after parturition is known as colostrum. Composition of colostrum is considerably different from composition of normal milk. Usually a period of 3 to 5 days immediately post-partum is needed for secretions to return to the normal composition of milk. During this period total solids, especially the globulin fractions (protein), are elevated. Newborn calves are practically devoid of gamma globulin, the blood fraction which contains antibodies against various disease organisms. Thus, calves must ingest gamma globulins from colostrum to acquire a passive immunity against common calfhood diseases. Feeding colostrum after birth is especially critical during the first 12 to 24 hours of a calf's life. After this time, enzymes in the digestive tract degrade the antibodies, and permeability of the gut to antibodies decreases. Thus, antibodies lose their effectiveness with time after birth. Heat or cold stress reduces rate of transfer of immunoglobulins to blood serum of newborn calves, which lowers resistance of calves to disease. Since newborn calves have poorly developed thermoregulatory mechanisms, protection from environmental extremes is warranted.

Lactose content is depressed, whereas fat and casein percentage in colostrum is variable. Since high lactose diets can cause scours in calves, reduced lactose content of colostrum helps to prevent this disease. Calcium, magnesium, phosphorus, and chloride are high in colostrum, whereas potassium is low. Iron is 10 to 17 times greater in colostrum than in normal milk. This high level of iron is needed to rapidly increase hemoglobin in red blood cells of the newborn calf. Colostrum contains 10 times as much vitamin A and 3 times as much vitamin D as that found in normal milk. Again, the newborn calf is practically devoid of vitamin A, and since it provides a degree of protection against various diseases every calf should be fed colostrum.

Shape of the lactation curve of the dairy cow is shown in Figures 18.2 and 18.7. At parturition milk production commences at a relatively high rate, and the amount secreted continues to increase for about 3 to 6 weeks. Higher-producing cows usually take longer than low-producing cows to achieve peak production. After the peak is attained, milk production gradually declines. The rate of decline is commonly referred to as persistency. In the average nonpregnant cow after peak production is attained, each month's production is about 94 to 96% of the preceding month's yield. Many nonpregnant cows will continue to secrete milk indefinitely, but at a reduced rate. Maintenance of high peak milk production should be every dairyman's goal, but this has never been achieved. In fact, there is a strong tendency for cows which achieve a high initial yield to be less persistent. During early stages of lactation, the stimuli to secrete milk are able to overcome many environmental or management problems (e.g., underfeeding or improper milking procedures). But as lactation progresses any adversity will reduce milk secretion to a greater extent than that expected in cows during early lactation.

Fat percentage in milk decreases slightly during the first 2 to 3 months of lactation and then increases as total production de-

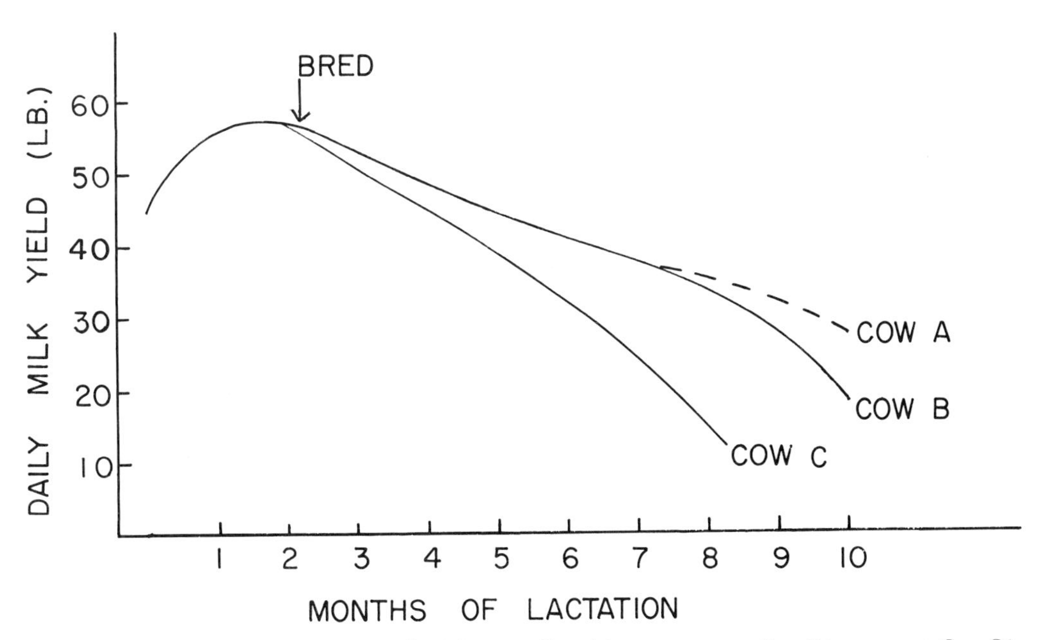

Fig. 18.7 Diagram of the lactation curve of a dairy cow. Cow A is not pregnant. Cow B is pregnant. Cow C is not as persistent as cow A or B.

creases with advancing lactation.[13] Milk protein content gradually increases with advancing lactation. Lactose decreases whereas mineral concentrations increase slightly during this time. These changes are illustrated in Figure 18.2. Most increases in SNF components of milk are associated with advancing stages of concurrent pregnancy rather than stage of lactation per se. Toward the end of lactation composition of milk tends to approach composition of blood.

Milk Secretion Rate

Milk secretion rate is rapid and relatively constant for 8 to 10 hours after milking and lowest just before and during milking. However, as milk accumulates during the interval between milkings, intramammary pressure increases and milk secretion rate per hour decreases (Figure 18.8). In general, increases in intramammary pressure in high-producing cows are less than in low-producing cows for the same quantity of milk.

Capacity of the udder to hold and secrete milk has a major influence on milk secretion rate. Usually larger udders produce milk at a greater rate than smaller glands. To give some idea of the capacity of the cow's udder, one study in mature Jersey cows at peak of lactation showed the maximal amount of milk that could be secreted and stored at one time was 54 lb, and it took almost 35 hours to achieve this production.[14] Another conclusion to be drawn is that if a cow is not milked, secretion stops about 35 hours after the last milking. Frequent removal of milk is conducive to increased milk secretion rates and decreased intramammary pressures.

Although much has been written about increased intramammary pressure reducing milk secretion rate, these studies have used milk accumulation to build up intramammary pressure. Therefore, the possibility exists that specific components of milk may act within the mammary cell to inhibit their own secretion, independent of intramammary pressure. In fact, elevated intramammary pressure does not appear to inhibit milk fat synthesis to the extent that it inhibits other milk constituents.

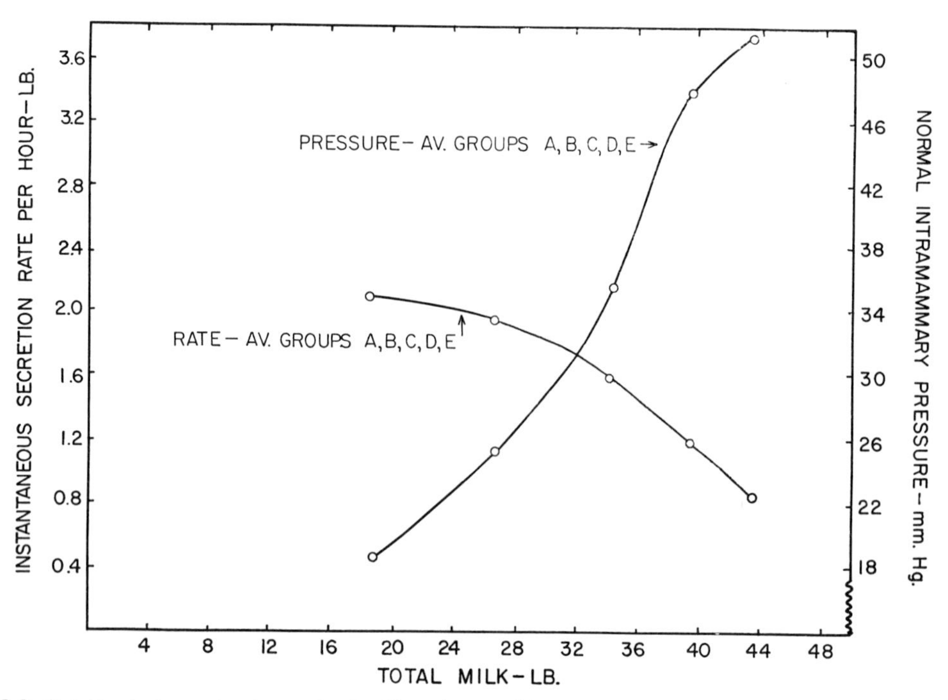

FIG. 18.8 Relation between total pounds of milk and normal intramammary pressure and total pounds of milk and instantaneous secretion rate. (Reprinted with permission from J. Dairy Sci. *44:* 1725, 1961)

Milking Practices

Cows are usually milked 2 × per day, and proper procedures to use at milking are discussed in Chapter 19. Increasing milking frequency to 3 × per day increases milk production by 10 to 25% and milking 4 × per day will stimulate milk yield another 5 to 15%. Whether these increases in production are worth the extra expense in labor, feed, and equipment depends upon conditions on a particular farm. Operations where cows are milked in parlors equipped with automatic teat-cup detachers or where extra labor and feed are available may find it economical to milk cows more frequently than 2 × per day. Milk yield response to 3 × milking is greater in later than in early stages of lactation, and feed requirements increase in proportion to milk produced.[15]

Milk first removed from the udder contains much less fat (as low as 1 to 2%) than milk removed at the end of the milking process (as high as 7 to 9%). The reason for this distribution of the fat globules is not known. It has been postulated that fat globules aggregate in the alveoli and are retarded in their passage toward the teat, whereas the fluid portion more readily passes around fat globules toward the base of the udder and teat. Thus, immediately preceding milking, milk in larger ducts of the gland has less fat than milk in alveoli.

Cows milked twice daily at 10- and 14-hour intervals produce about 1% less milk on average than cows milked at 12- and 12-hour intervals.[16] Higher-producing cows may show a greater inhibition in milk yield. Low-producing cows milked at 16- and 8-hour intervals produce only 1.3% less milk than similar cows milked at 12- and 12-hour intervals, but milk production losses of 4 to 7% occur in higher-producing cows and in heifers. Dairymen milking 80 to 200 ungrouped cows in a milking parlor may milk individual cows at markedly unequal intervals from day to day. However, by grouping cows according to milk yield or physiological state, high-producing cows and heifers can be milked close to a 12- and 12-hour interval.

Cows milked for 4 minutes for an entire lactation produce less milk, especially during early lactation, than similar cows milked for 8 minutes.[17] The 4-minute group was incompletely milked whereas the 8-minute group was somewhat overmilked. Milking times of most cows usually average just over 5 minutes to achieve maximal milk removal. Leaving 4 lb of milk in the udder after milking for 10 consecutive days permanently reduces milk yield for the entire lactation.[18] Machine-stripped cows do not produce significantly more milk than cows trained to milk without machine stripping, but machine-stripped cows require a longer time to milk. Thus, machine-stripping is not recommended. If practiced, machine stripping should be *brief.*

Age and Size of Cow

Milk yields increase at a decreasing rate until about the 8th year of age, depending upon the breed, and then decrease at an increasing rate. However, the decrease after the 8th year is much less than the increase before this age. Mature cows produce about 25% more milk than 2-year-old heifers. Increased body weight accounts for about 5% of this increase, whereas the remaining 20% is the result of increased development of the udder during recurring pregnancies.

Milk fat and SNF decrease about 0.2% and 0.4%, respectively, between the first and fifth lactation. There is little change thereafter. Lactose accounts for most of the decrease in SNF.

Heifers should be bred to calve at 24 months of age or earlier if they are of sufficient size to permit delivery of the calf. Although a heifer will produce more milk during the first lactation if breeding is delayed to the point where she calves after 30 months of age, total lifetime production will be reduced (Chapter 16).

In general, large cows produce more milk than small cows, but milk yield does not vary in direct proportion to body weight. Rather, it varies by the 0.7 power of body weight, which is an approximation of surface area of the cow. Thus, a cow which is twice as large as another usually produces only about 70% more instead of 100% more milk.

Estrous Cycle and Pregnancy

Estrus may temporarily depress milk yield, but experimental evidence is not consistent. As discussed in Chapter 15, higher-producing cows frequently delay their return to estrus after calving.

Cows with follicular cysts on the ovary produce more milk, adjusted for days not pregnant, than normal herd mates.[19] These same cows produced equivalent amounts of milk before the cystic condition was present. This suggests that circumstances associated with the cystic ovary increased milk production and that high milk production does not cause follicular cysts. Milk production of cystic cows is more persistent than production of normal herd mates. Anestrous cystic cows produce more milk than nymphomaniac cystic cows.

Pregnancy reduces milk production during concurrent lactation (Figure 18.7). For example, if a cow is bred at 90 days post-partum, she will produce 750 to 800 lb less milk in 365 days than if bred at 240 days post-partum. Most of this reduced yield occurs after the 5th month of pregnancy. By the 8th month of gestation milk production may be reduced 20% for that month in comparison with nonpregnant cows lactating the same length of time. Nonetheless, a regular calving interval is a major stimulant to high levels of milk production (Chapter 16). Factors such as feed, labor, advantages of fall calving, base period prices, and reproductive efficiency should be evaluated before a decision is reached on calving interval. Under most practical farm conditions it is usually best to rebreed cows at the first estrus occurring 45 to 50 days post-partum.

Dry Period

Cows should be given a rest period of 6 to 8 weeks between lactations. Either shorter or longer dry periods will reduce subsequent milk production. However, to maximize lifetime production of milk there must be a balance between production lost during the dry period and production gain in the subsequent lactation. During 2 consecutive lactations, optimum dry periods decline from 63 to 23 days as age at calving increases from 24 to 83 months.[20] However, cows with parturition intervals less than 340 days require dry periods of at least 55 days.

The proper procedure to dry off a cow is to withdraw all grain and reduce water supply several days before the start of the dry period. Then abruptly stop milking the cow. After milking is stopped intramammary pressure increases and inhibits further milk secretion. Sometimes if the udder becomes extremely congested, it may need to be remilked. However, this practice stimulates further milk synthesis because intramammary pressure is reduced and hormones are released. Perhaps more importantly re-milking removes leukocytes from the udder at a time when many are needed to combat infection. It is usually unnecessary to re-milk if production is reduced to 20 lb per day before milking is stopped. Some evidence suggests that *if* dry cow therapy for mastitis is *not* practiced, intermittent milking for a few days, as a means to dry off a cow, leads to less subsequent mastitis.

Environment

General relationships between environmental temperature, milk production, and feed consumption are illustrated in Figure 18.9. Increasing environmental temperatures increase the respiratory rate which is the primary mechanism whereby European-evolved breeds of dairy cattle dissipate heat. For example, respiratory rate increases about 5-fold when temperatures rise from 50 to 105°F. Heat produced by lactating animals is about double that of nonlactating cows. Milk production and feed consumption are reduced automati-

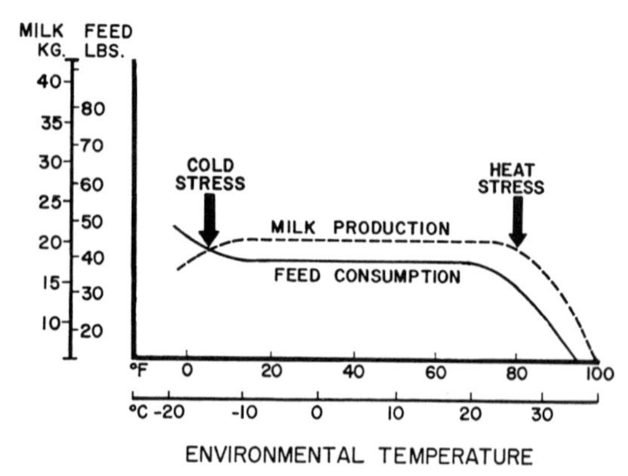

FIG. 18.9 When the environmental temperature is 75°F or above, reduced feed intake is followed by a decrease in milk production. At approximately 5°F, feed consumption increases and milk production decreases. (Courtesy of H. D. Johnson)

cally in an effort to curtail body heat production when temperatures become elevated. In fact, depressed appetite is the primary cause of reduced milk yields during heat stress.[21] Heat stress affects high-producing cows more than low producers, and heat stress is especially harmful at peak of lactation.

Milk production declines when environmental temperatures exceed 80°F for Holsteins and Brown Swiss, 85°F for Jerseys, and 90 to 95°F for Brahmans. Optimal temperature for European breeds of cattle is about 50°F. A rising temperature above 50°F is more detrimental than a similar fall below 50°F. High humidity adversely affects production only when temperatures exceed 75°F.

In general, milk fat and SNF percentages are greatest in winter and lowest in summer. Cows calving in autumn or winter produce more fat and SNF than cows freshening in spring and summer. At high temperatures (over 85°F) milk production is often reduced more than fat production which may result in a small increase in fat percentage in milk. At high temperatures there is an increase in chlorides and a decrease in lactose and protein content of milk. As temperatures decrease below 75°F, percent fat and SNF increase.

Provision of shade, use of fans, showers, or refrigerated air alleviate thermal stress. Air conditioning of cows in Florida stimulated milk yields almost 10%,[22] but costs of these elaborate systems prohibit their use on a commercial basis at present. It is perhaps more important to supply the right kind of feed and to select cows whose appetite is not readily decreased during heat stress. On the other hand, in hot, dry climates, use of relatively inexpensive evaporative cool shades in which air is blown over water reduces air temperatures as much as 12°F and stimulates milk yields 7% in comparison with unshaded cattle.[23] Insulated roof shades in subtropical humid climates, in which feed bunks and water are placed under the shades so that cows do not have to leave the shade during the hottest part of the day, increased milk yields 11% compared with cattle not shaded.[24] Major shade specifications are outlined in Table 18.2.

Milk yields for the entire lactation are usually greater when the cow calves in autumn or early winter. Yields decrease progressively if freshening occurs in winter, spring, and summer. Greater production in cows which calve in autumn is probably due to optimal temperatures, absence of flies, and more digestible feeds that are available in the autumn, winter, and early spring in comparison with summer conditions. Thus, when adverse conditions of summer are greatest, the autumn-freshening cow is usually in late stages of lactation or dry. Effect of season of calving is practically negligible when cows are fed stored feeds year round as in the dry lots of California.

TABLE 18.2 *Major Design Specifications for Shade Structures*

1. Orientation: Long axis parallel East–West.
2. Floor space: 50 to 65 ft^2/cow; concrete.
3. Height: 12 ft at lowest point.
4. Ventilation: Gable-roof with continuous open ridge.
5. Roof construction: White or aluminum; insulated.
6. Feeding and water facility: Under structure.
7. Waste management: Flushing system.
8. Adjoining dry lot: Required.

(Courtesy of R. J. Collier)

Heat stress during the last ⅓ of pregnancy reduces weight of calves at birth, alters endocrine function during pregnancy, and reduces subsequent yield of milk.[25] Provision of shade during pregnancy reduces these deleterious effects of heat stress.

Moderate exercise is conducive to high milk production, but too much or too little is detrimental. Thus, cows in stanchion barns should be turned out twice per day for exercise and heat detection. Animals on pasture require more energy than if feeds are brought to the cows. In fact, energy required to graze a poor summer pasture may double the maintenance requirement.

Disease and Drugs

Many diseases, especially mastitis, ketosis, milk fever, and digestive upsets adversely affect milk production and may alter composition of milk. These topics are discussed in Chapters 19 and 21.

Many drugs, including pesticides, used in treatment of cattle diseases are excreted into milk. Such milk should be discarded to prevent the drugs from entering the human food supply. Presence of antibiotics and pesticides in milk is illegal, and such milk is prohibited from sale. Dairymen should consult their veterinarian as to length of time milk should be withheld from market after a cow receives a drug.

18.8 Summary

The lactating mammary cell is a highly organized structure which can utilize up to 80% of available nutrients in blood for manufacture of milk. Most milk proteins and lactose are found only in milk. Milk fat of cows has a high proportion of short-chain fatty acids with a high degree of saturation. Free amino acids, glucose, acetate, and long-chain fatty acids are primary blood precursors of milk proteins, lactose, and short- and long-chain fatty acids, respectively. Limitation in any of these precursors reduces production and alters composition of milk.

In general, milk production and fat percent in milk are negatively related. Genetics, level of nutrition, stage of lactation, and environmental temperatures influence milk composition and yield. Large capacity udders and frequent and complete milking (at least 2 × per day) with attendant minimization of intramammary pressure are conducive to high milk secretion rates. Regular calving intervals (especially in autumn), dry periods of 6 to 8 weeks between lactations, older age and larger size of cows, and cool temperatures favorably influence milk yield.

Review Questions

1. Diagram cytological features of the mammary cell. Briefly outline function of each component of the cell.
2. List blood precursors of α-casein, lactose, and butyric acid in milk.
3. What do you think is the primary biochemical ingredient limiting synthesis of milk? Give reasons for your answer.
4. Speculate as to why milk fat percentage is so easily influenced whereas lactose percentage is very constant. List 5 factors that increase milk fat percentage.
5. What is colostrum? Why is colostrum important to the calf?
6. Outline a management program that maximizes milk production.

References

1. Patton, S.: Milk. Sci. Amer. *221:* 58, 1969.
2. Brown, R. E., Dept. of Dairy Science, Univ. of Illinois: Personal communication.
3. Keenan, T. W., Morre, D. J., and Huang, C. M.: Membranes of the mammary gland, in *Lactation: A Comprehensive Treatise,* B. L. Larson and V. R. Smith, eds. Vol. II. New York: Academic Press, 1974.
4. Larson, B. L.: Biosynthesis of the milk proteins. J. Dairy Sci. *48:* 133, 1965.
5. Aschaffenburg, R.: Genetic variants of milk proteins: their breed distribution. J. Dairy Res. *35:* 447, 1968.
6. Kronfeld, D. S.: Major metabolic determinants of milk volume, mammary efficiency, and spontaneous ketosis in dairy cows. J. Dairy Sci. *65:* 2204, 1982.
7. Ebner, K. E., and Brodbeck, U.: Biological role of α-lactalbumin: a review. J. Dairy Sci. *51:* 317, 1968.
8. Palmquist, D. L.: A kinetic concept of lipid transport in ruminants. A review. J. Dairy Sci. *59:* 355, 1976.
9. Bauman, D. E., and Davis, C. L.: Biosynthesis of milk fat, in *Lactation: A Comprehensive Treatise.* B. L. Larson and V. R. Smith, eds, Vol. II. New York: Academic Press, 1974.
10. Emery, R. S.: Biosynthesis of milk fat. J. Dairy Sci. *56:* 1187, 1973.
11. Peaker, M.: Recent advances in the study of monovalent ion movements across the mammary epithelium: relation to onset of lactation. J. Dairy Sci. *58:* 1042, 1975.
12. Laben, R. C.: Factors responsible for variation in milk composition. J. Dairy Sci. *46:* 1293, 1963.
13. Rook, J. A. F., and Campling, R. C.: Effect of stage and number of lactation on the yield and composition of cow's milk. J. Dairy Res. *32:* 45, 1965.
14. Tucker, H. A., Reece, R. P., and Mather, R. E.: Udder capacity estimates as affected by rate of milk secretion and intramammary pressure. J. Dairy Sci. *44:* 1725, 1961.
15. Pearson, R. E., Fulton, L. A., Thompson, P. D., and Smith, J. W.: Three times a day milking during the first half of lactation. J. Dairy Sci. *62:* 1941, 1979.
16. Schmidt, G. H., and Trimberger, G. W.: Effect of unequal milking intervals on lactation milk, milk fat, and total solids production of cows. J. Dairy Sci. *46:* 19, 1963.
17. Dodd, F. H., Foot, A. S., Henriques, E., and Neave, F. H.: Experiments on milking technique. 7. The effect of subjecting dairy cows, for a complete lactation, to a rigid control of the duration of milking. J. Dairy Res. *17:* 107, 1950.

18. Schmidt, G. H., Guthrie, R. S., and Guest, R. W.: Effect of incomplete milking on the incidence of udder irritation and subsequent milk yield on dairy cows. J. Dairy Sci. *47:* 152, 1964.
19. Johnson, A. D., Legates, J. E., and Ulberg, L. C.: Relationship between follicular cysts and milk production in dairy cattle. J. Dairy Sci. *49:* 865, 1966.
20. Dias, F. M., and Allaire, F. R.: Dry period to maximize milk production over two consecutive lactations. J. Dairy Sci. *65:* 136, 1982.
21. Collier, R. J., Beede, D. K., Thatcher, W. W., Israel, L. A., and Wilcox, C. J.: Influences of environment and its modification on dairy animal health and production. J. Dairy Sci. *65:* 2213, 1982.
22. Thatcher, W. W.: Effects of season, climate, and temperature on reproduction and lactation. J. Dairy Sci. *57:* 360, 1974.
23. Wiersma, F.: Research in Evaporative Cooling for Livestock, Semi-annual Meeting, American Society of Heating, Refrigerating and Air Conditioning Engineers, Columbus, Ohio, February, 1968.
24. Roman-Ponce, H., Thatcher, W. W., Buffington, D. E., Wilcox, C. J., and Van Horn, H. H.: Physiological and production responses of dairy cattle to a shade structure in a subtropical environment. J. Dairy Sci. *60:* 424, 1977.
25. Collier, R. J., Doelger, S. G., Head, H. H., Thatcher, W. W., and Wilcox, C. J.: Effects of heat stress during pregnancy on maternal hormone concentrations, calf birth weights and postpartum milk yield of Holstein cows. J. Anim. Sci. *54:* 309, 1982.

Suggested Additional Reading

Larson, B. L., ed.: *Lactation.* Ames: Iowa State University Press, 1985.

Larson, B. L., and Smith, V. R., eds.: *Lactation: A Comprehensive Treatise.* Vol. II. New York: Academic Press, 1974.

CHAPTER 19

The Milking Program

19.1 Introduction

The payoff in the dairy farm operation is harvesting the milk crop. While secretion of milk is a continuous process, harvesting usually occurs only twice daily. Characteristics of good milking habits include: milking at regular intervals; fast, gentle, and complete milking; use of sanitary procedures; and efficient use of labor. Persistent use of these procedures will result in higher milk yields and milk quality, less mastitis, longer life in the herd, and more profit per cow.

The milking operation requires more labor (approximately 55%) than any other single chore in the dairy barn.[1] Many improvements in milking equipment and design of milking facilities have been made in the past, but numbers of cows milked per person per hour in the U.S. have not changed appreciably since introduction of the herringbone milking parlor. However, introduction of automatic udder washing devices, automatic removal of the milking machine, and newly designed milking parlors increases milking efficiency.

19.2 Milk-Ejection Reflex

The small amount of milk present in cisterns and large ducts of the udder can be removed merely by overcoming resistance of the sphincter muscle surrounding the streak canal of the teat. However, the major portion of milk present in the udder must be forced from the alveoli and small milk ducts by the activation of a neurohormonal reflex called milk ejection or milk "let-down."

The milk-ejection reflex (Figure 19.1) involves activation of nerves in the skin of the teat which are sensitive to touch or temperature. Neural impulses ascend via the inguinal nerve (Figure 17.5) to the spinal cord then to the paraventricular and supraoptic nuclei in the hypothalamus and finally to the posterior pituitary (Figure 14.5). Milking- or suckling-induced activation of this afferent neural pathway causes discharge of oxytocin from the posterior pituitary into blood which carries oxytocin to the udder. Oxytocin diffuses out of capillaries in the udder, binds to receptors on the myo-epithelial cells, and causes contraction of myo-epithelial cells which surround the alveoli and smaller ducts (Figure 17.9). This squeezing action increases intramammary pressure and forces milk through the ducts to the gland and teat cisterns.

Contraction of myo-epithelial cells occurs within 20 to 60 seconds after stimulation of teats. A second discharge of oxytocin can be elicited late in the milking process, but it is more difficult to achieve than the first discharge, and it is usually not a fully effective response. After discharge

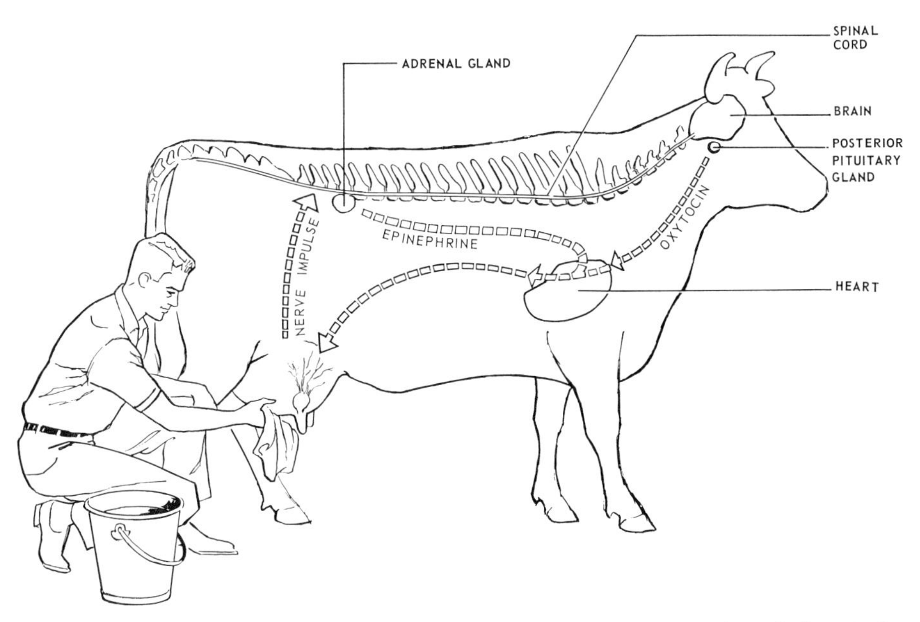

Fig. 19.1 Milk-ejection reflex. When the udder is stimulated, nerve impulses pass through the spinal cord to the brain. The nerve impulses from the brain stimulate the posterior pituitary gland to discharge oxytocin. Blood carries this hormone to the myo-epithelial cells which surround the alveoli. The myo-epithelial cells contract and force milk into the gland cistern. Excitement or pain will cause epinephrine and norepinephrine to be secreted. These hormones reduce efficacy of oxytocin. (Reprinted with permission from Univ. Calif. Agr. Ext. Serv. Pub. AXT-94, 1967)

of oxytocin flow of milk decreases with time, regardless of the amount of milk in the udder. This may be due to fatigue of myo-epithelial cells or inactivation of oxytocin. In fact, time required for one-half of the oxytocic activity in blood to disappear in the cow is only about 1 to 3 minutes,[2] and effective levels last for only 6 to 8 minutes. Thus, it is important to remove milk rapidly during the period when oxytocin causes contraction of myo-epithelial cells. Myo-epithelial cells contract in response to direct mechanical stimulation.[3] Thus, massage of the udder during machine stripping will express additional small quantities of milk from the alveoli.

External stimuli other than washing the udder will initiate the milk-ejection reflex. A potent stimulus to release of oxytocin is presentation of the calf to its mother. Other stimuli include noises associated with milking, feeding cows, presence of the milker, and coitus.

Reduced tone of smooth muscle components of the udder coincides with stimulation of oxytocin release at milking.[4] The reduced tone is associated with reduced secretion of norepinephrine in the udder and is conducive to rapid removal of milk from the udder.

The milk-ejection reflex can be inhibited also. When this occurs, only a small portion of milk can be removed from the udder. External unpleasant events at milking may cause sympathetic nerves to release epinephrine from the adrenal medulla into blood (Figure 19.1) or activate the sympathetic system in the udder which releases norepinephrine. Epinephrine and norepinephrine are potent vasoconstrictors (Chapter 14) which reduce blood supply to the udder and thereby prevent oxytocin from reaching myo-epithelial cells in sufficient amounts to cause their contraction. Injection of oxytocin at this time is ineffective. Some evidence also suggests that epinephrine can directly inhibit myo-epithelial cells from responding to oxytocin. Inhibition of the milk ejection reflex may also occur when the udder becomes engorged with milk. In this case capillary blood flow is reduced so much that oxytocin can no longer reach the myo-epithelium.

If the dairyman "keeps his cool," so will the majority of cows. A few cows do not respond to kindness, and they should be culled because they can upset other cows in the herd.

Internal emotional upsets which occur before activation of the milk-ejection reflex may prevent release of oxytocin from the posterior pituitary. Under these conditions, injection of oxytocin causes contraction of the myo-epithelial cells because vasoconstriction does not occur. This is an example of inhibition of the reflex at the level of the central nervous system. It is the most prevalent type of inhibition encountered in first-calf heifers when they first enter the milking herd. Injection of oxytocin for several milkings often corrects this problem. It is most important to remove this milk because production for the entire lactation can be reduced by incomplete milking.

19.3 Removal of Milk from the Udder

Although tone of the sphincter muscle around the streak canal relaxes during milking,[5] some external mechanism must be used to overcome the major resistance of this muscle.

Suckling

During nursing, the calf presses its tongue around the teat and against the palate and creates a negative pressure by separation of the jaws or retraction of the tongue. Positive pressure develops around the teat when the calf swallows. Between 80 and 120 alternating suck and swallow cycles occur per minute. In one experiment, calves created an average pressure differential across the streak canal of 535 mm Hg, whereas machine and hand milking created differentials of only 310 and 352 mm Hg.[6] Calf suckling is also the fastest method for evacuating milk from the udder.

Hand Milking

This practice is still employed in many countries of the world. Even in the U.S. hand milking may be performed on a short-term basis in special cases, usually associated with disease or injury, where it may be more convenient to milk by hand than to use a milking machine.

To hand milk a cow properly the teat is closed between the index finger and thumb. Then, milk in the teat is forced to the exterior as the other fingers compress against the teat. Next the index finger and thumb are relaxed to allow the teat cistern to refill, and the cycle is repeated. Superior hand milkers can usually obtain more milk from cows than milking machines.

Machine Milking

Milking machines of one sort or another have been available since 1859. Modern milking machines use alternating negative and atmospheric pressures, for which a double-chambered teat cup assembly is required (Figure 19.2). The chamber into which the teat is inserted is under continuous vacuum to retain the teat cup on the teat, and to create a pressure differential across the streak canal which forces milk from the teat cistern into the teat chamber of the teat cup. Negative pressure and atmospheric pressure are allowed alternately to enter the chamber between the rubber inflation (liner) and metal shell of the teat cup assembly. When atmospheric pressure enters the chamber, the rubber liner collapses around the teat. This assists blood and lymph to flow out of the teat. If atmospheric pressure is not introduced, the teat under continuous vacuum becomes congested. The amount of time the inflation is expanded compared with the time it is collapsed is termed the pulsation ratio. Ratios from 1:1 to 2.5:1 are most popular. Except when milk flow is rapid, milk does not flow through the streak canal when the rubber inflation is collapsed. When atmospheric air is evacuated, the inflation quickly returns to the resting state (left panel of Figure 19.2) and milk flow is maximum.

Complementary Milk

About 15 to 25% of total milk in the udder at the start of milking usually remains in the udder when milking is completed. This milk is termed complementary, or residual, milk. It can be obtained if oxytocin is injected and the cow milked a second time. Less complementary milk is obtained from first-calf heifers than from mature cows, and high-producing cows usually have less complementary milk than low-producing cows. Faster-milking cows

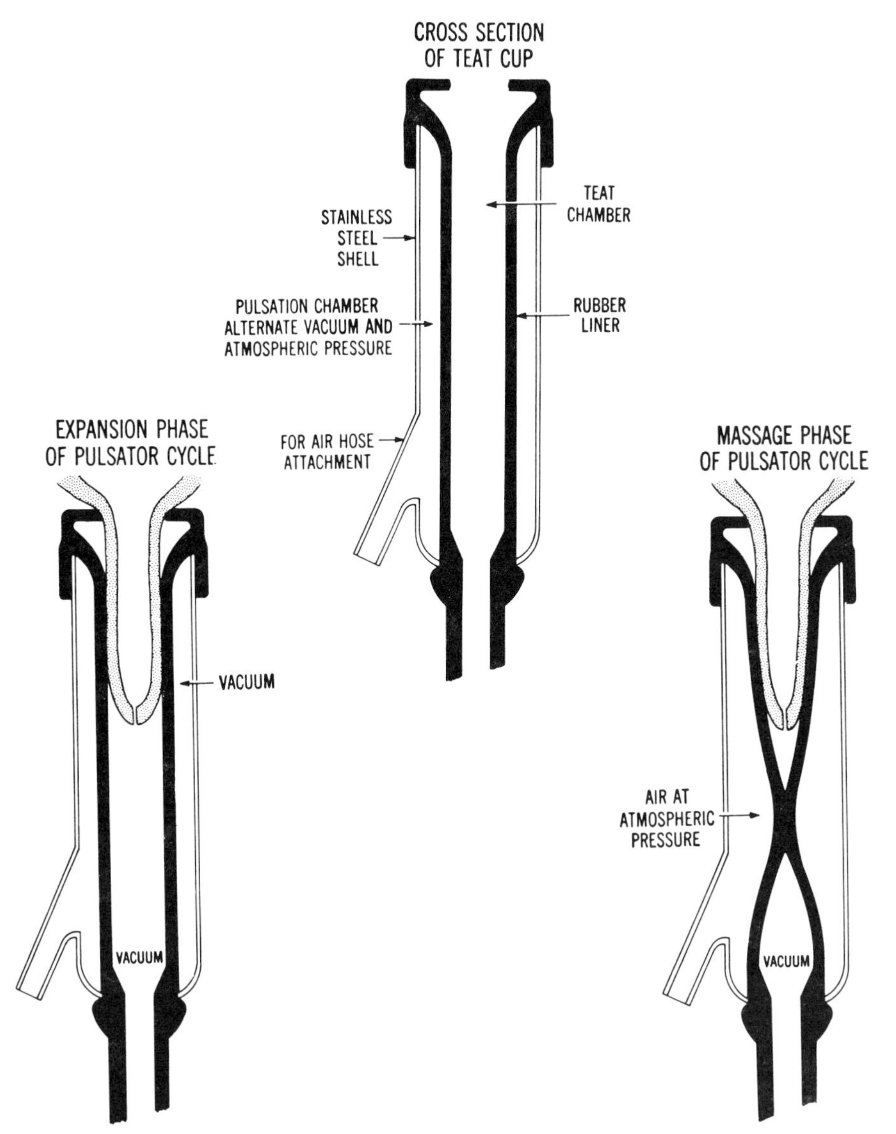

FIG. 19.2 Diagrams of the teat cup at various phases of milking. During expansion phase *(left)* the rubber liner is held by vacuum in close association with the metal teat cup shell, and milk flow rate is maximal. During the massage phase *(right)* atmospheric air enters and collapses the rubber liner around the teat, and milk flow is minimal. There is constant vacuum in the teat chamber of the teat cup. (Reprinted with permission from *The DeLaval Handbook of Milking*, 1963; courtesy of The DeLaval Separator Co.)

usually have less complementary milk, are more persistent, and thus have a higher yield for a total lactation.

19.4 Principles and Practices of Proper Machine Milking

Excellent or poor milking habits can be developed in both cows and dairymen.[7,8] If dairymen train themselves to maintain a calm atmosphere in the barn and to practice good milking habits, most cows will respond favorably. Recommended practices for machine milking are outlined in Table 19.1.

To achieve maximum milk yields dairymen should learn the individual habits of each cow. There is no such thing as a standard cow, especially when it comes to milking habits. With expanding herd sizes, group milking in herringbone parlors, and emphasis on milking more cows per person per hour, this goal of individual attention is

Table 19.1 *Recommended Practices for Machine Milking Cows*

1. Establish a regular routine and standard milking interval.
2. Maintain and operate the milking machine in accordance with manufacturer's directions.
3. Wash the udder and teats. Dry teats.
4. Remove 1 or 2 streams of milk into strip cup or paddle for cow side mastitis test.
5. Apply milking machine within 1 minute after start of wash.
6. Remove milking machine promptly when milk flow stops. Break vacuum first.
7. Apply teat dip to teats.
8. Record milk weights (Chapter 3).

becoming more difficult to achieve. In fact it is probably more economical in large commercial dairies to milk a uniform group of cows as rapidly as possible and to sell those cows that require too much individual attention.

Preparation of the Cow

The teats and lower portion of the udder should be washed with a warm sanitizing solution. A disposable paper towel should be used for each cow. Washing not only removes dirt, but also initiates the milk-ejection reflex. The sanitizing solution should be changed periodically because washing cows with contaminated disinfectant solutions can spread mastitis. Effectiveness of disinfectants decreases as organic matter content increases. Automatic washing devices which spray water on the udder are gaining in popularity, especially in larger dairy farms (Figure 19.3).

Teats should be dried with a disposable paper towel (Figure 19.4) and 1 or 2 streams of milk carefully drawn from each teat into a strip cup (Figure 19.5). Care should be taken to prevent formation of an aerosol. Formation of aerosols may transmit organisms from teat to teat. The first milk drawn is always higher in leukocytes and bacteria and should be discarded. Also, this practice of removing 1 or 2 streams of milk is a quick screening test for abnormal milk (Section 19.8). Milk containing flakes, strings, blood, or other signs of abnormality probably indicates mastitis and should be discarded. Forestrippings should not be squirted onto the floor in stanchion barns, because organisms can be transmitted to other quarters or even other cows. However, this practice is acceptable in milking parlors with grates in the floor

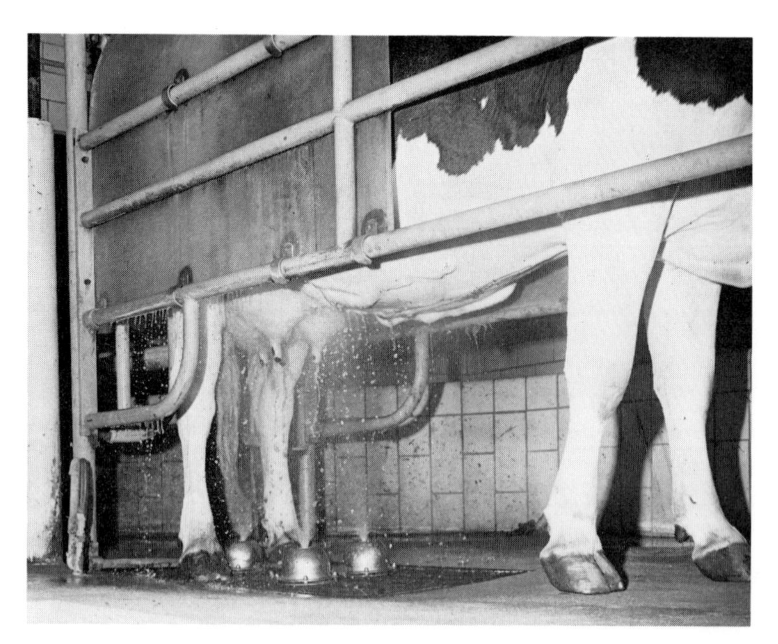

Fig. 19.3 Automatic udder washing device. (Courtesy of L. J. Bush; reproduced with permission from October 1976 issue of Hoard's Dairyman. Copyright 1976 by W. D. Hoard & Sons Company, Fort Atkinson, Wisconsin 53538)

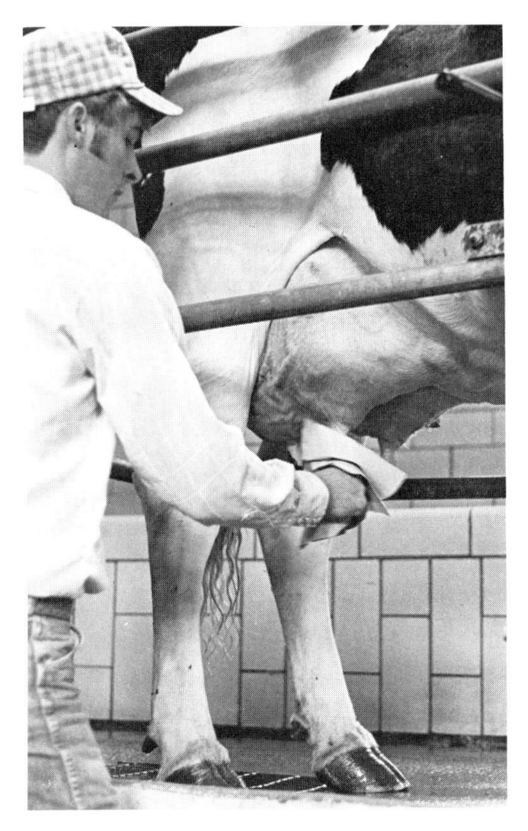

FIG. 19.4 After washing, dry teats and udder with individual paper towel. (Courtesy of L. J. Bush; reproduced with permission from October 1976 issue of Hoard's Dairyman. Copyright 1976 by W. D. Hoard & Sons Company, Fort Atkinson, Wisconsin 53538)

since infected milk can be washed away conveniently without contacting other cows. To achieve maximal release of oxytocin, washing, drying, and removing foremilk (stimulation time) requires 30 seconds. Cows stimulated for only 15 seconds or less require more time to milk.

The milking machine unit should be applied within 60 seconds after the start of the udder wash since effective levels of oxytocin remain in the blood for only 6 to 8 minutes. As shown in Table 19.2, as interval between washing and application of the milking machine increases, time required to milk the cows increases.

Physiological Factors Affecting Milking Rate

Teat cups should be carefully positioned on the teats. Speed with which a cow is milked is of economic importance in the

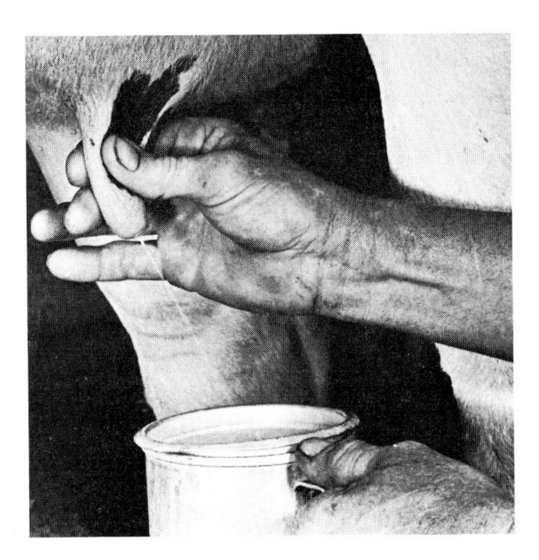

FIG. 19.5 Use of strip cup before milking provides a warning of abnormal milk. (Reprinted with permission from *Modern Mastitis Management,* 1970; courtesy of The Upjohn Company)

TABLE 19.2 *How Milking Time Was Affected by Time Lapse between Preparation and Attachment of Milker Unit*

Preparation lag time	*Average time milkers on cows*	*Number of farms*
30 seconds to 1 minute	4 min. 51 sec.	17
1 to 3 minutes	5 min. 31 sec.	24
3 to 6 minutes	6 min. 46 sec.	10
6 or more minutes	6 min. 12 sec.	5

Source: Fryman and Albright, Univ. of Illinois Circ. 851, 1962.

dairy farm operation. Physiological traits of the cow, certain physical properties of the milking machine, and management of the milking operation affect speed of milking. The average cow when milked correctly requires about 5 minutes to milk.

Size of the opening of the streak canal and tension of the teat sphincter muscles influence rate of milk flow. The ideal is to have just enough tonus on the streak canal to prevent milk from leaking before milking. Milk flow rates are slower in cows with longer teats. Since these characteristics are inherited, some selection may be warranted (Chapter 6).

In general, greater milk production induces faster milk flow from the udder per unit of milking time. However, cows producing large quantities of milk often require a longer total milking time since increased rate of milk flow cannot entirely compensate for increased level of milk production.

Machine Factors Affecting Milking Rate

Most commercial milking machines in the U.S. operate very effectively between 11 and 15 inches of Hg (a measure of vacuum). Operating milking machines at vacuum levels above those recommended initially causes faster milk flow. However, the teat cups tend to crawl up the teats and partially occlude the teat cistern. This increases the time required to obtain milk and may irritate the delicate tissue lining the teats.

Fluctuations in vacuum level in the teat cup decrease milk flow rates. The most common causes of fluctuating vacuum are teat cups flooded with milk, insufficient vacuum levels or lack of reserve vacuum, transporting milk above the udder, insufficient diameter of the air admission holes in the teat cup assembly, and simultaneous pulsation of all four teat cups rather than in pairs when the teat cup assembly is too small. Frequently the claw of the milking machine floods with milk before the teat cup floods.

Pulsation ratios, as previously stated, usually vary from 1:1 to 2.5:1, depending upon manufacturer and when the system was installed. Most new systems feature wider ratios because they milk cows faster. These machines should be removed sooner than machines set at the 1:1 ratio. However, the primary criterion as to when to remove the machine is always based upon cessation of milk flow.

The pulsation rate, which is the number of cycles of alternating negative and atmospheric pressure, is usually set around 48 to 60 cycles per minute. Pulsation rate has little effect on the speed of milking unless excessively slow or fast speeds are used. Then milk flow rates decrease.

Claw assemblies from most manufacturers weigh 4.5 to 5 pounds; if assemblies including inflations weigh less, slow milking may occur. Excessive weights or tension cause teat cups to fall off.

It is generally agreed that narrow-bore rubber liners ($<3/4''$ inside diameter) milk cows better than wide-bore liners and result in better udder health. Metal teat cup shells should match the bore of the rubber liners. Do not put narrow-bore liners in wide-bore shells.

The number of milking machines one person can handle successfully depends on type of barn or milking parlor, type of milking system, and ability of the milker. In a stanchion barn, more cows per hour are milked with 2 units per person than with 3 units per person. Even in present herringbone milking parlor systems in the U.S., one person usually can handle adequately no more than 4 units when no automation is present. Depending upon the individual parlor installation the number of cows milked per hour per person usually varies between 30 and 44. Addition of automatic milking machine detachers eliminates the second person in the pit of double 6 to 10 herringbone parlors. However, greater profits are to be gained if dairymen use practices that increase the number of pounds of milk harvested per person per hour rather than practices that affect only numbers of cows milked per hour.

Removal of Teat Cups

Cessation of milk flow and a collapsed, pliable udder are best indicators of when the udder is milked out. At this time the dairyman may briefly massage the udder and place some tension on the claw of the milking machine to remove strippings. Machine stripping should be brief or not practiced (Chapter 18). Vacuum supply to teat cups should be shut off and then teat cups gently removed. Use of automatic milking machine removal devices, which increases labor efficiency, will probably increase in the future, especially in larger herds.

After the milking machine is removed, teats should be dipped in an antiseptic iodophor, chlorhexidine, sodium hypochlorite, quaternary ammonium, dodecyl benzene sulfonic acid, or acrylic latex solution especially designed for such purposes (Figure 19.6). Avoid irritating solutions not designed for dipping teats, and do not use teat dips that contain more than 10% oil or lanolin bases. The dipping procedure disinfects teats and removes the film of milk which normally remains on the outside of teats after milking. This practice will significantly reduce new infections.[9] The germicidal dips are most effective against gram-positive organisms, whereas latex dips are most effective against coliform, pseudomonas and other "environmental" pathogens. Freezing of teat dips must be avoided because chemical composition of the dips may change and subsequent use may irritate the teat. Precautions must be taken in freezing weather to prevent freezing of dipped teats; either retain cattle in a warm environment until the teat dip dries or lightly wipe off excess dip before exposing cows to freezing weather.

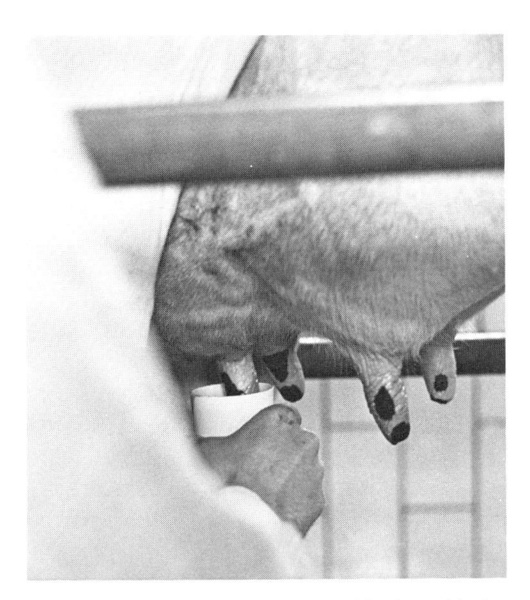

Fig. 19.6 Dip each teat in iodine, chlorine, chlorhexidine or other antiseptic solution immediately after removal of milking machine. (Courtesy of L. J. Bush; reproduced with permission from October 1976 issue of Hoard's Dairyman. Copyright 1976 by W. D. Hoard & Sons Company, Fort Atkinson, Wisconsin 53538)

Immersion of teat cups in lukewarm water in a pail followed by immersion in a sanitizing solution between the milking of every cow does not reduce spread of pathogenic organisms. However, use of an automatic backflush system between each cow essentially sterilizes teat cup liners and reduces spread of organisms.

19.5 Components of Milking Machine Systems

Complete discussion of the infinite variations in milking machine systems is beyond the scope of this book. Furthermore, many systems are customized for a particular dairyman's needs. Thus only major components of the milking system, i.e., vacuum pumps and lines, pulsators, and milking units will be described. More details can be obtained from books listed at the end of this chapter and from persons who service milking machines.

Vacuum Pumps and Lines

Most common types of pumps used in milking operations are piston, rotary vane, or centrifugal water displacement pumps. Piston pumps are dependable and operate at low to moderate speeds. Piston pumps were used extensively in the past, but rotary vane or centrifugal water displacement pumps are used on most new systems because they move more air than piston pumps of the same horsepower; thus they are more efficient.

Type of pump is not as important as capacity. Pumps in pipeline systems should have capacity to move 15 to 20 cubic feet per minute (CFM) of free air per milking unit in 2 to 4-unit systems (American Standard). Five milking units require 12 CFM, 6 to 12 units require 8 to 12 CFM, whereas 12 or more units require 8 to 10 CFM per unit. Fluctuating vacuum level is a more important contributory cause of mastitis than high vacuum level. Thus, the need for a large air flow reserve is evident.

A typical vacuum system is shown in Figure 19.7. Vacuum supply lines are usually made of polyvinyl chloride or galvanized iron. Vacuum supply lines (1¼ to 3 inch diameter) from the vacuum pump should enter the middle of a header tank, exit from top of the header, and enter top of a vacuum reserve tank. Vacuum pulsator lines (1½ to 3 inch diameter), should be as short as possible and form a complete loop back to reserve tank. Large size piping extends vacuum reserve throughout the system and minimizes vacuum fluctuations. Vacuum lines should be sloped 1½ inches per 10 feet toward the vacuum pump and have drain plugs for cleaning. Stall cocks should be tapped directly into the vacuum line.

In bucket systems the vacuum regulator, which admits air to prevent vacuum from exceeding a prescribed level, should be located between the pump and first stall cock. Vacuum regulators for pipeline systems should be installed between the header and vacuum reserve tank or between the reserve tank and sanitary trap. The area should be clean, dry and accessible for regular cleaning of the regulator. Vacuum level at the teat end should be maintained at approximately 11 to 12 inches Hg during peak milk flow. Diaphragm regulators stabilize vacuum better than dead weight or spring-loaded regulators. Capacity of the regulator must be at least equal to capacity of the vacuum pump.

In pipeline systems a vacuum gauge should be installed on the line that supplies vacuum to the sanitary trap. In bucket systems a gauge may be installed close to the vacuum pump.

A vacuum reserve tank with a minimum capacity of 30 gallons is recommended. With systems employing more than 6 milking units, capacity of the vacuum reserve tank should be 5 gallons per milking unit. The vacuum reserve tank should be equipped with a drain plug to facilitate removal of liquids.

The vacuum supply line to the milk pipeline should exit from middle of the vacuum reserve tank and connect to a sanitary trap. A stainless steel or glass milk pipeline should connect the sanitary trap to a milk receiver jar (stainless steel or glass) in high or low pipeline milking systems. The milk transport pipeline connects to the milk receiver jar.

Pulsators

The pulsator is a valve that alternately admits atmospheric air pressure to and then removes air from the chamber between the rubber liner and metal teat cup shell (Figure 19.2). Electrically controlled pulsators are more reliable and require less maintenance than master or mechanical-type pulsators.

Pulsation systems are available which

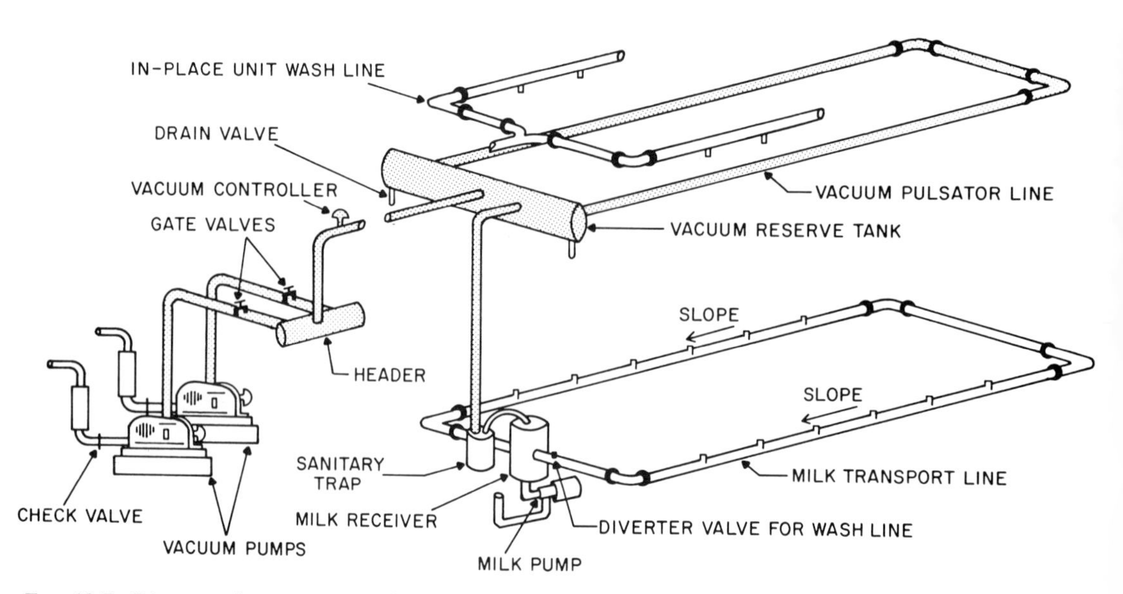

Fig. 19.7 Diagram of a vacuum supply system to the milking machine.

milk all 4 quarters simultaneously or which milk 2 quarters while resting the other 2 quarters. The first system works well if the teat cup assembly is designed to accept the higher milk flows from all 4 quarters and if vacuum reserves are adequate.

Milking Unit

Essential components of the milking unit are the teat cup assembly (Figure 19.2), air and milk hoses, and claw or bowl to hold or convey milk (Figure 19.8).

Milk Transport

Milk is transported via gravity or vacuum from the teat cup assembly to either: a suspended bucket, a bucket on the floor, a calibrated weigh jar or meter device and then into the milk line, or to the milk line directly. Use of a weigh jar or calibrated milk meter permits keeping of daily individual cow milk production records, and as stressed in Chapter 3, records are essential to a profitable dairy farm operation.

Permanent milk transport lines (Figure 19.7) may be constructed of stainless steel or glass, whereas portable systems use plastic hose. Milk lines are usually 1½ to 3 inches in diameter, and unless weigh jars are used, milk lines should be looped and contain no dead ends. Diameter of milk lines should be of sufficient size to prevent filling the line more than one-half full of milk. Double 6, 8, or 10 herringbone milking systems usually employ larger diameter pipelines. Milk and air should enter the top half of the pipe to avoid excessive vacuum fluctuations at the teat cup. Unless weigh jars are used, milk lines should be located below the udder and should be sloped a minimum of 1½ inches per 10 feet. Low lines are relatively easy to install in milking parlors, but difficult to install in stanchion-type barns.

With pipeline systems, milk is usually delivered to the side of a receiver jar, which in newer systems is located in the pit of the milking parlor. Milk is then transported by a positive pressure sanitary pump (Figure 19.7) from the receiver jar

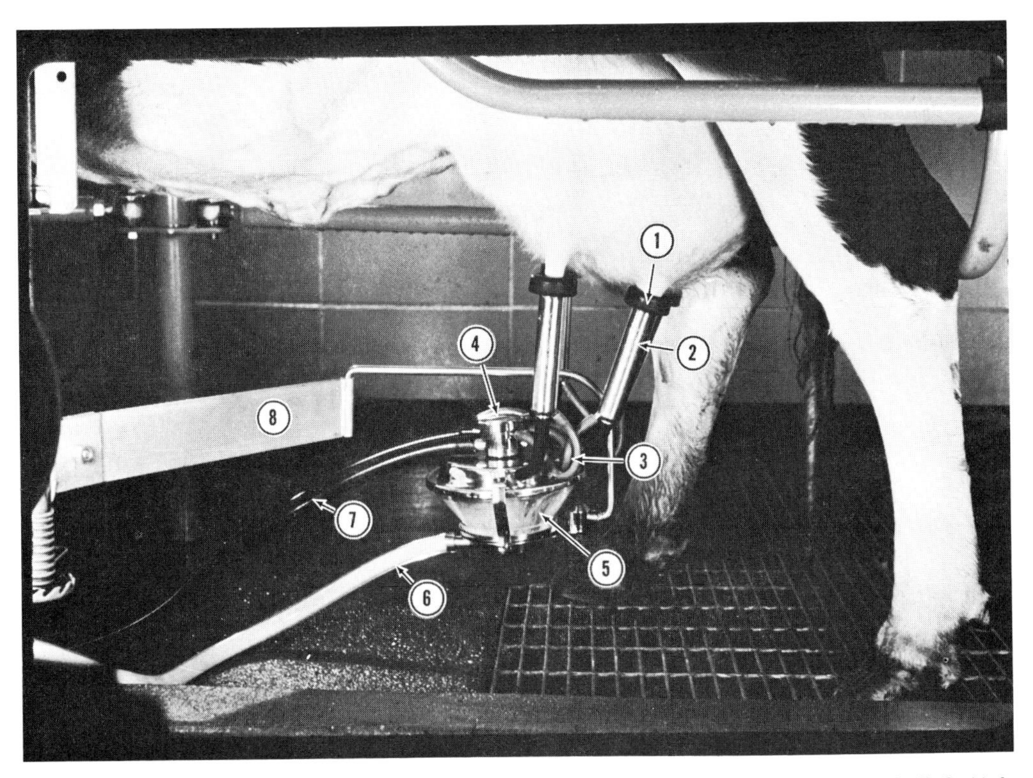

Fig. 19.8 Typical components of the milking machine. 1. Rubber inflation or liner; 2. Teat cup shell; 3. Air hose; 4. Pulsator; 5. Bowl; 6. Milk hose; 7. Vacuum hose; 8. Suspension arm of milking unit. (Courtesy of D. V. Armstrong)

through a stainless steel line, through a filtering device, and then into a bulk tank.

19.6 Sanitation of the Milking Machine System

Bacteria adversely affect milk quality. Methods and thoroughness of cleaning and sanitizing milking equipment are important factors that affect bacterial counts in milk.[7,8,10] Earlier, slow cooling of milk contributed to rapid bacterial growth, but use of bulk tanks largely eliminates this problem.

Rubber parts of the milking machine are especially troublesome to clean. Milk solids, especially fats, accumulate in pores of natural rubber and provide an excellent nutrient medium for bacteria. Mineral deposits are more likely to accumulate on synthetic rubbers. Silicone rubber is more resistant to accumulation of fats and minerals. Accumulation of fat or minerals and development of cracks in inflations eventually result in poor teat massage and slow milking. Molded inflations must be replaced after 1000 to 1200 individual milkings whereas stretched natural rubber inflations need to be replaced after 500 to 600 milkings. Silicone liners should be replaced after 6,000 to 10,000 milkings. Replace inflations in groups of 4, not one at a time.

Effectiveness of sanitizers is influenced by temperature, pH, sanitizer concentration, length of exposure, amount of contaminating organic matter, and milk solids. Effective procedures for stainless equipment requiring hand cleaning include rinsing with lukewarm water, brushing with hot alkaline detergent solution, rinsing, and then treating all utensils that come in contact with milk with a sanitizing solution. Use of soft water increases effectiveness and reduces quantity of cleaners and sanitizers needed. Rinsing equipment immediately after milking is especially important in effective sanitation of milking equipment.

Milk transport lines, if properly installed, are easily cleaned and sanitized in place (Figure 19.7). Pipeline milking systems are first flushed with lukewarm water (105°F) and then an alkaline detergent is circulated for 10 to 15 minutes. The pipeline should then be sanitized for about 5 minutes with a lukewarm sanitizing solution. Periodically a solution of organic acid detergent should be substituted for the alkaline detergent. The pipeline should be drained thoroughly before permitting entry of milk.

The vacuum or pulsator line is often overlooked in sanitation of the milking system. However, the pulsator or vacuum line should be cleaned 4 times per year for bucket milkers, 2 times per year for pipeline milkers, and every time milk is suspected of entering the line. Excessive foaming of milk may lead to contamination of the vacuum line in a bucket system. A foamless alkaline cleaner (1 to 2% lye) should be used. Insertion of one quart at several stall cocks is satisfactory. Be sure the vacuum pump does not become flooded with cleaning solution. A portable trap inserted in front of the pump reduces the chance of alkaline fumes and fluids entering the pump. Remove the dirty water from the vacuum reserve tank and rinse with an acid cleaner followed by hot water. Finally run the vacuum pump with stall cocks open to dry the pulsator or vacuum line.

19.7 Maintenance of the Milking Machine System

A properly operating milking machine should milk cows efficiently without irritating the udder. Nonetheless, problems associated with the milking machine often arise, and any milking system may wear out or become obsolete after a few years.

Signs of malfunctioning milking systems include: (1) teat cups falling off; (2) excessive vacuum fluctuations; (3) flooded milk lines and uneven milk flow; (4) slow return of vacuum level after an air leak; (5) slow milking; and (6) excessive teat erosion. These symptoms indicate that the system should be tested for malfunction.[8] However, some of these problems develop gradually, and dairymen are often unaware of them.

Performance of every milking system should be thoroughly checked at least twice yearly under a full operating load. Test equipment available includes air flow meters, vacuum recorders, and gauges.

An overloaded milking system is a common problem. There are 2 solutions. Either reduce the number of milking units to match air flow capacity of the system or increase pump and air flow reserve capacity of the system to handle the number of units employed.

19.8 Management Aspects of Mastitis

Mastitis is an inflammatory reaction of udder tissue to bacterial, chemical, thermal, or mechanical injury.[11] The inflammatory response consists of increased blood proteins and white blood cells in mammary tissue and in milk. The purpose of the inflammatory response is to destroy or neutralize the irritant, repair tissue damage, and return the udder to normal function.

Approximately 5 to 10% of all cows produce abnormal milk at any given time, and 40% of cows in the U.S. are infected with pathogenic bacteria in 2 or more quarters.

Mastitis reduces milk production and milk quality, and is a frequent reason for culling cows. In addition, mastitis increases labor costs and veterinary medical treatment costs. The average dollar loss to dairymen in the U.S. amounts to approximately $200 per cow per year. However, these losses can be reduced substantially by following sound management practices.

Bacterial Causes of Mastitis

Bacteria are by far the most common causes of mastitis and *Streptococcus agalactiae* and *Staphylococcus aureus* are the chief pathogens involved. But other bacteria, including *Streptococcus dysgalactiae, Streptococcus uberis,* Pseudomonas, Corynebacteria, and coliforms may be important causative agents in certain herds.

Actually it is not bacteria per se, but rather their toxins that produce inflammation. Persisting inflammation leads to tissue damage and replacement of secretory cells of the udder by nonproductive connective (scar) tissue. This loss of secretory tissue reduces milk production.

Yeasts also cause mastitis, and penicillin therapy may enhance growth of yeast within the udder. Overuse of antibiotics and poor sanitation, especially multiple use of syringes without proper sterilization, contribute to yeast mastitis.

Clinical Definitions of Mastitis

Subclinical mastitis is a form of the disease in which there is no swelling of the gland or gross abnormality of the milk, but certain changes in the milk can be detected by special tests outlined below. There are 15 to 40 cases of subclinical mastitis for each case of clinical mastitis.

Clinical mastitis indicates that gross abnormal conditions exist in the udder and milk. Milk may show flakes, clots, or a watery appearance. Severe forms (acute mastitis) may involve a sudden swelling of the infected quarter which may be hot, hard, and sensitive to touch. This type of mastitis may become systemic with signs of fever, rapid pulse, depression, weakness, and loss of appetite.

Chronic mastitis implies a persistent udder infection that exists continuously. On occasion these cases develop into the clinical form. Following these flareups there is a return to the subclinical form.

Diagnosis of Mastitis

For treatment purposes, the specific organism causing mastitis should be identified by a veterinarian from bacteriological cultures of the milk. Treatment aspects are presented in Chapter 21.

Presence of a pathogen in milk and inflammation of the udder signify infection. The inflammatory response which results in abnormal milk is usually detected by the dairyman or dairy plant personnel. The flakes in milk are congealed leukocytes, secretory cells, and protein. Use of a strip cup to determine presence of abnormal milk is an example of a test for inflammation.

Most screening tests for inflammation measure the number of leukocytes and mammary cells in milk. Noninfected quarters usually contain less than 500,000 cells per ml of milk, whereas higher counts may signify inflammation. However, noninfected cows subjected to high temperatures, such as those encountered in the South of the United States in the summer, and noninfected cows at the beginning and end of lactation frequently secrete milk which contains more than 500,000 cells per milliliter. Foremilk and strippings are especially high in cell numbers, and older cows have more cells in their milk than younger cows. Since these are normal responses they should not be considered to be inflammatory responses in the usual sense. These considerations emphasize that the mastitis screening tests outlined below must be interpreted with some caution. They are useful in quickly detecting early stages of inflammation, but they should not be used as the only basis for treatment. This should be reserved for the veterinarian.

Most common cow-side tests for udder inflammation, in addition to the strip cup, are the California Mastitis Test (CMT), Milk Quality Test (MQT), and the Whiteside Test. These tests involve mixing 1 or 2 streams of milk with an equivalent amount of their respective reagents in a 4-compartment paddle (Figure 19.9). Increased gel formation signifies higher numbers of white blood cells and other somatic cells in milk (Table 19.3).

Other tests for abnormal milk or inflammation, usually conducted in a laboratory or dairy plant, include the Wisconsin Mastitis Test (WMT), catalase test, and direct microscopic cell count of milk.[12] More recently, electronic cell counting devices have been developed, and many states have DHIA programs whereby somatic cells of milk from individual cows are quantified monthly. This is an excellent method to monitor general health of the udder. At

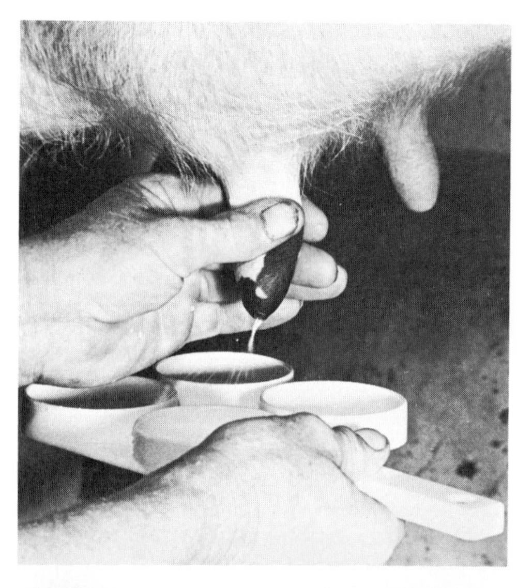

Fig. 19.9 Use of a typical cow-side test for abnormal milk. Squirt milk into test paddle *(top left)*. Add equivalent amount of mastitis test reagent *(top right)*. Evaluate amount of gelation of mixture to determine degree of inflammation *(bottom)*. (Reprinted with permission from *Modern Mastitis Management*, 1970; courtesy of The Upjohn Company)

Table 19.3 *Grading and Interpretation of the Cow-side Gelation Tests for Udder Inflammation*

Score	Description of reaction	Approximate cell count in milk
Negative	No visible gel	<200,000/ml
1	Slight gel is observed when paddle is tipped	≈500,000/ml
2	Heavier gel which does not move toward center when paddle is swirled	≈1,500,000/ml
3	Thick gel which moves quickly to center when paddle is swirled	≈5,000,000/ml
4	Thick gel which adheres to bottom of paddle	>5,000,000/ml

least one of these tests is routinely applied, on a herd basis, to all milk shipped interstate.

Factors Influencing Susceptibility to Mastitis

Many factors predispose the cow to an invading organism or they allow a preexisting low-grade infection to become more active.

Bacteria. Some organisms are more virulent than others in causing mastitis. Also some strains within a genus of bacteria show varying abilities to cause severe mastitis. Following invasion, establishment of the disease depends upon ability of the bacteria to grow in the udder and produce substances toxic to secretory tissue. Some infections will disappear spontaneously, some will be eliminated by antibiotic treatment, and others are never eliminated. Leukocytes in milk can phagocytize the invading organism and probably account for a large portion of cases that undergo spontaneous recovery. Introduction through the streak canal of as few as 35 colony forming units of *Streptococcus agalactiae* or 3 colony forming units of *Staphylococcus aureus* organisms can result in infection. In view of the large numbers of these pathogens on skin of the teat and in other places in the environment of the cow, it is surprising that cows are not continuously infected in all quarters.

Cow. Bacteria penetrate the streak canal to become established in the udder. There is a sebum-like mass of cells and lipid substances that line the streak canal which provide a seal on the teat and also exhibit bactericidal properties. Thus, the streak canal of the teat is the primary barrier to infection. Bacteria may be drawn inside the udder by the physical forces of milking, and some bacteria grow through the streak canal. Size and shape of the teat or length of the streak canal are of relatively little importance in predisposition of a cow to infection. However, cows with weak teat sphincter muscles tend to leak milk and have increased susceptibility to infection. Entry of bacteria often follows injury and local inflammation in the end of the teat.

Incidence of mastitis increases with increasing age of the cow. Nonetheless, it is possible for udders of first-calf heifers to be infected at parturition.

Within the udder neutrophile polymorphonuclear leukocytes mobilize to phagocytize microorganisms.[13] More than 50 million neutrophiles per milliliter of milk are commonly found in an infected quarter of the udder, yet pathogenic bacteria may remain viable. Indeed, phagocytic ability of leukocytes varies markedly among cows which may explain some of the variation in resistance to mastitis.

Level of milk production is not directly related to incidence of mastitis.[11] However, other factors which affect milk yield, such as milking rate and pendulous udders, may be related to mastitis incidence. These are discussed below. Although bacteria can infect an udder at any stage of lactation, the majority of new infections become established during the first 3 weeks of the dry period and during the first month after parturition. These facts further substantiate the idea that milk production is not directly related to mastitis because in these two physiological states the cow either secretes no milk (dry period) or production is maximal (first month of lactation).

It has been surmised that hormones, especially estrogens in the diet of the cow or produced in the cow herself, are involved in predisposing cows to mastitis. As yet, however, no hormone has been shown to render cows more susceptible or more resistant to mastitis.

Susceptibility to infection is an inherited

characteristic. Length of the leg in proportion to udder size and relative strength of udder attachments are examples of this. Large, pendulous udders tend to exceed capacity of the supportive ligaments, with a consequent breakdown of the udder, which subjects it to more physical injuries and makes it more susceptible to infection and clinical mastitis. Furthermore, milking rate, which depends in part on dilatability of the teat orifice, is an inherited characteristic that is related to susceptibility to infection. Selection of bulls which produce daughters with increased resistance to mastitis may become more widespread in the future.

Milking Machine. Improper use of the milking machine is related to tissue irritation[14] and incidence of mastitis. But some of the conclusions are based on presumptive evidence and have not always been substantiated in controlled experiments. For example, there is little conclusive evidence to indicate that milking with a high vacuum predisposes a cow to mastitis, although teat erosions, cyanosis, and edema of the teat ends develop more readily when vacuum levels reach 18 to 20 inches Hg. Teat erosions provide a place for organisms to grow, and this increases the chance for new udder infections to become established.

Irregular fluctuation of vacuum level is an important contributing factor to susceptibility of cows to mastitis. There is a tendency for increased mastitis with pulsation ratios of 70:30; optimum pulsation ratio for least mastitis is 62:38.[15] Pulsation rate is of little importance in etiology of mastitis provided vacuum levels and teat cup assembly meet manufacturer's specifications. Narrow-bore inflations reduce incidence of clinical mastitis.

The milking machine should be removed promptly when milk flow ceases, because overmilking may contribute to increased infection rates, clinical mastitis, and elevated leukocyte counts. Slippage of inflations on the teat create irregular fluctuations in vacuum at teat ends and contribute to a greater chance of establishment of a new infection. There is greater frequency of inflation slippage and bacterial impacts on the streak canal when milk flows are low; e.g., at the end of milking and during machine stripping. Thus, air leaks and machine stripping should be avoided. Vented inflations and low milk lines reduce retrograde impact of bacteria on teat ends. Incomplete milking also may increase incidence of mastitis. Some evidence suggests that cows milked 3 × per day have less mastitis than cows milked 2 ×.

Environment. Mastitis often increases when cows are initially turned onto pasture. Furthermore, chilling of the udder in spring or autumn increases clinical mastitis. Adequate bedding is recommended for protecting the udder from cold.

Housing, as it relates to degree of udder and teat injuries and to improper ventilation and dampness, influences incidence of mastitis. Fewer teats are stepped on in loose-housing or free-stall systems than in stanchion barns. Yet, there is insufficient data to indicate whether type of barn affects mastitis.

Nutrition. Feeding of high levels of grain or special components of grain does not influence mastitis. Fresh legume forages and silage are claimed by some to increase occurrence of clinical mastitis, but research evidence suggests this is not true. There is no evidence that adding vitamins or minerals over those amounts needed for normal lactation help to control mastitis.

Mastitis Control Program

Prevention is the key to mastitis control programs. A control program should emphasize those factors that reduce the rate of new infections.[16] To date, vaccines are of limited value in mastitis control programs, but a combination of preventive measures and therapeutic use of antibiotics markedly reduce incidence of mastitis.

Bacteria are most frequently transmitted to the uninfected udder from the contaminated hands of the milker, by wash cloths, or by teat cup liners. The milker's hands should be thoroughly washed in disinfectant soap before milking. Clinically infected cows should be milked last. The udder should be washed in a sanitizing solution with individual paper towels and after milking the teats should be immersed in an appropriate teat-dip solution. Backflushing systems are justified in larger herds. The teat cup assembly, milk pipelines, and other utensils should be cleaned and sanitized between each milking.

These hygienic procedures may reduce by 50% or more the incidence of new infections caused by streptococci and *S. aureus,* although new coliform infections are not reduced. Because of existing infections these preventive measures only reduce mastitis incidence by about 14% within a year.[17] Thus, preventive measures must be combined with treatment of existing infections, as prescribed in Chapter 21.

19.9 Summary

Efficient, rapid removal of milk from the udder should be the goal of every milking program because milking is the largest single labor-requiring job on the dairy farm. Udder and teats should be washed with a warm sanitizing solution for 30 seconds. Washing initiates discharge of oxytocin, which in turn causes contraction of myoepithelial cells. This is essential for expulsion of milk into ducts and cisterns. The udder and teats should be dried. A few streams of milk should be removed from each quarter and checked for flakes as an indication of mastitis before the machine is attached to teats. The milking machine should be attached to teats within 1 minute after starting to wash the udder. As soon as milk flow ceases the teat cup assembly should be carefully removed. Dipping teats in a proper disinfectant solution after milking effectively prevents mastitis.

Average milking times should be about 5 minutes. As with any mechanical device, milking machines wear out, and parts must be replaced. Signs of a malfunctioning milking system include: teat cups falling off, vacuum fluctuations, flooded milk lines, slow return of vacuum, slow milking, and severe teat erosions. Milking equipment should always be operated according to manufacturer's specifications. Routine service of milking equipment at least twice yearly is well worth the cost.

Mastitis is any inflammation of the mammary gland, and the most common cause is bacteria. Consultation with a veterinarian is indispensable for effective diagnosis and treatment of mastitis. Prevention is the key to the control of mastitis. Use of individual paper towels to wash the udder, milking clinically infected cows last, dipping teats after milking, and treating all quarters at beginning of the dry period are highly recommended to prevent spread of this disease.

Review Questions

1. Discuss activation and inhibition of the milk-ejection reflex. Why is each process of importance to the dairyman?
2. Outline the principles of proper milking.
3. What factors regulate rate of milking?
4. Diagram essential components of a milking machine. What is the function of each component?
5. Define mastitis. List some causes of mastitis. What is the relationship between the milking operation and mastitis? Outline a control program for mastitis.

References

1. MacLachlan, D. L.: A study of dairy chore labor under different systems of free-stall housing. Thesis, Mich. State Univ., 1967.
2. Gorewit, R. C., Wachs, E. A., Sagi, R., and Merrill, W. G.: Current concepts on the role of oxytocin in milk ejection. J. Dairy Sci. *66:* 2236, 1983.
3. Grosvenor, C. E.: Contraction of lactating rat mammary gland in response to direct mechanical stimulation. Amer. J. Physiol. *208:* 214, 1965.
4. Lefcourt, A. M., and Akers, R. M.: Is oxytocin really necessary for efficient milk removal in dairy cows? J. Dairy Sci., *66:* 2251, 1983.
5. Lefcourt, A. M.: Effect of teat stimulation on sympathetic tone in bovine mammary gland. J. Dairy Sci. *65:* 2317, 1982.
6. McDonald, J. S., and Witzel, D. A.: Differential pressures across the bovine teat canal during three methods of milk removal. J. Dairy Sci. *49:* 176, 1966.
7. The DeLaval Handbook of Milking. Poughkeepsie, N.Y.: The DeLaval Separator Company, 1963.
8. Milking management. AXT-94. Univ. Calif. Agr. Ext. Ser. 1967.
9. Wesen, D. P., and Schultz, L. H.: Effectiveness of a post-milking teat dip for preventing new udder infections. J. Dairy Sci. *53:* 1391, 1970.
10. Klenzade Dairy Farm Sanitation Handbook. St. Paul, Minn.: Klenzade Products, Division of Economics Laboratory, Inc., 1968.
11. Schultz, L. H., Natzke, R. P., Smith, J. W., Brown, R. W., Mellenberger, R. W., Thompson, P. D., Jasper, D. E., and Philpot, W. N.: Current concepts of bovine mastitis. Hinsdale, Ill.: The National Mastitis Council, Inc., 1978.

12. American Public Health Assoc.: Standard methods for the examination of dairy products. 13th ed. New York: Amer. Pub. Health Assoc., Inc., 1972.
13. Paape, M. J., Wergin, W. P., Guidry, A. J., and Pearson, R. E.: Leukocytes—second line of defense against invading mastitis pathogens. J. Dairy Sci. *62:* 135, 1979.
14. Peterson, K. J.: Mammary tissue injury resulting from improper machine milking. Amer. J. Vet. Res. *25:* 1002, 1964.
15. Mahle, D. E., Galton, D. M., and Adkinson, R. W.: Effects of vacuum pulsation ratio on udder health. J. Dairy Sci. *65:* 1252, 1982.
16. Natzke, R. P.: Elements of mastitis control. J. Dairy Sci. *64:* 1431, 1981.
17. Neave, F. K., Dodd, F. H., Kingwill, R. G., and Westgarth, D. R.: Control of mastitis in the dairy herd by hygiene and management. J. Dairy Sci. *52:* 698, 1969.

Suggested Additional Reading

Larson, B. L., ed. *Lactation.* Ames: Iowa State University Press, 1985.

Schultz, L. H., Natzke, R. P., Smith, J. W., Brown, R. W., Mellenberger, R. W., Thompson, P. D., Jasper, D. E., and Philpot, W. N.: Current concepts of bovine mastitis. Hinsdale, Ill. The National Mastitis Council, Inc., 1978.

Schalm, O. W., Carroll, E. J., and Jain, N. C.: *Bovine Mastitis.* Philadelphia: Lea & Febiger, 1971.

PART FIVE

Herd Management

CHAPTER 20

Raising Calves—Growing Heifers

20.1 Introduction

The future of any dairy operation depends upon a successful program for raising calves and heifers for replacements or upon purchased replacements which equal or exceed the current levels of milk production. Yet average mortality rates in calves under 3 months of age range up to 20% in many dairy areas and are substantially higher in some individual herds.

The average length of time that a cow stays in a milking herd varies between 3 and 4 years in most areas of the U.S. Therefore, from 25 to 33% of the milking herd must be replaced each year. It usually is more economical for a dairyman to raise his own heifers than to buy replacements, although there are exceptions. For example, dairy farms located on high-value land near metropolitan areas often sell all calves at birth and buy all replacements because of a lack of land for raising heifers. In other cases, heifers are raised by a second party for a predetermined fee and returned to the original owner a few weeks before first freshening. Still other dairymen sell their calves with an option to buy them back later as springing heifers. In any case, whether a dairyman raises his own calves or has someone else to do it, the first few weeks are critical in the life of a calf.

20.2 Care of the Cow and Calf at Calving

Care of the Dam

During the last one-third of gestation more nutrients are required for the development of the fetus than at any other time. The cow will require nutrients in addition to maintenance to satisfy the needs of the growing fetus. These needs are not large but must be met in order for healthy, vigorous calves to be delivered. A ration of good quality forage, a moderate level of grain during the last 2 to 4 weeks of the dry period and adequate mineral and vitamin supplementation will take care of these needs. If the forage fed is poor quality, vitamins A, D and E should be added to the ration. This may be accomplished by including a vitamin premix in the grain or by using an injectable vitamin preparation. Also, trace mineral salt and a calcium-phosphorus supplement should be available to the cow. When milk fever is a problem, a phosphorus supplement may be used.

It is inadvisable to feed the cow so that she becomes overly fat. Overconditioned cows may have difficulty in calving and also may be subject to metabolic diseases and other disorders.

At Calving

The choice of the calving area is an important item that should not be overlooked. An area that is clean, comfortable and, if possible, isolated from the rest of the herd should be selected. An ideal location for calving is an open area out of doors that has not been used extensively by cattle. During the winter months, however, a maternity pen is usually the best place for a cow to calve. This pen should be of ample size and should be well-bedded. A strong overhead beam or a hook suspended from the ceiling is desirable for attachment of a hip lift to raise cows afflicted with milk fever or the downer syndrome. Between calvings the pen should be thoroughly cleaned and sanitized with a good disinfectant. This will help reduce some of the problems with calf losses and infections of the reproductive tract of the cow. When losses of calves become abnormally high shortly after birth, it is wise to allow the area to "rest" for several weeks and to use another location for calving. Often, a calving area that is used repeatedly may be subject to a buildup of disease organisms that may infect the newborn calf very shortly after birth. Use of another calving area is a method of stopping these kinds of infections.

During parturition, the cow should be left alone unless she requires assistance. It is best to check the cow periodically for any calving problems. Do not be overly anxious to pull the calf unless assistance is really necessary. However, if trouble does occur, then the cow should be assisted. In many cases the calf may need its position within the uterus readjusted. If these troubles cannot be overcome within a reasonable length of time, a veterinarian should be called.

Care of the Calf after Birth

After calving the cow will usually be up and will begin to dry the calf. If, for some reason, the cow is unable to get up, then the calf should be dried with towels or other suitable material. Treatment of the navel with iodine is a good procedure and will help to reduce infection. The iodine solution should be introduced into the navel and not just around the outer portions. The calf should be on its feet and nursing within one-half hour after birth. Weak calves will require assistance in nursing or it may be necessary to obtain colostrum from the dam and to give it to the calf via nipple bottle.

It is generally best to remove the calf from its dam shortly after birth or within 24 hours after birth to reduce exposure to coliforms and other organisms and to reduce the difficulties with maternal instincts when the calf is left with the cow too long.

20.3 Feeding the Calf from Birth to 3 Months

Colostrum Feeding

Calves have no immunity against disease because antibodies or immunoglobulins (Ig) are not transferred across the placenta from the dam to the fetus. Colostrum is the only source of immunoglobulins (Ig) and is the method by which calves acquire passive immunity. Colostrum also provides a highly nutritious balanced diet for newborn calves. Colostrum usually includes the first six milkings following freshening and is unsaleable. The first milking of colostrum is the most important for calves as it is highest in immunoglobulins and nutritional value (Table 20.1).

The immunoglobulin content of colostrum is affected by age and breed of cows along with successive milkings following calving. Cows in second or later lactations produce larger quantities of colostrum and have higher immunoglobulin concentrations in colostrum than cows in first lactation. Older cows have been exposed to a wider range of diseases than young animals and therefore have produced more immunoglobulins against them. Thus, the first milking of colostrum from older cows can be fed to calves from first lactation heifers if livability problems are encountered. Cows which are not dried up, or have too short a dry period (less than 30 days) before their next lactation will not have sufficient levels of immunoglobulins in their first milk.

Time of Feeding

For maximum protection against infection, newborn calves should be fed colos-

Table 20.1 *Composition and characteristics of colostrum and normal whole milk.*

Item	*First milking*	*Second milking*	*Second day*	*Third day*	*Whole milk*
Specific gravity (gm/ml)	1.056	1.040	1.034	1.033	1.032
Total solids (%)	23.9	17.9	14.0	13.6	12.9
Fat (%)	6.7	5.4	4.1	4.3	4.0
Nonfat solids (%)	16.7	12.2	9.6	9.5	8.8
Protein (%)	14.0	8.4	4.6	4.1	3.1
Lactose (%)	2.7	3.9	4.5	4.7	5.0
Ash (%)	1.1	1.0	.8	.8	.7
Vitamin A (ug/100 ml)	295.0	190.0	95.0	74.0	34.0
Immunoglobulins (%)	6.0	4.2	1.0	—	.09

trum immediately after birth. The highest absorption of immunoglobulins occurs at this time and during the first 6 hours after birth, but by 24 hours most of the absorption capability is lost (Table 20.2). Feeding colostrum after intestinal absorption has ceased is of no benefit for passive immunity, but may provide some local protective action against disease-causing organisms within the intestinal tract. However, bacteria present in the intestine before the first feeding of colostrum block absorption of immunoglobulins and accelerate intestinal closure to absorption. Thus, the first milking of colostrum should be fed to calves within 15 to 30 minutes of birth and before the intestinal tract becomes innoculated with bacteria.

Amount of Colostrum to Feed

Calves should receive 6% of birth weight within the first 6 hours after birth.

This amounts to about 5.5 pounds or 2.5 quarts for a 90-pound calf. Research has shown calves fed 4.5 pounds of colostrum at birth had nearly double the blood IgG level (14.9 mg/ml) of calves fed 2.5 pounds (8.5 mg/ml) and triple calves fed 1 pound at birth (5.2 mg/ml). For protection against heavy disease challenges, calves should receive 3/4 to 1 pound of immunoglobulins during the first 24 hours of life. Calves should receive 10 and up to 12% of their birth weight in the first milking of colostrum during the first 24 hours after birth.

Table 20.2 *Effect of age at first feeding on immunoglobulin absorption.*

Age at first feeding (hours)	*IgG*	*IgM*	*IgA*
	Percent of calves absorbing Ig		
12	100	100	97
16	90	97	83
20	77	70	70
24	50	47	43

Source—Univ. of Arizona

Methods of Feeding

Milking colostrum from cows and hand feeding it to newborn calves is advisable. Calves which nurse have a faster rate and higher amount of immunoglobulin absorption, than hand-fed calves but less than 50% of optimal disease protection. If calves are allowed to nurse, they should be observed and assisted, if necessary, to achieve ample protection. If calves refuse to eat, colostrum should be fed using an esophageal feeder or stomach tube.

Emergency Sources

Sometimes colostrum is not available from cows. A frozen supply of first milking colostrum from home-raised older cows should be kept for use in these situations. Frozen colostrum is preferred, but sour or fermented colostrum can be used in emergencies. Sour colostrum should be buffered with sodium bicarbonate (1 teaspoon per quart) at time of feeding to enhance absorption of immunoglobulins.

Second and Third Day Feeding

Colostrum equal to 8% of birth weight should be fed to calves the second and third days of life. The immunoglobulins in this colostrum will not be absorbed but may provide some local protective action against bacteria and other organisms in the digestive tract.

Day 4 to Weaning

Some dairymen discontinue milk feeding at 3 to 4 weeks of age, especially with large vigorous calves. Others prefer to wean at 6 or even as late as 12 weeks. The plan chosen will depend upon labor and other cost factors, as well as the size, rate of growth, and general health of the calves.

Milk Feeding. Whole milk, excess colostrum, or milk replacer should be fed to calves for 3 to 8 weeks after birth. The amount to feed each day is dependent upon body size. A feeding schedule which has worked well in Illinois is outlined in Table 20.3.[1] This schedule allots milk to calves on the basis of birth weights. The daily allowance of milk which is usually fed in 2 equal feedings is equal to approximately 8, 9, 10, 8, and 5% of the birth weight for the first through the fifth week respectively.

If heart girth tapes or scales are not available for determining birth weights, calf sizes can be estimated as small, medium, or large; categories 2, 4, or 6 in the feeding schedule are then used. Amounts recommended are for use as guides only, and larger or smaller amounts of milk should be fed depending on the health of the calf and the desired rate of gain. In no case should milk be discontinued until the calf is eating at least 1 lb of dry calf starter mix per day. This may require more than 5 weeks of milk feeding for calves which were born weak or exceptionally small in size.

Milk replacers can be fed instead of whole milk when the latter is scarce or high priced. High-quality milk replacers usually contain a large proportion of milk products. An example of a suitable formula is shown in Table 20.4[2]

The best milk replacers contain at least 20% protein, all of it derived from milk. The protein level should be 22% when specially manufactured soy flours or soy concentrates are used, because plant proteins are less digestible than milk protein. Table 20.5 lists the various sources of protein according to their acceptability in milk replacers fed to calves from birth to 3 weeks of age.

The fat level in a good milk replacer powder should be at least 10% and may run as high as 20%. The higher fat level tends to reduce the severity of diarrhea and provides additional energy for growth. Good-quality animal fats are preferable to most plant fat sources. Soy lecithin, when homogenized, is a very acceptable fat source.

Carbohydrate sources that the calf can use efficiently include lactose (milk sugar) and dextrose. Two common carbohydrate sources that should be excluded from milk replacers are starch and sucrose (table sugar).

How to Mix Replacer Powder

Most milk replacers are formulated to be mixed and fed like normal whole milk. It is generally desirable to follow the manufacturer's directions in mixing the liquid diet from the dry powder. A complete liquid solution is usually most easily obtained by

TABLE 20.3 *Suggested Calf Feeding Schedule*

Feeding category	*Birth weight*	*First week*	*Second week*	*Third week*	*Fourth week*	*Fifth week*	*Total milk*
	lb	Pounds of milk per day*					lb
1	50–63	5	5½	6	5	4	178
2	64–73	5½	6	7	6	4	199
3	74–83	6	7	8	7	4	224
4	84–93	7	8	9	8	5	259
5	94–103	8	9	10	8	5	280
6	104–113	9	10	11	9	5	308
7	above 113	10	11	12	10	5	336

*Whole milk or milk replacer.

Source: Fryman, L. R., Harshbarger, K. E., and Muekling, A. J.: Feeding, managing and housing dairy calves. Ext. Cir. 992. College of Agriculture, Univ. of Illinois, December 1968.

Table 20.4 *Example of a Milk Replacer*

Item	*Amount in pounds*
Dried skim milk	70
Dried whey	18
Lecithin	2
Animal fat	10
Dicalcium phosphate	1.7
$CuSO_4$	+[a]
$FeSO_4$	+
$MnSO_4$	+
$CoSO_4$	+
Antibiotic	+

[a]+ = trace amounts
Source: Porterfield, R. A., McGrew, C. D., and Hibbs, J. W., Raising dairy herd replacements. Ext. Bul. 514. College of Agriculture, Ohio State Univ., November 1969.

placing the dry powder on top of the water, then mixing. Many milk replacers dissolve most easily in hot water. Consistency in the dry matter content (same proportion of dry replacer and water each feeding) is important.

Once a Day Feeding

While milk or milk replacers normally are given in 2 equal feedings per day, the daily allowance can be given in one feeding in order to save labor. When milk replacer is fed only once a day, less water is mixed with the total daily allowance of dry replacer. This more concentrated solution makes it easier for the calf to consume its daily allowance of replacer in one feeding.[3]

A milk replacer must be of high quality if it is offered only once daily. Ideally, a milk replacer fed once daily should contain 22 to 24% protein and 20% fat. The carbohydrate portion should be milk sugar (lactose) with no sucrose, starch, or fiber-containing ingredients. This extra-good quality product helps supply the necessary nutrition since the amount of replacer fed must be limited to control scours. Levels of dry milk replacer beyond one pound daily frequently cause diarrhea. Research suggests that only 0.8 pound of dry replacer in 7 pounds of water should be offered in a once-daily feeding program to the larger breeds. Small breeds need proportionately less.

Once-a-day feeding of calves may be desirable on large dairy farms to reduce labor costs. However, most dairymen prefer twice-a-day feeding because it ensures that calves get more attention, and signs of digestive upsets are noted more promptly.

Soured Colostrum

There has been much recent interest in feeding naturally fermented colostrum, sometimes called pickled milk. Reports from dairymen and recent results from experiment stations have been reasonably good. Many trials have been clearly successful, and only a few dairymen have since abandoned the practice. Here are points to consider in feeding soured colostrum:

1. Make sure that the first colostrum is given directly to the calf. Mix and save all excess colostrum (and colostrum-like milk) obtained from the cow in the first seven milkings. These seven milkings provide enough milk for one calf for about one month. If 2 or more cows freshen the same day, combine their colostrum. However, the number of calves fed from the combined container should not exceed the number of cows contributing to the total.
2. Use a covered container to store the fermenting colostrum. A container with a plastic liner is easier to clean before each re-use. Furthermore, a plastic liner prevents the soured milk (which be-

Table 20.5 *Protein Sources, Characterized by Suitability in Milk Replacers*

A. Optimum	*B. Acceptable*	*C. Questionable*
Skim milk powder	Chemically modified soy protein	Meat solubles
Buttermilk powder	Soy concentrate	Fish protein concentrate
Dried whole whey	Soy isolate	Soy flour
Delactosed whey		Distillers' dried solubles
Casein		Brewers' dried yeast
Milk albumin		Oat flour
		Wheat flour

comes quite acidic) from corroding a metal container and causing excessive intake of zinc or other minerals.

3. Most dairymen allow the fresh colostrum to ferment naturally even though the wrong kind of fermentation may occur. Some dairymen inoculate the fresh batch with 3 to 5 tablespoons of fermented colostrum from a previous batch to help start proper fermentation.
4. Do not hold the fermented colostrum for long periods before feeding, especially in summer. Research has shown that over time this product becomes very acidic and much of the protein is destroyed.
5. Stir the fermenting colostrum daily, preferably twice daily, during storage and prior to feeding. Stirring helps prevent scum from forming and minimizes large lumps.
6. Whether the sour colostrum should be diluted before feeding depends on its consistency. If it is quite concentrated, similar to much first-milking colostrum, a dilution might be desirable. Dilute 2 or 3 parts of colostrum with one part of water. For example, 4, 5, or 6 pounds of colostrum, depending on the size of the calf, can be mixed with 2 pounds of water. Using hot water will warm the colostrum and make it more acceptable to the calf.
7. Start feeding the fermenting colostrum on the fourth day after birth (after feeding the fresh undiluted colostrum directly from the cow the first 3 days). Doing this assures that the colostrum has not yet completed its fermenting process and will not be as acid tasting to the calf. This practice helps teach the calf to consume the soured product.
8. If 2 or more batches of soured colostrum have been collected for at least a week and have fully fermented, they may be mixed together to save space and shorten the time spent mixing. However, do not feed more calves than there are cows contributing to this mixture.

Preserved Colostrum

At times, undesirable fermentations may render soured colostrum unacceptable to the calf. Preservatives, primarily in the form of one or more organic acids, decrease protein breakdown and allow some control over the quality of a colostrum product stored at room temperature. Colostrum preservatives prevent or decrease microbial fermentation. Adding an appropriate acid to colostrum accomplishes artificially what a desirable natural fermentation would accomplish, an environment too acidic for growth of undesirable types of bacteria. Mixtures of organic acids are available commercially. The use of a single acid may be nearly as effective and more economical. For example, adding 0.7% propionic acid by weight has proven to be an effective means of preserving colostrum. Propionic acid, available for preserving high moisture corn, can be used for this purpose.

A colostrum preservative usually is not necessary during the cool months of the year, but may be advisable during the summer months when high temperatures may increase protein breakdown and spoil untreated colostrum.

Dry Calf Starter. Calves usually start eating a small amount of dry starter mix at about 1 week of age. To start them eating, rub a small amount of the mix in their mouths or put a little in the bottom of the pail after feeding milk. Fresh starter mix should be placed in the feed box every day in amounts that they will just clean up.

Complicated calf starters are not necessary.[4] The important characteristics of a starter mix are that it should be palatable, be high in energy, and contain 16 to 18% crude protein. One that has given good results contains 40% corn, 25% oats, 23% soybean meal, 8% molasses, 1.85% dicalcium phosphate, 1% trace mineralized salt, 1% Aurofac or similar antibiotic feed supplement, and 0.15% vitamins A and D supplement. In areas where corn or oats are not available or are high priced, barley, milo, or wheat can be substituted. Also, cottonseed meal, linseed meal, or other high-protein supplements can replace soybean meal if they are more readily available. Urea should not be included in the mix until a calf is ruminating and consuming at least 1 lb per day of roughage per 100 lb of body weight.

As soon as the calf is consuming at least 1 pound of starter daily, milk feeding can be discontinued. The time to reach this level of consumption will vary with the breed and the individual calf. Larger calves will reach it at 3 to 6 weeks. A complete calf starter containing dehydrated alfalfa is shown in Table 20.6.[5]

Quality Hay. Calves normally start eating small amounts of hay during the first week of life. High-quality, soft-textured hay (preferably high in legumes) should be available free-choice within the first week or two. Fresh hay should be fed each day, and old hay should be discarded or fed to older animals. Availability of high-quality hay encourages early rumen development with subsequent beneficial effects on health and economy of weight gains.

Feed Additives. Antibiotics such as Aureomycin in the diet of young calves stimulate appetites, increase growth rates, and may reduce the incidence and severity of scours. After 3 months of age, however, there is no benefit to routine feeding of antibiotics.

Although green roughages are good sources of carotene that can be converted to vitamin A by the calf, young calves may not eat enough roughage to fulfill their requirements. Therefore, most commercial calf starters contain a vitamin A supplement. Older calves eating more good-quality roughage usually obtain sufficient amounts of vitamin A from their diets without supplementation.

This is true, also, for vitamin D. Once the calf is eating about 2 lb per day of sun-cured hay, haylage, or silage, a vitamin D supplement is unnecessary. However, calf starters usually contain supplemental vitamin D for insurance until the calf is eating adequate roughage. All other vitamins usually are supplied in adequate amounts in normal calf rations.

Dicalcium phosphate, bone meal, or similar commercial mineral mixtures should be included in calf starters as 1 to 2% of the mix to provide adequate calcium and phosphorus after milk feeding is discontinued. Also, 1% of trace mineralized salt usually is included to insure adequate intakes of all other essential minerals.

Table 20.6 *A Complete Calf Starter Used at Iowa State University*

Ingredients[a]	%
Dehydrated alfalfa	20
Coarse ground shelled corn	32
Coarse ground oats	20
Liquid molasses	10
Soybean meal	15
Steamed bone meal or dicalcium phosphate	2
Trace mineralized salt	1

[a] Add per pound of starter
2,000 I.U. vitamin A
300 I.U. vitamin D
10 *mg* chlortetracycline (Aureomycin) or oxytetracycline (Terramycin)

Source: Nelson, K.: Calf survival—major link in herd improvement. New England Holstein Bulletin, May–June 1970. (Reprinted from Iowa Holstein Herald)

Weaning to 3 Months of Age

After weaning, calves can be fed up to 6 lb per day of calf starter in conjunction with good hay available free-choice. More than 6 lb is not recommended because of reduced hay intake at higher starter levels. Roughage intake should be encouraged at this time in order to hasten development and function of the rumen (Chapter 8).

Calves at this age usually are housed and fed together in pens which hold 5 to 10 calves each. Good-quality hay and clean water should be available free-choice at all times. The starter mix should be group fed at least twice a day. All calves in a pen should be about the same age and weight to minimize the possibility of greedy eaters consuming more than their share of the starter mix.

Silage, greenchop, and pasture can be fed to calves at this age, but these forages should not make up the entire ration. Dry hay and starter are needed in addition to provide adequate intake of all nutrients.

20.4 Housing for Dairy Calves

The basic requirements for housing calves from birth to weaning age are easy to list but more difficult to attain. The objective is to provide housing which is inexpensive to build and economical to operate and maintain while at the same time being easy to clean and disinfect to minimize calf mortality. During the milk-feeding period the calves should be housed and fed individually.

While the ideal temperature for young calves is probably between 50° and 60°F, they adjust readily to temperatures which are substantially higher or lower if they have a dry bed, are protected from drafts, and the system of ventilation is adequate to prevent dampness. The area should have adequate lighting. Stalls with solid floors and bedding are recommended if the environmental temperature drops below 40°F. See Section 22.6 for a discussion on ventilation requirements.

Housing for calves varies from simply tieing calves to the wall in the litter alley behind cows in stanchion barns to having a fully automated and integrated feeding and housing system in some larger dairies. The potential of young animals being exposed to diseases carried by older animals is often disregarded. A housing scheme growing in popularity, designed to minimize transfer of disease from one animal to another, utilizes naturally ventilated facilities, and: (a) individual hutches for calves from birth until 2 weeks after weaning, (b) small groups of calves (7 or 8 animals) until 4 or 5 months of age, and (c) larger groups in more labor-efficient facilities in later life (see Figure 20.1). See Section 22.12 for a more complete discussion on size and dimensions for separate calf housing facilities.

Methods of Feeding and Watering

As indicated in Section 20.3 calves from day 4 to weaning should be encouraged to consume dry calf starter and leafy hay. Therefore each calf stall or pen should have a small rack for hay and a removable feed box 8″ x 10″ x 6″ deep for starter. The top of the box should be 20″ from the floor. Unless a water bowl is installed in each pen, provision should be made for a water pail. Milk or liquid milk replacer can be fed in a bucket which is sterilized after use, by means of a nipple pail, or a plastic nursing bottle. All feeding utensils should be washed after use and sterilized.

Bedding and Manure

It is important that the calf be kept dry. Unless elevated stalls with slatted floors are provided, there should be enough bedding to absorb all of the urine. Dry sawdust or peat moss are excellent bedding materials. Other common bedding materials include shavings, straw, or chopped or long hay. However, these bedding ingredients

Fig. 20.1 This Minnesota dairy uses calf hutches (top) to provide semi-isolation. Then groups of 7 or 8 calves are moved to a super-hutch (center) where they become acclimated to group housing. When animals reach 400 or more pounds, they are moved to a larger, labor efficient unit (bottom).

may be eaten by the young calf and may cause digestive upsets. Manure should be removed regularly in warm barns but in cold barns, bedding should be added to build up a warm pack.

20.5 General Management Practices

With labor shortages, there is a tendency for dairymen to concentrate on the milking herd and to overlook or postpone some important details of calf raising. This happens particularly in permanently identifying heifer calves, in dehorning, and removing extra teats at the proper age.

Identification

The importance of accurate identification of dairy females was discussed in Section 3.6. It also listed most of the methods commonly used for both temporary and permanent identification. Placing a nylon cord with a numbered plastic tag around the neck of each calf when it leaves its dam will provide accurate and convenient temporary identification.

For herds enrolled in the DHIA program, the supervisor at each monthly visit will insert a zinc-coated metal tag in the ear of each grade heifer born since his previous visit. This number, 21-WZZ-9999 for example, or a portion of it will identify the calf throughout life and be used later as the DHIA cow index number when the heifer freshens. For purebred calves, if the owner prefers not to have an eartag inserted, the supervisor will report the date of birth and the name of the sire. When the calf has been registered, he will forward the registration number to the processing center. Registered Jerseys must be tattooed for registration (Figure 20.2).[6] Tattooing in the ear can also be used for permanent identification of registered calves in the other dairy breeds and for grade calves.

Accurate identification of youngstock is a must. Because of its importance, the National Cooperative DHI Program is now offering a new optimal service called Verified Identification Program (VIP). This nationwide program is described in Section 3.4.

Upon request, AI inseminators will eartag grade calves in herds not enrolled in a production testing program. Calves and heifers not previously identified will be eartagged by the veterinarian at the time of a tuberculosis test using the USDA series. The owner may also purchase various types, colors, and sizes of metal or plastic tags from agricultural supply companies and establish his own system of permanent identification. The important thing is that each heifer calf be accurately and permanently identified at an early age.

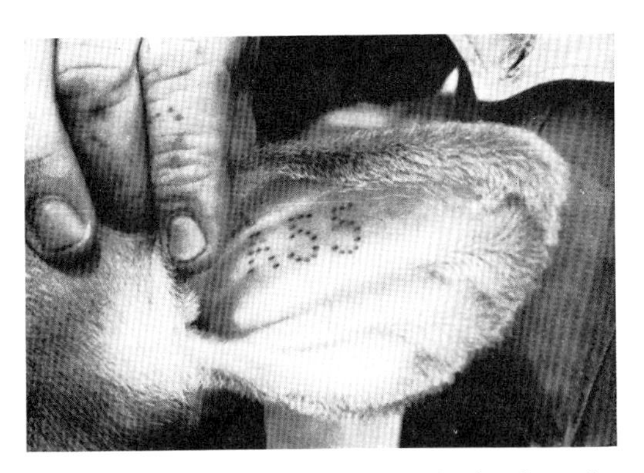

FIG. **20.2** A neat tattoo showing the herd number assigned to the calf provides an accurate means of identification throughout life. (Courtesy Dairy Science Department, Louisiana State University)

Dehorning

Most registered females are now dehorned even in those herds that still exhibit in the showring. There are no good reasons why heifers or cows should be horned and several reasons why they should be dehorned as young calves. The principal reason for dehorning, of course, is to eliminate the risk of injury to other animals in the herd or to people working with the cattle. Introducing a horned animal into a polled herd is poor judgment, and grade cows with horns often sell at a reduced price.

Dehorning calves under 30 days of age with a commercial electric dehorning iron is currently the most popular method. The length of time that the iron is applied will depend upon the age and breed of the calf. It usually requires from 10 to 20 seconds. When properly done, the area around the horn button should be a dark copper color[7] (Figure 20.3).

A stick of caustic potash works well in dehorning calves at 1 to 3 weeks of age. It is

FIG. 20.3 An example of the appearance of the horn button after electric dehorning. (Courtesy of Montana State University)

inexpensive, easy to do, and does a neat job with a minimal risk of scurs[8] (Figure 20.4).

For older calves, yearling heifers, and cows in milk, the veterinarian will anesthetize the area around the base of the horns and remove the horns with minimal bleeding and discomfort to the animal. However, the appearance of the poll will be less attractive in those animals which are dehorned at advanced ages.

Removing Extra Teats

Calves may be born with more than 4 teats. Usually the extra teats are located posterior to one or both rear teats but they may occur between the front and rear teats on one or both sides of the calf's udder. Since such extra teats detract from the appearance of the udder and may interfere with milking, they should be removed at 1 to 2 months of age. In most cases this can be done by the herdsman. With the calf lying on its side and the hind leg drawn forward, the udder should be wiped clean with an antiseptic solution. By placing the fingers of one hand under a fold of skin below the extra teat, it can be removed easily with a pair of curved, sharp, disinfected scissors. If necessary, trim the area around the base of the teat, and then apply tincture of iodine. When this minor operation is performed properly, there will be very little bleeding and when the heifer freshens, the scar will be scarcely noticeable. If there is a double teat or when the base of the extra teat is close to the base of the normal teat, it is recommended that a veterinarian perform the operation.

Exercise and Training to Lead

After weaning or even before, calves will benefit from exercise outdoors either in an individual outside run or by turning a pen of calves out into a paddock. As long as young calves have a dry bed and shelter from drafts, rain, and snow they adapt readily to subfreezing temperatures.

Calves should also be halter-broken and trained to lead at 2 to 3 months of age. Start by tying the young calf with a rope halter to the wall of the pen or to a fence to accustom it to being restrained. Then train it to lead by walking beside it assisting with a gentle push from the rear as necessary. Do not engage in a tug-of-war by dragging the calf with the halter rope. To anyone who has attempted to halter-break a 700- to 800-pound yearling heifer, the advantages of training young calves to lead are obvious.

20.6 Growing Heifers

Once the calf has been weaned, the high-risk period of calf mortality and the most expensive period of feeding have passed; the emphasis can then be shifted to greater economy in a ration which will provide for normal growth rates but not result in heifers becoming too fat (Figure 20.5).

Weights of heifers can be estimated from heart girth measurements. The measurement is made with a tape placed snugly around the body immediately behind the front legs and shoulders. Estimated weights of heifers of 4 breeds with varying heart girths are shown in Table 20.7. The ages at which these weights normally are reached also are shown in the table as a standard for comparison.

3 to 10 Months of Age

Heifers that are 3 months old can be shifted gradually from the starter to a concentrate mix with fewer ingredients. If good-quality legume hay is available free-choice, a concentrate mixture with 12 or

Fig. 20.4 Dehorning with caustic potash:
(A) Clip the hair covering the horn button and apply a ring of petroleum jelly to prevent caustic from injuring the eyes or burning the hair on cheek or jaw if there is bleeding.
(B) Apply the caustic to horn buttons.
(C) The caustic has destroyed all the horn cells and there will be no scurs to mar the appearance of the head.
(D) When the animal is dehorned correctly, the appearance of the head will resemble that of a naturally polled animal. (Courtesy of the University of California, Davis)

13% crude protein is adequate. However, a crude protein level of 15 or 16% may be needed if only fair-quality hay is fed. Corn, barley, milo, wheat, or oats can make up the major portion of the grain in the mix, depending on price and availability. A high energy level similar to the starter mix is desirable, and inclusion in the mix of 1% each of dicalcium phosphate and trace mineralized salt will fulfill mineral requirements.

The amount of concentrates to feed to heifers at this age depends on the amount and quality of roughage being fed. Heifers should receive enough concentrates to achieve rapid growth without becoming overly fat. A starting point for 4-month-old calves is about 1 lb concentrates per 100 lb body weight. This can be reduced as roughage consumption increases. Body condition should be the determining factor for the amount of concentrates to feed.

Fig. 20.5 These registered Guernsey heifers in Montana are being grown out economically on a home-mixed grain ration plus hay and corn silage. (Courtesy of Hoard's Dairyman)

10 Months to a Few Weeks before Calving

At about 10 months of age, heifers make satisfactory growth on high-quality roughage alone. When fed lower-quality roughages, however, they may need some concentrates up to 14 months of age. Rations for growing heifers and bulls should contain about 12% crude protein on a dry matter basis, as shown in Appendix Table III-C.

Any of the commonly fed roughages, such as pasture, hay, silage, haylage, or greenchop can be the major feed source for heifers during this period. The amount of concentrates to feed depends on roughage quality. Enough should be fed to keep the growing heifer in good condition, but not too fat. Heifers that are excessively fat may have more breeding and calving difficulties and lower lifetime milk production than those that are kept in normal condition (Section 16.5).

During the last few months before calving, concentrate feeding should be in-

Table 20.7 *Normal Heart-Girth Measurement and Weight of Calves and Heifers During the Growing Period*

Age in months	*Holstein*		*Ayrshire*		*Guernsey*		*Jersey*	
	Inches	*Pounds*	*Inches*	*Pounds*	*Inches*	*Pounds*	*Inches*	*Pounds*
At birth	31	96	29½	72	29	66	24½	56
1	33½	118	32	98	31½	90	29½	72
2	37	161	35½	132	34½	122	32½	102
3	40¼	213	38¾	179	38	164	35¼	138
4	43½	272	42¾	236	41¼	217	38¼	181
5	47	335	45½	291	44¼	265	41½	228
6	50	396	48¼	340	47	304	44½	277
7	52½	455	51¼	408	49¾	362	47¼	325
8	54¾	508	53	447	51¾	410	49¾	369
9	57	559	55	485	53¾	448	51¾	409
10	58¾	609	57	526	55	486	53¼	446
11	60½	658	58	563	56¾	521	55	481
12	62½	714	59	583	58¼	549	56½	520
13	63¼	740	60¾	630	59¼	587	57½	540
14	64¼	774	62	666	60½	615	58½	565
15	65¼	805	63	703	61¾	640	59	585
16	66¼	841	64	731	62½	674	59¾	611
17	67¼	874	65¼	758	63½	696	60½	635
18	68½	912	66	781	65	727	61½	660
19	69¼	946	66½	813	65½	752	62½	687
20	70½	985	67½	841	66¼	780	63	712
21	71½	1,025	68½	885	67½	816	64	740

Source: Farmers Bulletin No. 2176. Animal Husbandry Research Division, ARS, USDA, October 1961.

creased to fulfill the needs of pregnancy. Up to 8 lb per day may be needed with average-quality roughage, whereas 1 to 2 lb may suffice with excellent roughage. If the heifers receive the same roughage as the milking herd, the herd concentrate mix can be fed to both.

The Last Few Weeks before Calving

During the last 2 to 3 weeks before calving, heifers should receive at least 5 lb of concentrates to get them accustomed to the higher levels needed to meet their greater nutrient requirements upon freshening. This ration also allows their ruminal microorganism populations to adjust their proportions gradually in order to efficiently metabolize the high energy milking cow ration which the heifers will be fed when they enter the milking herd.

20.7 Summary

In calves which are normal at birth, most deaths from infectious scours, pneumonia, and related bacterial and viral diseases occur during the first month. The best method of providing some degree of immunity for young calves is a feeding of colostrum within an hour after birth with additional feedings during the next 72 hours.

Most management problems in raising calves from birth to weaning age can be controlled: (1) by an adequate feeding program of whole milk or milk replacer; (2) by the use of individual pens or stalls at least up to weaning age; and (3) by prompt treatment of calfhood diseases (see Chapter 21). Before 60 days of age, extra teats should have been removed, and the calves dehorned and permanently identified by eartagging, tattooing or freeze-branding (Appendix Table IIA).

At approximately 3 months of age the calves are shifted gradually from calf starter to a concentrate mix plus quality hay fed free-choice. From 3 months of age, emphasis should be placed upon normal growth rates, economical rations, and labor efficiency by group feeding and management. Groups of heifers of approximately the same age can be housed and fed as a unit.

Review Questions

1. What factors determine the age at which a dairy calf should be weaned?
2. How do milk replacers and calf starters differ in composition?
3. Which method of dehorning calves do you prefer? Give at least 3 reasons for your choice.
4. Would corn silage alone fed free-choice be a satisfactory ration for yearling heifers? If not, how would you supplement it?
5. How would you modify an all-silage ration for dairy heifers during the last 3 months of pregnancy?

References

1. Fryman, L. R., Harshbarger, K. E., and Muehling, A. J.: Feeding, managing, and housing dairy calves. Ext. Circ. 992. College of Agriculture, Univ. of Ill., December 1968.
2. Porterfield, R. A., McGrew, G. D., and Hibbs, J. W.: Raising dairy herd replacements. Ext. Bull. 514. Ohio State Univ. November 1969.
3. Willett, L. B., Albright, J. L., and Cunningham, M. D.: Once versus twice-daily feeding of milk replacer to calves. J. Dairy Sci. *52:* 390, 1969.
4. Gardner, R. W.: Acceptability and nutritional response comparisons between calf starters. J. Dairy Sci. *50:* 729, 1967.
5. Nelson, K.: Calf survival—major link in herd improvement. New England Holstein Bull. May-June 1970. (Reprinted from Iowa Holstein Herald.)
6. Frye, J. B., Jr., Anderson, H. W., Kilpatrick, B. L., and Waters, W. H.: Your guide for raising dairy calves. Agri. Ext. Pub. 1372. La. State Univ., December 1963.
7. Quesenberry, N. C.: Using the electric dehorner. Mont. Agr. Exp. Sta. Circ. 229. May 1960.
8. Mead, S. W., and Ronning, M.: Managing young dairy stock in California. Calif. Agr. Exp. Sta. Ext. Circ. 497, March 1961.

Suggested Additional Reading

Harrington, R. M., and Cristensen, R. L.: Raising dairy replacements. Depts. of Agricultural and Food Economics, Food and Agricultural Engineering, and Veterinary and Animal Sciences, Univ. of Mass. Coop. Ext. Serv., Publ. 64, November 1970.

Cole, H. H.: *Introduction to Livestock Production.* 2nd ed. San Francisco: W. H. Freeman & Co., 1966.

ASAE. *Dairy Housing II.* American Society of Agricultural Engineers. St. Joseph, MI. 1983.

CHAPTER 21

Herd Health

21.1 Introduction

Dairy herd health programs that emphasize prevention of disease rather than treatment play a central role in any attempt to increase production efficiency. Treatment will always be important in terms of survival of the individual sick animal. However, in terms of survival of the total production unit (profit versus loss) prevention is the more desirable method of disease control. Individual animal treatment should be viewed as a salvage operation since it occurs after varying amounts of production have already been lost. Under present economic conditions the proverb, "an ounce of prevention is worth a pound of cure," is more true than ever before. Emphasis in the following discussions is on prevention of diseases (an offensive approach) rather than on treatment (a defensive approach). Consequently, the selection of drugs and biologics and their use have been left to the discretion of the dairy manager in consultation with his veterinarian. Specific information on the treatment of various diseases can be found in the texts listed at the end of the chapter.

21.2 Herd Health Programs

The primary objective of a herd health program is to increase profit by limiting the occurrence of *economically significant* diseases. There is no justification for implementing a control program that costs more than the disease itself. Since each dairy is unique with respect to its management, physical facilities, and problems, there can never be a "universal dairy herd health program" containing a list of very specific steps that will automatically lead to success. Each program must be tailored to fit the needs of the individual dairy and must be modified continually as conditions on the dairy change. Herd health programs range from the absolute minimum of routine brucellosis vaccination of heifer calves to programs that include the use of computers for daily surveillance of a herd's health status.

The importance of herd health programs becomes evident when the cost of health problems incurred by the dairy industry is examined. In California alone, in 1974, the estimated cost of selected dairy cattle health problems was $155,590,300 (Table 21.1). Effective herd health programs can increase the production efficiency of the dairy industry by helping to reduce these losses.

In the discussions that follow, the reader should be aware that: (1) not all effective control techniques are discussed, and (2) certain strategies for controlling disease may be practical in one situation but may spell economic disaster in another. Thus, a general framework for a herd health program is presented, and it is the responsibility of the dairy manager in consultation with his veterinarian to fill in the details of the program if maximum success is to be achieved.

The most important element for a successful herd health program is the cooperation and dedication of everyone involved in

TABLE 21.1 *An Estimate of the Cost of Selected Health Problems to California Dairymen*

Health problems	*Magnitude of the problem*	*Annual cost per animal ($)*	*Annual cost to the industry ($)*
Mastitis	50% infected	100.00[a]	78,900,000[b]
Adult mortality	2%	12.00[c]	11,835,000[b]
Sterility	7%[d]	25.20[e]	19,882,800[b]
Delayed conception	Mean calving interval exceeds ideal by 45 days	40.50[f]	31,954,000[b]
Calf mortality (bulls)	30%	1.50[g]	1,183,500[b]
Calf mortality (heifers)	30%	15.00[h]	11,835,000[b]
Total		194.20	155,590,300

[a]A National Mastitis Council estimate.
[b]Based on a California dairy cattle population of 789,000.
[c]Based on a replacement cost of $600.
[d]From C. L. Pelissier, American Dairy Science Annual Meeting, 1971.
[e]Based on replacement cost of $600.
[f]Based on the cost of feeding 30 lb of $60/ton hay daily for 45 days.
[g]Based on a bull-calf value of $10 and a 50:50 sex ratio.
[h]Based on a heifer-calf value of $100 and a 50:50 sex ratio.

it. Any program will fail unless everyone maintains an interest in doing the best job possible. Incentive or bonus programs can be used to help maintain employee interest in the program.

21.3 Records

A vital part of every herd health program is its records. If a program is to be successful, people must be able to see where they have been, where they are, and where they are going. Dairy management has become so complex that most people cannot remember all of the pertinent information. A prime example is the area of calf mortality. A dairyman may be firmly convinced that calf mortality is less than 10%, yet after keeping adequate records for a number of months, he may find that calf mortality is actually somewhere between 25 and 30%. The same is true of most disease conditions encountered on the dairy. An unfortunate aspect of records is that most people find them uninteresting and inconvenient to use. The records always seem to be at another location when something needs to be recorded. As a result, information is often written on a piece of scrap paper that eventually is lost. Constant effort is needed to keep records current and accurate. Careful analysis of data provided by well-conceived and conscientiously maintained animal health records is essential for accurate definition of health problems.

Records can be of many styles; the important point is that the appropriate information be recorded in a form that can be used to assist in making management decisions (on an individual animal basis as well as on a herd basis). Records may range from a few simple notations to complex systems that are tabulated and summarized on computers. The dairy manager must evaluate the potential cost effectiveness of additional records, keeping in mind that there is little justification for keeping information if it is not analyzed and used. Proper utilization of detailed records of the health status of animals can be expected to increase dairy profits.

Calf Record Information

Individual calf records should include the calf's identification number, date of birth, sex, dam number, sire number, date of brucellosis vaccination and brucellosis ear tag number (for heifers only), and the date of and reason for death or disposal. From these data one can determine the time of year and age at which calves are dying, and appropriate changes can be made in the calf management programs. A more sophisticated record system might include information concerning: the size of the calf at birth (or the birth weight), difficulty of delivery, time of day of birth, and identification of the attendant present.

Heifer Record Information

Individual heifer records should include information concerning vaccination history, disease conditions, and treatments. More sophisticated records may include information about weights at various ages, breedings, identification of the bull used, and results of pregnancy examination. This information will allow a closer monitoring of the breeding, health, and general status of the heifers.

Cow Record Information

For maximum production efficiency, cow health record systems need to be closely integrated with production record systems. Some information about the breeding and reproductive status (e.g., heat dates, breeding dates, pregnancy status, and sire identification) is already included in the NCDHIP record system, along with milk production information. Additional information that should be recorded includes the status of the reproductive tract at the time of postpartum fertility examination, and disease occurrence and treatment. Unless problem areas can be precisely defined, programs to remedy health problems cannot be instituted.

A number of different record forms are available through the agricultural extension service of land-grant universities, commercial breeding firms, and agricultural supply companies. In most cases, personal preference will be the determining factor in selection of a specific system.

21.4 Physical Facilities

The importance of the following basic requirements for physical facilities should not be overlooked: (1) provision for efficient and safe handling and restraint of animals, (2) protection from the elements, (3) adequate lighting, and (4) readily available water. Although many procedures have the potential for yielding a several-fold return over investment, the same procedures are equally capable of causing a net loss if the animals involved cannot be handled efficiently or if the necessary procedure cannot be performed efficiently.

21.5 Diagnostic Procedures

Diagnostic procedures play an important role in the development and maintenance of an effective herd health program. In addition to the routine examination of cattle (both sick and well), serologic tests and necropsies deserve special mention.

Serologic tests are used to determine the level of antibodies present in the animal's serum (blood with the red blood cells and fibrinogen removed). Results of serologic tests are expressed as "titers." The titer is the last dilution of the serum that still shows a positive reaction in the test. For example, an animal may have a leptospirosis titer of 1:640, meaning that the last positive test occurred at a dilution of one part serum with 639 parts of diluent. Correct interpretation of titers requires a knowledge of the disease in question as well as of the specific test used, since several different types of tests are available. The presence of a titer does not mean that the animal currently has the disease; it merely indicates that at some (unknown) time the animal was exposed to the causative agent and responded by producing antibodies. Two blood samples should be obtained at a 3-week interval. If the animal had the disease at the time the first blood sample was taken, the second blood sample should have a higher titer than the first. It is important to be certain that both the acute (first) and convalescent (second) samples are obtained if a definitive diagnosis is to be made.

A necropsy should be performed on all animals that die from unknown causes. A number of diseases can be diagnosed on the basis of a necropsy, whereas the diagnosis of other diseases may have to await the results obtained from samples of tissues submitted to diagnostic laboratories.

21.6 Costs

Calculations of average costs for herd health programs are not very meaningful since there are so many variables (e.g., investment in physical facilities; records; secretarial help; professional fees for veterinarians, nutritionists, agricultural engineers; drugs and biologics; scope of the program).

Veterinary services usually account for a major part of the cost involved in the development and maintenance of herd health programs. Depending on the size of the dairy and the desires of management, veterinarians will have varying degrees of input. This input may range from occasional visits to care for emergencies or to vaccinate heifer calves for brucellosis, to a predetermined amount of time each week at the dairy and a significant input in management decisions.

Financial arrangements for professional services rendered by veterinarians vary widely. Examples of various kinds of financial arrangements include: (1) a flat fee per call or service rendered, (2) an hourly fee for time spent rendering service, (3) a monthly charge based on a per head assessment of a certain age or production group of cattle on the dairy, (4) a fixed yearly charge, or (5) a percentage of the gross income or net profits. In most studies where cost-benefit ratios for herd health programs have been determined, the ratio has been approximately 1:5 (i.e., a $5 return for each dollar spent on veterinary services).[1,2]

21.7 New Technology

As new technology and techniques become available, they are being incorporated into the daily operation of herd health management programs. Automated milk measuring devices that record production of each cow at each milking allow constant indirect monitoring of cow health. The use of computers on the dairy is now a reality. In addition to herd health and production data processing, maximum income over feed cost ration formulation, feed inventory, and accounting capability are now available right on the dairy. Owners who do not have a complete computer system can have access to these services via remote terminals. The potential uses of these capabilities are limited only by the imagination of the dairy's management team.

A relatively new technique that shows promise for herd health management is herd metabolic profile testing (MPT). Cow blood samples are submitted to laboratories for a series of specific biochemical tests (e.g., blood urea nitrogen, calcium, phosphorous, glucose, hemoglobin). This methodology has been used for a number of years on individual animals to aid in diagnosis. By testing blood samples from representative numbers of cows in various lactation and age groups, the metabolic status of entire groups of cows can be ascertained. The procedures not only aid in the identification of known problems but can also provide an early warning of impending metabolic disturbances such as low blood calcium levels (see Milk Fever, Section 21.15).[3,4,5,6] New developments will permit accelerated improvements in production efficiency in the future for those who are willing to make the necessary investments in time, effort, and money.

21.8 Basic Preventive Measures

The major factors that influence the occurrence of an infectious disease can be summarized by the following formula:

$$\frac{\text{Exposure Opportunity (number of disease organisms)} \times \text{Virulence (of disease organisms)} \times \text{Stress (on animals)}}{\text{Resistance of Animals}}$$

Factors in the numerator (Exposure Opportunity, Virulence, Stress) promote disease whereas the factor in the denominator (Resistance) prevents disease. Since manipulation of any of these factors can change the outcome (Disease), this same formula can be used to emphasize the importance and purpose of the basic preventive measures common to the control of many infectious diseases.

SANITATION reduces the number of disease organisms in the environment and thus the risk of exposure. Some diseases cannot be eradicated because the causative agent is so widespread in the environment. However, sanitation can reduce exposure to a level where clinical disease will be rare. Sanitation involves both physical removal of organisms (detergents and scrubbing) and chemical inactivation of organisms (disinfectants). Disinfectants are only effective when in direct contact with the disease organisms. They are not a substitute for a good cleansing. If the source of infection can be removed or reduced to a low level, disease incidence will decrease.

ISOLATION of sick or newly purchased animals will limit disease spread. Infected animals shed disease organisms into the environment via the breath, saliva, manure, urine, and abnormal secretions at various times during illness. If sick animals are isolated, the number of disease organisms to which other animals are exposed decreases. Additions to the herd should be isolated until they are shown to be free of disease (usually 2 to 3 weeks). It is senseless to eradicate the mastitis organism, *Streptococcus agalactiae,* from a herd only to reintroduce it with a newly purchased cow. Isolation of sick animals also helps to suppress the virulence of the disease organisms. As disease organisms pass from animal to animal, their ability to produce disease (i.e., virulence) increases. Consequently, the severity of the disease often increases as it spreads through the herd.

TESTING is a preventive measure, especially when done during an isolation period. The chances of introducing a specific disease organism into the herd can be reduced by confirming that new additions to the herd are free of infection. Tuberculosis and brucellosis are prime examples of major diseases that are controlled by this measure. Culturing the milk of newly purchased cows for the presence of *Streptococcus agalactiae* and other mastitis-causing organisms should be a routine part of a mastitis control program.

CULLING diseased animals from the herd can also be considered a preventive measure for the remainder of the herd. Although listed as a separate preventive measure, it is actually a form of isolation. For example, culling a cow that is chronically infected with staphylococcal mastitis is one of the best methods to decrease the exposure rate of other cows.

VACCINATION of animals is an effective preventive measure. Vaccination increases the animal's resistance to a particular disease by stimulating the animal to produce antibodies and/or increase the cell-mediated immune (CMI) response. Antibodies are the circulating protein molecules that help the body to combat invading disease organisms. In contrast, CMI refers to enhanced protective mechanisms at the cellular level.

The majority of vaccines now available are of excellent quality. When vaccination fails (a vaccine "break"), it is usually due to improper storage or administration of the vaccine. If the instructions provided in the package insert accompanying the vaccine are not followed, optimum results from vaccination cannot be expected. There are several important points to remember about vaccination.

1. A small percentage of animals are not capable of responding to certain vaccines, and will not be protected following vaccination. This lack of response is inherent in the animal and is not the fault of the vaccine. Consequently, vaccination cannot be expected to protect every animal.
2. Resistance is a relative matter. If a vaccinated animal is exposed to overwhelming numbers of disease organisms, the animal's resistance may be inadequate. Consequently, vaccination is an adjunct to, and not a substitute for, other preventive measures.
3. A number of days (up to 14) must elapse following vaccination before the animal's resistance is increased significantly.
4. Vaccination protocols often call for repeated administration of the vaccine at appropriate time intervals. These booster injections must be given if maximum protection is to be achieved.
5. Animals that are stressed or ill may not respond satisfactorily to vaccination. In spite of these limitations, vaccination plays a vital role in any herd health program.

A number of different types of vaccines are available. Before a vaccine is administered, the package insert should be read carefully. The duration of immunity (protection) following vaccination depends on the causative agent involved, the type of vaccine, and the method of its manufacture. Bacterins consist of suspensions of killed bacteria. They can be given to either pregnant or open cows. Toxoids consist of inactivated bacterial toxins. Modified live virus (MLV) vaccines consist of lyophilized, live, attenuated virus. Production of immunity relies on replication of the vaccine virus within the animal's body. Once reconstituted with the diluent supplied, these vaccines should be kept cool and out of direct sunlight, and they should be administered within an hour. MLV vaccines should not be given to pregnant animals since abortions may result. They should be administered at least 30 days prior to breeding in order to avoid delayed conception. The exception to this rule is the intranasal MLV vaccine for infectious bovine rhinotracheitis (IBR). This vaccine is applied to the nasal mucosa. Localized viral (vaccine) replication takes place rather than systemic replication. Killed virus vaccines are similar to bacterins except that they contain inactivated virus. The duration of immunity provided by these vaccines is short, since there is no viral (vaccine) replication after administration.

NUTRITION is often overlooked when herd health programs are discussed. Nevertheless, it plays a vital role in maintaining the animal's resistance to disease and in sustaining overall herd productivity. The nature of nutrition-disease interactions has not been adequately studied. However, a number of disease problems have been shown to have a nutritional basis; i.e., when the nutritional imbalance is corrected, the disease disappears.[7] The reader is referred to the chapters on nutrition for more information about this important aspect of herd health management.

21.9 Calf Management from Birth to 12 Months

The newborn calf has virtually no immunity (circulating antibodies) to infection and yet the calf must survive in a highly contaminated environment. As the calf takes its first breath or licks its muzzle for the first time, organisms, harmless and/or pathogenic, start entering the body. Although the calf is born with a sterile intestinal tract, within a few days after birth the intestinal tract will be colonized with organisms and certain areas will contain billions of organisms per gram of contents.

The risk of exposure to pathogenic organisms can be markedly reduced by having the cow calve in a clean environment. The ultimate in sanitation is to place the cow in a clean maternity pen a day or two before calving after clipping and washing her udder and escutcheon.

The calf's navel should be promptly dipped in 7% tincture of iodine. When the umbilical cord separates, the exposed vessels provide a pathway for organisms to gain entry into the calf's body. The tincture of iodine cauterizes the exposed tissues as well as killing any organisms already present. Organic iodines (e.g., teat dips) are not suitable for this purpose. Long, dangling umbilical cords should be severed close to the body before being treated with iodine. If the cut cord bleeds, the stump should be tied with a piece of string.

Nature helps protect the calf until its own immune system begins producing antibodies by transferring the immunity of the dam to the calf via antibodies secreted in the dam's colostrum. The maternal antibodies persist for varying lengths of time after ingestion by the calf. One complication caused by the presence of maternal antibodies is interference with the calf's own response to vaccination. Consequently, some vaccines are not recommended for use prior to a specific age, at which time maternal antibodies, if present, will have decreased to a negligible level.

The calf is capable of absorbing antibodies from the colostrum for only a short period of time. Maximum absorption occurs immediately after birth and decreases with time. By 24 hours after birth very little absorption of antibodies occurs through the intestinal wall,[8] although the colostrum will still be an excellent source of nutrients. In addition, any antibodies present in the colostrum will still have a local protective effect in the intestine. In cattle, there are three distinct classes of antibodies, which differ in their molecular weights and specific biological properties. Recently it has been found that they also differ in the length of time following birth that they can be absorbed from the intestine. The heaviest class of antibodies seems to be most protective against *Escherichia coli,* the bacterium most commonly associated with diarrhea in calves. Unfortunately, this class of antibodies appears to be the one that is absorbed for the shortest period of time following birth.

Colostrum should be provided as soon after birth as possible and in adequate amounts. The sooner the calf receives maternal antibodies (and nutrients), the better will be its chance for survival. Colostrum feeding is so important that colostrum is sometimes milked from the cow immediately after calving and given directly to the calf by bottle or esophageal tube. This technique has some additional benefits: (1)

early detection of cows freshening with mastitis (in which case the milk should not be given to the calf) and (2) detection of cows that "leaked" milk before calving. Some cows lose so much colostrum before calving that inadequate amounts remain for the calf. In either of these cases, colostrum from other cows should be given to the calf. The colostrum used should come from cows that have been on the dairy for a number of months so they will have had the opportunity to become exposed and respond (produce antibodies) to the pathogenic organisms endemic on the dairy. The calf should receive the first colostrum rather than second or third milking colostrum since the antibody concentration is much higher in the first colostrum.

Two other important points about colostrum feeding should be mentioned. Calves receiving adequate amounts of colostrum will have a significantly lower death rate than those that do not receive colostrum, since the antibodies present in the colostrum help protect the calf from developing a septicemia (i.e., invasion of the organisms through the mucous membranes into the bloodstream). However, proper colostrum feeding will not, by itself, reduce the incidence of calf scours.

The second point relates to absorption of antibodies. Although calves may receive adequate amounts of colostrum, a small percentage of calves will not absorb appreciable amounts of the antibodies present. The exact mechanism for this absorption failure is not known. Nevertheless, the net result of feeding adequate amounts of colostrum early in life will be an overall decrease in calf death losses. The reader should consult the chapter on calf rearing for more information about amounts of colostrum to be fed, methods of colostrum storage, calf nutrition and housing.

Every calf management program should include brucellosis vaccination of all heifer calves. In many areas, especially if animals have or will have access to pasture, blackleg and malignant edema vaccination are imperative. Local conditions will determine which other vaccines may be necessary for inclusion in the program.[9]

Calf Diseases

The key to low mortality and high profits in raising dairy calves is a sound program of feeding and management carried out by individuals who are both experienced and interested in calves. Although tender loving care is difficult to define, it still plays an important role in calf disease prevention.

Infectious Calf Scours (Diarrhea, Enteritis)

The term calf scours, rather than indicating a specific disease, simply describes the major clinical signs of a disease that can have many causes. In scours the intestinal lining either fails to absorb fluids or intestinal secretions are greatly increased, resulting in a loss of essential body fluids. Dehydration is usually the cause of death. The infectious agent responsible for the diarrhea is important, however, from the standpoint of disease prevention.

Infectious calf scours is the most frequent cause of death in calves between birth and 10 days of age. While a variety of bacteria as well as viruses have been implicated, *Escherichia coli* is the major causative agent. *E. coli* is widely distributed in the environment and is normally present in the lower portion of the small intestine and in the large intestine. The majority of strains of *E. coli* are harmless. However, certain strains are enteropathogenic and produce potent toxins. Under certain conditions, enteropathogenic strains multiply rapidly in the digestive tract and release the toxins that result in diarrhea or sudden death. Other causative agents of diarrhea in calves include *Clostridium perfringens, Salmonella spp.,* bovine virus diarrhea, reovirus,[10,11] corona virus,[10,11] and coccidia (in older calves).

The first symptoms noted may be a severe diarrhea, loss of appetite, and general dullness, followed by progressive dehydration. The color and consistency of feces are of little value in determining the precise causative agent. The temperature of the calf may be below or above normal. The age of the calf has a marked influence on the survival rate; the younger the calf, the greater the mortality rate.[12]

Preventive measures include colostrum feeding as soon after birth as possible, proper sanitation (of calving facilities, calf pens, and calf feed and feeding equipment), proper housing, isolation, and treatment of affected calves.

Nutritional Scours (Noninfectious Scours)

In contrast to infectious calf scours, the mortality rate with nutritional scours is low if the calf receives prompt attention. Overfeeding is a primary factor in the development of nutritional scours in young calves. Even though nutritional scours is a noncontagious disease, affected calves should be treated promptly since nutritional scours may predispose the calf to infectious scours. In addition, if scours persist, resistance to other diseases is lowered (see Section 20.3 for a discussion of calf nutrition).

Pneumonia

Pneumonia is a major cause of death in calves between 3 and 16 weeks of age, especially during the late summer and winter months. The disease may occur suddenly in apparently normal calves, or it may be associated with infectious scours or any other disease that lowers the resistance of the calf. The causative agent may be bacterial or viral. Among the bacteria, *Pasteurella multocida* and *Corynebacterium pyogenes* are involved most frequently.

Signs include temperatures of 103° to 106°F, rapid breathing, coughing, and a nasal discharge that may contain pus. Antimicrobial therapy should begin immediately and should continue for several days after the temperature returns to normal. Prevention involves the use of proper housing (avoiding overcrowding, draft control, temperature modulation, and humidity control), sanitation, and feeding management.

Other Calf Diseases

In addition to scours and pneumonia, calves may contract a variety of other diseases, the majority of which are discussed in the latter part of this chapter.[13]

21.10 Heifer Management from 12 Months to Calving

Heifers undoubtedly receive the least management consideration of any age group of cattle on the dairy. Regular inspection for internal and external parasites is often overlooked. Internal parasite control will require more attention when animals have access to pasture than when animals are housed under drylot conditions, since drylot conditions do not provide a favorable environment for the larval stages of internal parasites. The exception to this rule is coccidiosis, which can occur under a wide variety of management situations.

Although not routinely employed, regular weighing of individual heifers or groups of heifers allows accurate monitoring of their nutritional condition and general health status. Proper feeding during the heifer's growing period will do much to prevent disease and management problems at the time and later in life. See Section 20.6 for further discussion of heifer nutrition.

Oral administration of a magnet is a cheap and effective way of minimizing the incidence of hardware disease. All animals should receive a magnet by the age of 1 year. If animals are held off feed for 18 hours prior to administration, the magnet will be more likely to reach the reticulum. A hand compass can be passed along the left side of the cow, in the area behind the elbow, to determine whether the animal already has a magnet in the reticulum. An obvious deflection of the compass needle will occur if a magnet is present.

Vaccination programs for heifers (and new cows entering the herd) may include infectious bovine rhinotracheitis (IBR), bovine virus diarrhea (BVD), and leptospirosis. Development of immunity not only protects the cow from disease but also protects her calf from being aborted (a common result of these diseases). In addition, appropriate antibodies are passed to the newborn calf via the colostrum. The MLV vaccines (IBR and BVD) should be administered at least 30 days prior to breeding.

Breeding heifers (at the appropriate body weight for the breed) to bulls known to produce small calves will reduce trauma at calving and subsequent reproductive tract problems.

21.11 Cow Management

Health programs for cows are concerned primarily with reproduction and mastitis. Conditions at calving have a profound influence on a cow's subsequent reproductive performance. If assistance at calving is necessary, strict attention should be given to the cleansing of equipment (i.e., obstetrical chains), the attendants' hands and arms, and the cow's vulva and adjacent skin. Organisms carried into the cow's reproductive tract may subsequently cause infection and delayed conception. The use of clean calving facilities will improve calf health and reduce the number of postpartum intrauterine infections and peracute mastitis cases. An added benefit of a special calving area is that cows can be observed more closely and assistance provided, if necessary. If special maternity pens are used, adequate protection from sun, wind, and rain should be provided.

Nutrition, especially during the dry period, plays an important role in the reproductive health of the cow. For example, as the incidence of milk fever increases, so does the incidence of retained placentas. In turn, cows with retained placentas are more likely to have subsequent intrauterine infections.

Thirty days after calving the cow's reproductive tract should be examined to see whether the uterus has involuted properly, the ovaries are functioning properly, and any intrauterine infection is present. Early initiation of treatment for intrauterine infections will result in faster and more successful control.

Cows should be checked for pregnancy by approximately 40 days after breeding. If cows are open, the reproductive tracts can be reexamined to see whether corrective measures should be taken.

If a mastitis control program is in effect on the dairy, culturing the milk of all new additions to the herd will identify cows that are capable of introducing new infections. Infected cows can then be treated or culled. Routine culturing of clinical mastitis cases will aid in the design of effective mastitis control programs. Until the incidence of mastitis is reduced to a very low level, all cows in the herd should be treated at drying off with an approved intramammary dry cow infusion product.

All cows added to the herd should be tested for brucellosis and tuberculosis by a federally accredited veterinarian. Appropriate vaccines should be administered and the animals should be checked for the presence of a magnet. As stated previously, MLV vaccines (except the intranasal IBR vaccine) should not be given to pregnant cows, since abortion may occur. If cows have already been vaccinated for IBR and BVD, only special circumstances warrant revaccination since the immunity persists for a long time. If MLV vaccines are used, they should be administered at least 30 days prior to breeding. The ideal time to administer these vaccines is at the time of the postpartum reproductive examination. If leptospirosis is a problem in the herd, cows should be vaccinated annually. In problem herds, vaccination may have to be repeated at 6-month intervals. Since the vaccine (a bacterin) consists of killed organisms, it can be given to either pregnant or open cows without deleterious effects.

If calf scours caused by *Salmonella spp.* are a problem, cows should be vaccinated when they are dried off at the end of lactation and again 3 weeks later. The resistance of the calves will then be augmented by the antibodies they receive in the colostrum.

Since conditions vary so much from dairy to dairy, no general deworming program can be recommended. Ideally, groups of animals should be checked periodically to determine their parasite load. Internal parasites cannot be economically eradicated, so the objective of a parasite control program is to maintain the level of infection below the point where production is appreciably impaired. The general rule is to check fecal samples from a minimum of 10 cows or 10% of a group (e.g., dry cows or heifers), whichever is larger. By having a veterinarian follow the change in parasite egg counts over a period of several years, an optimal control program can be devised.

21.12 Outline of a Basic Herd Health Program[a]

A. Calves
   1. Birth—7% tincture of iodine on navel—feed colostrum immediately
   2. First week—dehorn, remove supernumerary teats, castrate bulls
   3. Two months—brucellosis vaccination (heifers only)

[a]Depending on the local conditions, the basic program may be expanded or contracted. Only general vaccination recommendations are included. Specific vaccination programs for individual herds should be made in consultation with a local veterinarian.

4. Six months—blackleg, malignant edema vaccination (especially if cattle will be on pasture)

B. Heifers
   1. Twelve months—IBR-BVD-leptospirosis vaccination—magnet (orally)
   2. Fifteen months—breed (breed earlier if animals have sufficient body weight, e.g., Holsteins—750 pounds)
   3. Twenty-four months—calve

C. Cows and Heifers
   1. Teat dip after each milking.
   2. Thirty days post partum—reproductive examination, leptospirosis vaccination.
   3. Forty-five to 60 days post partum—breed
   4. Forty to 60 days post breeding—pregnancy examination
   5. Drying off—dry treat udder with intramammary infusion approved specifically for dry cow treatment

21.13 Infectious Diseases

The importance of the various infectious diseases varies both regionally and seasonally throughout the country.

Mastitis

Mastitis refers to an inflammation of the mammary gland. The most common causes of mastitis are bacteria that gain entry to the udder via the teat canal opening (see Table 21.2). Therefore, a mastitis prevention program should be focused on reducing the number of disease organisms impinging on the teat end. Proper washing of the teats with an approved sanitizer, drying of the teats before milking and the use of an effective post-milking teat dip results in a reduction of organisms on the teat end. The normal resistance of the teat canal opening can be protected by following approved milking techniques, by avoiding excessive vacuum and overmilking, and by performing regular milking equipment checks (at least twice a year).

Cows should not be milked wet. Water draining off the udder onto the teats contains many coliform organisms, which flow down the teat during milking and collect on the teat end following milking. After milking, the sphincter muscle surrounding the teat canal remains dilated for a varying period of time, facilitating invasion of the teat canal by bacteria. Thus, teat dips are most effective when applied immediately after the milking machine is removed. Providing feed for cows after they leave the milking parlor keeps them standing so that the sphincter muscle has time to contract before the cows lie down in soiled bedding.

Cows are also exposed to mastitis organisms via the milking machine when milked after a cow affected with clinical or subclinical mastitis. Systems that sanitize the milk cluster between cows are being developed, and preliminary results appear very promising. Cleansing and sanitizing of the milking equipment between milkings are also very important.

Dry treatment of cows, with an intramammary infusion of an antibiotic approved specifically for dry cow treatment, is also an integral part of a mastitis control program. The decision to treat all cows at drying off, in contrast to treating selected cows, is influenced by the type of organisms causing the mastitis and the extent of the problem in the herd. It is extremely important that strict sanitation be employed in both dry cow treatment and treatment of clinical mastitis cases.

The teat end should be cleansed thoroughly with alcohol swabs before infusion. A single infusion tube should never be used on more than one quarter. Careless infusion has often resulted in udder infections with microorganisms that did not respond to treatment with routine antimicrobial agents.

Since the appropriate control method varies with the specific causative organism, it is imperative that the organisms be identified. No program can be effective unless the problem is carefully defined. In the initial stages of control, rigorous culling may be necessary.

Brucellosis

The control and eventual eradication of brucellosis are important because of the danger of human infection and the economic loss to dairymen from infertility and abortions.

After years of calfhood vaccination, blood testing, and slaughter of positive reactors, the goal of complete eradication of

TABLE 21.2 *Major Mastitis Organisms*

The following basic preventive measures are recommended at all times:

1. Milking sanitation
2. Teat dipping after milking
3. Proper installation, adjustment, maintenance and use of mechanical milking equipment
4. Culling of nonresponsive cases
5. Dry treatment with an approved intramammary dry cow infusion product

Organism	*Primary source of infection*	*Prevention*	*Comments*
Staphylococcus aureus	1. Infected cows 2. Environment	1. Dry treatment 2. Identify and segregate infected cows—milk last	1. Forms tiny abscesses in udder; the longer the cow has been infected, the more difficult to eliminate the infection
Streptococcus agalactiae	1. Infected cows	1. Identify infected cows a. Intramammary infusion of antibiotic during lactation and/or at drying off	1. Can be eradicated from the herd
Other Streptococcus species	1. Environment 2. Infected cows		
Nocardia and yeast	1. Organisms carried into udder at time of infusion a. Teat surface contamination b. Infusion product contamination	1. Sanitize teat before infusion 2. Use single dose medication 3. Use approved (commercial) medication	
Coliform bacteria	1. Environment 2. *Klebsiella* associated with raw sawdust 3. Infected cows	1. Do not milk wet cows 2. Reduce contamination of udder a. Keep housing in sanitary condition b. Provide feed after milking	
Mycoplasma	1. Infected cows	1. Eradicate: segregate; cull as soon as possible	

brucellosis in the United States has not yet been achieved.[14] In recent years, when many dairymen have neglected to vaccinate heifer calves with the Strain 19 vaccine, an upsurge in the incidence of brucellosis has occurred. Vaccination only protects approximately 65% of the recipients against a normal field exposure to brucellosis. Nevertheless, vaccination is far better than no protection at all. A new reduced dosage vaccine for calves is now in use. This vaccine does not stress calves as severely and simplifies the interpretation of subsequent blood test results. Until other control techniques are developed, the importance of calfhood vaccination to brucellosis prevention cannot be overemphasized. MLV vaccines should not be administered at the same time as the brucellosis vaccine.

The other important aspect of a brucellosis prevention program is serologic testing and isolation of all new animals entering the herd. This testing (as well as the calfhood vaccination) must be done by a federally accredited veterinarian.

Bovine Virus Diarrhea (BVD)

Bovine virus diarrhea is characterized by erosions of the digestive tract from the mouth to the anus. Clinical signs include diarrhea, lameness, abortion, and congenital defects (in calves). Infected animals have an initial temperature of 104° to 106°F and a clear nasal discharge. In mild cases the temperature returns to normal within several days and the animal recovers promptly. In severe cases, bloody diarrhea, loss of appetite, loss of weight, and dehydration may occur. Diagnosis is confirmed by serologic tests.[15]

The virus is spread from animal to animal by direct contact. MLV vaccines are available. Animals should not be vaccinated before the age of 8 months due to the persistence of maternal antibodies (absorbed from colostrum). Vaccination confers lifelong immunity. Consult your veterinarian concerning the correct vaccine to use.

Foot Rot (Necrotic Pododermatitis)

Most dairy herds have occasional cases of foot rot. In some herds foot rot may reach epidemic proportions. *Fusobacterium necrophorum (Sphaerophorous necrophorus)* and *Bacteroides melaninogenicus* are the causative agents, although other bacteria may become involved in secondary infections. The bacteria are present in manure and thus are prevalent in the cow's environment. Any injury between the claws or at the junction of horn and skin will permit entrance of the organisms. Wet conditions in pastures or barnyards, the presence of abrasive materials (sharp stones), and overgrown hooves are predisposing factors for infection.

The first sign of infection is redness and a slight swelling of the area between the claws. Lameness is generally the first clinical sign noted. If not treated, the infection spreads deeper into the tissues and may penetrate a joint, resulting in an unfavorable prognosis. In milking cows the pain reduces both appetite and milk yield.

If cases are detected early, a thorough cleansing of the hoof and treatment with astringents (copper sulfate) may be sufficient. In many cases systemic antibiotic therapy is necessary. Low level feeding of organic iodides has been advocated for the prevention of foot rot. Theoretically the organic iodides reduce the population of the causative organism in the manure and thus in the environment. The effectiveness of this procedure has not been well substantiated. Excessive amounts of iodine can produce harmful effects. Cows often have access to iodine from a variety of sources (rations, supplements, mineral mixes) so the potential for excessive intake can occur.

Improved sanitation, removal of abrasive objects from the areas where cattle walk, and regular foot trimming do much to decrease the incidence of the disease. Foot troughs containing copper sulfate or lime lower the incidence of the disease in affected herds. Troughs should be placed so that cows must walk through them on the way to or from the milking parlor.

Hardware Disease (Traumatic Reticuloperitonitis, Traumatic Pericarditis)

Hardware disease is caused by the perforation of the wall of the reticulum by sharp foreign objects (nails, wire, staples). As a result, any of the bacteria present in the ingesta may escape into the surrounding tissues and be responsible for the disease. The clinical signs include a reluctance to move, lack of appetite, reduced milk yield, fever, and reduced rumen motility. Treatment of the disease involves the use of antimicrobials, coupled either with surgery to remove the foreign object or the administration of a magnet (unless already present), followed by immobilization of the cow in a stanchion for several weeks. The magnet prevents the foreign object from penetrating further (unless the object is already out of range or is of nonferrous material). The infected areas can then be walled off by the cow's defense mechanisms.

Prevention involves the elimination of wire, nails, staples, and other sharp objects from the feed. The majority of cases can be prevented by prior administration of a magnet. The use of twine, rather than baling wire, reduces the incidence of the disease, as does the feeding of cubed hay (unless the cuber picks up baling wire left in the field). See Section 8.4 for a further discussion of hardware disease.

Infectious Bovine Rhinotracheitis (IBR, Rednose)

Infectious bovine rhinotracheitis (IBR) is an acute, contagious disease caused by a herpesvirus, resulting in a variety of clinical manifestations, of which inflammation of the upper respiratory tract is the most common. The incubation period is less than a week, and the initial body temperature ranges from 104° to 106°F. The nasal discharge may be clear or purulent. Most animals recover within a period of 2 weeks. However, if the disease develops into a bronchopneumonia, recovery is slow and death may occur.[15]

The other forms of disease caused by the virus are: a pustular inflammation of the vulva and vagina (infectious pustular vulvovaginitis-IPV), conjunctivitis, encephalitis, and abortions. Abortions usually occur 1 to 3 months after the initial viral infection of the upper respiratory tract, and are most common in the last trimester of gestation. The disease is spread from animal to animal by contact (via the respiratory route). Vaccination is used to prevent the disease. An intranasal vaccine can be administered to pregnant animals and has been very effective in calves, since the presence of maternal antibodies does not interfere with the development of active immunity.

Johne's Disease (Paratuberculosis)

Johne's disease is caused by the bacterium, *Mycobacterium paratuberculosis.* The organism localizes in the intestine and causes a thickening of the intestine wall. The clinical signs include chronic weight loss and chronic diarrhea. Organisms are passed in the manure and other animals become infected by ingestion of the bacteria. Young animals seem to be most susceptible and infections usually occur early in life. The disease has a long incubation period and clinical signs often do not appear until after 2 years of age.

The disease is frustrating to diagnose in individual animals because of: (1) the lack of dependability of most of the available tests, and (2) the time delay between the collection of fecal samples and the culturing and positive identification of the organism.

Culturing of feces is the method of choice for herd testing. Depending on the number of animals involved, positive animals can be disposed of or else segregated. Calves from infected cows should be removed at birth and reared in a separate facility to break the chain of infection. An experimental vaccine is being tested, however, its major drawback is that its use interferes with TB testing results.

Leptospirosis

Leptospira pomona, L. hardjo, and *L. grippotyphosa* are the most common serotypes associated with leptospirosis in cattle. Following an incubation period of 4 to 10 days, initial infection usually causes a fever for 6 to 48 hours. Unless other signs of illness are evident, the disease may go unnoticed. Occasionally a milking cow may drop in milk production and produce yellow, clotted milk in all four quarters. In more severe infections animals may develop anemia, hemoglobinuria (red-tinged urine), jaundice (yellow color of the nonpigmented skin areas), listlessness, and pneumonia. Fatal infections are more common in calves than in adult cattle. Abortion and stillbirth, the most commonly recognized clinical evidence of leptospirosis in cattle, occur 1 to 4 weeks following the initial bacteremia. Infertility has been associated with *L. hardjo* infection. Recovered animals may shed organisms in the urine for several months. Contamination of the environment by infected animals is the source of infection for other cows. The disease can be diagnosed by serologic means.

Vaccination with the appropriate bacterins results in decreased abortion and stillbirth rates in enzootically infected herds. Very little cross protection occurs between the serotype bacterins so the specific serotype(s) involved should be determined. Vaccination at 6-month intervals is recommended in problem herds.[16,17]

Pasteurella; Parainfluenza-3 (Shipping Fever)

Shipping fever is a pneumonia complex involving bacteria *(Pasteurella sp.),* parainfluenza-3 (PI-3) virus, and stress (e.g., shipping and severe weather changes). The disease is characterized by high body temperatures (104° to 107°F), rapid breathing, accelerated pulse, coughing, and nasal discharge. Severe systemic infection in pregnant animals may cause abortion. Any factor that stresses the animal may be a predisposing factor in the development of the disease. Cattle returning to the herd from shows and sales are often the source of an outbreak. Early recognition of the disease and administration of antimicrobials are essential for successful treatment.

A number of different vaccines are available. The use of Pasteurella bacterins is advocated by some veterinarians. The PI-3 vaccine is often incorporated with the IBR-BVD-leptospirosis vaccines.

Tuberculosis (TB)

Tuberculosis is important for public health reasons as well as for its adverse effects on animal production. The disease is most commonly seen in the respiratory form in cattle although the organism may localize in the intestinal tract, uterus, or udder. Tuberculosis is contracted by the inhalation or ingestion of the bacterium, *Mycobacterium bovis.* Since the infected animal is the primary source of infection, any type of management which results in crowding of animals will favor the spread of the disease. Disease organisms can be excreted in exhaled air, manure, milk, or body discharges. The primary mode of transmission to man is via milk. Fortunately, pasteurization of milk almost completely eliminates this possibility.

The federal bovine tuberculosis eradication program has been responsible for bringing the disease under control in the United States. However, sporadic outbreaks continue to occur so the disease must always be kept in mind. Control at the herd level is based on the identification and removal (slaughter) of infected animals and avoidance of further introduction of the disease. In addition to regular herd tests, all new animals entering the herd should be TB tested. The intradermal TB test is administered by injecting tuberculin into the skin and then examining the injection site 72 hours later for evidence of a reaction. The testing must be done by a federally accredited veterinarian.

Nonspecific Infections of the Reproductive Tract

Nonspecific infections of the reproductive tract are classified according to the tissues involved. Endometritis refers to an inflammation restricted to the endometrium (lining of the uterus). Since no systemic involvement or signs exist, diagnosis can be made only by rectal (or vaginal) examination or by the observation of pus discharging from the vulva. The infection is usually mild and can be eliminated by appropriate intrauterine treatment.

Metritis refers to a much more serious infection involving all the tissues of the uterus. Systemic signs of illness are present, and immediate systemic as well as intrauterine antimicrobial treatment is necessary.

Pyometra refers to an accumulation of pus in a uterus with a tightly closed cervix. If endometritis cases are not treated promptly, they may become pyometra cases, with a much less favorable prognosis.

TABLE 21.3 *Other Infectious Diseases*

Disease	*Causative agent*	*Source of infection*	*Signs*	*Prevention and comments*
Anaplasmosis[18]	*Anaplasma marginale*	Transmitted by blood from affected animals. Vectors are ticks, biting insects, man (during dehorning, vaccinating, etc.)	1. Anemia; pale nonpigmented skin areas 2. Jaundice; yellow color of nonpigmented skin areas	1. All ages susceptible, clinical signs in animals over 18 months 2. Vaccine available
Blackleg	*Clostridium feseri (chauvoei)*	Contaminated soil and feed; ingested	1. Sudden death 2. Lameness 3. Swelling on hip, shoulder, neck, chest or back; gas in tissues	1. All animals on pasture should be vaccinated by 6 months of age; lifelong immunity 2. Vaccination may not be necessary in dry lot 3. Most common in cattle 4 months to 2 years of age
Cancer eye (squamous cell carcinoma)	Virus	Infected animals	1. Cauliflower-like growth on third eyelid, cornea on eyelid	1. Lack of pigment around the eye increases incidence 2. Have growth removed as soon as noticed; surgery or cryosurgery 3. Seldom seen in cattle less than 3 years old
Johne's disease[19]	*Mycobacterium paratuberculosis*	Contamination of environment by infected animal feces	1. Diarrhea 2. Progressive loss of weight	1. Current diagnostic tests are inadequate for eradication[20] 2. Use fecal culture to identify infected cows 3. Animals become infected during first 6 months of life; don't show symptoms until several years later 4. Remove calves at birth and raise in an area away from adult cows; feed milk replacers to reduce the spread of the disease in an infected herd 5. Jersey and Guernsey breeds are most susceptible
Lymphosarcoma[21,22]	Virus	Infected animals; not highly communicable	1. Tumors of lymphoid tissues 2. Signs depend on locations of tumors	1. Blood test to identify animals exposed to the virus is available; used to prevent entry of exposed animals into a clean herd 2. Can affect animals of any age

edema	*ticum*		1. Fluid accumulation in area of wound 2. Sudden death 3. Uterine or birth canal infections	1. Incidence not age related 2. Vaccinate annually if a postpartum problem
Pinkeye (infectious keratoconjunctivitis)	*Moraxella bovis*	Infected animals; spread by flies, contact	1. Ulcer on cornea 2. Excessive tearing	1. Wind, dust, sharp plant awns, sunlight aggravate the condition 2. Face fly is an important vector—dust bags aid control
Ringworm	Trichophyton species (fungus)	Infected animals; direct contact	1. Heavy, grey-white crust raised above the skin	1. Cases regress in the spring when animals are exposed to sunlight
Salmonellosis	*Salmonella dublin, S. typhimurium*	Carrier cows contaminate environment via feces	1. Diarrhea with blood 2. 2–6 weeks of age	1. Usually occurs subsequent to stress 2. Pneumonia is a common complication 3. Vaccine (bacterin) available[23]
Tetanus	*Clostridium tetani*	Contaminated soil enters via wound	1. Erect ears 2. Stiff legs	1. Can be a problem when elastrator band is used to castrate bulls 2. Vaccine (toxoid) available
Trichomoniasis	*Trichomonas foetus*	Infected bull; transmitted at mating	1. Infertility 2. Abortion in early pregnancy 3. Uterine infection	1. Potential problem with natural service; not with artificial insemination 2. Cull infected bulls; use artificial insemination
Tuberculosis	Mycobacterium species	Contamination of environment by infected animal	1. Internal tubercles; usually found at slaughter 2. Progressive weight loss	1. Prevent entry of infected animals by skin testing all new additions to herd 2. Federal eradication program (test and slaughter) 3. Can cause tuberculosis in humans
Vibriosis	*Vibrio fetus*	Infected bull; transmitted at mating	1. Repeat breeding 2. Irregular estrous cycles 3. Abortion in late gestation	1. Potential problem with natural service; not with artificial insemination 2. Vaccinate cows at least 30 days before breeding; annual booster 3. Research reports indicate that vaccination of bulls will control the disease[24]
Warts (bovine papillomatosis)	Virus	Contact with infected animals	1. Whitish cauliflower-like growths on head, neck, and shoulders	1. Cattle over 2 years old are resistant 2. Commercial vaccine available 3. Best results with autogenous vaccine 4. Usually recover spontaneously

Table 21.4 *Parasites*

Parasite	*Life cycle of parasite (and location of various stages)*	*Importance and signs*	*Prevention and comments*
Internal Parasites			
A. NEMATODES 1. Intestinal roundworms (numerous genera and species)	Adults —stomach, intestine —lay eggs Eggs —passed in feces Larvae —eaten along with feed —some migrate in body, others remain in stomach and/or intestine —develop into adults	1. Interfere with digestion 2. Unthriftiness 3. Diarrhea	1. Regular deworming needed on pasture; schedule depends on local conditions 2. Drylot conditions not conducive to perpetuation of life cycle 3. Must have constant exposure to maintain adult population 4. Reduce fecal contamination of feed, if possible 5. Avoid feeding on ground or else change feed area daily 6. Avoid overgrazing 7. Segregate young animals from mature animals
2. Lungworms *(Dictyocaulus viviparus)*	Adults —lungs —lay eggs Eggs —hatch Larvae —coughed up and swallowed —pass out in feces —eaten along with feed —migrate to lungs —develop into adults	1. Pneumonia 2. Unthriftiness	1. Require moisture for survival; prevent access to wet areas 2. Regular deworming; schedule depends on local conditions
B. PROTOZOA Coccidiosis (Eimeria species)	Adults —intestinal wall Oocysts —passed in feces —eaten with feed —complete life cycle in intestinal wall	1. Diarrhea, often with some fresh blood	1. Reduce fecal contamination of feed and water 2. Avoid feeding on ground or else change feed area daily 3. During periods of high exposure, feed coccidiostats 4. More likely to be a problem in calves

C. TREMATODES Liver flukes *(Fasciola hepatica)*	Adult —bile ducts —eggs passed via bile into feces Eggs —hatch into intermediate stages Intermediate stages —enter snails —passes through several more stages —leave snail —encyst on forage —eaten along with feed —migrate to bile ducts, mature	1. Unthriftiness 2. "Clay-pipe" tracts in liver; liver condemned at slaughter	1. Requires snail as an intermediate host; preventing access to snail habitat (wet areas) will reduce exposure 2. No drugs currently registered for snail control 3. Only one drug (which has restricted approval) is available for use in animals 4. Clinical disease usually seen in cattle under 2 years of age 5. Older animals may show lowered productivity
D. DIPTERA Grubs (heel flies, warbles; *Hypoderma* species)	Adults —lay eggs on legs Eggs —hatch Larvae —burrow through skin —eventually reach back —cut breathing hole —drop out on ground in spring Pupae —emerge in late spring	1. Adult flies bother cattle; lower milk production 2. Abscesses along back lower hide value	1. Do not treat for grubs after date recommended for your area; if treated after this date, larvae may be in area of esophagus or spinal column resulting in severe reaction when the larvae die 2. Adult fly has no mouth parts; sole function is to reproduce 3. No systemic grubicide currently approved for use on lactating cattle
External Parasites			
Lice (Various genera and species)	Spend entire life on animal	Biting lice 1. Irritation, animals rub Sucking lice 1. Suck blood 2. Animal may become anemic	1. Most severe in winter (especially in housed animals) 2. Most noticeable around head and tail head 3. Spray with approved insecticide 4. Use dust bags with approved insecticide 5. Dust manually with approved insecticide

NOTE: The list of drugs approved for use in parasite control in various groups of cattle (lactating, dry, beef) changes so frequently that current recommendations should be obtained from a local veterinarian, extension agent, or drug supplier.

Reproductive tract infections occur as a result of bacterial contamination of the uterus, usually at the time of calving. Proper sanitation of calving facilities, cleanliness when assisting at calving time, and proper nutrition will reduce the incidence of infections. Early detection and treatment of reproductive tract infections are important if the goal of high fertility is to be achieved.

A number of other infectious diseases are summarized in Table 21.3.

21.14 Parasitic Diseases

Parasites, both external and internal, present only minor problems in well-managed dairy herds where symptoms are noted early and animals are treated promptly. The relative importance of the various species of parasites varies both regionally and seasonally throughout the country. Each dairy manager, in cooperation with his veterinarian, should develop a control program best suited to the needs of the herd. The principal parasites affecting dairy cattle are listed in Table 21.4.

21.15 Metabolic Diseases

Although the underlying mechanisms of the important metabolic diseases of dairy cattle are not fully understood, proper nutrition will drastically reduce their incidence.

Milk Fever (Parturient Paresis)

The most significant characteristic of milk fever is hypocalcemia (low blood calcium). The name "milk fever" is a misnomer since affected cows have a subnormal or normal body temperature. Initiation of lactation places a severe strain on the calcium balance of the cow due to the amount of calcium secreted in the milk. All cows are slightly hypocalcemic at the time of calving, but some cows become so hypocalcemic that clinical signs of milk fever develop. Since the calcium ion is necessary for normal muscle function, a severely hypocalcemic cow initially has an unsteady gait, trembles, or has difficulty in rising. If untreated, the animal goes down, often lying with the head turned back toward the flank, and is unable to rise. Delayed treatment often results in death or a very slow response to treatment. Milk fever cases should be regarded as emergencies and the affected animal should be treated as soon as possible. Relapses may occur in treated cows.

The majority of cases occur within 3 days after calving; however, cases have been reported to occur during all stages of lactation and the dry period. The incidence of the disease increases with increasing age of the cow and is higher in the Jersey and Guernsey breeds. Once a cow has milk fever, she is more prone to have it in succeeding lactations. Cows with a history of milk fever should be watched carefully at calving for early signs of the condition.

The standard treatment is intravenous infusion of calcium borogluconate as soon as the first signs are seen. Cows that already are recumbent will normally stand up within 1 to 2 hours following treatment. If the cow does not rise or if a relapse occurs, the treatment can be repeated. Since overdoses of calcium salts can result in acute heart damage, intravenous administration should be performed slowly. If more than one retreatment is deemed necessary, the diagnosis should be reassessed and professional advice sought.

The incidence of milk fever can be reduced to a very low level by proper feeding during the dry period. Feeding low calcium (less than 100 grams per day)—high phosphorus (more than 40 grams per day) rations during the dry period is important. Excellent results have been reported by feeding specific calcium-deficient diets in the last 2 to 3 weeks before calving. See Sections 9.6 and 9.7 for further information on the use of both vitamin D and calcium for the prevention of milk fever.

Ketosis (Acetonemia)

The characteristics of ketosis are hypoglycemia (low blood sugar), ketonuria (ketone bodies in the urine), loss of weight, and reduced milk production. Most cases occur in well-conditioned cows, 2 to 6 weeks after calving. Rapid utilization of body reserves and impaired carbohydrate metabolism are involved in the development of the disease.

Affected cows refuse to eat grain, then silage, and finally hay. They become gaunt and milk production drops rapidly. Listlessness and the odor of acetone (a ketone) in the urine, in the milk, and on the cow's breath are other indications of the disease. A positive diagnosis can be made by testing the milk or urine for the presence of ketone bodies.

Treatment of ketosis usually involves intravenous administration of glucose solutions and glucocorticoids (to temporarily increase blood sugar levels) as well as the oral administration of propylene glycol. Numerous other treatments are used.

None of the current recommendations for the prevention of ketosis are totally effective. Currently accepted feeding recommendations include: gradually increasing grain intake after calving to avoid indigestion and subsequent disease, using hay in preference to high silage rations, and preventing cows from becoming too fat during the dry period. The feeding of propylene glycol or sodium propionate has been successful in herds where ketosis is a major problem.

Three other metabolic diseases of dairy cattle are summarized in Table 21.5.

21.16 Poisoning

When poisoning occurs in dairy cattle, it is usually the result of human error. Poisoning cases are often very difficult to diagnose since (1) many of the clinical signs are nonspecific, (2) a number of different toxic substances produce the same clinical signs, and (3) the number of toxic substances to which animals may be exposed is enormous. Since the prognosis is generally unfavorable, and death occurs in many cases before the condition can be diagnosed and treated, prevention is the best way to avoid losses. The prevention of poisoning can be simply summarized by saying that animals should not be allowed access to anything that they should not consume.

The poisonings listed in Table 21.6 have been chosen because their occurrence in cattle has been well documented. However, on a specific dairy, other poisonings may be much more important.

21.17 Summary

A herd health program, developed and executed jointly by the dairy manager and his veterinarian, for the prevention and

TABLE 21.5 *Other Metabolic Diseases*

Disease	*Cause*	*Signs*	*Prevention and comments*
Displaced abomasum	Exact cause uncertain	Off feed, reduced milk yield	1. Highest incidence at calving time 2. More common in cows fed high level of grain or low forage rations 3. Provide proper nutrition; see Section 8.4
Retained placenta	Varied 1. Nutritional; vitamin, mineral or energy deficiency 2. Other; twinning, induced parturition	Placenta not shed within 12 hours after calving	1. Provide proper nutrition; see Section 8.4
Udder edema	Excessive accumulation of fluid in udder tissue	Enlarged and swollen udder	1. More severe in first lactation cows. 2. Limit access to both sodium and potassium salts during dry period 3. Avoid excessive concentrate consumption prepartum

TABLE 21.6 *Poisoning*

Poison	Signs	Prevention and comments
Nitrates, nitrites	1. Labored breathing 2. Unsteady gait 3. Brownish color of non-pigmented skin soon after ingestion of nitrates or nitrites	1. Excessive intake of nitrates or nitrites 2. Nitrites convert hemoglobin to methemoglobin, reducing blood's ability to transport oxygen 3. Fertilizers are a common source of poisoning 4. Some plants under stress conditions (drought) accumulate nitrates or nitrites
Lead	1. Unsteady gait 2. Unable to rise 3. Convulsions 4. Blindness	1. Most common sources are old batteries, lead-base paint on barns 2. More common in calves
Plants (several hundred)	1. Depends on plant	1. Animals consuming adequate amounts of other feeds seldom eat enough of a poisonous plant to do any harm 2. When feed is scarce, animals may consume enough of a toxic plant to produce toxic or even fatal effects 3. Become familiar with plants on farm that can be poisonous; eradicate if possible[25]
Moldy feed toxicity (toxins produced by fungi)	1. Depends on fungus involved 2. Often affects liver 3. Lowered performance	1. Chronic nature; by time problem is noticed, source cannot be found 2. Avoid feeding moldy feeds
Chemicals Pesticides Herbicides	1. Depends on chemical	1. Contamination of feeds 2. Always follow directions on label of chemicals 3. Properly dispose of chemical containers and unused portions of chemicals

treatment of infectious and metabolic diseases, internal and external parasites and other common ailments, as well as accidental injuries or poisoning, will pay generous dividends in terms of increased production efficiency. Herd health is the responsibility of everyone involved in the herd operation. A daily health check of all animals in the herd, along with regular analysis of the dairy records, will reveal health problems in the early stages when treatment is most effective, recovery is most rapid, and proper measures can be taken to prevent their spread.

Review Questions

1. What are the differences between herd health management programs and individual animal treatment protocols?
2. Outline the basic information that should be included in animal records.
3. What are the components of a successful herd health program?
4. What are the roles of dairy managers and veterinarians in herd health programs?
5. Outline a calf health management program (birth to 1 month of age).
6. Discuss the cause and prevention of 5 infectious diseases of dairy cattle.
7. What factors influence the incidence of parasitic diseases?
8. Discuss the cause and prevention of one metabolic disease of dairy cattle.

References

1. Frederick, G.: Herd health investment boosts milk 3,000 pounds per cow. Hoard's Dairyman *122:* 78, 1977.
2. McCauley, E. H.: The contribution of veterinary service to the dairy enterprise income of Minnesota farmers: production function analysis. J. Am. Vet. Med. Assoc. *165:* 1094, 1974.
3. Norman, B. B.: Metabolic profile testing-problems of putting MPT to work. Anim. Nutr. Health *31:* 12, 16, 1976.
4. Payne, J. M., Dew, S. M., Manston, R., and Faulks, M.: The use of a metabolic profile test in dairy herds. Vet. Rec., *87:* 150, 1970.
5. Stevens, J. B.: Metabolic and cellular profile testing: an aid to dairy herd health management. Anim. Nutr. Health *30:* 14, 1975.

6. Stevens, J. B., Anderson, J. F., Perman, V., and Schlotthauer, J. C.: Metabolic and cellular profile testing—a modern approach to herd health management. 35th Minnesota Nutrition Conference, pp. 175–190, 1974.
7. Morrow, D. A.: Nutritional health program for high producing dairy herds. Bovine Practitioner, No. 11: 16, 22, 1977.
8. Bush, L. J., Aguilera, M. A., Adams, G. D., and Jones, E. W.: Absorption of colostral immunoglobulins by newborn dairy calves. J. Dairy Sci. *54:* 1547, 1971.
9. Smith, P. C.: Proposed calfhood immunization program for the commercial dairy herd. J. Dairy Sci. *60:* 294, 1977.
10. McClurkin, A. W.: Probable role of viruses in calfhood diseases. J. Dairy Sci. *60:* 278, 1977.
11. Mebus, C. A.: Viral calf enteritis. J. Dairy Sci. *59:* 1175, 1976.
12. White, R. G.: Scours—the calf killer. Anim. Nutr. Health *31:* 14, 16, 1976.
13. White, R. G.: Bovine pediatrics. Bovine Practitioner, No. 10: 34, 1975.
14. Becton, P.: Brucellosis status report. J. Dairy Sci. *59:* 1163, 1976.
15. Newman, L. E.: Infectious bovine rhinotracheitis and bovine virus diarrhea. J. Dairy Sci. *59:* 1179, 1976.
16. Hanson, L. E.: Bovine leptospirosis. J. Dairy Sci. *59:* 1166, 1976.
17. Hanson, L. E.: Immunology of bacterial diseases, with special reference to leptospirosis. J. Am. Vet. Med. Assoc. *170:* 991, 1977.
18. McCallon, B. R.: Anaplasmosis. J. Dairy Sci. *59:* 1171, 1976.
19. Larsen, A. B.: Johne's disease (paratuberculosis) in 1976. Proc. 9th Ann. Mtg. Am. Assoc. Bovine Practitioners, pp. 20–23, 1976.
20. Moyle, A. I.: Culture and cull procedure for control of paratuberculosis. J. Am. Vet. Med. Assoc. *166:* 689, 1975.
21. House, J. A., Glover, F. L., and House, C.: Current aspects of bovine leukemia. Proc. 8th Ann. Mtg. Am. Assoc. Bovine Practitioners, pp. 147–150, 1975.
22. Olson, C., and Baumgartener, L.: Lymphosarcoma (leukemia) of cattle. Bovine Practitioner, No. 10: 15, 18, 22, 1975.
23. Rankin, J. D., Taylor, R. J., and Newman, G.: The protection of calves against infection with *Salmonella typhimurium* by means of a vaccine prepared from *Salmonella dublin* (Strain 51). Vet. Rec. *80:* 720, 1967.
24. Clark, B. L., Dufty, J. H., Monsbourgh, M. J., and Parsonson, I. M.: Immunization against bovine vibriosis—vaccination of bulls against infection with *Campylobacter fetus subsp. venerealis.* Aust. Vet. J. *50:* 407, 1974.
25. Fowler, M. E.: Poisonous plants in harvested feeds. Proc. 9th Ann. Mtg. Am. Assoc. Bovine Practitioners, pp. 30–34, 1976.

Suggested Additional Reading

Amstutz, H. E., editor, *Bovine Medicine and Surgery,* 2nd ed., Santa Barbara: American Veterinary Publications, 1980.

Blood, D. C., Radostits, O. M., Henderson, J. A.: *Veterinary Medicine,* 6th ed., London: Bailliere Tindall, 1983.

Buck, W. B. and Osweiler, G. D.: *Clinical and Diagnostic Veterinary Toxicology,* 2nd ed., Dubuque: Kendall Hunt, 1982.

Freeman, A., ed.: Bovine infectious diseases report. J. Am. Vet. Med. Assoc. *163:* 777, 1973.

Freeman, A., ed.: Colloquium on bovine mastitis. J. Am. Vet. Med. Assoc. *170:* 1115, 1977.

Gillespie, J. H. and Timoney, J. F.: *Hagan and Bruner's Infectious Diseases of Domestic Animals,* 7th ed., Ithaca: Cornell University Press, 1981.

Howard, J. L., editor, *Current Veterinary Therapy: Food Animal Practice,* Philadelphia: W. B. Saunders Co., 1981.

Kingsbury, J. M.: *Poisonous Plants of the United States and Canada.* Englewood Cliffs, New Jersey: Prentice-Hall, 1964.

Radeleff, R. D.: *Veterinary Toxicology,* 2nd ed. Philadelphia: Lea & Febiger, 1970.

Roberts, S. J.: *Veterinary Obstetrics and Genital Diseases,* 2nd ed. Ann Arbor, Michigan: Edwards Brothers, 1971.

Schalm, O. W., Carroll, E. J., and Jain, N. C.: *Bovine Mastitis.* Philadelphia: Lea & Febiger, 1971.

Schultz, L. H., Brown, R. W., Jasper, D. E., Mellenberger, R. W., Natzke, R. P., Philpot, W. N., Smith, J. W., and Thompson, P. D.: *Current Concepts of Bovine Mastitis,* Washington National Mastitis Council, 1978.

CHAPTER 22

Facilities, Equipment, and Technological Developments

22.1 Introduction

The dairy housing system should serve several functions:

1. Provide a healthy, comfortable environment for cows.
2. Provide a desirable working condition for the operator.
3. Be integrated with the feeding, milking, and manure handling systems.
4. Comply with applicable sanitary codes.
5. Optimize labor efficiency in terms of cows handled per man and milk produced per man.
6. Be economically feasible.

The type of housing best adapted to a particular farm depends on many factors. Among the major considerations are: (1) climate, (2) size of herd, (3) condition and layout of the present housing system, (4) cost, and (5) personal preference. Recent and projected trends in housing were outlined briefly in Chapter 1. This chapter contains a discussion of housing systems, ventilation requirements, waste handling, feeding systems, milking facilities, and housing of youngstock.

Definitions of the various systems discussed in this chapter are as follows:

Cold housing refers to buildings kept cold in winter. Natural air movement removes moisture and keeps inside temperatures near outside temperatures. These buildings may be uninsulated, but light insulation is frequently used in the Lake States and Northeast regions in order to reduce winter condensation and summer heat buildup.

Warm housing refers to buildings kept warm in the winter. They are equipped with mechanized ventilation and environmental control. Insulation helps retain animal heat to prevent freezing within the structure in winter and keep out excessive heat in summer.

Stall barns are those where each cow is confined in a stanchion or tie-stall. In most instances, stall barns are insulated, mechanically ventilated, and maintained at a 40° to 50°F winter temperature.

Free-stall barns are those where cows have access to individual resting stalls, but are free to move among resting, feeding, and watering areas.

Loose housing barns provide a manure pack in a "cold" barn. The cows are free to move among resting, feeding, and watering areas.

Open lots are outside exercise areas. They are used by themselves for resting, or in combination with loose housing, freestalls, or stall barns for feeding and watering. In warm climates, sun shades may be provided.

22.2 Stall Barns

Stall barns are by no means standardized units. They differ in arrangement (face out or face in), type of stalls (stanchion versus tie), and equipment associated with feeding, waste handling, and milking. With current emphasis on haylage or silage for forage, most new units are one-story structures with adjacent silos for storing feed.

Stall barns are the most common form of housing for the milking herd in the Northeast and Lake States regions, where protection from adverse weather must be provided. This housing is the most economical and practical for smaller herds (less than 60 cows). Feeding of cows and handling of wastes can be mechanized as much in stall barns as in free-stall barns. The main advantage of stall barns is the opportunity for greater individual attention to cows and maximum comfort for the operator. For the breeder of registered cattle, stall barns enable him to let prospective buyers see his cattle to their best advantage at all times of the year. Problems in stall barns include: tying and untying cows, distribution of feed and bedding, difficulty in installing modern pipeline milking systems that result in stable milking vacuum, stooping to milk, and difficulty in controlling moisture in old, uninsulated buildings.

Arrangement of Barn

The face-out system is generally considered more desirable than the face-in system. While both systems have advantages and disadvantages, facing the cows out generally reduces the labor required. Approximately 60% of the operator's time is spent behind the cow, 15% in front, and 25% in other parts of the barn. A particular advantage of the face-out arrangement is that it is much easier and more economical to install a pipeline milking system that permits the operator to milk cows on either side simultaneously. Also walls adjacent to litter alleys are more difficult to clean than those adjacent to feed alleys.

Those who prefer the face-in system believe the cows are easier to get into their stalls, and that feeding of the cows is made easier. Drive-through face-in barns, once popular in the Northeast to facilitate feeding of hay and bunker stored silage, have lost their popularity, probably because of the trend to larger herds utilizing freestalls and the difficulty in controlling temperature and moisture during the winter months.

New barns are being built with a 1-in. slope per 10 ft of barn length toward the milk room. The slight slope is not noticeable but minimizes the height of the milk

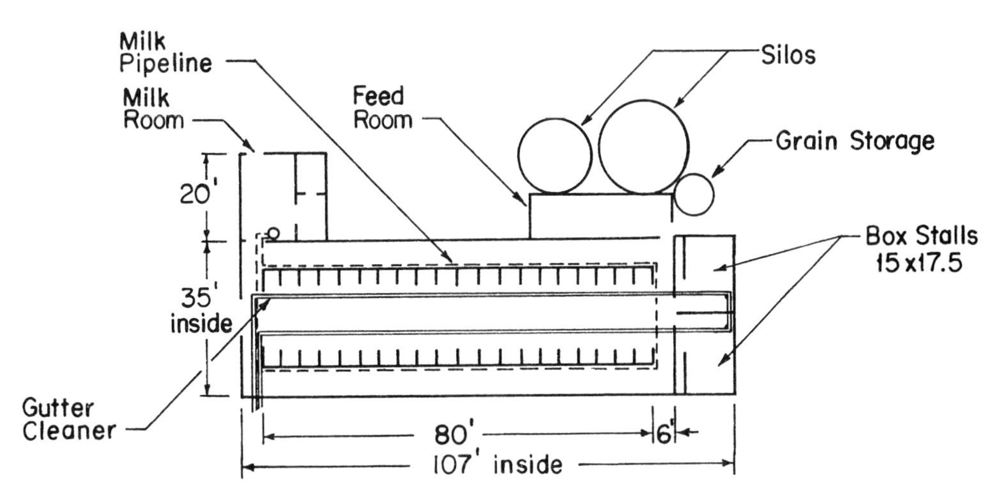

Fig. 22.1 Plan for 40 cow tie stall barn.

FIG. 22.2 A portion of a typical midwestern tie-stall barn. The gutters on either side of the center alley have grates over the gutter.

pipeline in long stall barns. Modern designs call for both the manure loader and milk room to be located at opposite sides of one end of the barn; cow exits and entries, and feed storage, are at the other end. Figure 22.1 shows the design graphically, Figure 22.2 pictures a portion of a typical stall barn. It has the advantage of minimizing cross-overs and permits lower (less height) installation and proper slope of the milk pipelines.

One box stall for each 25 cows is recommended. Box stalls should be at least 12 ft wide and either 12 or 13 ft long to permit adequate work space. Arrangements should include watering device, feeding space, a lock-in stanchion, and an overhead hoist for elevating downer cows. A more complete discussion of hospital area facilities is found in Section 22.3.

Type of Stalls

The two general types of stalls in use are: stanchion and tie. Stanchion stalls, with a yoke-fastened top and bottom that is free to swing and with each cow released individually, are seldom put into new barns. Lever stanchions, which secure or release all cows simultaneously, are more common in the "flat milking barns" of the Southwest. Lever stalls substantially reduce the time and labor required in fastening and releasing cows, but fail to provide the same

TABLE 22.1 *Recommended Cow Stall Platform Sizes, Combined with Electric Cow Trainers*

	Stanchion stalls		*Tie stalls*	
Cow weight	*Width*	*Length*	*Width*	*Length*
Under 1,200 lb	4′0″	5′6″	4′0″	5′9″
1,400 lb	4′6″	5′9″	4′6″	6′0″
Over 1,600 lb	Not recommended		5′0″	6′6″

Source: Dairy Housing and Equipment Handbook. Midwest Plan Service Publ. No. 7, 1976.

TABLE 22.2 *Stall Comparisons*

Stall type	*Cost*	*Easy to tie*	*Cow comfort*	*Cow access to manger*	*Can be home made*
Regular stanchion[a]	Med	Med	Least	Least	Hard
Lever stanchion	High	Best	Least	Least	Hard
Regular tie	Med	Med	Med	Least	Med
New York tie	Low	Least[b]	Best	Best	Easy
Comfort tie	High	Least[b]	Best	Best	Med
Inverted V tie	Med	Med	Best	Best	Hard

[a]Stanchions without a swivel yoke can easily be made of wood, but restrict cow head movement.
[b]If the cow backs up while being tied.
Source: Dairy Housing and Equipment Handbook. Midwest Plan Service Publ. No. 7, 1976.

degree of cow comfort. Regardless of the stall front, platforms are best kept clean with proper platform size (Table 22.1) and by use of cow trainers. A comparison of stanchions and various types of tie-stalls are shown in Table 22.2.

Tie-stalls offer more freedom than stanchions. The four types are: regular tie, New York tie, comfort, and inverted V. Each consists of a neck tie (chain or strap) fastened to the curb or stall front to prevent the animal from backing out, and a restraint in front to prevent the animal from walking into the manger. More labor is required to fasten and release cows in tie-stalls than in stanchions.

The *regular tie-stall* has the advantage of permitting the cow to be down either with her neck through the opening or with her head over the platform area (Figure 22.3). The *New York stall* is lower in cost since the horizontal pipe can double as a water pipe. Its disadvantages, however, are that cows have access to feed intended for adja-

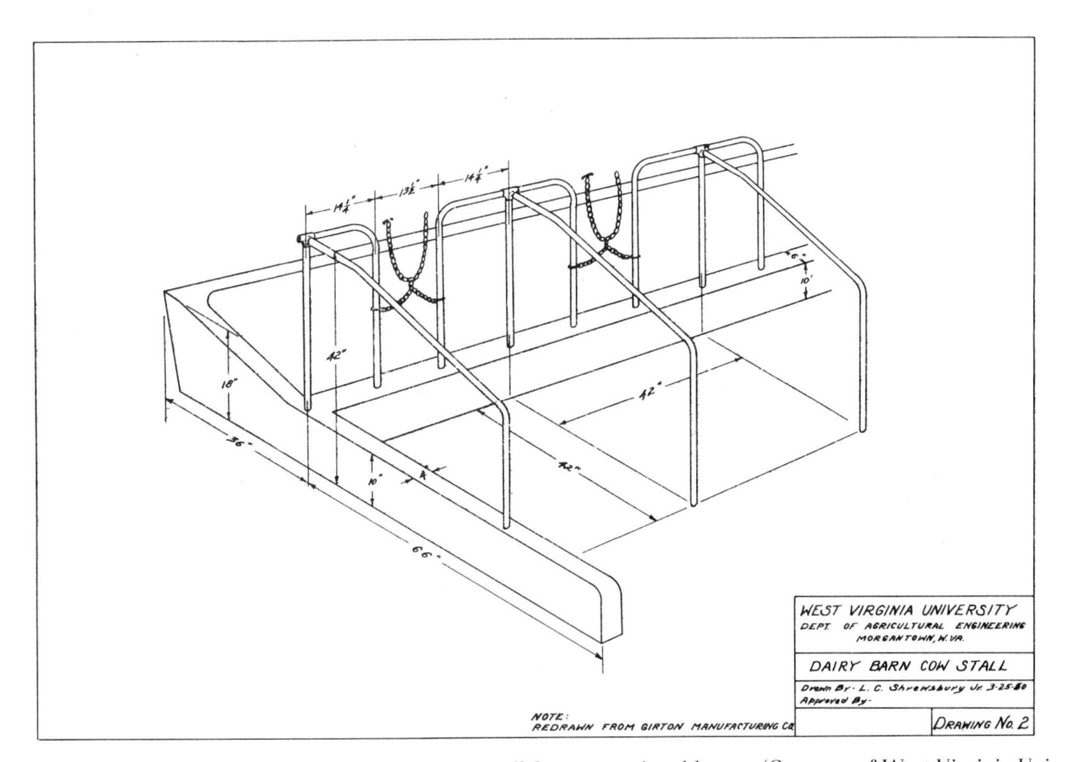

FIG. 22.3 Design and dimensions of a typical tie stall for conventional barns. (Courtesy of West Virginia Univ., Morgantown, W. Va.)

cent cows and untrained cows more frequently attempt to enter the manger area. The 3-horizontal pipe stall is known as the *comfort stall.* These pipes are positioned to force the cow to stand near the rear of the platform, but they still allow her to move forward in arising and provide free head movement over the curb. Usually the top pipe is used as a vacuum pipe and the bottom one is a water line. *Inverted V stalls* restrain forward movement when the cow stands, yet they do not restrict the animal's head movement when she is lying down.

Stall Size

Today's dairy cow is a big animal. Holsteins weigh from 1,200 to 1,800 lb. Small stalls in older stanchion barns are directly responsible for many injuries that have led to unjust criticism of stall barns. Be sure that platform widths are consistent with the length. As a general rule, platform widths should be about 75% of the length (Table 22.1). Wide stalls permit the operator freedom of movement at milking time. Some dairymen prefer stalls of two sizes, 30% small for heifers and 70% large for mature cows.

Stall Mats

Because of the shortage of bedding materials and the need to reduce bedding where liquid manure systems are used, dairymen are seeking a replacement for straw that, in many areas, has been difficult to find. Straw is highly absorbent and has good insulating properties. Options include: (1) nearly bare cement with reduced amounts of chopped straw or shavings, (2) rubber mats embedded in concrete, (3) rubber mats mechanically held in place, (4) rubber mats bonded to the concrete, (5) poured-in-place synthetic resin mats, and (6) indoor-outdoor carpeting.

Some dairymen believe that barns should be equipped so that some stalls have rubber mats and others are left as plain concrete. Cows with poor foot structure (shallow heel) are definitely more prone to excessive toe growth when housed continuously on rubber mats or other softer surfaces. Foot trimming at regular intervals may become a necessity. A disadvantage of most mats is that water, manure, urine, and bedding eventually get beneath them or moisture moves up through the concrete. Such conditions are favorable to rapid bacterial growth; cleaning the underside is difficult and time consuming. Indoor-outdoor carpeting wears out quickly and is irritating to open sores; it is not recommended.

None of the systems described appear to be completely satisfactory. Dairymen continue to experiment in search of the ideal stall surface. You must recognize that no mat alone can keep cows clean. Properly adjusted cow trainers and continued housekeeping are necessary. Because the mat acts as an insulator under a cow's feet, be sure to use metal tie chains, rather than leather straps, in order to provide an electrical ground when trainers are used.

Barn Width

A 36-ft outside barn width is usually recommended in new construction. If the combined feed alley and manger width can be reduced from 6 ft 7 in. to 5 ft 10 in., a 34-ft wide barn can be satisfactory. A combined flat manger and feed alley is now common, necessitating less incline for cross alleys; it allows more room for mechanical feed carts and permits easier cleaning. Service alleys, between cows in a face-out barn, that are 6 ft wide are adequate when using gutter cleaners or grates for storing liquid manure underneath them. Typical cross sections are shown graphically in Figure 22.4; recommended stall barn dimensions are provided in Table 22.3.

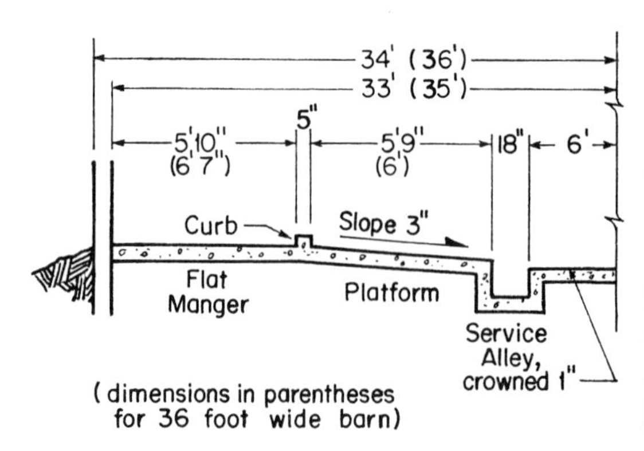

FIG. 22.4 Cross section for barn with flat manger and feed alley.

Table 22.3 *Recommended Stall Barn Dimensions*

Alley width	
Flat manger-fed alley	5′8″–6′6″
Feed alley with step manger	4′0″–4′6″
Service alley with barn cleaner	6′0″
Cross alley[a]	4′6″
Manger width	
Cows under 1,200 lb	20″
Cows 1,200 lb or more	24″–27″
Gutters	
Width[b]	16″ or 18″
Depth, stall side	11″–16″
Depth, alley side	11″–14″

[a]Taper the end stalls inward 6 in. at the front for added turning room for a feed cart.
[b]Or as required for barn cleaner.
Source: Dairy Housing and Equipment Handbook. Midwest Plan Service Publ. No. 7, 1976.

Gutter Design

While cows appear larger when they are slightly elevated, steel grates over gutters of the same height on either side permit easier movement by the operator when milking. Furthermore, grates help keep bedding out of the gutter when liquid manure systems are used and prevent cows from stepping in gutters.

Remodeling

Reasons for remodeling a structurally sound building include:

1. Replacement of worn out equipment.
2. Provision of more comfortable stalls.
3. A more convenient arrangement to reduce labor.

Stall sizes recommended for new construction should be used. Similarly, gutters should be at least 16 in. wide. Consider these two dimensions fixed; others are flexible. Flat manger-feed alleys can be narrower than recommended in Table 22.3 if cows are turned out while feed is distributed. Do not consider remodeling barns less than 30 ft wide inside.

22.3 Enclosed Free-Stall Barns

These barns are most common in herds of 150 or more cows, employing 3 or more men in the dairy operation (Figure 22.5). However, numerous total confinement,

Fig. 22.5 A modern enclosed free-stall barn located in Minnesota. Only one side is shown. Three rows of free-stalls provide a resting place for cows that eat at a mechanized feeder (portion of feed bunk shown at extreme right).

free-stall operations exist in the 60 to 100 cow range. Production potential frequently is limited because of management problems associated with system design errors. A list of such problems includes: (1) poor free-stall design, (2) inability to group cows, (3) inability to segregate cows, (4) inability to move cows to the milking center without crossing another herd, (5) dead-end alleys with access limited because one cow lies down in the alley, (6) feed alleys too narrow for adequate cow movement, (7) unprotected overhead door tracks or watering devices, subject to damage by vehicles, (8) turns required when scraping manure, (9) excessive number of gates to open while scraping or feeding, (10) gates hinged to swing in the wrong direction, and (11) absence of man passes (narrow openings in fences to permit the operator to move between groups of cows without opening and closing gates).

Before the dairyman accepts a design proposal, he should visit several existing systems to acquaint himself with the mistakes others have made. Further, problems that can be managed in small housing systems can become magnified in larger herds. Finally, a free-stall system should be designed to permit further expansion at a later date.

Free-Stall Design

Free-stalls are meant to provide a comfortable resting space, away from the feeding area. They must be durable, and should provide for cow cleanliness, minimal wastage of bedding, and freedom of injury to the cow. Stall dimensions are important. Table 22.4 provides free-stall dimensions for cows of differing average herd size. The suggested length is for stalls filled to the top of the back curb, and the length includes the curb itself. Cleanliness is maintained by frequent removal of manure within the stall. Front partitions are usually 8 to 12 in. higher than side partitions to prevent cows from standing with their heads over the partition.

TABLE 22.4 *Free-Stall Dimensions*

Size of cow, lb.	*Width*	*Length*	*Height of side-partition[a], in.*
1,000	3′6″	6′10″	40
1,200	3′9″	7′0″	42
1,400	4′0″	7′3″	45
1,600	4′0″	7′6″	48

[a]Height from curb to top of upper rail.
Source: Dairy Housing and Equipment Handbook. Midwest Plan Service Publ. No. 7, 1976.

Partition construction varies greatly. Most commercial stalls are 2 or 3 horizontal pipes. Because the space between rails is greater than in those with wooden 2 × 6 in. construction, a fourth rail may be added to prevent cows from inserting their head through the partition; when this situation occurs, it is easier to turn small cows around within the stall. Most recommendations call for a 15-in. clearance between the stall floor and lower rail.

More recently, suspended free-stall partitions have gained in popularity (Figure 22.6). Because no alley post exists, partitions can be installed after the concrete work is complete. The partitions are 1 ft shorter than the stalls and permit maintenance of the stall platform with a tractor-mounted blade. Further, less opportunity exists to damage the stall when scraping alleys with a tractor.

Many curbs lack sufficient height to hold the daily manure collection, and spill-over into the stalls is common during the scraping process. When the scraping alley is quite long or when manure is scraped infrequently, the height of the curb should be increased. Curb heights of 10, even 12 in., are desirable.

Bedding is expensive and bothersome. Sawdust, shavings, straw, and/or dried manure are commonly used bedding materials. Compacted clay or stone dust material may be used without bedding. These materials require regular maintenance to prevent holes from developing in the stall. Sand, gravel, clay, or stone dust may cause undue wear of liquid manure pumps.

Concrete floors are less comfortable for cows; thus, some cows make only minimal use of them. Rubber mats set into concrete have worked satisfactorily (see discussion in Section 22.2). One stall is usually provided for each cow. When complete-feed rations are available *ad libitum*, cows do not always lie down at the same time. Some dairymen have had success with 10 to 30% more cows than stalls.

FIG. **22.6** Suspended free-stalls. (Courtesy of Hoard's Dairyman)

Barn Design

Free-stall barn design should permit expansion without alteration of the milking center, provide for 3 or more groups of cows, have convenient access to hospital facilities, and permit a smooth flow of cow traffic.[3] The layout shown in Figure 22.7 contains all of these features.

Free-stall barns are commonly built with 2, 3, or 4 alleys. Most designs use some alleys for both feed and stall access. Barn width is directly related to alley widths. See Figure 22.8 for dimensions. When mechanical manure scrapers are used and when alleys are not excessively long, solid floor alleys may be reduced to a 6-ft width. Slope solid floor alleys to storage areas or cross conveyors to facilitate drainage. In solid floor barns, make curbs straight and alleys of uniform width. Locate fences, waterers, and any cow cross alleys in areas elevated from and sloped to the alleys. These elevated areas are usually cleaned by hand.

Cow Traffic

The design shown in Figure 22.7 reduces the size of the holding area and the time cows are retained prior to milking. Fifty-four free-stalls are built for each group of cows. If the milking parlor is a double-6 herringbone, for example, the holding area need be only large enough to hold 42 cows (54 cows if 25 to 30% more cows than free-stalls are present in a given group).

Hospital Area

A convenient place to handle cows either on an individual basis (treating sick cows, breeding cows, or freshening cows) or in groups (pregnancy examinations, vaccinations) is probably the most neglected area in free-stall barns. A good veterinary hospital area should contain (1) separation area just off the return alley(s) between the milking parlor and the cow feeding and resting area, (2) treatment stalls, (3) maternity stalls, and (4) a loading chute.[4] These facilities minimize the labor required in maintaining a well-managed herd.

The separation and treatment area should be located in the milking center where hot and cold water, heat, and dry and refrigerated storage are available. By using power gates controlled from within the parlor, specified cows can be shunted

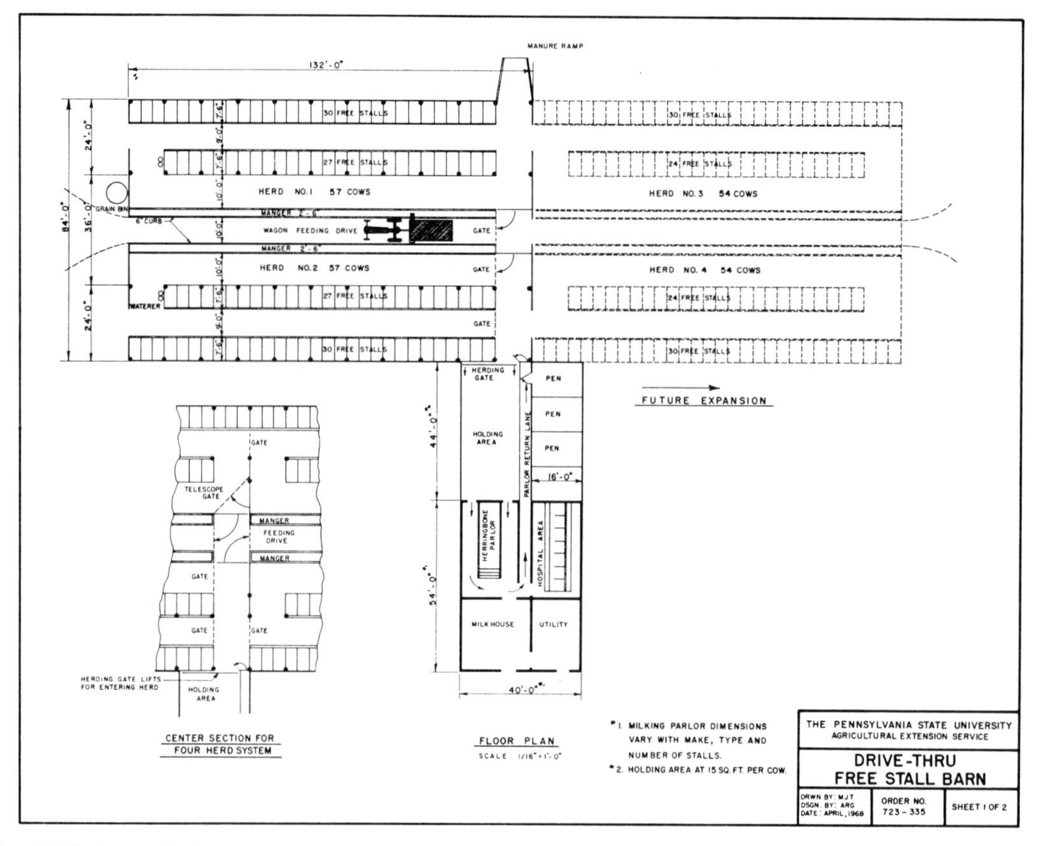

Fig. 22.7 Desirable free-stall barn layout. (Courtesy of Pennsylvania State University Agr. Ext. Ser.)

into the separation area temporarily without seriously deterring the milking operation. When milking is completed, cows are either treated immediately and returned to their group, or placed in a box stall if their stay is to be more prolonged.

The floor of the separation and treatment area should be a rough finish, sloped ¼-in./ft toward a drain with doors located so manure can be scraped directly into outside alleys and removed mechanically. Unless fence-line lever-operated stanchions are provided in one or more cow-lots, it is suggested that the number of treatment stalls be equal to the number of cow-stalls on one side of the milking parlor—an 8-stall treatment area for a double-8 herringbone parlor. Do not crowd these stalls close to a wall; the veterinarian requires adequate room on all sides of the animal to perform a complete examination.

Some treatment rooms also contain box-

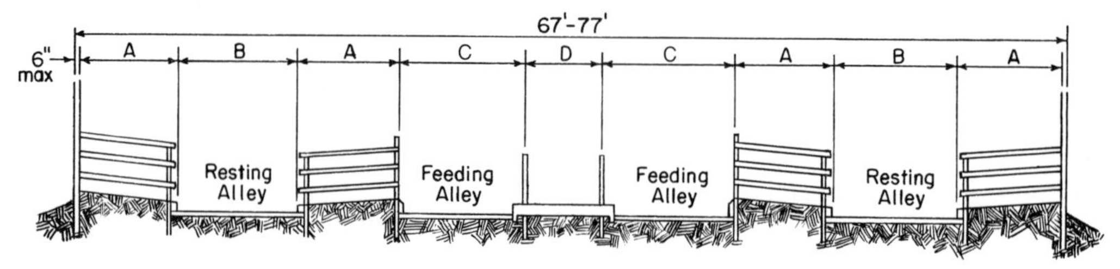

Fig. 22.8 Cross section of a typical free-stall barn.
Stall and Alley Widths:
A = 6′10″ to 7′6″; B = 8′ to 10′ with solid floor, 6′ to 9′ with slotted floor; C = 9′ to 10′ as a feeding alley, 10′ to 12′ as a combination feeding and resting alley; D = 5′ to 6′ with mechanical bunk, 15′ to 18′ with feeding fences and drive-through alley.

stalls, separate from the maternity stalls. These stalls should be located close to the feed source and convenient to the milking parlor. Dirt floors are best, but they should be elevated to prevent moisture draining into the pens. Overhead supports for rope or chain hoists to elevate "downer" cows should be standard equipment.

Loading chutes located outside the building proper, but adjacent to the holding pen and connected by a door that is normally closed, are ideal. The chute should not be over 32 in. wide so animals cannot turn around while being loaded.

22.4 Open or Partially Open Free-Stalls

Open free-stalls are used in regions with limited snowfall and reasonably warm average winter temperatures. While the free-stall area is protected by a roof, both the cows and feed in the bunk in outside yards are exposed to the weather. Cattle are usually less confined than in the enclosed system, but more concrete and fencing are desirable for moving animals among the housing, feeding, and milking areas.

22.5 Open Corrals

In the warmer and dry climatic regions, most dairies group cattle in corrals, providing about 350 sq ft per animal. These corrals may be either rectangular or pie-shaped. To provide serviceability, flexibility, safety, sanitation, and economy, corrals should include:

1. Concrete paving in areas of heaviest cow traffic such as at the feeding platforms, around drinking water tanks, and near the outlet gates to the milking parlor.
2. An arrangement that minimizes the distance between cows and the milking and service areas.
3. Perimeter feed bunks along fencelines. Mechanical conveyors to feed bunks are satisfactory, but rather expensive in large herds when portable (truck or tractor and wagon) feeding is practical.
4. Sheltered areas (i.e., loose housing or shades) at least 1 foot above grade to keep cows clean and dry during wet seasons.[5]

Most of the larger, commercial dairies in the Southwest climatic region utilizing rectangular corrals use one of four basic arrangements (Figure 22.9). Type A has a common lane for feeding and moving cows; types B, C, and D have separate facilities. The various items that dictate choice of corral arrangement include:

1. *Continuous flow* of cows to and from the parlor. Only types B and C provide continuous flow of cattle when 2 different groups are milked simultaneously.
2. *Drainage* away from the feeding area is essential. When cow-movement lanes

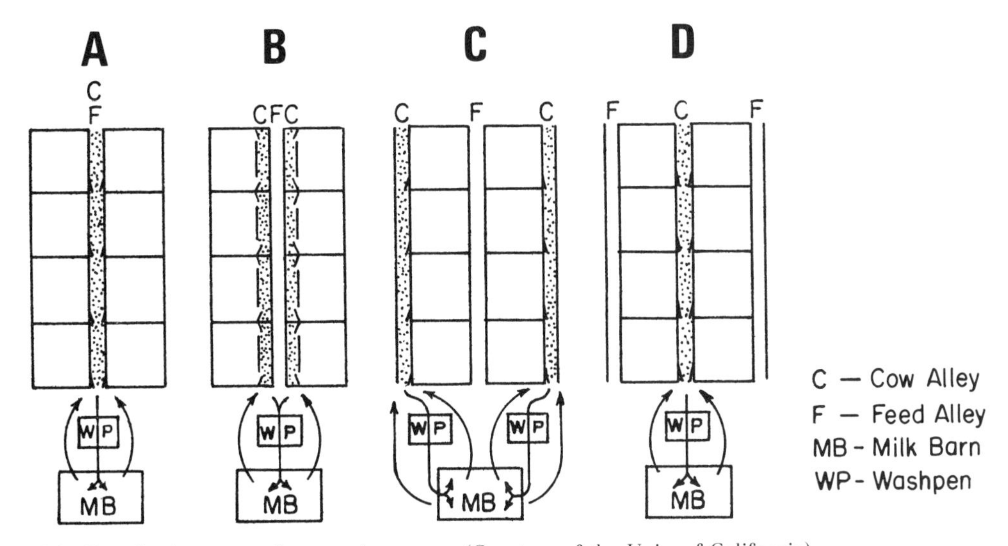

FIG. 22.9 Four basic rectangular corral systems. (Courtesy of the Univ. of California)

are on one side of the corral and the feeding area is on the other side, cross drainage through the corral and away from the milking center may become necessary.

3. *Hay storage* close to the feeding area increases feeding efficiency. However, high stacks cast long shadows and have the disadvantage of slowing air movement and retarding corral drying to the north or east of the stack.
4. *Portable feed wagons* require considerable space to turn around. Only type D arrangement eliminates the necessity of opening and closing gates or vehicle turnarounds.
5. *Lock stanchions* are required in all corral types except type B if cows are to be isolated from feed prior to milking. With type B arrangement, only 1 or 2 corrals need be equipped with lock stanchions to facilitate clipping, vaccination, and pregnancy examination.
6. *Mangers* with high fronts are usually recommended in type A corrals. In addition, manger space is lost to a gate in each corral, increasing the distance to each succeeding corral. Noncontinuous mangers hamper feed delivery from self-unloading wagons.

In herds of more than 500 cows, the distance from the holding corral to the farthest rectangular pen becomes great. Pie-shaped corrals have the advantage that all corrals are approximately the same distance from the milking center. When there is heavy dependence on feeding of green chop or silage via self-unloading wagons, continuous perimeter feeding may be a labor-saving attribute. Hay storage is frequently provided at an interior fenceline dividing 2 groups of cows. Feedbunk space and desired square footage per cow determine the angle and length of pie-shaped corrals (Figure 22.10).

Cattle shades to protect dairy cattle from intense summer sun are sometimes needed to maintain milk production and reproductive efficiency. When metabolic rates become too high because of cattle being overly warm, their appetite and feed intake decline and milk production suffers. Properly constructed summer shades can reduce the radiant heat load of cattle by as much as 50%.

Shades should be over an elevated area, from 12 to 14 ft high at the eaves, and up to twice as wide as the height. They should be north-south oriented, to promote drying of wet spots and control of flies. From 30 to 50 sq ft of shade per animal is recommended, with the larger figure used when air movement is likely to be minimal.

Many types of roof materials are satisfactory. Flat roofs are satisfactory in areas of low rainfall; other regions favor a sloped roof. Using corrugated aluminum covered shades as a reference point (shade effectiveness value of 1.00), a cover of 6 in. of hay is valued at 1.20; galvanized steel with the topside painted white and the underside painted a dark color to absorb radiant heat has a value of 1.05. Saran shade cloth and single layers of snow fence are valued at only 0.84 and 0.59, respectively.

Wind breaks are sometimes provided in dry climates that receive some snow along with cold winds. Fences, 10 ft high, with approximately 80% solid board, provide better snow protection than solid fence. Windbreak fences are usually constructed of: (a) 1 x 10's—2½ in. apart, (b) 1 x 8's—2 in. apart, or (c) 1 x 6's—1½ in. apart in the vertical position. When a fence is built inside a lot on a mound for wind protection, the maximum board spacing is 2 in. to reduce drafts on animals lying close to it.

22.6 Ventilation

Ventilation is a continuous process to remove moisture from inside the building, provide fresh air for animals, remove excess heat in warm weather, and remove odors and gases from animal waste. A "system" is required to bring fresh air into the building, distribute it evenly, and remove it. This system is completely different for the 2 types of housing environments, "cold" and "warm."[1]

In "cold" housing, wind and natural convection forces move the air, and properly located adjustable inlets provide distribution and volume control. In "warm" housing, a mechanical ventilation system, either pressure or exhaust, is used. Fans force air into or out of the building, and distribution is provided by properly located inlets or outlets.

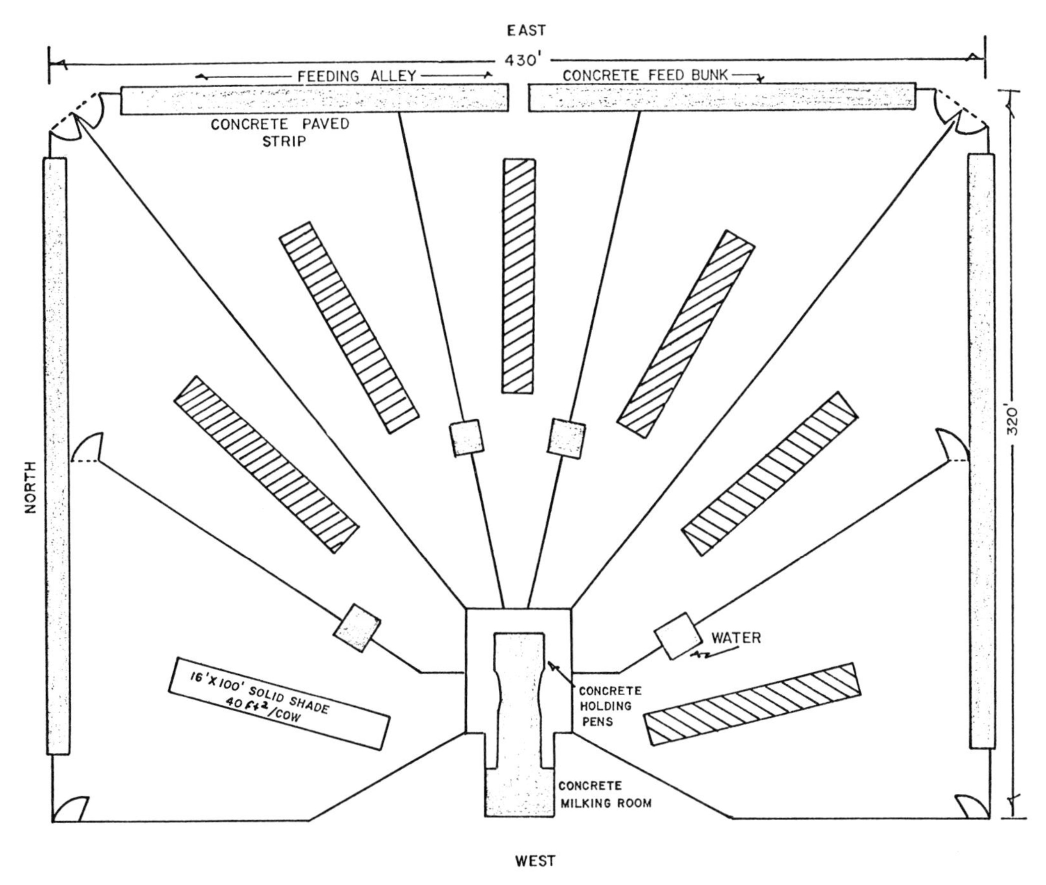

Fig. 22.10 The fan-type corral layout substantially reduces travel time and distance for cows and cow handlers. (Courtesy of the Univ. of Arizona)

Ventilation in Cold Housing

In many parts of the United States light insulation in the ceiling is needed to control condensation and frost formation on inside surfaces in winter and to reduce heat buildup in summer. Figure 22.11 illustrates the regions where some insulation is needed.

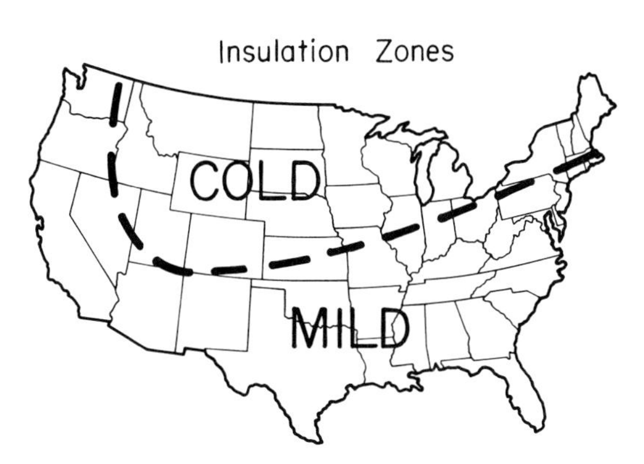

Fig. 22.11 Cold region of U.S. requiring insulation.

An open ridge outlet is common in cold loose housing and free-stall barns. Table 22.5 lists recommended opening widths. Open ridges are less effective when: (1) cows have access to outside areas, (2) high silos, trees, or other buildings are close to the barn, (3) too few cows are in the barn, and (4) the roof has little slope (less than 1-ft elevation in 3-ft span).

Roof ventilators are common on livestock buildings, but they do tend to fill with

Table 22.5 *Recommended Minimum Size for Open Ridges*

Building width (ft)	24	30	40	50	60	70	80
Open ridge (in.)	4[a]	5[a]	6	8	10	12	14

[a] Minimum of 6 in. in climates where frost may build up and block the ridge opening.
Source: Dairy Housing and Equipment Handbook. Midwest Plan Service Publ. No. 7, 1976.

dust, birds' nests, or frost. While they have the advantage of keeping snow and rain out of the building, they seldom provide adequate air movement and are not as satisfactory as an open ridge.

Inlets in the wall of the building need to be of at least 2 sizes, large openings for summer ventilation and much smaller ones to provide air movement in winter. Summer air inlets are often 3 x 6 ft or 4 x 8 ft doors, which may be adjusted during changing weather. Winter air inlets are commonly under overhangs and may be equipped with hinged doors that can be closed during snowstorms.

Ventilation in Warm Housing

Tightly constructed insulated buildings with mechanical ventilation are referred to as warm housing. Inside temperatures in winter are maintained in the 40 to 70°F range. To control temperature and moisture, the following items must be provided: (1) insulation in the walls, ceiling, and foundation, (2) vapor barriers, (3) adjustable air inlets or outlets, (4) limited door and window openings, (5) fans, and (6) supplemental heat if needed.

The amount of insulation needed depends on climate, type of insulation, size and type of building, number and size of animals housed, and supplemental heat requirements. Insulating capabilities are indicated by "R" values, illustrated in Appendix Table V-B. Materials having high numbers are good insulators. The insulating value of a wall or ceiling is the total "R" value of its components, including siding and air spaces, for example. In the colder regions, an R value of 14 in the walls and 23 in the ceiling is recommended. Lower values are appropriate in the milder climates, but should not be less than 9 in the walls and 12 in the ceiling in order to reduce summer heat gain.

Exhaust ventilation systems used in "warm" buildings usually have an attic chamber to supply "tempered" air during cold weather, an inlet system to distribute the air, and exhaust fans to remove moisture and stale air in winter and excess heat in summer. In addition, special summer inlets to bring air directly in from the outside are recommended. The diagram of a 2-fan exhaust system in a building less than 50 ft wide, with slotted floors and manure pit is shown in Figure 22.12. Center baffled ceiling inlets are needed in buildings more than 50 ft wide.

Some important considerations in designing an exhaust ventilation system include:

1. Allowing 1 sq ft of inlet space for each 600 cubic feet per minute (cfm) of fan capacity.
2. Adding an adjustable baffle to the slot inlet if control of direction and velocity of air is desired.
3. Two or more fans, one smaller than the other(s) and running constantly, are needed to exhaust different amounts of air under varying weather conditions. The smaller fan should be located in the manure pit of buildings so equipped to exhaust toxic fumes away from livestock. In solid floor buildings, all fans should be located in a wall opposite the air intake ducts.
4. In solid floor buildings, a duct around the small exhaust fan located just below an eave permits it to exhaust cooler air from near the floor during winter and thus serves to conserve heat.

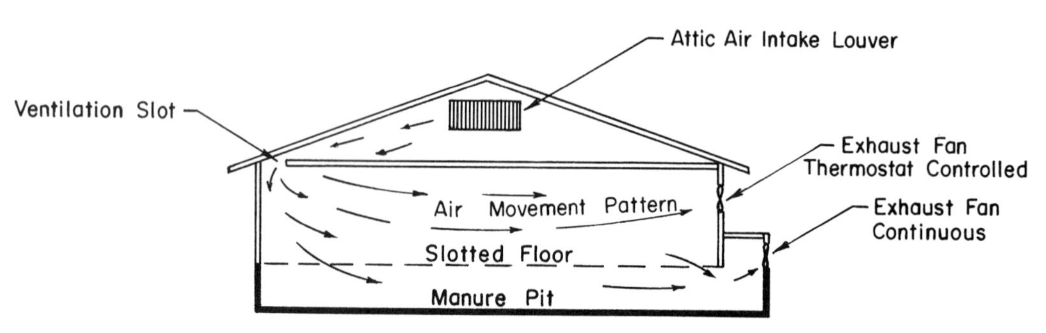

FIG. 22.12 Ventilation scheme for buildings with manure pit.

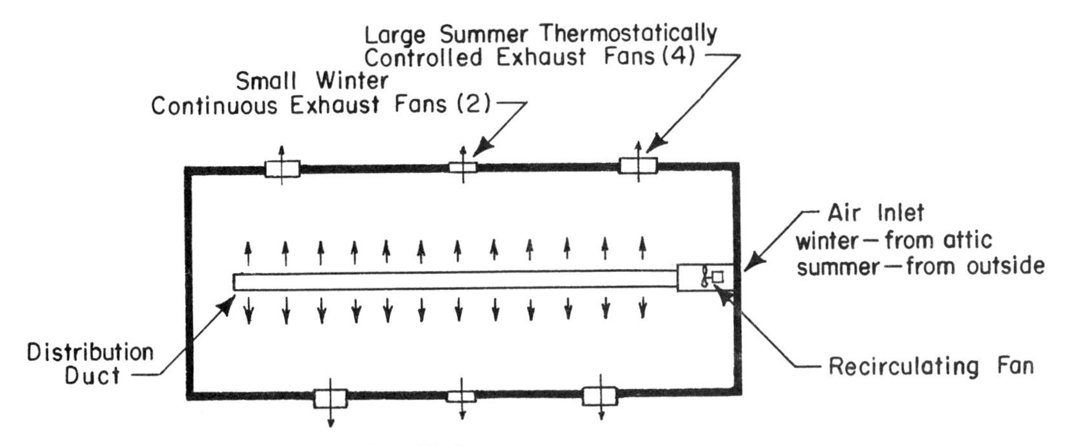

FIG. 22.13 Pressurized duct system for wide barns.

Pressure ventilation systems differ from exhaust systems in that fans are used to force air into a building. Milkhouses almost always use pressure systems to avoid drawing in dusty air and odors from the milking parlor or barn. Air distribution is provided by perforated inlet ducts and/or outlets spaced evenly around the building (Figure 22.13). It is important that at least 1 sq ft of outlet per 600-cfm fan capacity be provided. Outlets need anti-backdraft devices to prevent high winds from forcing some outlets closed and disrupting the desired air flow. Pressurizing fans equipped with a thermostatically controlled heating element are frequently a part of this system when added heat is desired.

Ventilation fan capacity recommendations are 100 cfm/1,000 lb of body weight in winter; double this rate in summer (Table 22.6). About one-fourth of the winter requirements should be exhausted continuously, even during extremely cold weather.

Calf barns frequently require additional heat. A heater capable of supplying 1,000 Btu/hr per calf is usually recommended. Heaters and exhaust fans should be installed on opposite sides of the barn.

TABLE 22.6 *Recommended Ventilation Rates*

Milkhouse, pressurizing fan	600–800 cfm
Milking parlor	
Winter	100 cfm/stall
Summer	400 cfm/stall
Warm housing	
Calves	
Continuous "winter," min	0.1 cfm/lb or 4 air changes/hr
Summer	20 air changes/hr
Cows	
Continuous "winter," min	25 cfm/1,000 lb
Normal winter	100 cfm/1,000 lb
Summer	200 cfm/1,000 lb

Source: Dairy Housing and Equipment Handbook. Midwest Plan Service Publ. No. 7, 1976.

22.7 Handling Manure

The collection, storage, transportation, and land application of manure must be compatible with sanitary milk production, the housing system, and pollution control regulations. Runoff and seepage must be controlled. Minimizing odors, controlling flies, and guaranteeing the safety of people and animals are also essential.[2]

Principles

Any system starts with the source of wastes—an animal, the milkhouse, and rain or melting snow. Almost all systems end *in* the soil; no pollutant is intentionally released into a stream. Only a few systems currently involve refeeding, selling of fertilizer, or specialized treatment. All systems are compromises between convenience, labor, investment, and esthetics. No system is "best." Each component, facility, or process has advantages and disadvantages.[6]

Regulations

Federal, state, and local regulations control air, water, and noise pollution. These regulations are enforced to minimize or eliminate "pollution." *A pollutant is a resource out of place,* and manure is a resource—its fertilizer value can help crops grow.

Federal regulations apply to all states. Regulations are subject to change, frequently becoming more restrictive. At this time, any operation that: (1) discharges waste into navigable water, (2) has navigable water traversing the feeding or holding area, or (3) meets capacity standards of 700 dairy animals and discharges measurable wastes from a major storm (defined as a 24-hour storm that occurs only once every 25 years) MUST APPLY FOR A NATIONAL POLLUTANT DISCHARGE ELIMINATION SYSTEM (NPDES) PERMIT.[7] An estimated 32,000 dairy operations are affected by these provisions. Over 70% of the dairies judged to be affected have fewer than 70 cows.

Individual states or localities can impose additional or more stringent regulations regarding runoff, land application, or ground water protection. Some agencies attempt odor control by regulating the distance between public housing and livestock units. Seek out the regulations that apply to your facilities while you are planning—and *before you start new construction.*

Consistency and Amount of Manure

When all manure produced by dairy animals is collected and not altered by the addition of bedding, it contains about 87% water (13% solids). Manure that has had bedding added or liquid separated can be handled as a solid product when its dry matter content exceeds 20%. To handle as a liquid slurry, manure must be less than 15% solid matter; for irrigation, added water is necessary to bring the dry matter content down to about 5%. Thus, the manure handling system of choice is dependent on what is added to, removed from, or bypassed in the collection of manure.[2]

The 24-hr production of manure by dairy animals of varying sizes is shown in Table 22.7. These data assume total collection, nothing lost and nothing added. A cubic foot of manure contains 7.5 gal and weighs about 62 lb. A gallon of manure weighs approximately 8¼ lb.

TABLE 22.7 *Dairy Animal Daily Manure Production and Composition*

Animal size lb.	Total daily manure production lb.	cu. ft.	gal.
150	12	0.19	1.5
250	20	0.33	2.4
500	41	0.66	5.0
1,000	82	1.32	9.9
1,400	115	1.85	13.9

Source: Dairy Housing and Equipment Handbook. Midwest Plan Service Publ. No. 7, 1976.

Bedding

A typical herd of 100 Holsteins and 82 youngstock maintained in loose housing bedded with straw for 6 months requires nearly 150 tons of straw. Since straw is in short supply and expensive, dairymen are searching for alternative bedding sources. Corn stover or mature hay, when chopped, is a satisfactory bedding material. Some shells or hulls, such as peanut or cottonseed, absorb nearly as much liquid as most straws. The effectiveness of wood bark, chips, shavings, or sawdust depends on the hardness of the source material. Oats straw, for example, has a water holding capacity rating of 2.5; i.e., to raise the dry matter content of 100 lb of pure manure, including urine, from 13 to 20% so that it can be handled easily as a solid, nearly 11 lb of oats straw must be used. The water holding capacity of several common bedding materials is shown in Appendix Table V-C. Storage space requirements and estimated daily bedding requirements are shown in Appendix Tables V-D and V-E.

22.8 Solid Manure

Manure can be handled as a solid if it is mixed with bedding or if the liquids are allowed to evaporate or drain away. Stall barn manure is usually loaded directly into a spreader with a barn cleaner. In free-stall barns, a tractor scraper and front-end loader are most common, although mechanical scrapers are sometimes used (Figure 22.14).

Daily hauling has the advantages that labor requirements are spread evenly throughout the year, little accumulation is

Fig. 22.14 A tractor scraper pushing manure into a recession where it is pumped to an adjacent storage area (left), and a mechanical scraper moving manure to a cross-alley not in view (right).

present and odor problems are minimized, and the investment is small. The disadvantages are that manure must be hauled in all types of weather, the work is time consuming, and many dairymen consider it to be a drudgery. If 80% of the manure is collected and hauled along with 8 lb of bedding materials per cow daily, the manure from a herd of 100 cows weighing 1,400 lb each will require 608 trips to the field annually if a 125-bushel capacity spreader is heaped full each trip.

Short-term storage is desirable when field conditions are poor or equipment breaks. Tractors and equipment used in cold weather wear out faster. Pollution runoff may be excessive if manure is spread on steep slopes or snow-covered ground.

Storage of solid manure needs to be convenient for loading and away from waterways; surface water should be diverted. Stacking works best with manure containing bedding. High stacks built during freezing weather flatten out when the spring thaw arrives, putting tremendous pressure on storage structures. While stacking manure overcomes several of the daily haul disadvantages, it can cause offensive odors and create a fly problem. Where minimal bedding is used, drains leading to an approved holding pond may be located in the low corner of the storage area.

22.9 Liquid Manure

Year-long liquid storage permits incorporating manure into the soil at the best time and provides a place to put milkhouse and parlor wastes. A separate system for human waste is usually required. The investment is generally high, and odors can be a problem, especially when agitating and spreading. Labor requirements may also interfere with field work.

Several types of storage are used for liquid manure: (1) storage tank under the barn, (2) outside below-ground storage tanks, (3) earthen basins, and (4) silos. Manure is moved to storage by: (1) dropping through slots into storage below, (2) tractor scraper, (3) barn cleaner, and (4) pressure pumps from a hopper or tank to the storage structure.

Slotted floor alleys (or gutters in stall barns) eliminate the labor and cost of scraping equipment. Since manure does not build up on the floor, cows remain comparatively clean.

Slotted floors provide rapid segregation of an animal from its manure. Concrete slats are the most common, but are the heaviest and require the strongest support. Steel or aluminum are more uniform than wood or concrete, but are usually more expensive. Tapered slats, greater top width than bottom width, tend to pass wastes bet-

TABLE 22.8 *Suggested Slat Size and Spacing*

	Slat spacing		
	Narrow slats	*Wide slats*	*Expanded metal*
Cows	Not recommended	1½ to 1¾ in.	Not used
Calves	¾ in. between 1 × 2's on edge	1¼ in.	9-gauge (flattened)

Source: Livestock Waste Facilities Handbook. MWPS-18. Midwest Plan Service, Iowa State Univ., Ames, Iowa.

ter than uniform-width slats, especially if slat depth is more than 1 in. Approximate recommendations for slat size and spacing are listed in Table 22.8.

Manure scrapers work very well but the labor invested is rather high and daily manure management becomes a high priority item to prevent buildup. In cold buildings during the winter months, freezing manure can be a problem, which may be partly overcome by frequent or continued scraper operation or by installing electric (20 watts/sq ft) or hot water (75 Btu/sq ft) heat in the floor.

Pumping manure slurries with up to about 15% solids can be achieved. Piston pumps with solid pistons can move manure with fibrous bedding up to 300 ft to storage. Other types with large hollow pistons are more suitable for pure manure free of bedding material or wasted hay. More recently, tanks equipped with securely fastened and airtight doors in combination with an air pressure pump have been used to move manure, either liquid or solid, to holding basins.

Storage capacities for liquid systems should allow for 1⅓ cu ft per 1,000 lb animal per day plus extra wastes. This capacity is about 2 cu ft per day for Holstein cows. Milkhouse and milking parlor wastes and water used in high pressure cleaning of these facilities require additional storage capacity.

Earthen basin storage is more economical than concrete storage tanks, but it is important to invest enough monies to facilitate mixing and removing of the manure. Side slopes of earthen storage basins should be 2:1 to 4:1 (4 ft of run per foot of rise). A depth of 10 to 12 ft is common. They need to be equipped with either a dock (for vertical loading pumps) or a concrete ramp and floor (for horizontal pumps). If liquids are to be drained away so that remaining manure can be handled as a solid product, a porous drain or perforated pipe and spillway leading to a liquid irrigation holding pond is recommended. See Figure 22.15 for an example of how a ramp, dock, or porous dam is utilized.

Earthen basins must be: (1) at least 100 ft from a water supply, (2) above the water table, (3) convenient for filling and emptying, (4) located so as not to be a receptacle of excess "clean" water, and (5) constructed in soil that does not allow seepage. Such storage basins SHOULD NOT BE CONSTRUCTED on creviced bedrock.

An alternative to the earthen basin is the above-ground silo. When compared to the earthen basin, it has the advantages of being safer for both humans and livestock, and of being more attractive in appearance. The major disadvantages are cost, although it is usually more economical than the slotted floor, below-building storage of manure.

Screening can be accomplished in several different ways. Grass is the natural filter used in settling channels, but should be used only in slurries lightly loaded with organic solids, such as milkroom wastes. A screening system provides for handling and disposing of both solids and liquid. The liquid portion is suitable for pumping, and the solids will contain from 20 to 30% dry matter (Figure 22.16). About half of the original volume is in each of the liquid and solid fractions. Screened solids can be dried for bedding in dry climates or further processed for livestock feed.

Dehydrators can remove most manure moisture, but this method has several disadvantages: high initial cost, mechanical problems, high energy requirements, unpleasant odors, and severe corrosion of dryer parts.

Milkhouse and parlor wastes can add

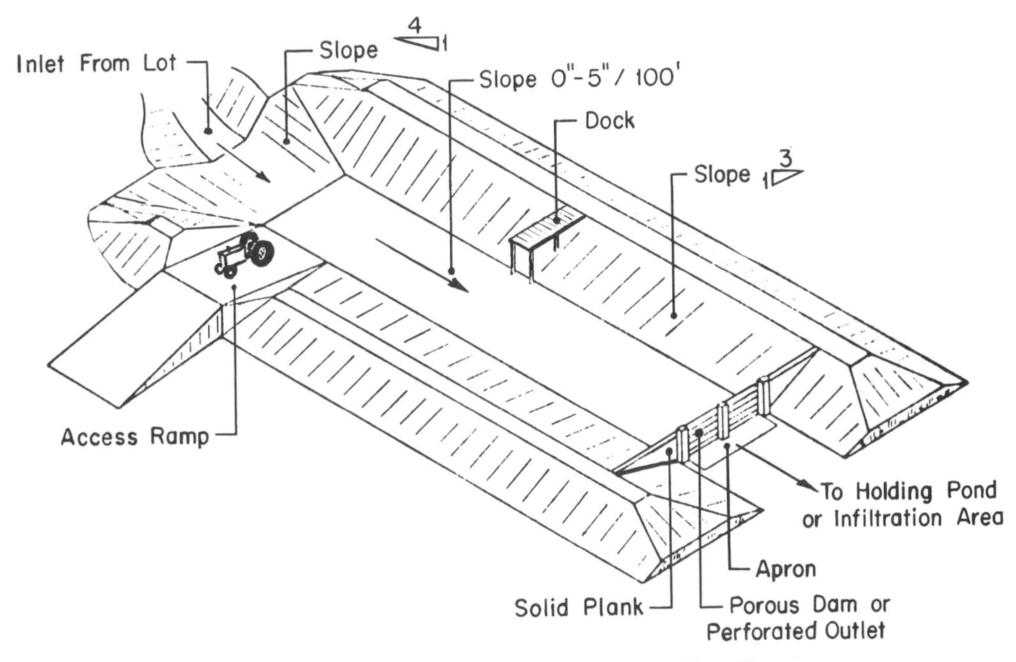

FIG. 22.15 Earthen and concrete settling basins and perforated pipe with spillway.

considerable volume to the manure disposal process. Management and equipment used greatly affect the volume produced from the milking and cleaning operations. If cows' udders are washed using paper towels and disinfectant, little waste occurs; if cows are washed in an automatic prepstall, up to 9 gal/day for each cow milked may go down the drain. Floors can be washed with relatively little water and a stiff-bristle broom or with a high pressure hose and a relatively large amount of water. The volumes of milkhouse and parlor wastes produced are estimated in Appendix Table V-F.

FIG. 22.16 Separation of liquid and solid portion of manure.

Small units tend to use less water daily, but more per cow for cleaning floors and equipment. A 100-cow unit with automatic washing equipment can easily use over 800 gal daily. Milk parlor waste can resemble dilute liquid manure when it contains large amounts of feed and bedding. Milk room cleaners and sanitizers do not settle easily, can cause severe odors in anaerobic lagoons, and plug leaching fields. It is much more convenient to dispose of milk solids by field spreading—another advantage for liquid manure systems.

22.10 Feeding Systems

Mechanical feeders are used to reduce the labor required in feeding and minimize the drudgery involved. Further, they should be designed to limit the opportunity for cows to select those feeds that they like and discriminate against those that they do not like. The more widely feedstuffs vary, particularly in protein content, the more serious the problem becomes because grain mixtures are usually formulated to supplement average forage intake.

Stanchion and comfort stall barns provide an environment in which individual feeding is possible, but this type of feeding requires much labor unless the system is mechanized. A feeding system involves more than the simple installation of mechanical feeders; it involves the arrangement of the entire dairy facility, the physical form of the forages to be handled, and the type of storage.

A herd of 100 Holstein cows fed in the barn or dry lot throughout the year with an average daily consumption per cow of 10 pounds of hay, 80 pounds of corn silage, and 14 pounds of grain mixture requires approximately 1,900 tons of feed, which must be stored and fed.

Storing and Feeding Forages

The comparative advantages and disadvantages of the more common types of silos were discussed in Section 10.4. Both haylage and silage are well adapted to mechanized feeding. Current trends are to feed more alfalfa haylage and corn silage, depending less on baled hay. In the Lake States region, where the farms are generally smaller, upright silos are the most popular, the primary reasons being (1) convenience, (2) ease of mechanization, (3) reduced labor in handling bulky feeds, and (4) fewer problems during freezing weather. Dairymen storing more than 1,000 tons of silage annually, especially corn silage, find that bunker silos are more economical and are frequently more reliable in that they are less vulnerable to power failure or breakage of mechanical feed delivery systems.

A front-end tractor loader and side-unloading wagon work well in distributing silage or haylage from horizontal silos to cattle in drive-through stanchion barns, in

Fig. 22.17 This inside view of a Vermont open front barn shows the concrete bunk for feeding silage with hay racks in the background. The side-unloading wagon parked at end of covered runway distributes silage from both the upright silo and a horizontal silo. (Courtesy of Hood Dairy Farm Engineering Service, Middlebury, Vt.; photo by Dick Smith, North Conway, N.Y.)

corrals with fence-line bunks, and in some enclosed housing and feeding systems (Figure 22.17). Silage and haylage can be self-fed from horizontal silos or incorporated into a complete ration (Section 12.3).

Many dairymen prefer to include some dry hay in the ration. The more common kinds of hay fed to dairy cattle and the various methods of harvesting, storing, and feeding it were discussed in Sections 10.6, 10.7, and 10.8. Harvesting and storing of baled hay may be highly mechanized and quite efficient. The amount of manual labor required in moving baled hay from storage to feeding areas is high, but it can be reduced in several ways: (1) if available, hay cubes or wafers can be distributed by augers to mangers and feed bunks (Section 10.8); (2) chopped hay can be handled mechanically, but many dairymen object to its dustiness and low palatability; and (3) pelleted hay is ideally adapted to mechanized feeding but may cause a decrease in fat test (Section 10.8).

Alternative hay handling methods have been developed recently. Comparative advantages of each method are shown in Table 22.9. The large, round bales and 3-ton stacks are not easily moved many miles, require special equipment, and are seldom utilized on farms with fewer than 60 cows or in stanchion barns.[8]

Estimated harvest and storage losses when legume-grass forages are harvested at varying moisture levels and by alternative methods are illustrated in Figure 22.18.

Feeding Grain Mixtures

In tie-stall barns, feeding cows individually is a time-consuming but relatively easy job. The grain mixture may include all ingredients offered except the forage (complete grain mix), or each ingredient may be fed separately (top dress feeding of protein supplement and/or mineral supplement). Mechanical feeding equipment available includes simple pushcarts; motorized and mechanized feeding carts; and augers delivering feed to small, lever-operated storage boxes between pairs of cows.

Feeding the desired amount of the grain mixture to individual cows in a milking parlor is difficult to achieve because of the limited time each cow is in the stall. Dairymen seldom are willing to slow the parlor throughput (cows per hour) to the extent that cows may consume all feed allotted or required. Intake in parlors varies as follows: finely ground grain—0.5 lb/min; coarse ground grain—0.9 lb/min; and pelleted feeds—1.3 lb/min.

Several alternative programs are rapidly gaining acceptance. The first is the use of computerized feeders or other mechanical feeding devices that allow only specified cows to obtain supplemental grain beyond what is offered in the parlor (Figure 22.19). These systems are common where herd size or building arrangement prevents grouping cows by production level. Each cow intended to have access to supplemental feed carries a recognition key suspended from her neck and is allotted a portion of her daily grain ration at each visit.

The feeding of a constant amount of supplemental grain mixture by top-dressing silage is very popular, especially in areas where high-moisture grains are stored in a separate upright silo adjacent to the forage silos. The primary disadvantage is that dry cows and low producers tend to consume more feed than required. Thus, separation of cows into different groups is advised.

Separation into different milking strings is particularly appealing to those dairymen with large herds and the physical facilities to allow such a grouping (Section 12.3).

22.11 Milking Barns and Parlors

The production of high-quality milk depends upon clean, healthy cows milked in a clean, efficient manner. Factors to consider in the selection of a milking barn or parlor include: (1) type of housing, (2) number of cows, (3) number of operators, (4) mechanization desired (now or in the future), (5) capital resources available, and (6) personal preferences. Similarly, many choices exist among the types of systems available, and careful consideration of the alternatives is essential.

Most dairymen housing cattle in tie-stall barns prefer to milk their cows without moving them to a parlor. This practice immediately limits the alternatives to one of three choices: (1) milking in a bucket and carrying milk to the milkroom, (2) using a bucket system in combination

TABLE 22.9 *Comparative Analysis of Various Methods of Feeding Dry Forages*

		Losses expected			Hours labor required per ton to		
Method of storage	*Method of feeding*	*Harvesting*	*Storage*	*Feeding*	*Store*	*Feed*	*Tons of hay required*[a]
A. Conventional bales	Individually in stanchions or group fed at rack	20	4	5	2.00	1.11	137
B. Large round bales	Stored inside; fed in rack	20	4	4	0.10	0.55	135
	Stored outside; fed on ground	20	11	22	0.10	0.40	166
C. 3-ton stack	Stored outside; fed in rack	20	4	4	0.08	0.24	135
	Stored outside; fed on ground	20	9	28	0.08	0.17	171
D. Cubed or Pelleted	Individually in stanchions or group fed at bunk	Usually purchased	2	2	(Depends on facilities and equipment)		130

[a]Tons of hay required for 50-cow herd consuming 2.5 ton per cow annually (13.7 lb daily). Additional forage from another source assumed.
Source: Adapted from 1977 Minnesota Dairy Report, Univ. of Minnesota, St. Paul.

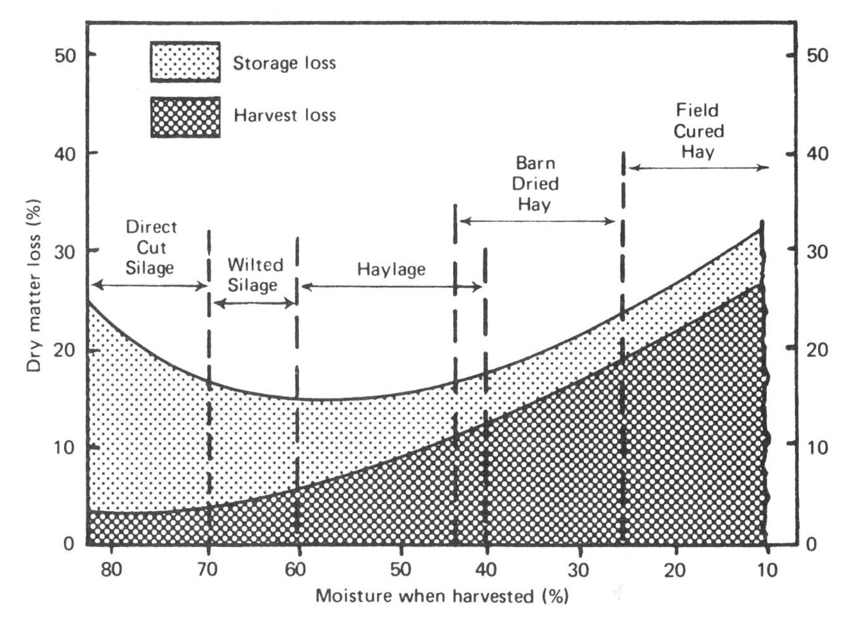

FIG. 22.18 Losses in legume-grass forages at various moisture levels. (Courtesy of the Agr. Ext. Ser., Michigan State University)

with a step-saver or dump station, and (3) using a pipeline milking system.

When milking parlors are used, the choices of systems are much broader: (1) flat milking barns similar to small stanchion barns, (2) side-opening parlors, (3) herringbone parlors, (4) one of several types of rotary parlors, and (5) the polygon parlor. Pipeline milking systems are almost standard equipment in each of these, but the degree of mechanization varies tremendously.

It is clear that no one type of milking parlor is superior in all respects. Thus, in choosing the type of barn or milking parlor, consider these points:[5]

1. *Labor efficiency* or milk per manhour is influenced most by the ability of the operator to use a maximum number of milking units with minimum idle time of the units, within the boundaries of good milking technique. Herringbone and polygon milking parlors are generally superior for minimizing lost time.
2. *Adaptability to mechanization and automation* includes the opportunity to use automatic unit detachment, power gates, pre-milking washing of the udder, and mechanical grain feeding devices. The fewer milking stations in a barn consistent with the number of units an operator can use properly, the quicker it becomes economically feasible to install these labor-saving devices. The tie-stall is the least suitable for mechanization because equipment must be moved from one location to another.
3. *Comfort and convenience,* indicated by the operator's ability to keep clean and dry and be neither too cool or too warm, plus the ability to have all switches and controls within easy reach to reduce fatigue. The design of the parlor, including space allotted, location of passageways, amount of illumination, and ventilation provided, is more important than the type of parlor.
4. *Time available for cows to consume grain* has become an important consideration as level of production and grain consumption increases. Conventional tie-stall barns clearly have the advantage in this respect. Four-sided polygon parlors afford 2 to 3 times the "feeding time" permitted in most milking parlors, because of the number of cows in stalls at any given time.
5. *Ease of cow identity* is essential in herd management. The herringbone, polygon, and rotary parlors present problems in this respect when neck chains

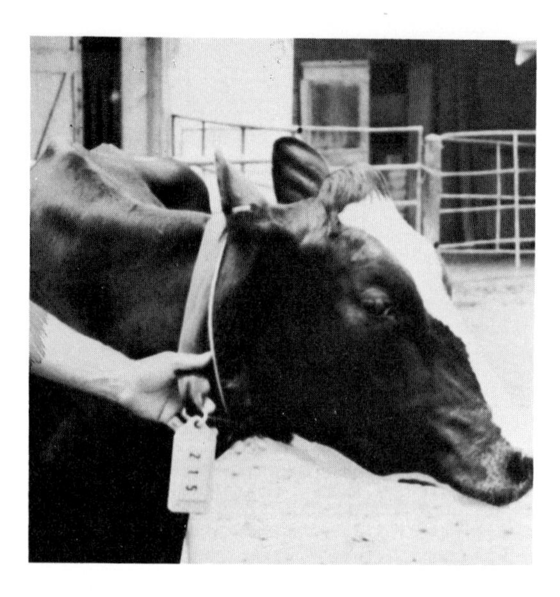

Fig. 22.19 Cow accessing grain at a feeder controlled by a computer (top). The special "key" around the cow's neck (center) identifies the cow as she enters the feeding stall. All adjustments of feeding schedules may be controlled at the computer located in the dairy office (bottom).

are used. Both the tie-stall and side-opening parlor allow for easy and accurate identification. Ankle straps are satisfactory where cows remain on concrete, but are difficult to read when cows legs become dirty from muddy lots. No method of identification is considered equal to a "freeze-brand" placed on either side of the hips or flanks.

6. *Initial cost* of the different systems is difficult to determine because of the wide variety of accessory equipment that may be installed. Facilities that yield the most milk per man-hour of labor usually have the lowest cost per cow milked.

Types

A brief description of each type of milking system common in the United States follows. An analysis of their relative efficiency is shown in Table 22.10.[9]

Tie-stall or flat milking barns are the most widely used milking facility. While they have the advantage of individual cow handling and observation, they usually result in lower efficiency ratings (milk per man-hour of labor) than elevated parlors. Their biggest disadvantage is the amount of stooping and kneeling required of the operators. When one person milks 40 cows, typical of many Midwest and Lake States, stall barn operators typically perform 200 kneebends and 82 backbends daily while milking cows and moving equipment. Usually less kneeling and bending are needed in the flat parlors where no gutters exist and a milking unit is attached between each pair of cows for the duration of milking.

Side-opening parlors have from 2 to 4 stalls head to tail in a row, and most parlors have a row on both sides of a pit to reduce operator walking distances. Because cows are handled individually, a cow having a long milking time does not hold up the entire operation as it can with other parlor systems. Besides power-operated gates and crowd gates, the most common mechanization is for udder washing and stimulation. Automatic milking machine detaching units allow one man to operate from 6 to 8 milking units and still minimize the possibility for overmilking (Figure 22.20).

Herringbone parlors vary in size from double-3 parlors to double-10's and even

Table 22.10 *Standard for Milking in Different Types of Milking Facilities at Two Different Levels of Production*

Item	Tie-stall: Step-saver	Tie-stall: Pipeline	Parlors: Side-opening	Parlors: Herringbone	Parlors: Fully mechanized parlor[a]
A. Average cows (20 lb/milking)					
No. of units	2	3	4	4	5
Chore time/cow, min	2.00	1.67	1.55	1.45	1.00 or less
Cows/man-hour	26	36	38	41	60 or more
Milk/man-hr, lb	520	720	760	820	1,200 or more
B. High producing cows (30 lb/milking)					
No. of units	3	4	5	5	6 or more
Chore time/cow, min	2.00	1.67	1.55	1.45	1.00 or less
Cows/man-hour	28	36	38	41	60 or more
Milk/man-hr, lb	840	1,080	1,140	1,230	1,800 or more

[a] Parlor includes push-button gate control, crowd gates, feed-bowl covers, and automatic milking unit detachers. Fully mechanized parlors may be 12 to 24 units per operator resulting in 80 or more cows being milked in an hour's time.

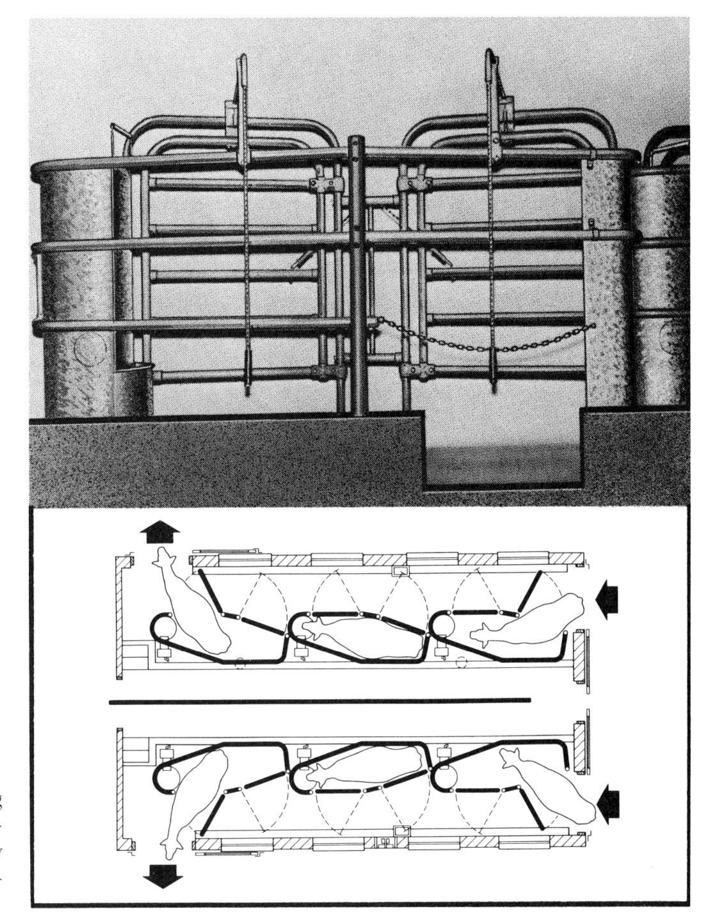

Fig. 22.20 Photo of a side-opening stall and diagram of double-3 side-opening stall milking parlor. (Courtesy of the DeLaval Separator Co., Poughkeepsie, N.Y.)

larger. This system is quite adaptable to mechanization and the distance between udders is minimized, thus minimizing operator walking distances and overall parlor length (Figure 22.21).

Polygon parlors are a 4-sided herringbone parlor (Figure 22.22). More sides with fewer cows per side reduces delays due to slow-milking cows. Thus, throughput is generally greater than if the same number of stalls were used in a double herringbone parlor. The polygon was originally conceived to maximize laborsaving capabilities of equipment used in mechanizing parlors. Its value appears greatest in herds of more than 400 cows.

Rotary parlors are of three general types—tandem, herringbone, and turnstile. Rotary *tandems* generally have 8 stalls, although systems with 5 to 22 stalls have been built. Similar to a side-opening parlor, they generally operate on a start-stop basis, rotating one stall every 25 to 45 seconds. This method of operation fixes the amount of time the operator can spend with a particular cow. The rotary *herringbone* usually has from 13 to 17 stalls, with some built up to 40 stalls. The rotary *turnstiles* in the United States have 17 stalls; cows walk onto the platform to enter and back off to exit. Cows are milked between the hind legs. Operators (usually 2) work on the outside of the solid platform, rather than on the inside of a circle as in the tandem or herringbone rotary parlors.

Fixed time milking is a problem in most rotary parlors. Unless equipped with automatic unit detachers, nearly all operators have a tendency to remove the unit as the cow approaches the exit, not necessarily when she has finished milking. Another problem is improper stimulation prior to unit attachment. Once the cow enters the platform, the operator typically washes the

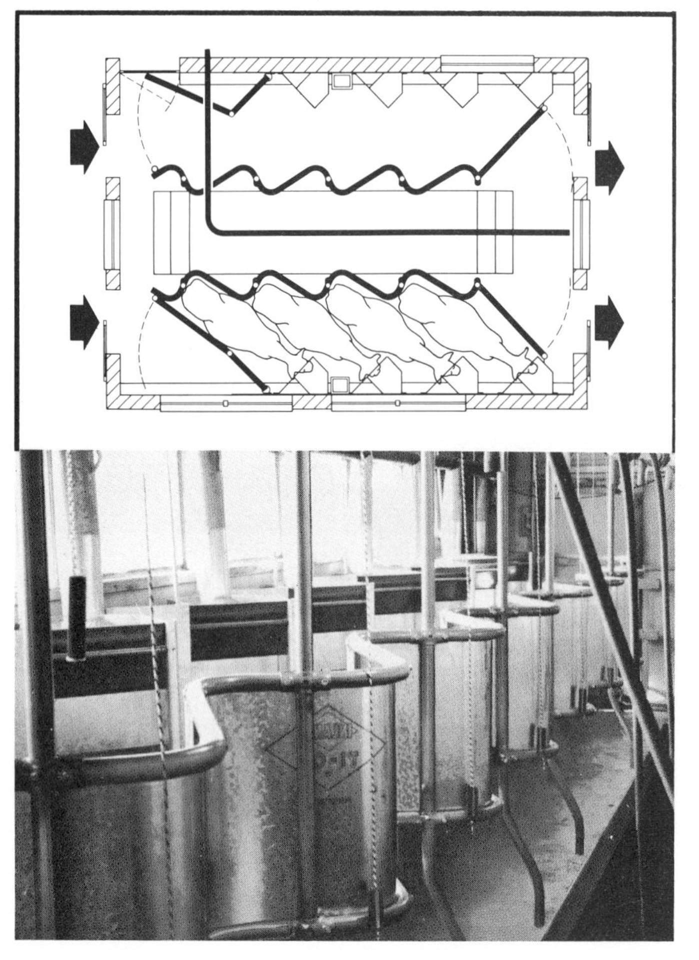

Fig. 22.21 Diagram of typical double-4 herringbone milking parlor and photo of an installation. (Courtesy of DeLaval Separator Co., Poughkeepsie, N.Y.)

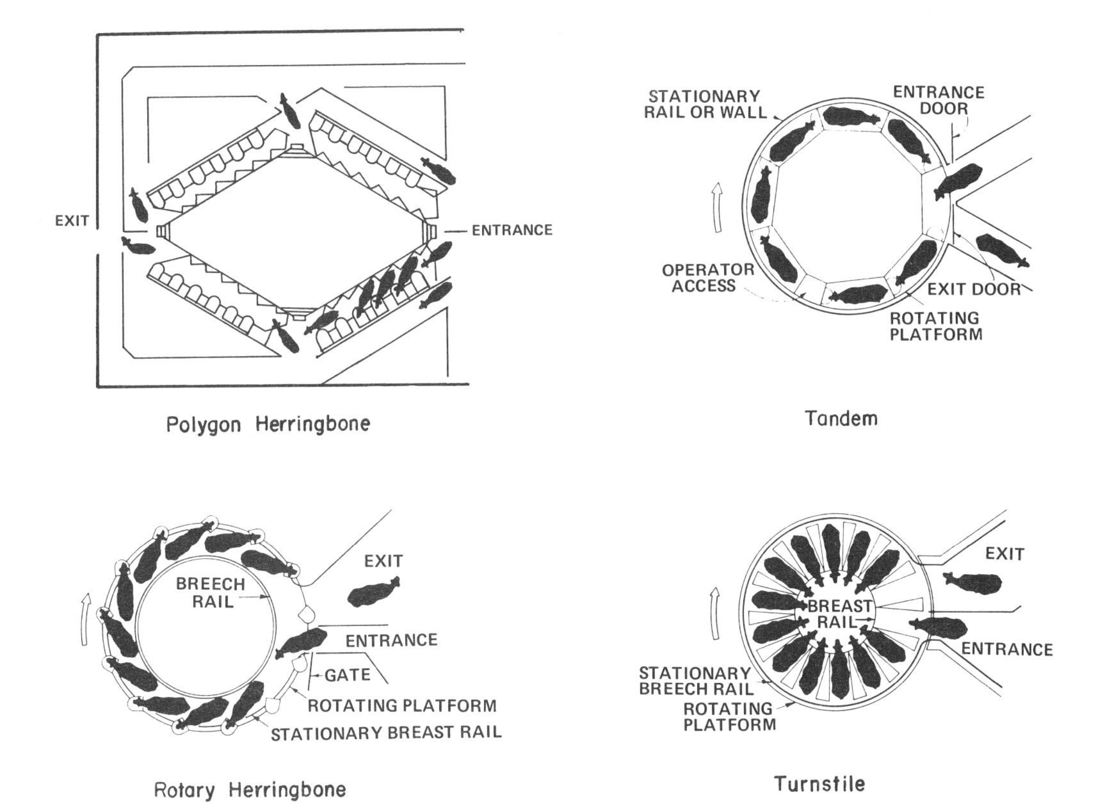

FIG. 22.22 Diagrams of the polygon herringbone parlor and 3 different types of rotary parlors.

udder, sometimes dries and foremilks the udder, then immediately attaches the milking unit. Prep stalls located in front of the entrance to the platform tend to hinder cow movement; thus, cow throughput efficiency is reduced. Most researchers have found that the pounds of milk obtained per man-hour of labor is inferior to that obtained in side-opening parlors.

Mechanized parlors may be installed to improve cow movement, provide stimulation to enhance milk let-down, reduce drudgery and speed up the operator in completing his routine, or automatically detach and remove the milking unit. Some factors to consider in choosing these labor-saving devices are:

1. Crowd gates should move forward in the holding pen, stop when they meet resistance from an animal, then restart after a preset time delay.
2. Power gates consist of pneumatic operation of cow entry and exit gates and doors. Controls should be located at both ends of a parlor and at the center of a parlor in multiple-operator installations.
3. Feedbowl covers facilitate cow movement, prevent cows entering from sampling feed in each bowl, and signal cows at rear of parlor when the exit gate opens.
4. Automatic detaching units should be loosely coupled to the claw, detect end-of-milking, have a positive vacuum shut-off prior to unit removal, and be removed from under the cow upon completion of milking.

See Chapter 19 for a discussion on the components of the milking machine system, its maintenance, and the principles and practices involved in proper machine milking.

22.12 Stray Voltages on the Dairy Farm

Stray voltages have caused serious problems on many dairy farms. Dairymen have lost milk production and experienced cow health problems due to small currents of electricity passing through the cows' bodies. The concept of stray voltage is relatively simple electrically, although the sources can be varied and complex.

As dairy farm operations increase in size and sophistication, as farmstead electrical wiring systems become obsolete or deteriorate, and as electrical loads on rural distribution systems increase, it is likely that stray voltage problems also will increase. There are many causes of stray voltages. They can arise off the farm or on the farm. Several sources can exist simultaneously.

Voltage is the product of current times resistance. The algebraic equation is:

E = IR, where E = voltage potential (volts)
I = current flow (amperes)
R = circuit resistance (ohms)

It is important that people making stray voltage measurements understand this relationship. At a given voltage, one of two interpretations can be made: (1) high current flow and low pathway resistance, and (2) low current flow and high pathway resistance. The first is much more likely to be a major problem, resulting in management problems on the dairy farm.

Cow Pathway Resistance

The pathway with the lowest average resistance, thus the pathway with the highest current at a given voltage measurement is the mouth-all hooves pathway. This pathway would be typical of that found in most milk barns, where the mouth is in frequent contact with the waterer or metal feeder while the hooves are on concrete conceivably at a much different voltage potential. The average resistance is 361 ohms.

In milking parlors, cattle may experience a front-rear hoof voltage drop sufficient to cause a problem when they enter the parlor or as they enter the parlor stall. The average resistance of this pathway is 734 ohms, approximately twice that of the mouth-four hooves pathway.

Since poor milk letdown, poor milking performance, and more mastitis are symptoms common to stray voltage problems, many dairymen feel that the teats and udder are involved in a pathway causing stray voltage problems. Recent research has shown that the pathway resistance involved with the milking system is so high it is unlikely this is the path of problem currents.

Current Sensitivity

Mouth-all hooves shocks as low as 1 mA cause a response about 14% of the time. Some adaptation occurs. That is, cows receiving these current levels learn to expect them and fail to respond when additional shocks occur. It generally requires 3 mA or more before cows have a significantly altered behavior.

Combining resistance and current sensitivity data provides an approximation of expected cow response to applied voltages. By Ohm's Law, a 1.0 mA mouth-all hooves shock is equivalent to 0.36 V. Minimal cow responses are expected to occur at this voltage drop across the cow. But, voltages of 1.0 across two cow contact points are sufficient to create behavior problems on dairy farms.

The corrective procedure employed depends on the source of the problem voltage. The contribution due to an on-farm problem will be minimized by maintaining good electrical wiring systems that meet the requirements of the National Electric Code. Properly balanced 120 volt circuits and conversion of larger 120 volt motors to 240 volts will reduce the effect of secondary neutral voltage drops at the farm service entrance.

The effectiveness and practicality of corrective procedures intended to minimize neutral-to-earth voltage problems have been determined. These methods have been used successfully.

Isolation transformers provide a physical separation of the neutral conductor at some point on the farm. They can be installed either for the entire farmstead at the central distribution point, or for a single building near the building service entrance. Isolating transformers must be installed by a licensed electrician and must meet the requirements of the State Board of Electricity. Other isolation techniques developed and being used include the saturable core reactors.

Equipotential planes may be used to control voltage differences. In an existing facility this may be done by cutting slots in the concrete floor, placing bare copper wire in the slots, and grouting the wires to the neutral at the electrical service panel. If all animal contact points are bonded together, a voltage differential cannot develop. If a neutral-to-earth voltage exists on the farm, then one problem is that animals may react as they enter onto the equipotential plane. To reduce the extent of this voltage differential, a gradual voltage gradient must be established between the remote floor or earth and the equipotential plane. One method of accomplishing this is to run a bare copper wire in the concrete out through the entry doors of the milking parlor a distance of 20 feet with the wire embedded deeper (to a final depth of 24 inches) as one becomes more distant from the equipotential plane.

It is recommended that an equipotential plane be established in all new milking facilities and other areas where animals make contact directly with grounded equipment. The equipotential plane should consist of No. 6 gauge or larger steel mesh with a grid dimension of 6″ × 6″. Sheets of reinforcement mesh should be welded together and welded to any steel stanchions embedded into the concrete. This mesh must be bonded to the neutral of the electrical service panel. Again, transitions on and off the plane are critical.

22.13 Calf and Youngstock Housing

The most critical period in the life of a dairy animal is the first 2 months. Mortality frequently runs 20% or more during this period. Good housing goes a long way in preventing these losses.

The most common and universally accepted youngstock facility for housing all youngstock, from birth to freshening, is one that includes: (1) a nursery or hutch for baby calves, (2) group pens for the weaned calves, 6 to 10 animals in each pen, and (3) a shelter, as required by the severity of the weather, for the older heifers.

In designing a youngstock facility, whether it is a structure specifically designed for this purpose or simply a remodeled barn previously used to house adult cattle, it is important to know how many youngstock of each age or body size grouping are likely to be on the farm. Guidelines are shown in Table 22.11.

TABLE 22.11 *Housing Needs, by Age Groups, Satisfactory for Providing 30 Replacements Annually in a 100-Cow Herd*

Description	*Age, months*	*Probable no. of heifers*	*Recommended facilities when male calves are:* Sold at birth	Kept for 1 year
Milk fed calves	0–2	7	10	18
Weaned calves	2–6	11	12	22
Senior calves	6–12	17	18	28
Junior yearlings	12–18	16	18	18
Bred heifers	18–26	18	20	20
Steers (dairy beef)	10–13	—	—	18
Total		69	78	124

"Cold" Calf Housing

Several building types are being used successfully to raise calves in a "cold" environment. These include: (1) open front hutches, (2) an open front barn, and (3) an enclosed barn.[1]

Enclosed and open front barns are most successfully ventilated by eave openings and an open ridge. See Section 22.6 for a discussion on ventilation principles. Pens are frequently 4 x 6 ft with 1 open side and 3 solid sides to prevent drafts. Calf hutches (Figure 22.23) need to be located in an area that drains well; a good sand base is desirable. Moving them between each occupancy onto clean soil helps prevent disease outbreaks. Calf hutches may be as small as 3 ft wide and 5 ft deep when used in regions with mild climates. In the colder regions, hutches are usually 4 ft wide and 8 ft deep to provide more protection during severe weather.

"Warm" Calf Housing

Warm calf nurseries provide uniform temperatures throughout the winter. It may be a part of the cow or heifer barn or a

Fig. 22.23 Calf hutches in winter in Minnesota. (Courtesy of Successful Farming)

separate building. When in the same barn with older animals, the calf area should be partitioned separately, and have a separate ventilation, insulation, and heating system. See Section 22.6 for a discussion on ventilation and insulation.

Advantages of a "warm" barn are dairyman's comfort, more efficient labor utilization, and no freezing of water pipes. Calves often do as well or better in a "cold" barn.

Calves should be penned separately until 2 weeks after weaning. The pen may be either a 2 x 4 ft stall elevated or floor level, with an additional 1 ft on the front for a feed box, or a floor pen (minimum 4 x 4 ft) with 4 ft high spaced boards or wire sides to permit adequate air movement. When calves have become used to solid feed, stress conditions can be reduced by putting them in group pens within the "warm barn" for several weeks before moving them to group pens in a cold building. A well-designed "warm" calf facility, with free stalls and slotted floors, is illustrated in Figure 22.24.

Older Animals

A loose housing resting area with a manure pack is adequate for the older heifers. Open front should be located to maximize the influence of the sun, especially in the colder climates.

A Virginia plan for housing youngstock

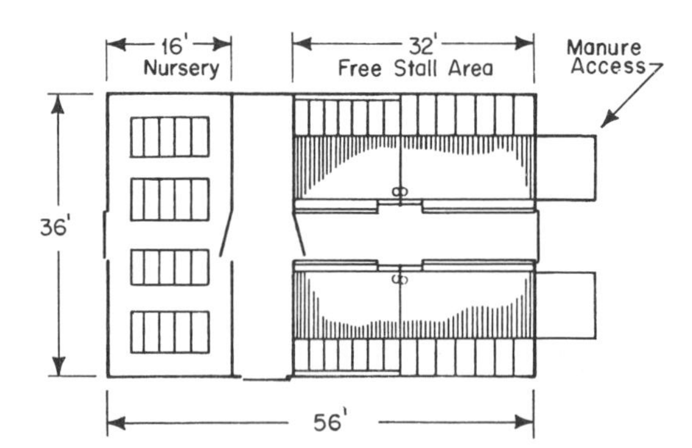

Fig. 22.24 Warm calf barn with free stalls.

incorporates self-cleaning concrete floors; group pens are approximately 75% roofed, with the resting shelter portion oriented in a southeasterly direction to utilize winter sun in drying animal droppings on the sloped floor (Figure 22.25).[10] A tractor-scraped litter alley lies between the resting area and the feeding slab. Pens 12 ft wide housing from 12 to 17 heads per pen provide from 20 to 30 sq ft of total area per animal, depending on number of animals in each pen. Approximately 60% of this area is resting area, 20% is litter alley, and 20% is feeding slab.

Free stalls are well suited to calves over 2 months of age. Recommended free-stall dimensions are shown in Table 22.12.

22.14 Summary

As dairymen replace worn-out and inefficient buildings and equipment or expand their facilities in order to enlarge their herd, increase their annual milk output per man, and obtain higher labor incomes, the various methods of housing, feeding, and milking cows should be studied in detail. How feed is distributed and manure disposed of should be carefully analyzed also, so that new equipment or a complete new facility can make a maximum contribution

Table 22.12 *Free-Stall Dimensions for Calves and Heifers*

Age	*Width*	*Length*
6 weeks to 4 months	2′0″	4′6″
5 to 7 months	2′6″	5′0″
8 to 18 months	3′0″	5′6″
18 months to freshening	3′6″	6′3″

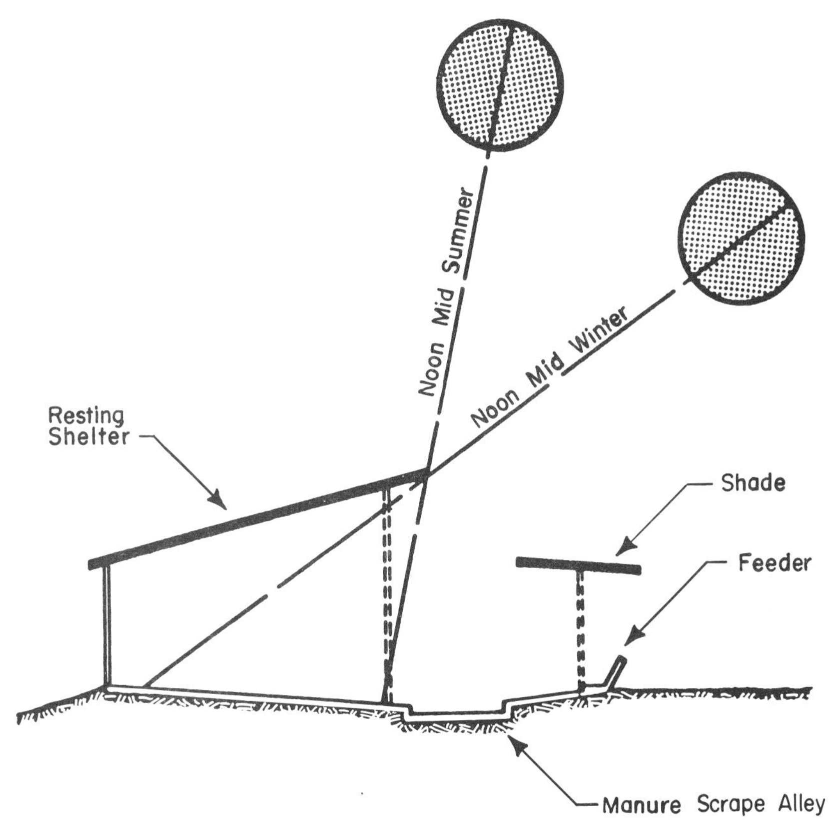

FIG. 22.25 Schematic of Virginia youngstock housing facility showing mid-day sun exposure. (Courtesy of Agr. Ext. Ser., Virginia Polytechnic Institute, Blacksburg.)

to an efficient and profitable dairy enterprise.

Review Questions

1. Indicate your choice between stall-barn housing or free-stall housing and give reasons for that choice. Also list the apparent disadvantages to the system you chose.
2. Describe your concept of an "ideal" free-stall. Include dimensions, type of partition, height of curb adjacent to manure-scraping alley, type of floor surface, kind and amount of bedding.
3. Describe the ventilation system best suited for your choice of housing system (question 1). Indicate the number and size of cattle you would be housing and calculate the ventilation fan capacity required, if forced-air ventilation is needed.
4. Suppose you have a herd of 100 cows, 40 yearling heifers, and 50 smaller calves. If all manure and urine were collected and stored for 6 months, what size storage basin would be required?
5. If you were to build a new milking parlor for a herd of 150 cows, what type of parlor would you choose and how would it be equipped? Give reasons for your decision.

References

1. Bates, D. W., Brevik, T. J., Johnson, D. W., Keith, R. K., and Pedersen, J. H.: *Dairy Housing and Equipment Handbook.* MWPS-7. Midwest Plan Service, Iowa State University, Ames, 1976.
2. Livestock Wastes Subcommittee: *Livestock Waste Facilities Handbook.* MWPS-18. Midwest Plan Service, Iowa State University, Ames, 1975.
3. Mellor, C. W., and Davidhizar, W. J.: Design of a 150 cow free stall dairy system. Proceedings, National Dairy Housing Conference, American Society of Agricultural Engineers, St. Joseph, Michigan, p. 46, 1973.
4. Boyd, J. S.: Veterinary hospital areas in free-stall housing systems. Proceedings, National Dairy Housing Conference, American Society of Agricultural Engineers, St. Joseph, Michigan, p. 76, 1973.
5. Bishop, S. E., Fairbank, W. C., Oliver, J. C.,

Smith, F. F., and Cleaver, T.: *Dairy Design.* AXT-250. Agricultural Extension Service, University of California, Berkeley, 1968.
6. Guest, R. W., Klausner, S. D., and March, R. P.: Guidelines for dairy manure management in New York State. N. Y. Agr. Engr. Ext. Bull. 400. Cornell University, Ithaca, New York, 1973.
7. Effluent limitations guidelines for existing sources and standards of performance and pretreatment standards for new sources. Federal Register, V. 38, No. 173, Part II, 1973.
8. Hutjens, M. F., Otterby, D. E., and Appleman, R. D.: Feeding dairy cattle. Agr. Ext. Ser. Bull. 218. University of Minnesota, St. Paul, 1976.
9. Bickert, W. G., and Armstrong, D. V.: Herringbone parlors, polygon parlors, side-opening parlors, rotary parlors, flat milking barns. AEIS publication No. 329 through 333. Michigan State University, East Lansing, 1975.
10. Collins, W. H., and Murley, W. R.: Countersloped, self-cleaning cattle confinement structures. Proceedings, Southeast Regional Meeting, American Society of Agricultural Engineers. University of Virginia, 1976.

Suggested Additional Reading

Bates, D. W., Brevik, T. J., Johnson, D. W., Keith, R. K., and Pedersen, J. H.: *Dairy Housing and Equipment Handbook.* MWPS-7. Midwest Plan Service, Iowa State University, Ames, 1976.

Proceedings, National Dairy Housing Conference. American Society of Agr. Engineers. St. Joseph, Michigan, 1973.

The Way Cows Will Be Milked on Your Dairy Tomorrow. Babson Brothers Dairy Research Service, Oak Brook, Illinois, (Revision in print, 1985).

Dairy Housing II. SP-4-83. American Society of Agricultural Engineers. St. Joseph, MI., 1983.

Cloud, H. A., Appleman, R. D., and Gustafson, R. J.: Stray Voltage Problems With Dairy Cows, University of Minnesota Ext. Folder 552, 1980.

Wilcox, C. J. et al. *Large Dairy Herd Management,* University of Florida Press, 1978.

CHAPTER 23

Business Management Decisions in Dairy Farming

23.1 Introduction

Dairy farming today is a dynamic, highly competitive industry. In the past the traditional dairy farm rested on land and labor as major resources. Now both land and labor availability is decreasing and costs are increasing. Thus, the use of more capital and improved management is increasing rapidly. These adjustments have fostered much larger dairies with higher capital investments per farm.

The traditional "labor oriented" dairyman is on a small farm, and utilizes only family labor (Figure 23.1). He has little desire to hire labor and thereby shifts his role to that of a personnel manager in the "capital-oriented league." His earnings may be high for the resources he controls, but his limited use of capital and hired labor puts a ceiling on his potential earnings.[1]

Labor-oriented dairymen not effectively combining their long hours of manual labor and limited capital with a high level of management soon become *low resource* dairymen. These dairymen are more and more finding themselves facing one of two choices:

1. Become a more skilled laborer and, at the same time, improve their management ability (1 in Figure 23.1).
2. Leave the dairy farm altogether (2 in Figure 23.1).

Many dairy farm operators have shifted their role from supplying skilled labor to managing capital and hired labor (3a in Figure 23.1). When successful, they receive a comfortable management income as well as a good return on investment. However, attempts to join the skilled money/labor manager group can result in a creditor's nightmare (3b in Figure 23.1) if the managerial requirement is beyond their capacity. This factor is the basis for the often-used statement: "Get better before getting bigger." Finally, staying where they are (4 in Figure 23.1) and supplementing dairy farm income with non-farm income has become a less workable option.

Employing large amounts of capital does not ensure high earnings. If a dairyman is inept at handling the larger capital resources, his earnings may actually fall because he has more units on which to lose money. Many an expansion-minded dairyman has become a *creditor's nightmare.*

Profitable expansion opportunities continue to exist for today's skilled managers. These dairymen can operate with lower costs per unit, and the large volume produced gives them above-average earnings. Keep in mind that many dairymen possess

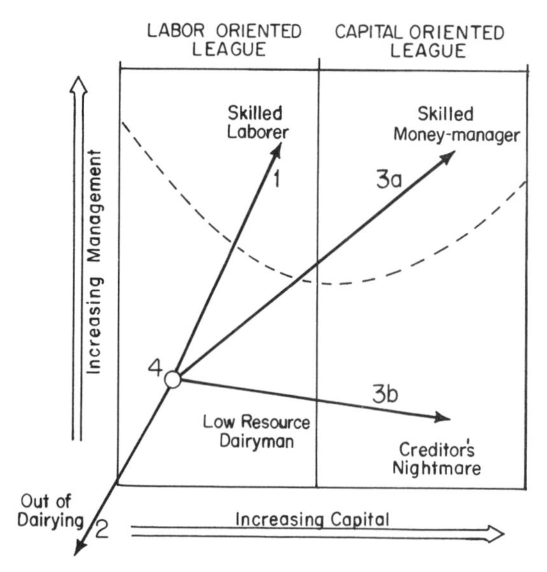

Fig. 23.1 Direction today's traditional dairyman may take.

only average amounts of management skills and available capital. Their degree of success depends on how effectively they employ both resources. Successful and vulnerable dairymen exist in both the labor- and capital-oriented leagues.

Every dairyman must regularly ask himself 3 basic questions: (1) Where am I? Am I capital oriented, labor oriented, or on the way out of business? (2) Where do I want to be? What status do I want to achieve? (3) How can I best get there from here?

This chapter is intended to assist in this decision-making process. Special attention is given to getting started, selecting business arrangements, and managing hired labor.

23.2 Getting Started

Modern dairy farm technology has caused the capital and management requirements to be very high. Most beginning dairymen recognize that they may have to begin dairying by working on an existing dairy farm, at least for a while. At the same time, many parents want to see their business continue, and have as their objective the passing on of their farm to the next generation. The problem is to determine how to get a son or son-in-law established and bring about a successful transfer of both property and management control to the next generation. A similar problem exists for the older dairyman with no apparent heirs who hires an unrelated young dairyman to enter the business.[2]

In most cases, the decision involves the economic security of the parents, the future career potential of the young dairyman, and overall family goodwill. Participants must realize that each family and farm situation is different and that each family has a different set of goals and objectives.

The individually operated farm business traditionally exhibits a life cycle that parallels that of the farm operator, often with a second generation superimposed on top of the one already existing (Figure 23.2). The typical dairyman passes through at least 3 stages during his career: entry, growth, and exit. In the *entry* stage, he evaluates his opportunity in dairying and decides whether it is to be his career. In the *growth* stage, he extends his resource base through purchase and lease, usually increasing his debt load, then tends to consolidate his gains and stabilize his income. Finally, in the *exit* stage, he considers retirement and transfer of the business.

It is important to recognize how each farm family situation differs. The first factor to consider is the *level of achievement* of the parents. Some put together a substantial business that easily accommodates a second generation of dairymen (situation 1 in Figure 23.2). Others have provided adequately for their family, but find themselves with a 1 to 1½ man business (situation 2).

The second factor is the *timing of the son's entrance* into dairying relative to the parent's stage in their life cycle. It makes quite a difference whether the parents have only 2 to 5 years until retirement or whether they have 20 years of activity remaining.

The third factor is the *number of children* with an interest in the farm. If only one child exists, the parents have a good business, and they are only 5 years from retirement, the problem is much simpler than if the parents are only 45 years old, the business is inadequate for 2 families, and there are 5 children, 2 of them wanting to become farmers.

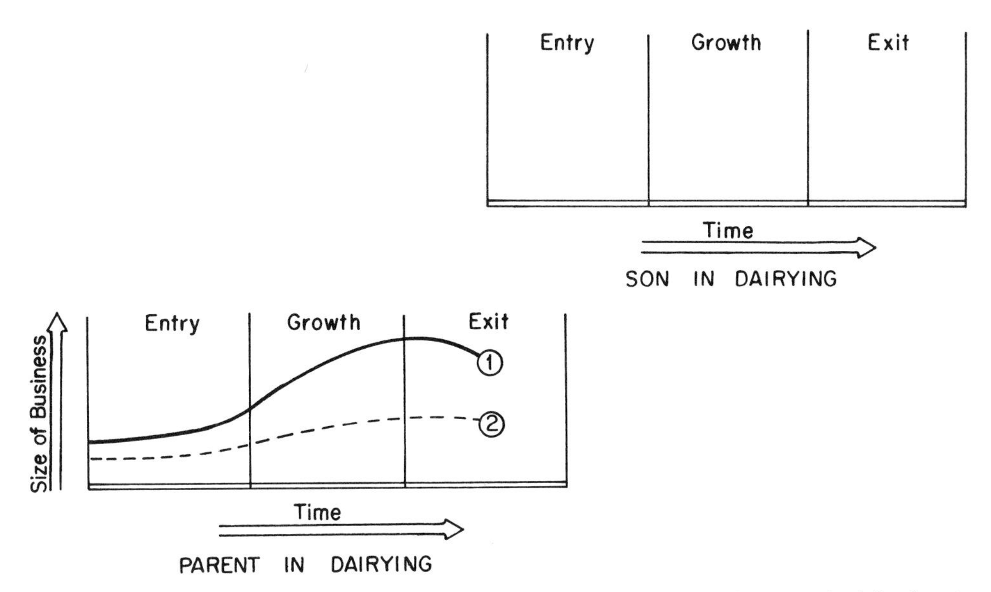

FIG. 23.2 Merging life cycle of parent and son, daughter, or unrelated young dairyman in dairy farming.

Testing Stage for a Business Arrangement

Whenever a single proprietorship is expanded to a 2-man operation, the relationship should go through a *testing stage.* This has a twofold purpose: (1) helping the son decide whether he really wants to dairy, and (2) helping both parties determine whether they can make a joint business venture work from a family and business point of view.

The common types of pre-farm partnership arrangements are: (1) wage, (2) wage-incentive program, (3) wage-share arrangements, and (4) enterprise working agreements. It is usually best to start out with a wage, wage-incentive, or wage-share arrangement. Examples of wage-incentive programs are discussed in Section 23.10.

As the junior member becomes more involved, an enterprise working agreement provides a means for him to build equity. These pre-partnership arrangements usually last 2 to 5 years before they are disbanded (parties go separate ways) or they formalize their working arrangement into a more permanent business arrangement (partnership or corporation).

In a *working agreement,* the junior member of the firm (or employee) may furnish some personal property (i.e., livestock and machinery) and some management in addition to labor. The employee would not normally make whole farm decisions, but may make most of the major decisions in one enterprise (i.e., the dairy enterprise but not the cropping or other livestock enterprises). This agreement should be in writing and signed by all parties involved. Be sure the writing covers such topics as: (1) job responsibilities of each party, (2) a clear description of all assets owned by each party, (3) a procedure for settling disputes, and (4) conditions under which the agreement can be dissolved.[3]

An example of enterprise working agreement is shown in Table 23.1. In this example, the employee is the herdsman and will contribute 25 cows and 25 replacement heifers, as well as the labor required in the dairy enterprise. In return, he is guaranteed $1,500 per month plus house and meat. If his combined management and labor capabilities result in sufficient spendable income (line 18), he may be able to reduce his debt load assumed in purchase of animals or, if they are already free of debt, acquire additional livestock or machinery.

Working agreements, in a legal sense, may or may not be considered as a legal partnership. In general, if the working agreement involves "sharing of income but

Table 23.1 *Example Dairy Enterprise Working Agreement*

			Example	
I. General description of working agreement			Employee will be herdsman and will contribute 25 cows, 25 heifer repl.	
II. Fixed contributions to enterprise by each party.			Employer	Employee
Livestock, number and value*			75 cows @ $1200	25 cows @ $1200
			75 heifers @ $300	25 heifers @ $300
Building and equipment	depreciation	(1)	$ 24,350	$ 0
	interest	(2)	$ 14,800	$ 0
	repairs	(3)	$ 3,480	$ 0
Taxes and insurance		(4)	$ 870	$ 0
Labor and management charge or wage		(5)	$ 0	$ 18,000
Other labor hired		(6)	$ 7,500	$ 0
Total fixed contribution		(7)	$ 51,000	$ 18,000
Grand total fixed contribution**		(8)	$ 69,000	
Value homegrown feed fed*		(9)	$ 75,000	
Subtotal (8 + 9)		(10)	$144,000	
Desired share arrangement(%)		(11)	75%	25%
Allocated fixed and feed contrib. (10 × 11)		(12)	$108,000	$ 36,000
Total fixed contribution (line 7)		(13)	$ 51,000	$ 18,000
Value homegrown feed contrib. (12 − 13)		(14)	$ 57,000	$ 18,000
III. Other enterprise costs* (share in desired percent: line 11)				
Purchase feeds			$ 6,750	$ 2,250
Breeding costs			$ 1,350	$ 450
Veterinary, medicine			$ 2,700	$ 900
Production testing			$ 945	$ 315
Marketing costs and commission			$ 4,200	$ 1,400
Other enterprise costs			$ 1,800	$ 600
Total other enterprise costs		(15)	$ 17,745	$ 5,915
IV. Expected income and debt servicing capacity of enterprise (share in desired percent: line 11)			14,000 cwt @ $12.60/cwt	
Milk			$132,300	$ 44,100
Cull sales			$ 15,750	$ 5,250
Total income, each party		(16)	$148,050	$ 49,350
Less selected expenses (3 + 4 + 6 + 14 + 15)		(17)	$ 86,595	$ 23,915
Approximate spendable income (16 − 17)		(18)	$ 61,455	$ 25,435
Sched. principal and interest payment		(19)	$ 39,150	$ 9,200
Amount left for living, etc. (18 − 19)		(20)	$ 22,305	$ 16,235
V. Guaranteed wage and fringe benefits			Employee guaranteed $1,500/mo. plus house plus (1) 1,000# steer/year	

*Livestock and other enterprise cost contributions and enterprise income will be shared in desired percentages (line 11). Adjustment of homegrown feed contribution by each party will be made to make arrangement equitable (lines 12 through 14).
**Total employer and employee fixed contributions.

not a sharing of losses," it is *not* a partnership. It is, however, considered a partnership if it is both profit and loss sharing.

23.3 Alternative Business Arrangements

The major alternatives for a more permanent form of dairy farm business organization are depicted in Table 23.2. They are organized in ascending degrees of complexity (left to right) and are to be implemented after the "working agreement" phase has been passed. In many cases, the overriding business concern is finding a way to permit the second party (son) to become established without jeopardizing the financial interests of the senior party.[2]

Whenever business size and family interest suggest the development of multi-ownership on a permanent and continuing basis, business development and "transfer of the business" concerns become more complex and important. If the owner-operators *supply all the capital,* no particular business arrangement has a major advantage over the others in terms of capital acquisition. However, if an untimely death occurs and the farm is transferred to heirs, some differences may occur. Frequently, the dairy farm operator inherits the personal property and other children take ownership of real estate, leasing it to the farm operator. This situation is where disagreements may arise over management policies.

In regards to *limited liability* considerations, one form of farm business organization has no particular advantage over another if the owner-operators have all their capital invested in the farm business. When personal investments exist outside the farm business, a corporation has an advantage. It should be understood, however, that limited liability, in an absolute sense, is nonexistent whenever shareholders assume personal liability for debt obligations.

Management and control of the business do differ, depending on how the business is organized. A sole proprietor makes his own decisions; unless an agreement specifies otherwise, partnerships usually imply that management decisions are made jointly. Neither partner has the freedom to be his own boss. Corporations are more structured, and the corporate bylaws specify what authority shareholders or board of directors have in management control. In a small farm corporation, the same persons may serve as shareholders, directors, officers, and workers; thereby, management control may function similar to a partnership organization.

Tax considerations become important in considering how the business is to be organized. Social security taxes are higher under the corporate structure. Further owner-operator employees of a corporation may be subject to workmen's compensation charges on their salary and entitled to benefits under the act; such coverage does not apply to sole proprietorships or partnerships. Because income tax in a corporation can be split between an owner-operator's salary and corporate profits, a direct comparison between rates cannot be made. For most levels of capital gain income, however, the corporation pays a larger tax than an individual.

Farm corporations are not numerous; only 1% of the farm businesses in the 1969 census were incorporated. A corporation is an artificial being created under state law; it is a separate business entity distinct from its owners. A *tax-option corporation* (subchapter 5) is a creation of the federal tax law. It is a corporation in all respects except that the corporate entity pays no income tax; instead, each shareholder-owner reports a share of corporate income for income tax purposes. In general, farm corporations are most advantageous when the business is relatively large and it is desirable to make provisions for transfer of ownership.

Transfer Considerations

The business arrangements described in this section and the previous section are designed to assist in transferring at least a part of the personal property and some of the management control of the business to a junior member of the firm. Eventually, the family must face the problem of setting up an effective transfer plan for the overall business.

The first step here is to determine their present situation. What assets do they have? How are they held? What gifts or commitments have been made? What is the current status and desire of the family?

TABLE 23.2 *Comparison of Farm Business Organizations*

	Sole proprietor	*Partnership*	*Corporation*
Nature of entity	Single individual	Aggregate of two or more individuals	Legal person separate from shareholder-owners
Life of business	Terminates on death	Agreed term; terminates at death of a partner	Perpetual or fixed term of years
Liability	Personally liable	Each partner liable for all partnership obligations	Shareholders not liable for corporate obligations
Source of capital	Personal investment; loans	Partners' contributions; loans	Contributions of shareholders for stock, sale of stock, bonds and other loans
Management decisions	Proprietor	Agreement of partners	Shareholders elect directors who manage business through officers elected by directors
Limits on business activity	Proprietor's discretion	Partnership agreement	Articles of incorporation and state corporation law
Transfer of interest	Terminates proprietorship	Dissolves partnership; new partnership may be formed if all agree	Transfer of stock does not affect continuity of business—may be transferred to outsiders if no restrictions
Effect of death	Liquidation	Liquidation or sale to surviving partners	No effect on corporation. Stock passes by will or inheritance
			Regular corporation
Income taxes	Income taxed to individual; 50% deduction for long-term capital gains	Partnership files an information return but pays no tax. Each partner reports share of income or loss, capital gains and losses as an individual	Corporation files a tax return and pays tax on income; salaries to shareholder-employees deductible Capital gains offset by capital losses; no 50% deduction for capital gains Rate: 22% on first $25,000, 48% on excess Shareholders taxed on dividends paid
			Tax-option corporation
			Corporation files an information return but pays no tax. Each shareholder reports share of income, operating loss, and long-term capital gain

Source: Thomas, K. H., and Boehlje, M.: Farm Business Arrangements: Which One For You? North Central Regional Extension Publication 50, 1976.

As parents proceed to set goals and develop a plan, they should remember that their No. 1 priority should be to provide themselves financial security in their retirement years. If income appears to be insufficient for the retirement years, they should be exploring ways of increasing income or reducing retirement needs rather than being concerned about the heirs and their problems. If income appears to be barely adequate for retirement, then plans to transfer property (e.g., wills and trusts) at time of death should receive priority. If probable income is more than adequate, then sale and/or gift of at least some of the property before death may be in order.

With their own situation clearly in mind, the parents should then focus on a second priority—developing a transfer plan that will treat all heirs fairly and that will minimize transfer costs. Professional legal help will be needed to develop an effective plan. If done right, it will be money well spent, in terms of both satisfaction gained and taxes saved.

23.4 Evaluation of Current Status

Where am I? To answer this question, the dairyman needs a good set of records (measurement of past performance) and a set of standards (by which to judge his performance). The kind of information needed depends on the answer to these 2 questions: What information is of value to me? How much time, effort, and cash am I willing to spend to get it?[1]

The more information and analysis obtained, the higher will be the cost. Therefore, 2 points that should always be kept in mind when deciding upon what records to keep are:

1. Keep only those records that will be used.
2. Keep only those records that will be kept well enough to be useful.

Table 23.3 can be used as a guide to obtain any of 4 objectives.[4] Records have 4 main uses. The first is a *legal requirement;* that is, they provide tax information for federal and state income tax returns and social security reports. Every dairyman must keep these records, but it is his option as to how many of the remaining objectives he wants to achieve from the use of records.

The second use of records is as a *diagnostic* tool. They can help determine the profitability of a business by identifying strengths and weaknesses. A good dairy manager will capitalize on his strong points, and after recognizing his weak points, he will take steps to correct them.

A third use of records is as an *indicator of progress,* from both a production management and financial management stand-

TABLE 23.3 *Records Needed to Evaluate the Dairy Farm Business*

Records needed	*Income tax*	*Cash flow*	*Net worth*	*Farm earnings*	*Enterprise analysis*	*Family living*
Inventory						
Depreciation schedule	x		x	x		
Other farm assets			x	x	x	
Non-farm			x			x
Production records						
Milk production (DHI)					x	
Other livestock production					x	
Livestock cash expenses					x	
Crop yields					x	
Feed records					x	
Financial transactions						
Farm receipts and expenses	x	x		x	x	
Non-farm receipts and expenses		x				x
Possible tax deductions	x	x				
Debt payments		x	x			

Source: Hasbargen, P. R.: Farm Record System: What Records Should I Keep? Univ. of Minn. Agr. Ext. Ser. Publ. FM-100, 1971.

point to measure change in size, productivity, efficiency, and organization. Actual performance can be measured in comparison with planned performance or standards of performance for the dairy farm business.

Finally, records can be used for both short- and long-term *forward planning.* Cash flow, credit needs, and repayment capacities can be projected. In addition, the dairyman can determine the profit generation capacity of different proposed alternatives. In these volatile economic times, forward planning is crucial.

Many good record books and record plans are available to farmers. Your local county agent or farm advisor, vocational-agricultural instructor, or banker can provide a good farm records book or suggest a local source. Mail-in electronic data processing (EDP) systems have the advantage of providing timely information useful for cash flow or financial control purposes.

23.5 Using Records as a Diagnostic Tool and Indicator of Progress

The following troubleshooting guide highlights factors that farm records and experience have shown are closely related to success in dairying.[5] A 2-step procedure is used: the first step is a financial analysis of the overall business (Table 23.4). Step 2 isolates potential problems via an enterprise analysis (dairy enterprise in Table 23.5 and cropping enterprise in Table 23.6).

Profitability and *financial soundness* (liquidity and solvency) are the chief financial concerns of a dairy farm operator. The returns to labor and capital are the best measures of profitability. To determine the degree of success obtained, these earnings can be compared either with the income reported by comparable businesses or with other employment opportunities.

Because of the wide variation in quality and quantity of labor and management resources used on dairy farms, no goal is specified for labor and management earnings in Table 23.4. Returns to capital invested of 8 to 10% and returns to equity capital of 10 to 15% are indicated because dairymen should not only recover their interest on borrowed money but also make a profit on their own money, comparable to what might be expected on an investment elsewhere. Since purchased inputs represent a growing proportion of total production expenses, the income per dollar of operating expense is increasing in importance as an indicator of profitability.

Table 23.4 *Conducting a Financial Analysis of the Overall Business*

Profitability

Item	*Profitable goal*	*Your farm*
1. Return to labor and management	________	________
2. Labor and mgt. return/operator	________	________
3. Rate of return to total capital invested	8–10%	________
4. Rate of return to equity capital	10–15%	________
5. Operating income per $1.00 of operating expenses	$1.90–$2.00[a]	________

Liquidity/Solvency

Item	*Acceptable standard*	*Your farm*
1. Ratio of current liabilities to current assets	0.5	________
2. Ratio of current and intermediate liabilities to current and intermediate assets	0.6	________
3. Ratio of long-term liabilities to long-term assets	0.6	________
4. Ratio of total liabilities to total assets	0.5–0.6	________

[a]Applicable in midwestern U.S. Figure may be inappropriate in regions where more feedstuffs are purchased.

TABLE 23.5 *Analysis of the Farm "Dairy" Enterprise, Including Youngstock*[a]

Symptoms of unprofitable returns

Item	*Profitable goal*	*Your goal*	*Your farm*
1. Cash receipts per man (devoted to dairy herd)	$62,000		
2. Milk sold per cow (lb)[b]	14,000		
3. Cattle sales per cow	$175		
4. Total returns per cow	$1,950		
5. Dairy returns per $100 feed[b]	$220		

Causes of unprofitable returns

Item	*Profitable goal*	*Your goal*	*Your farm*
1. Number of cows			
2. Milk produced per man[b] (thousand pounds)	500–600		
3. Feed cost per 100 lb milk	Under $6.00		
4. Percent of cows in milk	86		
5. Average calving interval (months)	12.5		
6. Percent of calves lost	10		
7. Price received for milk			

Symptoms of unprofitable costs

Item	*Profitable goal*	*Your goal*	*Your farm*
1. Feed cost per cow	Under $1,000		
2. Non-feed cash costs per cow	$350		
3. Value of cow and replacement	$1,500		
4. Building and equipment cost per cow	Under $450		
5. Labor cost per cow	$300		

Causes of unprofitable costs

Item	*Profitable goal*	*Your goal*	*Your farm*
1. Grain fed per cow (lb)	5,600		
2. Cows per man (devoted to the dairy herd) P=parlor, S=stall	P=45, S=35		
3. Veterinary and medicine expense per cow	$30		
4. Building and equipment investment per cow	Under $2,000		
5. Percent of housing capacity used	95–100		

[a]The youngstock may be considered a separate enterprise. In this example, however, they are included with the milking herd.
[b]These key factors have been found to have an especially close relationship with returns to operator labor and management income.

TABLE 23.6 *Analysis of the Dairy Farm "Cropping" Enterprise*

Symptoms of unprofitable returns

Item	Profitable goal	Your goal	Your farm
1. Crop value per crop acre	$250–350		
2. Hay yield per acre (in tons of hay equivalent)[a]	4–6		
3. Corn silage yield per acre (tons)[a]	15–25		
4. Corn yield per acre (bu)[a]	90–120		
5. Other grain yield per acre			

Causes of unprofitable returns

Item	Profitable goal	Your goal	Your farm
1. Total crop acres[a]			
2. Percent of tillable acres cropped	98–100		
3. Percent of tillable acres in a high value crop	80–90		
4. Fertilizer-lime expense per crop acre	$45–60		
5. Insecticide-herbicide expense per crop acre	$15–30		

Symptoms of unprofitable costs

Item	Profitable goal	Your goal	Your farm
1. Labor expense per crop acre			
Hay	$30–35		
Corn silage	$30		
Feed grains	$20		
2. Machine overhead expense per crop acre			
Hay	$22–26		
Corn silage	$30–35		
Feed grains	$30–35		
3. Total expense per crop acre[b]			
Hay	$200–230		
Corn silage	$225–300		
Feed grains	$225–300		

[a]These key factors have been found to have an especially close relationship with returns to operator labor and management income.

[b]"Total expenses" per acre include, besides labor and machinery, cash expenses for seed, fertilizer, chemicals, and interest and taxes on land. Any of these could be a cause for excessively high costs.

Note: Figures listed in the "profitable goal" column are applicable to midwestern U.S. and may be inappropriate in other regions.

Profitability and financial soundness tend to have a competitive as well as a complementary relationship. When measuring financial soundness, *liquidity* refers to the short-run (next 12 months) ability of the dairy farm to service current debts. *Solvency* measures indicate the intermediate and long-term financial stability of the dairy farm, reflecting the ability to pay all liabilities with existing assets at any point in time. Liquidity and solvency measures look at the business in a pessimistic light in that they consider the outcome of foreclosure by creditors or the ability of the business to survive periods of economic stress.

A larger ratio of liabilities to assets suggests a weaker liquidity position, but a smaller ratio is not necessarily more desirable. Productivity of highly liquid assets, such as large inventories of supplies, is generally less than assets in a less liquid form. Thus, a strong liquidity position (value considerably less than 0.5) tends to lower farm profitability.

Solvency measures also indicate the control the owner has over his business and the degree of risk assumed. As the ratio approaches 1.0, the business is operating in a high risk situation; additional credit may be hard to obtain. On the other hand, a low ratio suggests that profitability may be hampered by limited use of credit. A standard of 0.5 to 0.6 for the ratio of all liabilities to all assets is considered ideal for stable dairy farm businesses.

Given a profitable business, liquidity problems may result from poor timing in outflow of dollars, insufficient liquid assets, excessive family living expenses, or an inappropriate debt repayment schedule. When solvency problems occur, the dairyman may be faced with selling off some assets. Care should be taken, however, not to jeopardize long-run profitability; in other words, get rid of only the least productive assets.

The key to good liquidity and solvency is strengthening profitability. Thus, good farm managers devote much time to analyzing their business and identifying those areas that are less profitable.

Enterprise Analysis

The enterprise analysis is conducted in 2 phases. First, the dairyman searches for *symptoms;* then he attempts to identify *causes.* For most dairymen a two-enterprise analysis (dairy and crops) is meaningful. Some may choose to add a "replacement heifer" enterprise. The list of analysis factors provided in Tables 23.5 and 23.6 is not exhaustive; however, it contains the more important determinants of dairy farm profitability.

Managers tend to view a particular factor as the problem when it is merely a symptom of the difficulty. For example, low dairy cow productivity may seem to be the cause for low returns when actually the cause is poor reproductive performance. Searching for symptoms and then causes in the manner indicated should result in a more effective pursuit of business weaknesses.

Keep in mind that the primary causes for unprofitable dairy operations identified include the following:

1. Low production per cow.
2. Low production per man-year of labor expended.
3. High expenditures for feed per cow.
4. Low yield per acre or poor quality feed, whether it be hay, corn silage, or feed grains.

A critical analysis of the dairyman's Dairy Herd Improvement Association (DHIA) records can be extremely helpful in locating and correcting the cause of unprofitable dairy operations. See Chapter 3 for a discussion on this topic, and study Chapter 12 on computer-formulated rations for maximum profit.

Profitable goals vary depending on location, climate, soil productivity, and other factors. For that reason, columns are provided for each dairyman to list his own goals. Keep in mind that goals should be both realistic and attainable. As soon as a goal is reached, the standard is usually raised. It is by this means that the level of management improves, as indicated in Figure 23.1. Also see Appendix Table V-G.

Because of the extreme variation in herd sizes common to different regions of the country, and because of the variation in prices received for milk, no attempt is made to provide a suggested or guideline goal for either cause of unprofitability (i.e., dairying or cropping).

23.6 Setting Goals

To set future goals, the dairyman should carefully appraise his options and determine the benefits and consequences of each. After considering alternatives, he should eliminate those that are clearly unattractive or infeasible. To analyze the remaining alternatives he should: (1) gather information, (2) use appropriate analytical procedures, and (3) weigh the evidence.[1]

When evaluating feasible alternatives, the dairyman must look beyond the initial costs and a salesman's claims. He should attempt to obtain the best estimates of investment costs, operating expenses, and labor demands. See Sections 23.8 and 23.11, as well as the appropriate appendix tables, for assistance in estimating these inputs. Above all, the dairyman must remember that the *number* of arguments for or against a given proposal is not as important as the *weight* or *importance* of each argument. The dairyman should make the decision. Postponing action only perpetuates the existing situation, which may be the worst possible decision.

One of the alternatives frequently considered is expansion of the dairy business. Unfortunately, some dairymen have expanded without going through the process described previously, and found to their chagrin that nothing was gained other than more work. This is not to suggest that expansion is not a viable alternative. For many it is, and they should proceed. Expansion is most likely to be a viable alternative when one or more of the resources (land, labor, capital, and management) are not fully employed and additional amounts of the other resources can be obtained at a reasonable cost. Adding more cows upsets the existing balance of these resources and it does require resource realignment.[6]

Get better before you get bigger is a phrase that bears repeating. Figure 23.3 clearly illustrates that milking more cows provides no cure-all for dairymen with income problems. Management becomes more important as herd sizes become larger. Few dairymen producing less than 14,000 pounds of milk per cow annually (Holsteins) will profitably benefit from a major expansion of the dairy enterprise.

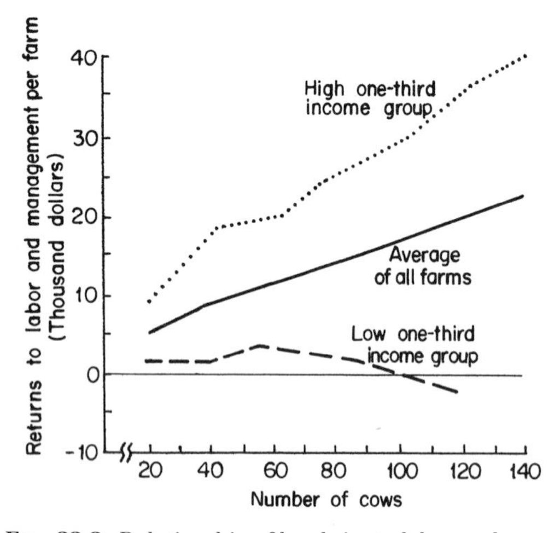

Fig. 23.3 Relationship of herd size to labor and management income. (Source: Willett, G. S., Howard, W. T., and Weigle, R. N.: Planning for Dairy Farm Expansion. Univ. of Wis. Agr. Ext. Ser. Publ. A2449, 1972)

23.7 Factors to Consider Before Expanding

Dairy farms throughout the United States vary greatly in size of herd, housing, and other characteristics. Dairy herds in Minnesota average only 33 cows compared with 400 cows in Florida. Many factors have influenced this development.[7]

Tradition and nationality have been major factors influencing the size and organization of dairy farms. In the early days Europeans established small dairy herds similar in size to the ones with which they were familiar in their home countries. Now the attitude of the individual has become more business oriented. The progressive dairyman is more interested in developing an operation that minimizes drudgery and hard physical work. Thus, the trend has been toward more specialization and mechanization, requiring more cattle over which to spread the added investment.

Attitude of credit agency personnel is another important element in the development of larger dairy farms. Midwestern and northern states bankers have tended to be more conservative while western lenders have been more oriented to large-scale operations. One reason is major differences in the types of credit needed; feed and livestock account for 50% of the total investment in California dry lot operations but only 33% in Minnesota. Thus, a higher percentage of the loans on California dairy farms are short-term (1-year) loans.

Climate and land resources in a given area are also important factors influencing the opportunity for expansion. Topography and land productivity, especially segmentation of crop land, make it difficult to acquire sufficient land base close to building facilities. Further, the cost of acquiring more crop acres is almost prohibitive in some areas near urban developments. Climatic problems, particularly in the northern portions of the Lake States and northeast regions, make it more difficult to harvest and store large quantities of high-quality forages. On the other hand, almost unlimited quantities of high-quality hay in Arizona, California, Washington, and Texas have encouraged the development of large-scale dairy farming. A major concern at the present time in these areas is the high cost of this purchased forage, due to the favorable prices received for cotton and other cash crops.

Location and kinds of markets have a definite influence on the size of herd and technology supplied. Most fluid milk production is adjacent to or reasonably near large metropolitan centers, so most larger herds are concentrated in those areas. Most large-scale operations are in high cost, and often irrigated, areas.

The desire of some farmers to include dairying as a supplementary enterprise is another consideration. This practice serves as a hedge against disastrous losses in income from droughts or early freezing of small grains and forage crops.

The impact of *pollution control regulations* has become an important factor. Some dairymen who would otherwise consider expansion are maintaining present herd size and facilities until pollution control measures are spelled out more definitely for their area.

When an individual is deciding whether to expand, he should look at the alternatives. The decision is a major one; he may live with it for 20 years. Thus, he should ask himself these questions:

1. *Is dairying the best alternative?*
   a. What is the future demand/price situation?
   b. What are the comparative advantages of dairying in the area?
   c. Do I like cows well enough?
   d. Am I young enough?
2. *Should I get bigger or just better?*
   a. What gives me more satisfaction, having a big job or doing a good job?
   b. Do I rank in the top 25% of dairymen in my size group?
   c. Do I have the managerial and financial capability to handle the larger herd?
   d. Do I have the labor supply? Can I manage labor?

23.8 Steps in Analyzing Expansion Alternatives

If adding more cows to the herd seems the best way to improve the dairy farm, these questions should be considered: (1) How many cows should be added? (2) How much additional feed will be required? (3) Where will the additional feed come from? (4) How much additional labor will be needed and where will it come from? (5) How much capital will be needed for new and/or expanded buildings, new equipment, more land, more replacement youngstock, and higher operating expenses? (6) Will expansion be profitable? (7) Can additional credit be obtained at an acceptable cost and with a reasonable repayment schedule? (8) Will the added debt still permit an overall acceptable debt level for the dairy farm business?[6]

The planning will be more effective if it is organized in a logical manner to deal with the problems encountered by milking more cows. Dairymen and their creditors should work through the following 7-step planning procedure before committing themselves to a major expansion of the dairy herd.

Step 1: Construct an Inventory

Specify the number of dairy cattle for the expanded herd, including both the number of milking cows and number of youngstock necessary to accompany the expanded herd (see Table 22.11).

Step 2. Determine the Feed Requirements

Feed requirements per head, by age group, need to be specified for all feeds in the ration. If you have kept good feeding records, use them for budgeting new feed needs. If you have no records, consult your state agricultural university or commercial

Table 23.7 *Example of Annual Feed Requirements for a 100-Cow Herd, Including Storage Losses and Wastage*

Animals		Corn or other feed grains (bu)		Protein		Corn silage (tons)		Hay or low moisture silage (tons of hay equivalent)		Pasture (acres)	
Age	No.	Required per head	Total	Required per head	Total	Required per head	Total	Required per head	Total	Required per head	Total
Milk cows	100	82	8,200	700	35	5.0	500.0	4.3	430.0		
Yearling heifers	32	8	256	80	1.6	—	—	2.4	76.8	1.5[a]	48
Sr calves	17	8	136	80	0.7	—	—	1.0	17.0		—
Jr calves	22	6	132	70	0.8	—	—	0.4	8.8		—
Total	171		8,724		38.1		500.0		532.6		48

[a]Pasture for 4 months, equivalent to 1.3 tons of hay equivalent (HE).

feed company. An example of the livestock feed requirements for the Lake States region is shown in Table 23.7. Allow for storage losses and waste, which may be about 5% for concentrates and as high as 15% or even more on forages (see Table 22.9). Keep in mind that animal size and production level are closely related to feed requirements. If you are adding cows substantially different from the rest of the herd in size or production, this change should be reflected in the budget feed requirements.

Step 3. Identify the Sources of Feed

Feed may be obtained by any one or a combination of 3 methods: (1) changing the cropping program on the existing land base, (2) purchasing or renting additional land, or (3) buying additional feed. If you expect to raise some or all of your feed needs, use conservative yields to take care of poor years. To calculate the yield potential of your existing acres, assess what adjustments in the cropping program will be necessary, determine what you can raise yourself, and then divide the feed shortage (if any) by the expected per acre yield to determine the change in acreage needed to match feed supply with demand. Try several different cropping plans and feed purchase plans to determine which overall plan is likely to be the more profitable. Mistakes on paper are much less costly than finding out later that another alternative would have been more profitable.

Step 4. Determine the Labor Requirements

Labor requirements depend on: (1) the number of cows milked and youngstock maintained, (2) the type of milking, housing, waste handling, and feeding system used, and (3) the amount and type of crops grown. Adding cows will usually lower the per head labor requirements because fixed or overhead labor is spread over more cows. Family farm operators that expect to utilize hired labor should not plan on hired labor working as many hours per year or being as productive as the operator himself. Many of the disappointments in the performance of hired labor have arisen from an underestimation of the contribution the operator's immediate family (spouse and children) have made in the ongoing operation.

Guidelines for determining annual labor needs are provided in Table 23.8. The lower numerical values are more likely to occur only in the larger, more mechanized operations. Be sure to consider the availability of labor during seasons of peak work loads such as corn planting or harvesting and hay harvesting. The estimated annual labor requirements for a 40-cow stall barn herd and a 125-cow free-stall herd are shown in Appendix Table V-H. The detailed procedures for calculating the labor required are given in Appendix Table V-I.

Step 5. Estimating the Additional Capital Required

While elaborate plans are not necessary to obtain usable investment estimates, it is important to be careful in choosing the figures ultimately used for planning. Experience has shown that investments commonly run higher than anticipated; thus, one should estimate toward the higher rather than the lower extreme. Section 23.11 provides guidelines for estimating the cost of new buildings and equipment. In addition, do not fail to include the investments necessary for additional livestock,

TABLE **23.8** *Suggested Annual Labor Requirements for Dairy and Crop Enterprises*

Dairy	*Annual hours per animal*	*Crop*	*Annual hours per acre*
Milk cows		Corn	3–5
Stanchion barns	55–60	Oats	2–4
Parlor	35–50	Corn silage	4–6
Youngstock	9–15	Hay	6–10
		Hay crop silage	5–7

including youngstock, and feed and labor necessary to maintain them.

Some added investments invariably result in other additional investments, frequently unplanned. For example, a new, taller upright silo can cause one to purchase a larger blower, which in turn results in the need for a bigger tractor, larger capacity field wagons, and more chopper capability.

The experience of 47 Michigan dairy farmers that have gone through a major expansion program is that immediately following expansion, serious animal health problems have occurred with large increases in calf losses and more reproductive problems, resulting in nearly a 1,000-lb drop in production per cow. Sixty-eight percent of all farms expanding have experienced a cash-flow problem lasting for about 2 years. Several returned to their creditors to obtain additional credit.[8]

Step 6. Estimating Income and Expenses

The next step is to itemize expected income and expenses. It is easy to be overly optimistic on the former and forget one or more critical items on the latter. Dairy or farm management specialists may have computer capabilities to help in this important aspect.[9] If so, they can create a report similar to that shown in Table 23.9. If not, use this one as a guide and develop your own by use of a table calculator. Milk sales depend on production per cow, number of cows milked, and milk price. Dairy returns should include cull cow, bull calf, and breeding animal sales. Sale of surplus feed crops should be included. Farm expenses are best estimated by using records from previous years as a resource. Livestock expenses (e.g., veterinarian and medicine, breeding fees, bedding, and supplies) per cow will normally not vary substantially as more cows are added. Similarly, crop expenses (e.g., fertilizer, seed, herbicides, and insecticides) per acre will remain relatively constant when additional acres are cropped.

Step 7. Performing a Financial Analysis

The final step considers: (1) projected profitability, (2) projected length of the payback period, and (3) expected solvency position of the business at the time of peak indebtedness. Alternative No. 1 in Table 23.9 is that of a *status quo* situation, in other words, what the long range performance is expected to be if no major expansion is undertaken. Alternative No. 2 depicts what is expected if a major expansion does occur.

Projected profitability is the single most important financial issue confronting the dairyman. The best measure of profitability on the projected expanded operation is the rate of return on the added capital investment (Table 23.9, line 15). Creditors will be especially interested in this factor since it provides an indication of how well their contribution (i.e., capital) to the venture will perform.

Projected length of time required to pay back or retire the additional debt is a second important issue to consider. An expansion program can be profitable and still not have an acceptable payback period. When expansion occurs, the business must generate enough cash to (1) pay operating expenses, (2) meet income tax and social security liabilities, (3) pay interest on existing and added debt, (4) retire the principal on current debt, and (5) support the operator's family at an acceptable level. In addition, enough cash surplus must be available to cover normal capital replacements and retire at least a portion of the principal on the added debt.

In general, the payback period should not exceed the economic life of the investment in question. A dairy cow should pay for herself (and/or her replacement) in 3 to 5 years; the cost of buildings and equipment should be recaptured in less than 10 years since the economic life of an asset is determined as much by obsolescence as by deterioration.

Additional Considerations

Most dairymen and many creditors are reluctant to do the kind of planning and budgeting discussed in this chapter. Foremost among the reasons for their reluctance are the time and effort involved in doing a good job of planning. Hopefully, more widespread use of computers will eventually relieve the dairyman of the drudgery of "number" manipulations. Another reason commonly cited for not

TABLE 23.9 *A Financial Analysis of a Projected Expanded Farm*[a]

Item	*Alternative 1*	*Alternative 2*
Plan Description		
Number of dairy cows	75	125
Table tillable acres	295	295
Total hours of labor	6,131	8,220
Total farm investment ($)	785,509	985,509
Corn equivalent balance (bu)	1,254	−10,388
Hay equivalent balance (tons)	1	5
Projected Profitability	($)	($)
1. Gross farm income	175,948	254,125
2. Cash operating expense	108,379	181,950
3. Net cash farm income (1–2)	67,568	72,175
4. Depreciation	18,000	28,400
5. Profit or loss (3–4)	49,568	43,775
6. Interest on farm net worth	34,111	34,111
7. Labor and management earnings (5–6)	15,458	9,664
8. Real estate (RE) and non-real estate interest paid	18,971	44,340
9. Value of operator's labor and management	28,797	32,705
10. Return on farm investment	39,742	55,408
11. Total farm investment	785,509	985,509
12. Return per $100 farm investment	5.1%	5.6%
13. Added return to added investment		15,666
14. Added capital invested		200,000
15. Return per $100 added investment		7.8%
Projected Payback Period		
16. Net cash farm income	67,568	72,175
17. Net non-farm income	0	0
18. Net cash available	67,568	72,175
19. Family living (2 families)	30,000	30,000
20. Income tax and social security	7,270	3,100
21. Annual RE principal payments	10,927	11,576
22. Cash available for non-RE debt	19,371	27,499
23. Amount of non-RE debt	43,044	118,044
24. Years to pay back non-RE debt	2.2 yr.	4.3 yr.
25. Added cash available for non-RE debt		8,128
26. Added non-RE debt		75,000
27. Years to pay back added non-RE debt		9.2 yr.
28. Total debt	222,998	422,998
29. Percent in debt	27.3%	41.6%
30. Net worth change per year	12,298	10,675

[a]The situation: This hypothetical dairy farm with 75 cows producing at the 15,000-lb level has been quite profitable. The two brothers, in a partnership, have several young children interested in working part-time on the farm. Their plan is to employ this help and, at the same time, reduce some of the drudgery involved in hauling manure daily; working conditions are to be improved further by building a new milking parlor.

Their plans call for an additional 50 milk cows replacing the dairy beef enterprise, which has not been too profitable in recent years. In order to have sufficient feed available, they plan to convert some of the land used to raise corn as a cash crop to corn silage and alfalfa hay, thereby producing their entire forage needs; they will purchase any feed grains necessary.

Current prices received for milk sold have been averaging $12.60 per 100 lb. Corn has been available for $2.50 per bushel. In order to provide some protection in the event that prices change radically, this projection was made using a planning price for milk of only $12.00 per 100 lb and the purchased corn was charged at $2.65 per bushel. Forages were valued at $50.00 per ton of hay equivalent.

doing this planning is that "things never go as planned, so why bother." However, this reason calls for more planning, not less. Everyone (dairyman and creditor) needs to know the consequences of not having things go as planned. This prediction can be formulated by changing some of the important assumptions used in the base budget (i.e., increasing the investment required or decreasing the price received for milk).

23.9 Evaluating Alternatives for Attaining Goals

Having decided on his goals, the dairyman must then determine how to attain them. He must appraise alternative development routes and come up with a plan that is technically feasible, and compatible with his managerial capacity and the present financial position of his business.[f] The following 4 tests may be helpful in evaluating the different alternatives:

Test 1. Is the proposed development route technically reasonable and desirable? Is it desirable from the standpoint of facility development, frequency of changes in equipment, and machinery lines, for example?

Test 2. Are the management requirements of the proposed development route consistent with available management capabilities? Is management capable of producing milk at the level of efficiency required?

Test 3. Will the proposed development plan exhibit reasonable profit and repayment potentials, particularly during the early development stage? For example, if an expansion is being planned, are enough cows in the herd producing at a level to meet all loan obligations when they first become due?

Test 4. Are the risks of the proposed plan consistent with the risk-bearing ability of the business? If the current business is in a weak financial position, then plans involving production and financial risks must be avoided.

If a major expansion (or addition of a dairy enterprise) is being planned, careful projections regarding production and investment must be made on an annual basis. These projections will prove beneficial to the dairyman in first convincing himself, and then in convincing others, that he has a workable plan.

Since additional outside resources are often needed to implement a plan, someone outside the business must become convinced to go along with the plan. At this point, the *when, where,* and *how* to present a plan becomes important.

In deciding *when* to contact a lender, remember to plan ahead. For sizable loans, as occur in expansion programs, start your request well in advance of signing a purchase order. *Where* do you sell the plan? If possible, invite the creditor to visit the farm and show him how the plan will fit into the ongoing business. In considering *how* to present the plan, be prepared to provide a summary of past production and financial performance as well as a detailed plan of your proposal.

Be prepared to explain how the loan will permit you to achieve your goals, the length of loan desired, the source of payment, and the timing of payments. Once the plan is implemented, keep your lender completely informed. Be punctual with payments, and if you are going to encounter difficulty, let your lender know. Meet with him at least once yearly, show him the year's results, and present a plan for next year's credit needs.

23.10 Managing Hired Labor

The key to a person's success in any endeavor is to be *qualified* and *motivated* to do the job. With dairy farms emerging as larger units, more dependence falls on hired labor. Many dairy farmers are relatively inexperienced in supervising hired labor and fail to recognize the motivational strategies necessary for good working relationships. Thus, turnover is large; supply is short. This situation has caused some dairymen to eliminate expansion as a possible alternative even when the profit potential appears attractive.

Satisfied dairy farm employees, when asked why they like to work for their present employer, provide many reasons. The gist of what they say is: "He treats his employees as human beings." This concept was expressed in many different fashions in several papers presented at the 1976 Large Dairy Herd Management Symposium held in Florida.[10] The points made were:

1. Employees appreciate being treated with dignity and understanding by their employers.
2. Each employee likes to know how he is getting along and appreciates this recognition.
3. Supervisors should make it clear what is expected of the employee; hired labor want to understand the rules.
4. Employees want training; they appreciate the boss pointing out ways to im-

prove, especially if also told why a different approach is a better one.

5. People like to be requested rather than ordered; they want to be informed now, not a month later, when a crisis has arisen.
6. Employees like to make use of their abilities; it is important that supervisors recognize each person's interest. It gives employees a sense of belonging.
7. Employees will not tolerate harsh words.
8. Employees should be encouraged to make suggestions, but most hired labor do not like to make decisions as they do not consider themselves a part of the management team.
9. Finally, they like to be productive, they like to be on a winning team, and they like to feel that they can advance.

On smaller farms with fewer hired workers, the situation seems to encourage a feeling of personal responsibility and a high level of independence. These workers, however, cannot be expected to work the same long hours, with feverish dedication, that the operator does if they receive no reward. Thus, many dairymen use an *incentive plan.*

Some basic principles in any wage incentive plan are: (1) the plan should be aimed at rewarding the employee for working in the interest of the employer; (2) the measures used should be influenced by the employee (i.e., percentage of profit is a poor incentive in that it is affected by too many factors not under the control of the employee); (3) computations used in determining the bonus should be simple and understood by both parties; (4) indicators of performance are preferably determined by a disinterested party or by objective criteria; and (5) bonus payments are made soon after the employee has met the specified requirements.

A list of incentive ideas is provided. These examples should serve only as a guide. Specific incentive plans must be adapted to each individual dairy farm.

A. Based on *milk shipped*
   1. ____¢ per 100 lb of milk shipped, with a base salary of $____ per month.
   2. ____¢ per 100 lb of milk sold per month above ____ hundredweight (specified amount).
   3. ____¢ per 100 lb milk sold, with the rate increased by ____¢ per each additional year employee stays, up to a maximum of ____¢ per hundredweight of milk.

B. Based on *production records*
   4. Annual bonus of $____ for each pound of herd average milkfat above ____ lb (i.e., 30¢ per lb for each pound above 400 lb average).
   5. Annual bonus of $____ for each 305-day record above ____ lb milk (i.e., $15 for each cow under employee's control exceeding 15,000 lb milk).

C. Based on *quality of milk* produced
   6. $____ per month each month the bacteria count is below ____ (specified level).
   7. $____ per month each month the somatic cell count is below ____ (specified level) or percent of cows positive to the California Mastitis Test is below ____% (specified percentage).

D. Based on youngstock program
   8. $____ for each heifer calf raised to 8 weeks of age.
   9. $____ annual bonus if percent calves born alive and successfully weaned exceeds ____% (specified level).

E. Based on reproductive performance
   10. ____¢ for each cow or heifer detected in heat and subsequently bred by technician.
   11. $____ for each cow diagnosed pregnant from a breeding performed between 40 and 120 days after calving.

One of the more difficult managerial duties for the farm operator employing hired labor is deciding when he should delegate duties, and when he should undertake the job himself. Some general guidelines are provided:

When to delegate

1. Delegate anything hired labor can do as well or better than the manager.
2. Delegate when management might do poorly because of a lack of time.

3. Delegate jobs that are repetitive, when a routine has been established.
4. Delegate when it costs too much for management to do the job.
5. Delegate low-risk jobs as a means of developing employees.
6. Delegate when management is spending too much time on operations and not enough time on managing.

When not to delegate

1. When no one else can do the job satisfactorily or when management's prestige is needed to get the job done.
2. When it is confidential beyond what subordinates can handle.
3. When a morale booster is needed and the manager sets the pace or pattern.
4. When it takes less time for management to do the job than it takes to delegate it, unless it is a recurring situation that hired labor will perform regularly.
5. Whenever management feels it is necessary to see problems or trends.

23.11 Investment in Facilities and Equipment

The major purpose of this section is to assemble, analyze, and project cost and price data when alternative housing, milking, waste handling, feeding, and youngstock raising systems are used. The authors are especially grateful to C. R. Hoglund for providing much of the base data to develop this material.[11]

Very few dairymen will be building a completely new dairy facility. To survive in the dairy business, however, many dairymen are faced with decisions regarding reorganization and modernization of their operations. It is important that they consider the impact of these improvements on labor use, investments, and costs and returns.

The planning information found in Appendix Table V-J was derived from studies conducted in the Lake States, the Northeast, and in the northern areas of the Corn Belt and Northern Plains regions, but is thought to be generally applicable in all parts of the United States. The input costs and product prices shown are approximately correct for 1983. A major problem in selection of specific cost and price data is that they become obsolete due to changing economic conditions. A slowing down of price increases appears to occur as a result of reduced demand and lower raw product costs. If one is using these data for planning purposes in the year 1984 or later, the user is advised to incorporate an appropriate annual inflation factor. Historically, inflation has resulted in an approximately 5% increase in prices annually.

An example of the investment, and annual costs associated with that investment, of a completely new 125-cow free stall operation is shown in Table 23.10. The input data used to derive these estimates were obtained from Appendix Table V-J.

The bottom line indicates that the investment totals $2,691 per cow. The estimated annual costs include interest, depreciation, repairs, and insurance. If the value of milk sold is $12 per 100 lb, it would take approximately 3600 lb of milk per cow to cover these costs plus the additional feed required for the higher production. Dairymen with cows producing 15,000 lb or more milk have a much better chance to pay off this kind of investment than do the dairymen obtaining only 10,000 lb milk per cow.

23.12 Summary

Three major problems facing the dairy industry today are: How do young dairymen get in business? How can the business grow in an orderly fashion to support 2 families? How do older dairymen get out of business? In most cases, these decisions involve the future career potential of the young dairymen and the economic security of those retiring from the dairy business.

Students hoping for a career in dairying may experience an entire cycle of: (1) being employed as a hired laborer, (2) developing a "working agreement" with only limited capital invested, (3) becoming a sole proprietor, and then (4) forming a partnership or (5) creating a corporation. The latter 2 will probably be multi-family operations that could involve expansion of facilities, more capital invested, and more cows in order to spread the added investment over more units. This adjustment usually calls for increased use of credit and a higher level of management skills.

TABLE 23.10 *Calculated Investment and Annual Costs for Buildings and Equipment Needed in a 125-Cow Dairy Enterprise*[a]

Item	*Investment*		*Annual cost*	
	$	$	$	$
Free-stall housing outside feeding				
Barn structure	44,560		6,016	
Concrete lot	17,610		2,202	
Free-stalls	9,805		2,256	
Water system and wiring	7,165		966	
		79,140		11,440
Milking parlor and milking system				
Milkroom and parlor	39,885		6,182	
Milking system	14,540		3,490	
Bulk tank	15,020		3,229	
		69,445		12,901
Waste handling				
Earthen basin	16,245		2,355	
Scraper, tractor	2,820		727	
Pump and agitator	4,425		1,185	
Underground pump system	6,440		1,726	
Liquid spreader	9,825		2,535	
		39,755		8,528
Feed storage and handling				
Bunker silo (15T per cow plus 20% loss and wastage)	34,790		4,348	
Mixing wagon, without scales	11,135		2,284	
Feed bunk, single side	5,920		1,006	
Hay storage (3T hay per cow plus 50T bedding)	30,190		4,076	
		82,035		11,714
Miscellaneous facilities				
Holding corral	7,410		1,000	
Powered crowd gate	1,775		373	
Hospital area, cold barn	3,810		515	
		12,995		1,888
Calf and youngstock facilities				
Naturally ventilated calf barn and pole barn for older heifers		53,000		7,950
Total		336,370		54,421
Investment per cow and replacement		2,691		
Annual cost per cow and replacement				435

[a]Example developed for a 125-cow free-stall barn with outside feeding, standard herringbone parlor without add-ons, 6-month storage of manure in an earthen basin with daily scraping, corn silage stored in bunker silo and fed by a mixing wagon, hay stored in pole barn adjacent to feed bunk, and youngstock raised in naturally ventilated calf barn with older heifers raised in an open pole barn.

Source: Adapted from Hoglund, C. R.: Dairy Systems Analysis Handbook. Mich. State Univ. Agr. Econ. Report No. 300, 1976; and adjusted to 1983 prices by correcting for inflation, using source data from: Agricultural Prices, Statistical Reporting Service, USDA, Publication Pr 1–3, 1983.

Profitable dairy enterprises are characterized by high milk yields per cow, above average volume of business, and lower than average costs for feed, labor, and facilities. In achieving these goals, the owner or operator can set the goals and plan the programs, but his success depends not only on his understanding of the principles involved in breeding, feeding, and milking the dairy herd, but also on the daily application of sound business management principles.

Review Questions

1. Outline a 10-year plan that reflects how you would hope to establish your dairy farm business. Set goals at periodic intervals and show

that your plan is reasonable and attainable.
2. List the factors that would be important to you if you were a hired employee on a dairy farm. Also list specific incentive plans that might be helpful in making you a more productive hired employee.
3. Describe the thought process required for you to determine whether dairy farming is your best career alternative and how you would determine what herd size and production level are necessary to assure success.
4. Cite an example from your experience, if possible, of the strengths and weaknesses of the dairy farm business. Include in your diagnosis the: (a) milking herd enterprise, (b) replacement youngstock enterprise, (c) cropping enterprise, (d) use of labor, (e) physical facilities used, and (f) capital investment required.
5. Describe a dairy farm from your experience, if possible, or a hypothetical farm, if need be, and draw plans "on paper" for a herd 50% larger. Then, use Appendix Tables V-H, V-I, and V-J to determine the labor requirements and added capital required to permit the expanded business to be successful.

References

1. Thomas, K. H., Weigle, R. N., and Hawkins, R. O.: Farm financial decision making: Part 1, A perspective; Part 2, Where am I?; Part 3, Where do I want to be?; Part 4, How can I best get there from here? North Central Regional Extension Publication 34, 1973.
2. Thomas, K. H., and Boehlje, M.: Farm business arrangements: which one for you? North Central Regional Extension Publication 50, 1976.
3. Thomas, K. H., and Freeman, M. L.: Prefarm partnership arrangements, Univ. of Minnesota Agr. Ext. Ser. Publ. FM-220PFP, 1975.
4. Hasbargen, P. R.: Farm record system: what records should I keep? Univ. of Minnesota Agr. Ext. Ser. Publ. FM-100, 1971.
5. Willett, G. S., Howard, W. T., and Weigle, R. N.: Guidelines for analyzing the dairy farm business. Univ. of Wisconsin Agr. Ext. Ser. Publ. A2504, 1973.
6. Willett, G. S., Howard, W. T., and Weigle, R. N.: Planning for dairy farm expansion. Univ. of Wisconsin Agr. Ext. Ser. Publ. A2449, 1972.
7. Hoglund, C. R.: The U.S. dairy industry: today and tomorrow. Michigan State Univ. Agr. Exp. Sta. Rept. 275, 1975.
8. Speicher, J. A.: Personal communication.
9. Thomas, K. H., Cuykendall, C. H., Hawkins, R. O., and Penning, W. S.: User's guide for FINLRB: a computerized long-range budgeting procedure. Univ. of Minnesota Agr. Ext. Ser. Econ. Info. Rept. R72-10, 1973.
10. Speicher, J. A., Appleman, R. D., Murrill, F. D., Holt, J., Wells, R. C., and Holtz, E. W.: Personnel management—6 papers. Proceedings, Large Dairy Herd Management Symposium. Univ. of Florida, Gainesville, 1976.
11. Hoglund, C. R.: Dairy systems analyses handbook. Michigan State Univ. Agr. Econ. Rept. No. 300, 1976.

Suggested Additional Reading

Harl, N. E.: *Farm Estate and Business Planning.* Agri Business Publication, Skokie, Illinois, 1977.

Wilcox, C. J. *et. al. Large Dairy Herd Management.* University of Florida Press, Gainesville, Fla. 1978.

Dairy Housing II. American Society of Agricultural Engineers, St. Joseph, Mich. 1983.

Appendix Tables

Table I-A *Dairy Cattle Breed Registry Associations in the United States*

1. Ayrshire Breeders Association, 2 Union Street, Brandon, Vermont 05733
2. The Brown Swiss Cattle Breeders Association, Box 1038, Beloit, Wisconsin 53511
3. The American Guernsey Cattle Club, P.O. Box 27410, Columbus, Ohio 43227
4. Holstein-Friesian Association of America, South Main Street, Brattleboro, Vermont 05301
5. The American Jersey Cattle Club, P.O. Box 27310, Columbus, Ohio 43227
8. Purebred Dairy Cattle Association, Peterborough, New Hampshire 03458
6. American Milking Shorthorn Society, 1722-JJ South Glenstone Avenue, Springfield, Missouri 65804
7. Red and White Dairy Cattle Association, Crystal Spring, Pennsylvania 15536

TABLE I-B *Participants in the DHIA Program*

Private Sector	*Government* — *State*	*Government* — *Federal*
A. Dairymen 1. National Dairy Herd Improvement Association, Inc. 2. State Dairy Herd Improvement Associations a. Central Testing Labs for Milk Components and Somatic Cell Counts 3. Local Dairy Herd Improvement Associations a. DHIA Supervisors B. Dairy Records Processing Centers C. Breed Organizations 1. Purebred Dairy Cattle Association 2. Breed Associations D. Artificial Insemination Industry 1. National Association of Animal Breeders 2. Artificial Insemination Organizations E. American Dairy Science Association	A. Land-Grant Universities 1. Cooperative Extension Service a. State Extension Dairymen b. Local Extension Specialists 2. Research 3. Teaching 4. Dairy Records Processing Centers	A. USDA—Federal Extension 1. Program Leader—Dairy Production B. USDA—Agriculture Research Service 1. Animal Improvement Programs Laboratory

The NCDHIP Policy Board

Purpose	*Membership*	
To provide a stronger program base for formulating policy, rules, and regulations and to administer their effective coordination and application to the DHIA program.	National Dairy Herd Improvement Association, Inc.	- 5
	USDA—Agricultural Research Service	- 1
	USDA—Extension	- 1
	State Cooperative Extension Service	- 2
	Purebred Dairy Cattle Association	- 1
	National Association of Animal Breeders	- 1
	Extension Committee on Organization and Policy	- 1

Table I-C *Freeze Branding as a Method of Animal Identification*

I. Objective

The basic goal of freeze branding is to provide permanent identification with a minimum of discomfort and damage to the animal.

II. Equipment and Materials Needed

1. Coolant (either A or B)
   A. Dry ice in either 95% ethyl, methyl, or isopropyl alcohol.
   B. Liquid nitrogen.
2. Insulated container for the coolant
3. Set of copper or high-quality bronze alloy branding irons, with sufficient mass and facing. Four-inch irons are recommended for yearlings and older cattle, 3-inch irons for cattle 6 to 12 months of age, 2-inch irons for calves up to 6 months. The irons should have a rounded face.
4. Goggles and gloves for protection of eyes and hands while handling liquid nitrogen or dry ice.
5. Set of cattle clippers with blades to clip either coarse or fine.
6. Squeeze bottle of alcohol.
7. Chute or other adequate means of restraining the animal.
8. Grooming brush.

III. Procedures for Branding with Dry Ice Plus Alcohol

1. Pour enough alcohol into an insulated container to completely immerse the head of the branding irons. Slowly add chipped dry ice until the alcohol is chilled to −79°C. To maintain the correct temperature, it is necessary to have ice in the solution at all times.
2. Place the irons in the solution. At first vigorous boiling will occur, and will continue until the temperature of the irons and the coolant solution is equal. When this occurs, the irons are ready for use. When the above conditions are met, proceed as follows:
   A. Restrain the animal.
   B. Select the brand site and clip the hair from the area to be branded as *close* as *possible* (fine clip).
   C. Brush the clipped area to remove loose hair, dirt and dandruff. If the skin is dirty, wash with an alcohol-soaked cloth.
   D. Soak the area to be branded with alcohol immediately before applying the iron. Repeat for each digit applied.
   E. Select the proper iron, shake off excess coolant solution and apply to the brand site. Exert enough pressure to assure firm contact between the face of the iron and skin. Maintain this pressure for the desired exposure time.
   F. The following times for each of the specified age groups have resulted in legible brands on dairy and beef cattle.

Age	Time	
	Dairy	Beef
Birth through 1 month	10 seconds	15 seconds
2–3 months	15 seconds	20 seconds
4–8 months	20 seconds	25 seconds
9–18 months	25 seconds	30 seconds
over 18 months	30 seconds	35 seconds

   G. For white animals, allow an additional 10 to 15 seconds of exposure. This will destroy the hair and result in a "bald" or "bare" brand.
3. Advantages of the Dry Ice method:
   A. Results are more consistent than those obtained with liquid nitrogen.
   B. Overexposures of 2 to 5 seconds are not critical.
   C. Digits applied are legible from branding until the regrowth of white hair.

TABLE I-C *Freeze Branding as a Method of Animal Identification (Continued)*

4. Disadvantages:
   A. Dry Ice is difficult to obtain in many areas.
   B. It is difficult to maintain the proper temperature.
   C. The coolant solution has a tendency to absorb moisture from the atmosphere; this reduces the effectiveness of the solution.
   D. Requires 60 to 90 seconds to re-chill the iron after use.
   E. Requires close clipping, which increases time requirements.

IV. Procedures for Branding with Liquid Nitrogen on Coarse Clipped Areas

1. Place liquid nitrogen in a *well*-insulated container with an open top just large enough to accommodate the irons. Maintain enough nitrogen in the tank to cover the head of the branding irons at all times.
2. Carefully place the irons in the tank to avoid being splashed with boiling nitrogen. As with dry ice, there will be vigorous boiling until the temperature of the irons is the same as the nitrogen. When this occurs the irons are ready for use. Then proceed as follows:
   A. Steps A, C, D, E, and G are similar to those quoted for the dry ice plus alcohol procedure.
   B. Select the brand site and clip the hair from the area to be branded with a coarse clip blade (83–84 AU) thus preventing the removal of most of the underfur. The additional hair will prevent excessive follicular and skin damage.
   C. The following exposure times for each of the specified age groups are recommended for liquid nitrogen.

Age	Time: Dairy	Time: Beef
Birth through 1 month	5 seconds	10 seconds
2–5 months	7 seconds	12 seconds
6–9 months	10 seconds	15 seconds
10–12 months	12 seconds	17 seconds
13–18 months	15 seconds	20 seconds
>18 months	20 seconds	25 seconds

3. Advantages of the liquid nitrogen (coarse clip) method:
   A. Liquid nitrogen is more readily available than dry ice in most areas.
   B. It is more convenient and easier to use.
   C. Irons can be re-chilled in a matter of seconds.
   D. Special clippers and/or blades are not required to prepare the brand site.
   E. Less time consuming than dry ice method because of shorter exposure times.
   F. Digits applied are legible from branding until the regrowth of white hair.
4. Disadvantages:
   A. Time exposures in excess of those recommended will result in excessive follicular and skin damage.
   B. Liquid nitrogen containers designed for freeze branding are expensive.
   C. In some areas, the cost of liquid nitrogen is high.
   D. Liquid nitrogen can be dangerous if not handled with care.

V. Procedures for Branding with Liquid Nitrogen on Unclipped Areas

1. Repeat steps A and B as outlined in the previous liquid nitrogen procedure.
2. Then proceed as follows:
   A. Select brand site and thoroughly brush the hair coat to remove as much loose hair, dirt and dandruff as possible.
   B. Thoroughly soak the area to be branded with enough alcohol to penetrate the hair and underfur immediately before the iron is applied. Repeat for each digit applied.
   C. Apply the iron, exerting considerably more pressure than would be used on a clipped sur-

Table I-C *Freeze Branding as a Method of Animal Identification (Continued)*

face and maintain for the desired exposure time. The additional pressure is necessary to assure an efficient transfer of heat through the hair and underfur.

D. The recommended exposure times by age based on current knowledge are as follows:

Age	*Time*[1] *Dairy*
Birth through 1 month	*
2–3 months	15 seconds
4–6 months	20 seconds
7–12 months	25 seconds
>12 months	30 seconds

[1]Time exposures for beef have not been firmly established.
*This approach is not recommended for calves through one month of age.

E. Allow an additional 15 to 20 seconds on white animals to produce a "bald" brand.

3. Advantages of the liquid nitrogen (unclipped) method:
   A. Clipping the brand site before branding is not required.
   B. Regrowth of white hair is more profuse, thus increasing legibility.
   C. Exposure times in excess of those recommended are not as critical as in methods requiring clipping.
4. Disadvantages:
   A. Exposure times required on unclipped areas are more than doubled, thus decreasing the number of animals that can be branded in one day.
   B. It requires twice as much alcohol to soak the brand site prior to branding as in methods where clipping is required.
   C. The digits applied are not legible until the white hair grows back in 6 to 8 weeks.
   D. The additional pressure required over a longer period of time can affect operator performance.

VI. Series of Events Which Follow a Freeze Brand

1. Skin is frozen and indented in the form of the brand applied.
2. Skin thaws out in 2 to 3 minutes after branding.
3. The branded area begins to redden, and edema develops.
4. The edema will persist for 24 to 48 hours, depending on exposure time.
5. When the reddening and edema recede, the area will appear dry and scruffy.
6. A scab will form and persist for 3 or 4 weeks.
7. When the scab sloughs off, varying amounts of hair and skin are lost.
8. White hair will begin to appear 6 to 10 weeks later, depending on the stage of the natural hair growth cycle when the brand was applied.
9. The brand will remain legible until white hair appears.

VII. Advantages of Freeze Branding

1. Less pain to the animal.
2. Less hide damage.
3. Improved legibility.

Source: N. W. Hooven, Jr., USDA.

TABLE II-A *Factors for Projecting Lactation Records with Less than 305 Days in Milk to a 305-day Basis and Weighting Projected Records in Genetic Evaluations*
TABLE II-A-1 *Factors for Projecting Lactation Records with Less than 305 Days in Milk to a 305-day Basis*
Milk Factors for First Lactation Holsteins in the Northern Region

Calving Season	*Days in Milk*	*Sample-day yield*		*Herd average*	
		Intercept	*Slope*	*Intercept*	*Slope*
DEC–FEB	7– 55	0.453	0.00296	1.314	−0.01538
	56–105	0.570	0.00085	0.599	−0.00238
	106–155	0.536	0.00117	0.787	−0.00416
	156–205	0.547	0.00148	4.866	−0.03418
	206–255	0.376	0.00232	9.545	−0.05700
	256–305	0.502	0.00182	−8.621	0.01424
MAR–MAY	7– 55	0.367	0.00364	1.487	−0.01787
	56–105	0.511	0.00101	0.578	−0.00134
	106–155	0.355	0.00250	0.981	−0.00518
	156–205	0.420	0.00219	12.811	−0.06840
	206–255	0.314	0.00271	14.562	−0.07694
	256–305	0.797	0.00082	−10.551	0.02154
JUN–AUG	7– 55	0.437	0.00312	1.268	−0.01269
	56–105	0.466	0.00260	0.844	−0.00499
	106–155	0.583	0.00149	0.699	−0.00360
	156–205	0.734	0.00088	5.054	−0.03293
	206–255	0.508	0.00198	12.628	−0.06988
	256–305	0.755	0.00101	−3.981	−0.00475
SEP–NOV	7– 55	0.494	0.00286	1.234	−0.01355
	56–105	0.578	0.00134	0.661	−0.00314
	106–155	0.620	0.00094	0.619	−0.00274
	156–205	0.737	0.00039	7.600	−0.03672
	206–255	0.278	0.00263	24.020	−0.11682
	256–305	0.403	0.00214	−12.652	0.02699

Example of procedure to project to 305 days in milk (DIM) the milk yield of a first lactation Holstein cow calving in the northern region in January. The cow gave 8,000 lb milk in 120 days, milking in a herd that averaged 16,000 lb. Her last test day milk yield was 40 lb.
Basic Formula:
Projected 305-day yield ($\hat{Y}_{305}$) = Actual yield (Y_{DIM}) + [Estimated average daily yield] · [Days remaining to 305]
Application of formula:

$$\hat{Y}_{305} = Y_{DIM} + \left[(\text{intercept} + \text{slope} \cdot \text{DIM}) \cdot \text{Test day yield} + (\text{intercept} + \text{slope} \cdot \text{DIM}) \cdot \frac{\text{herd average}}{1{,}000}\right] \cdot (305 - \text{DIM})$$

$$= 8{,}000 + \left[(0.536 + 0.00117 \cdot 120) \cdot 40 + (0.787 - 0.00416 \cdot 120) \cdot \frac{16{,}000}{1{,}000}\right] \cdot (305 - 120)$$

$$= 8{,}000 + [(0.6764 \cdot 40 + (0.2878) \cdot 16] \cdot 185$$

$$= 8{,}000 + [27.056 + 4.6048] \cdot 185$$

$$= 8{,}000 + [31.6608] \cdot 185$$

$$= 8{,}000 + 5{,}857$$

$$= 13{,}857$$

TABLE II-A-2 *Factors for Weighting Projected Records Used in Genetic Evaluations**

Lactation number	*Minimum days in milk* 15	46	76	107	137	168	198	229	259	290
	All milkings weighed on test day									
1.....	.47	.58	.67	.75	.81	.87	.92	.96	.99	1.00
≥2.....	.40	.53	.64	.74	.82	.89	.94	.98	.99	1.00
	AM-PM test plans									
1.....	.36	.49	.60	.69	.78	.85	.90	.94	.97	.98
≥2.....	.30	.44	.57	.68	.77	.84	.90	.95	.97	.98

*Terminated incomplete records for which all milkings are weighed on test day will have a weight of at least .70 and those from AM-PM test plans, a weight of at least .65 to insure that records from culled cows retain enough weight to impact genetic evaluations appropriately.
Source: Wiggons, G. R.: Length weights for AP and Standard DHI. Mimeograph, 1984.

TABLE II-B *Factors for Reducing 305-day, Age-corrected Records to a Twice-a-day Milking Basis*

	Factor for 3-times-a-day milking			*Factor for 4-times-a-day milking*		
Number of days milked	*2 to 3 years of age*	*3 to 4 years of age*	*4 years of age and over*	*2 to 3 years of age*	*3 to 4 years of age*	*4 years of age and over*
5 to 15	.99	.99	.99	.98	.99	.99
16 to 25	.98	.99	.99	.97	.98	.98
26 to 35	.98	.98	.98	.96	.97	.97
36 to 45	.97	.98	.98	.95	.96	.96
46 to 55	.97	.97	.97	.94	.95	.96
56 to 65	.96	.97	.97	.93	.94	.95
66 to 75	.95	.96	.96	.92	.93	.94
76 to 85	.95	.95	.96	.91	.92	.93
86 to 95	.94	.95	.96	.90	.91	.93
96 to 105	.94	.94	.95	.89	.91	.92
106 to 115	.93	.94	.95	.88	.90	.91
116 to 125	.92	.93	.94	.87	.89	.90
126 to 135	.92	.93	.94	.87	.88	.90
136 to 145	.91	.93	.93	.86	.88	.89
146 to 155	.91	.92	.93	.85	.87	.88
156 to 165	.90	.92	.93	.84	.86	.88
166 to 175	.90	.91	.92	.83	.85	.87
176 to 185	.89	.91	.92	.82	.85	.86
186 to 195	.89	.90	.91	.82	.84	.86
196 to 205	.88	.90	.91	.81	.83	.85
206 to 215	.88	.89	.90	.80	.83	.85
216 to 225	.87	.89	.90	.79	.82	.84
226 to 235	.87	.88	.90	.79	.81	.83
236 to 245	.86	.88	.89	.78	.81	.83
246 to 255	.86	.88	.89	.77	.80	.82
256 to 265	.85	.87	.88	.77	.79	.82
266 to 275	.85	.87	.88	.76	.79	.81
276 to 285	.84	.86	.88	.75	.78	.80
286 to 295	.84	.86	.87	.75	.78	.80
296 to 305	.83	.85	.87	.74	.77	.79

Source: Kendrick, J. F.: Standardizing dairy herd improvement association records in proving sires. ARS-52-1. January 1955.

Table II-C *A Partial Set of National Holstein Age and Month-of-Calving Milk Adjustment Factors for Cows in the United States*
Linear interpolation between ages at calving is satisfactory when using this table. The complete national set of factors is contained in the source reference.

Age (Months)	Jan.	Feb.	Mar.	Apr.	May	Jun.	Jul.	Aug.	Sep.	Oct.	Nov.	Dec.
22	1.31	1.32	1.32	1.33	1.35	1.36	1.39	1.39	1.35	1.32	1.31	1.30
24	1.27	1.28	1.28	1.29	1.30	1.32	1.34	1.34	1.31	1.28	1.27	1.26
26	1.23	1.24	1.25	1.25	1.27	1.28	1.31	1.30	1.27	1.25	1.23	1.23
28	1.21	1.21	1.22	1.22	1.24	1.25	1.28	1.27	1.25	1.22	1.21	1.20
30	1.19	1.19	1.19	1.20	1.21	1.23	1.25	1.25	1.22	1.20	1.19	1.18
32	1.17	1.17	1.17	1.18	1.19	1.21	1.23	1.23	1.21	1.18	1.17	1.16
34	1.15	1.15	1.15	1.16	1.17	1.19	1.22	1.22	1.19	1.16	1.15	1.15
36	1.13	1.13	1.13	1.14	1.15	1.18	1.21	1.21	1.18	1.15	1.13	1.13
38	1.10	1.10	1.11	1.12	1.13	1.16	1.19	1.19	1.16	1.13	1.11	1.11
40	1.09	1.08	1.09	1.10	1.11	1.14	1.18	1.18	1.14	1.11	1.09	1.09
42	1.07	1.07	1.07	1.08	1.10	1.13	1.16	1.16	1.13	1.09	1.08	1.07
44	1.05	1.05	1.06	1.07	1.09	1.12	1.15	1.15	1.11	1.08	1.06	1.06
46	1.04	1.04	1.05	1.06	1.08	1.10	1.14	1.13	1.10	1.06	1.05	1.04
48	1.03	1.03	1.04	1.05	1.07	1.09	1.12	1.12	1.09	1.05	1.04	1.03
50	1.02	1.02	1.03	1.04	1.06	1.08	1.11	1.11	1.08	1.04	1.03	1.02
52	1.01	1.01	1.02	1.03	1.05	1.07	1.10	1.10	1.07	1.03	1.02	1.01
54	1.00	1.01	1.01	1.02	1.04	1.06	1.09	1.09	1.06	1.02	1.01	1.00
56	1.00	1.00	1.00	1.02	1.03	1.06	1.08	1.08	1.05	1.02	1.00	1.00
58	0.99	.099	1.00	1.01	1.03	1.05	1.08	1.08	1.04	1.01	1.00	0.99
60	0.99	0.99	0.99	1.01	1.02	1.05	1.08	1.07	1.04	1.01	0.99	0.99
62	0.98	0.98	0.99	1.00	1.02	1.04	1.07	1.07	1.04	1.00	0.99	0.99
64	0.98	0.98	0.98	1.00	1.01	1.04	1.07	1.07	1.03	1.00	0.98	0.98
66 TO 71	0.97	0.97	0.98	0.99	1.01	1.04	1.06	1.06	1.03	1.00	0.98	0.98
72	0.97	0.97	0.98	0.99	1.01	1.03	1.06	1.06	1.03	1.00	0.98	0.98
74	0.97	0.97	0.98	0.99	1.01	1.03	1.06	1.06	1.03	0.99	0.98	0.97
76 TO 80	0.97	0.97	0.97	0.99	1.00	1.03	1.06	1.06	1.02	0.99	0.97	0.97
82	0.97	0.97	0.97	0.99	1.00	1.03	1.06	1.06	1.03	0.99	0.98	0.97
84	0.97	0.97	0.98	0.99	1.01	1.03	1.06	1.06	1.03	0.99	0.98	0.97
86	0.97	0.97	0.98	0.99	1.01	1.03	1.06	1.06	1.03	1.00	0.98	0.98
87 TO 93	0.97	0.97	0.98	0.99	1.01	1.04	1.06	1.06	1.03	1.00	0.98	0.98
94 TO 99	0.97	0.97	0.98	0.99	1.01	1.04	1.07	1.07	1.03	1.00	0.98	0.98
100 TO 110	0.98	0.98	0.99	1.00	1.02	1.05	1.07	1.07	1.04	1.01	0.99	0.99

Source: Norman, H. D., Miller, P. D., McDaniel, B. T., Dickinson, F. N., and Henderson, C. R.: USDA-DHIA factors for standardizing 305-day lactation records for age and month of calving. ARS-NE-40. September 1974.

Table II-D *Calculation of a Simplified Selection Index*

Assume that a dairyman has decided to emphasize 3 traits in his selection program; milk yield, milking speed, and dairy character (sharpness). He has determined that improvement of these 3 traits will result in the greatest overall profit to him. He has checked in Table 6.3 and found that the genetic correlations of milking speed and dairy character with first lactation yield are such that it is reasonable to include them together in an index, and he has mastitis well under control. Therefore, he wishes to evaluate the cows in his herd for these traits using a selection index. He does this using the following 5-step procedure:

(1) He determines the heritability for each trait from Table 6.2.

Heritability of milk yield expressed as a deviation from herdmates	= 0.25
Heritability of milking speed	= 0.30
Heritability of dairy character	= 0.20

TABLE II-D *Calculation of a Simplified Selection Index (Continued)*

(2) He determines the phenotypic standard deviation (S.D.) for each trait from reference material.

S.D. of milk yield expressed as a deviation from herdmates = 2400 lb
S.D. of milking speed on basis of a 3-point scale
(1 = slow, 2 = average, 3 = fast) = 0.6
S.D. of dairy character on basis of a 3-point scale
(1 = thick, 2 = moderate, 3 = sharp) = 0.5

(3) He decides on the economic value that he should place on one standard deviation's worth of improvement in each trait.

One S.D. of milk (deviation from herdmates) at $10.00 per cwt is worth $240.00 per lactation.

If, on the average, the difference in milking time between a slow and medium or between a medium and fast cow is 2 minutes per milking, one S.D. would be worth about 2.4 minutes per day, or 732 minutes over a 305-day lactation. At $4.00 per hour for labor, therefore, one S.D. in milking speed would be valued at $48.80 in labor cost per lactation.

He believes that it is worth $20 per cow to him to improve the dairy character of his herd by one standard deviation, which is 0.5 of the way between a thick and moderate cow or a moderate and sharp cow.

(4) Now he needs to determine each cow's P value for each trait in the index. He does this by expressing each cow's phenotypic value as a deviation from the breed mean and dividing this deviation by the standard deviation from 2 above. This expresses the cow's phenotypic value in standard deviation units above or below the breed mean. Let's assume that a hypothetical cow to be evaluated has the following phenotypes for the traits to be included in the index: A USDA-DHIA Cow Index Value (or an estimated transmitting ability by some other method) of +600 lb milk; is categorized as a fast-milking cow; and is categorized as being thick in dairy character.

$$\text{Therefore } P_1 \text{ is } \frac{600}{2400} = .25;$$

$$P_2 \text{ is } \frac{3-2}{0.6} = 1.67, \text{ and;}$$

$$P_3 \text{ is } \frac{1-2}{0.5} = -2.0$$

(5) The final step in the selection index evaluation of each cow is to solve the index equation using the values determined in steps 1 to 4. Note that the heritability of milk yield was accounted for in the calculation of the USDA-DHIA Cow Index. Therefore, the heritability is not used in that term in the selection index equation, and the first term is simply v_1P_1.
The index equation for the hypothetical cow is:

$$\begin{aligned} I &= v_1P_1 + v_2h_2^2P_2 + v_3h_3^2P_3 \\ &= (240.00)(0.25) + (48.80)(0.30)(1.67) + (20.00)(0.20)(-2.0) \\ &= 60.00 + 24.45 - 8.00 \\ &= 76.45 \end{aligned}$$

Therefore, the hypothetical cow's index value is 76.45. Note that since the hypothetical cow is above average in milk yield and milking speed, these traits receive plus weighting in the index, but dairy character receives a minus weight because the cow is below average. Other cows in the herd would be evaluated in the same manner, and then those cows with the highest index scores would be selected for breeding.

TABLE II-E *Formula for Calculating $PD_\$$ and $CI_\$$ Based on Yield of Milk and Milk Components*

The formula is shown in terms of $PD_\$$. The same formula can be used for $CI_\$$ by substituting CI for PD.

The formula is:

$$PD_\$ = PD_M(M_\$ - F_\$B_F - P_\$B_P - S_\$B_S) + PD_F F_\$ + PD_P P_\$ + PD_S S_\$$$

where subscripts M, F, P, and S = milk, fat, protein, and SNF.

$M_\$$ = price of a pound of milk at base tests (3.5% fat, 3.2% protein, 8.5% SNF) ($.1240).
$F_\$$ = price for a pound of fat ($1.71).
$P_\$$ = price for a pound of protein ($1.26).
$S_\$$ = price for a pound of SNF ($.88).
B = base test for each component designated by the subscripts (0.035, 0.032, and 0.085).

Example:

Assume milk is paid for on the basis of volume with differentials for fat and SNF as follows: $12.40 per cwt ($.1240/lb) for milk with 3.5% fat and 3.2% protein; a fat differential of $.171 per 0.1%, and a protein differential of $.126 per 0.1%. The bull whose $PD_\$$ is being calculated has the following PDs for yield: $PD_M = +1{,}000$ lb; $PD_F = +32$ lb; $PD_P = +30$ lb. Note that this bull has no PD_S and the SNF in the milk has no direct market value in this pricing formula. Then,

$$\begin{aligned} PD_\$ &= 1{,}000\ (.1240 - 1.71 \times .035 - 1.26 \times .032 - 0 \times .085) + 32 \times 1.71 + 30 \times 1.26 + 0 \\ &= 1{,}000\ (.1240 - .05985 - .04032 - 0) + 54.72 + 37.80 + 0 \\ &= 1{,}000\ (-.02383) + 92.52 \\ &= \$68.69 \end{aligned}$$

TABLE III-A *Daily Nutrient Requirements of Dairy Cattle*

Body weight (lb)	Breed size, age (wk)	Daily gain (lb)	Feed DM (lb)	Feed energy			Crude protein (lb)	Minerals		Vitamins	
				NE_m (Mcal)	NE_g (Mcal)	TDN (lb)		Ca (lb)	P (lb)	A (1000 I.U.)	D (I.U.)
Growing Dairy Heifer and Bull Calves Fed Only Milk											
55	S-1	0.7	1.00	0.85	0.53	1.20	0.25	0.013	0.009	1.0	165
65	S-3	0.8	1.15	0.94	0.63	1.38	0.28	0.014	0.010	1.2	200
93	L-1	0.9	1.38	1.25	0.70	1.66	0.33	0.018	0.011	1.8	280
106	L-3	1.1	1.65	1.36	0.90	1.98	0.40	0.020	0.013	1.9	300
Growing Dairy Heifer (F) and Bull (M) Calves Fed Mixed Diets											
100		0.6	2.7	1.35	0.52	2.08	0.31	0.018	0.013	1.9	300
100(F)	S-10	0.8	2.8	1.35	0.69	2.25	0.36	0.020	0.013	1.9	300
100(M)	S-10	1.0	2.8	1.35	0.87	2.42	0.40	0.022	0.014	1.9	300
100(F)	L-3	1.2	2.8	1.35	1.05	2.55	0.44	0.024	0.015	1.9	300
100(M)	L-3	1.4	2.8	1.35	1.22	2.67	0.48	0.027	0.015	1.9	300
100		1.6	2.8	1.35	1.40	2.80	0.58	0.028	0.016	1.9	300
150		0.8	4.0	1.82	0.70	2.89	0.49	0.024	0.015	2.9	450
150(F)	S-19	1.0	4.1	1.82	0.88	3.08	0.55	0.026	0.015	2.9	450
150(M)	S-18	1.2	4.1	1.82	1.06	3.27	0.59	0.028	0.016	2.9	450
150(F)	L-9	1.4	4.1	1.82	1.24	3.42	0.64	0.031	0.017	2.9	450
150(M)	L-8	1.6	4.1	1.82	1.42	3.58	0.68	0.033	0.018	2.9	450
150		1.8	4.1	1.82	1.59	3.70	0.73	0.034	0.018	2.9	450
200		0.8	5.3	2.26	0.73	3.53	0.65	0.031	0.017	3.8	600
200		1.0	5.4	2.26	0.92	3.76	0.71	0.032	0.018	3.8	600
200(F)	S-26	1.2	5.4	2.26	1.10	3.94	0.75	0.034	0.018	3.8	600
200(M)	S-24	1.4	5.4	2.26	1.29	4.12	0.80	0.036	0.019	3.8	600
200(F)	L-14	1.6	5.4	2.26	1.47	4.27	0.85	0.038	0.020	3.8	600
200(M)	L-12	1.8	5.4	2.26	1.66	4.45	0.90	0.040	0.021	3.8	600
200		2.0	5.4	2.26	1.84	4.64	0.94	0.041	0.022	3.8	600
Growing Dairy Heifers											
300		0.8	7.9	3.07	0.83	4.82	0.88	0.034	0.021	5.8	900
300	S-38	1.0	7.9	3.07	1.04	5.06	0.92	0.036	0.022	5.8	900
300		1.2	7.9	3.07	1.25	5.25	0.96	0.037	0.023	5.8	900
300		1.4	7.9	3.07	1.46	5.48	1.00	0.039	0.024	5.8	900
300	L-23	1.6	7.9	3.07	1.66	5.70	1.04	0.041	0.025	5.8	900

300		1.8	7.9	3.07	1.87	5.88	1.09	0.043	0.026	5.8	900
400		0.8	10.5	3.81	0.96	6.09	1.15	0.040	0.026	7.7	1,200
400	S-52	1.0	10.5	3.81	1.20	6.35	1.19	0.042	0.028	7.7	1,200
400		1.2	10.5	3.81	1.44	6.63	1.22	0.043	0.029	7.7	1,200
400		1.4	10.5	3.81	1.68	6.88	1.26	0.044	0.030	7.7	1,200
400	L-32	1.6	10.5	3.81	1.92	7.09	1.30	0.045	0.030	7.7	1,200
400		1.8	10.5	3.81	2.16	7.34	1.33	0.046	0.031	7.7	1,200
500		0.8	12.7	4.50	1.08	7.19	1.35	0.046	0.032	9.6	1,500
500	S-67	1.0	12.7	4.50	1.35	7.52	1.38	0.047	0.033	9.6	1,500
500		1.2	12.7	4.50	1.62	7.82	1.41	0.048	0.034	9.6	1,500
500		1.4	12.7	4.50	1.89	8.10	1.44	0.049	0.035	9.6	1,500
500	L-41	1.6	12.7	4.50	2.16	8.38	1.47	0.050	0.036	9.6	1,500
500		1.8	12.7	4.50	2.43	8.66	1.50	0.051	0.037	9.6	1,500
600		0.8	14.7	5.16	1.18	8.23	1.52	0.049	0.034	11.5	1,800
600	S-81	1.0	14.7	5.16	1.47	8.60	1.53	0.049	0.035	11.5	1,800
600		1.2	14.7	5.16	1.76	8.89	1.56	0.050	0.036	11.5	1,800
600		1.4	14.7	5.16	2.06	9.22	1.59	0.051	0.037	11.5	1,800
600	L-50	1.6	14.7	5.16	2.35	9.56	1.61	0.052	0.038	11.5	1,800
600		1.8	14.7	5.16	2.65	9.85	1.64	0.053	0.039	11.5	1,800
700		0.8	16.2	5.79	1.26	9.07	1.64	0.050	0.036	13.5	2,100
700	S-95	1.0	16.4	5.79	1.57	9.51	1.67	0.051	0.037	13.5	2,100
700		1.2	16.4	5.79	1.88	9.84	1.69	0.052	0.038	13.5	2,100
700		1.4	16.4	5.79	2.20	10.18	1.72	0.053	0.039	13.5	2,100
700	L-59	1.6	16.4	5.79	2.51	10.50	1.74	0.054	0.040	13.5	2,100
700		1.8	16.4	5.79	2.83	10.84	1.76	0.054	0.041	13.5	2,100
800	S-109	0.6	16.2	6.40	1.00	9.07	1.58	0.048	0.036	15.4	2,400
800		0.8	17.6	6.40	1.33	9.86	1.75	0.054	0.038	15.4	2,400
800		1.0	18.0	6.40	1.66	10.40	1.80	0.055	0.039	15.4	2,400
800		1.2	18.0	6.40	1.99	10.73	1.82	0.056	0.040	15.4	2,400
800	L-68	1.4	18.0	6.40	2.32	11.12	1.83	0.056	0.041	15.4	2,400
800		1.6	18.0	6.40	2.66	11.48	1.85	0.057	0.042	15.4	2,400
900	S-133	0.4	15.9	6.99	0.70	8.90	1.51	0.045	0.035	17.3	2,700
900		0.6	17.4	6.99	1.04	9.74	1.68	0.050	0.039	17.3	2,700
900		1.0	19.2	6.99	1.74	11.14	1.88	0.056	0.042	17.3	2,700
900		1.2	19.2	6.99	2.09	11.52	1.90	0.057	0.043	17.3	2,700
900	L-78	1.4	19.2	6.99	2.44	11.94	1.90	0.057	0.044	17.3	2,700
900		1.6	19.2	6.99	2.78	12.29	1.92	0.058	0.045	17.3	2,700

TABLE III-A *Daily Nutrient Requirements of Dairy Cattle (Continued)*

Body weight (lb)	Breed size, age (wk)	Daily gain (lb)	Feed DM (lb)	Feed energy NE_m (Mcal)	Feed energy NE_g (Mcal)	Feed energy TDN (lb)	Crude protein (lb)	Minerals Ca (lb)	Minerals P (lb)	Vitamins A (1000 I.U.)	Vitamins D (I.U.)
1,000		0.4	17.1	7.57	0.73	9.58	1.60	0.049	0.040	19.2	3,000
1,000		0.6	18.7	7.57	1.09	10.47	1.78	0.054	0.042	19.2	3,000
1,000		1.0	20.2	7.57	1.82	11.82	1.95	0.060	0.045	19.2	3,000
1,000		1.2	20.2	7.57	2.18	12.22	1.96	0.060	0.046	19.2	3,000
1,000	L-88	1.4	20.2	7.57	2.55	12.63	1.97	0.060	0.046	19.2	3,000
1,000		1.6	20.2	7.57	2.91	13.01	1.98	0.060	0.046	19.2	3,000
1,100		0.4	18.3	8.13	0.76	10.25	1.70	0.051	0.042	21.2	3,300
1,100		0.8	20.9	8.13	1.53	11.91	1.98	0.060	0.045	21.2	3,300
1,100	L-98	1.2	20.9	8.13	2.29	12.85	1.99	0.060	0.046	21.2	3,300
1,100		1.6	20.9	8.13	3.06	13.69	2.00	0.060	0.046	21.2	3,300
1,200		0.4	19.4	8.68	0.79	10.86	1.79	0.053	0.042	23.1	3,600
1,200	L-110	0.8	21.6	8.68	1.58	12.53	2.01	0.060	0.044	23.1	3,600
1,200		1.2	21.6	8.68	2.38	13.39	2.02	0.060	0.044	23.1	3,600
1,200		1.6	21.6	8.68	3.17	14.26	2.04	0.061	0.046	23.1	3,600
1,300		0.4	20.5	9.21	0.82	11.48	1.88	0.054	0.040	25.0	3,900
1,300		0.8	21.9	9.21	1.63	12.92	2.01	0.058	0.040	25.0	3,900
1,300		1.2	21.9	9.21	2.45	13.91	2.01	0.058	0.042	25.0	3,900
1,300		1.6	21.9	9.21	3.26	14.78	2.02	0.058	0.042	25.0	3,900
Growing Dairy Bulls											
300		1.0	7.9	3.07	1.01	5.02	0.93	0.036	0.022	5.8	900
300	S-34	1.4	7.9	3.07	1.41	5.43	1.02	0.039	0.024	5.8	900
300		1.8	7.9	3.07	1.82	5.81	1.11	0.043	0.026	5.8	900
100	S-10	1.0	2.9	1.52	0.87	2.54	0.39	0.022	0.014	1.9	300
100	L-2	1.2	2.9	1.52	1.04	2.65	0.43	0.024	0.015	1.9	300
100		1.4	2.9	1.52	1.22	2.78	0.48	0.027	0.015	1.9	300
150		1.0	4.1	1.90	0.89	3.14	0.51	0.026	0.015	2.9	450
150	S-18	1.2	4.1	1.90	1.07	3.31	0.55	0.028	0.016	2.9	450
150		1.4	4.1	1.90	1.25	3.46	0.60	0.031	0.017	2.9	450
150	L-8	1.6	4.1	1.90	1.42	3.60	0.64	0.033	0.018	2.9	450
150		1.8	4.1	1.90	1.60	3.75	0.69	0.035	0.018	2.9	450
200		1.2	5.6	2.28	1.10	4.00	0.74	0.034	0.018	3.8	600

200	S-24	1.4	5.6	2.28	1.29	4.18	0.79	0.036	0.019	3.8	600
200		1.6	5.6	2.28	1.47	4.35	0.83	0.038	0.020	3.8	600
200	L-12	1.8	5.6	2.28	1.66	4.52	0.88	0.040	0.021	3.8	600
200		2.0	5.6	2.28	1.84	4.70	0.92	0.042	0.022	3.8	600
300		1.0	7.9	3.07	1.01	5.02	0.93	0.036	0.022	5.8	900
300	S-34	1.4	7.9	3.07	1.41	5.43	1.02	0.039	0.024	5.8	900
300		1.8	7.9	3.07	1.82	5.81	1.11	0.043	0.026	5.8	900
300	L-20	2.0	7.9	3.07	2.02	6.00	1.15	0.045	0.027	5.8	900
300		2.2	7.9	3.07	2.22	6.19	1.20	0.047	0.028	5.8	900
400		1.0	10.5	3.81	1.10	6.30	1.20	0.041	0.027	7.7	1,200
400	S-44	1.4	10.5	3.81	1.54	6.73	1.29	0.044	0.029	7.7	1,200
400		1.8	10.5	3.81	1.98	7.17	1.37	0.047	0.031	7.7	1,200
400		2.0	10.5	3.81	2.20	7.38	1.41	0.048	0.032	7.7	1,200
400	L-28	2.2	10.5	3.81	2.42	7.59	1.46	0.050	0.033	7.7	1,200
500		1.0	12.7	4.50	1.18	7.34	1.41	0.048	0.032	9.6	1,500
500	S-55	1.4	12.7	4.50	1.65	7.85	1.49	0.051	0.033	9.6	1,500
500		1.8	12.7	4.50	2.12	8.36	1.56	0.053	0.035	9.6	1,500
500		2.0	12.7	4.50	2.36	8.57	1.60	0.054	0.036	9.6	1,500
500	L-34	2.2	12.7	4.50	2.60	8.80	1.64	0.056	0.037	9.6	1,500
600		1.0	14.8	5.26	1.27	8.41	1.59	0.053	0.036	11.5	1,800
600	S-65	1.4	14.8	5.26	1.78	9.03	1.64	0.055	0.038	11.5	1,800
600		1.8	14.8	5.26	2.29	9.55	1.71	0.057	0.040	11.5	1,800
600		2.0	14.8	5.26	2.54	9.80	1.75	0.058	0.041	11.5	1,800
600	L-41	2.2	14.8	5.26	2.79	10.03	1.79	0.060	0.042	11.5	1,800
700		1.0	17.0	6.02	1.38	9.57	1.76	0.056	0.040	13.5	2,100
700	S-75	1.4	17.0	6.02	1.93	10.20	1.82	0.057	0.042	13.5	2,100
700		1.8	17.0	6.02	2.48	10.81	1.87	0.059	0.043	13.5	2,100
700		2.0	17.0	6.02	2.76	11.08	1.91	0.060	0.044	13.5	2,100
700	S-47	2.2	17.0	6.02	3.04	11.36	1.94	0.061	0.045	13.5	2,100
800		1.0	18.7	6.78	1.48	10.51	1.90	0.058	0.043	15.4	2,400
800	S-85	1.4	18.7	6.78	2.07	11.24	1.93	0.058	0.044	15.4	2,400
800		1.8	18.7	6.78	2.66	11.91	1.98	0.060	0.045	15.4	2,400
800		2.0	18.7	6.78	2.96	12.21	2.01	0.061	0.045	15.4	2,400
800	L-54	2.2	18.7	6.78	3.26	12.49	2.04	0.062	0.046	15.4	2,400
900		1.0	20.0	7.55	1.62	11.50	1.96	0.060	0.045	17.3	2,700
900	S-95	1.4	20.0	7.55	2.27	12.26	1.99	0.061	0.047	17.3	2,700
900		1.8	20.0	7.55	2.92	13.00	2.02	0.062	0.049	17.3	2,700

Table III-A *Daily Nutrient Requirements of Dairy Cattle (Continued)*

Body weight (lb)	Breed size, age (wk)	Daily gain (lb)	Feed DM (lb)	Feed energy			Crude protein (lb)	Minerals		Vitamins	
				NE_m (Mcal)	NE_g (Mcal)	TDN (lb)		Ca (lb)	P (lb)	A (1000 I.U.)	D (I.U.)
900		2.0	20.0	7.55	3.24	13.40	2.04	0.063	0.050	17.3	2,700
900	L-60	2.2	20.0	7.55	3.56	13.60	2.08	0.064	0.050	17.3	2,700
1,000		1.0	21.0	8.33	1.73	12.33	2.00	0.061	0.046	19.2	3,000
1,000	S-106	1.2	21.0	8.33	2.08	12.73	2.01	0.062	0.047	19.2	3,000
1,000		1.6	21.0	8.33	2.77	13.52	2.04	0.063	0.048	19.2	3,000
1,000	L-67	2.0	21.0	8.33	3.46	14.22	2.08	0.064	0.050	19.2	3,000
1,000		2.2	21.0	8.33	3.81	14.55	2.10	0.065	0.050	19.2	3,000
1,100		0.8	22.0	8.94	1.45	12.58	2.06	0.062	0.049	21.2	3,300
1,100	S-118	1.2	22.0	8.94	2.17	13.46	2.07	0.062	0.049	21.2	3,300
1,100		1.6	22.0	8.94	2.90	14.30	2.09	0.063	0.050	21.2	3,300
1,100	L-74	1.8	22.0	8.94	3.26	14.70	2.10	0.064	0.050	21.2	3,300
1,100		2.0	22.0	8.94	3.62	15.03	2.12	0.064	0.051	21.2	3,300
1,200	S-129	0.6	22.6	9.55	1.13	12.66	2.08	0.063	0.049	23.1	3,600
1,200		1.0	23.0	9.55	1.88	13.69	2.13	0.064	0.050	23.1	3,600
1,200	L-82	1.4	23.0	9.55	2.63	14.61	2.13	0.064	0.050	23.1	3,600
1,200		1.8	23.0	9.55	3.38	15.41	2.16	0.065	0.051	23.1	3,600
1,300		0.6	23.7	10.14	1.16	13.27	2.16	0.064	0.049	25.0	3,900
1,300		1.0	23.7	10.14	1.94	14.29	2.16	0.064	0.050	25.0	3,900
1,300	L-92	1.4	23.7	10.14	2.72	15.26	2.16	0.065	0.050	25.0	3,900
1,300		1.8	23.7	10.14	3.49	16.12	2.17	0.066	0.051	25.0	3,900
1,400		0.2	21.5	10.71	0.40	12.04	1.89	0.064	0.049	26.9	4,200
1,400		0.6	24.3	10.71	1.19	13.80	2.19	0.065	0.050	26.9	4,200
1,400	L-102	1.0	24.3	10.71	1.99	14.87	2.18	0.065	0.050	26.9	4,200
1,400		1.4	24.3	10.71	2.79	15.80	2.18	0.065	0.051	26.9	4,200
1,500		0.2	22.5	11.28	0.41	12.60	1.97	0.064	0.050	28.8	4,500
1,500	L-116	0.6	24.9	11.28	1.22	14.37	2.21	0.065	0.050	28.8	4,500
1,500		1.0	24.9	11.28	2.03	15.39	2.20	0.066	0.051	28.8	4,500
1,500		1.4	24.9	11.28	2.84	16.31	2.20	0.066	0.051	28.8	4,500
1,600		0.2	23.6	11.84	0.41	13.22	2.05	0.064	0.050	30.8	4,800
1,600	L-140	0.6	25.5	11.84	1.22	14.87	2.23	0.066	0.051	30.8	4,800
1,600		1.0	25.5	11.84	2.04	15.94	2.22	0.066	0.051	30.8	4,800
1,700	L-163	0.2	24.6	12.40	0.41	13.78	2.14	0.064	0.051	32.7	5,100

1,700		0.6	26.1	12.40	1.23	15.32	2.27	0.066	0.051	32.7	5,100
1,700		1.0	26.1	12.40	2.05	16.39	2.26	0.066	0.051	32.7	5,100
Growing Veal Calves Fed Only Milk											
75	—	1.10	1.5	0.97	0.90	1.8	0.38	0.015	0.011	1.4	225
100	L-1.0	1.75	2.3	1.37	1.51	2.8	0.57	0.018	0.013	1.9	300
125	L-3.0	2.00	2.7	1.59	1.75	3.2	0.65	0.024	0.015	2.4	375
150	L-4.8	2.20	3.0	1.82	1.96	3.7	0.72	0.029	0.018	2.9	450
175	L-6.4	2.30	3.3	2.05	2.09	4.0	0.74	0.033	0.020	3.4	525
200	L-8.0	2.40	3.6	2.26	2.23	4.3	0.77	0.035	0.021	3.8	600
225	L-9.5	2.50	3.8	2.47	2.38	4.6	0.81	0.037	0.022	4.3	675
250	L-10.9	2.65	4.1	2.67	2.58	5.0	0.86	0.039	0.023	4.8	750
275	L-12.3	2.75	4.4	2.87	2.74	5.3	0.89	0.040	0.024	5.3	825
300	L-13.6	2.80	4.6	3.07	2.86	5.6	0.92	0.041	0.025	5.8	900
325	L-14.9	2.85	4.9	3.26	2.97	5.8	0.94	0.042	0.026	6.3	975
Maintenance of Mature Breeding Bulls											
1,200	—	—	18.3	9.98	—	10.3	1.58	0.042	0.036	23.1	—
1,400	—	—	20.6	11.20	—	11.5	1.76	0.049	0.040	26.9	—
1,600	—	—	22.7	12.38	—	12.7	1.93	0.057	0.045	30.8	—
1,800	—	—	24.9	13.53	—	13.9	2.09	0.064	0.049	34.6	—
2,000	—	—	26.9	14.64	—	15.1	2.25	0.071	0.053	38.5	—
2,200	—	—	28.9	15.72	—	16.2	2.41	0.079	0.057	42.3	—
2,400	—	—	30.8	16.78	—	17.3	2.56	0.086	0.061	46.2	—
2,600	—	—	32.7	17.82	—	18.3	2.71	0.093	0.064	50.0	—
2,800	—	—	34.6	18.84	—	19.4	2.85	0.099	0.068	53.9	—
3,000	—	—	36.5	19.84	—	20.4	3.00	0.107	0.072	57.7	—

Source: *Nutrient Requirements of Dairy Cattle*, Fifth Revised Edition, 1978, National Academy of Sciences—National Research Council, Washington, D.C.

TABLE III-B *Daily Nutrient Requirements of Lactating and Pregnant Cows*

Body weight (lb)	Feed energy NE$_1$ (Mcal)	Feed energy TDN (lb)	Crude protein (lb)	Calcium (lb)	Phosphorus (lb)	Vitamin A (1,000 IU)
Maintenance of mature cows[a]						
700	6.02	5.84	.71	0.028	0.023	24
800	6.65	6.45	.77	0.032	0.026	28
900	7.27	7.05	.83	0.035	0.028	31
1000	7.86	7.63	.89	0.038	0.030	35
1100	8.45	8.19	.95	0.040	0.032	38
1200	9.02	8.75	1.01	0.043	0.034	41
1300	9.57	9.29	1.06	0.046	0.037	45
1400	10.12	9.82	1.12	0.048	0.039	48
1500	10.66	10.34	1.17	0.051	0.041	52
1600	11.19	10.85	1.22	0.053	0.043	55
1700	11.71	11.36	1.27	0.056	0.045	59
1800	12.22	11.86	1.32	0.059	0.047	62
Maintenance plus last 2 months' gestation of mature dry cows						
700	7.82	7.60	1.32	0.047	0.033	24
800	8.65	8.40	1.45	0.053	0.038	28
900	9.45	9.17	1.57	0.059	0.042	31
1000	10.22	9.93	1.69	0.064	0.045	35
1100	10.98	10.66	1.80	0.070	0.050	38
1200	11.72	11.38	1.92	0.075	0.053	41
1300	12.44	12.08	2.03	0.080	0.057	45
1400	13.16	12.78	2.13	0.085	0.060	48
1500	13.85	13.45	2.24	0.090	0.064	52
1600	14.54	14.12	2.34	0.095	0.067	55
1700	15.22	14.78	2.44	0.100	0.071	59
1800	15.88	15.42	2.54	0.105	0.075	62
Milk production—nutrients per pound of milk of different fat percentages						
Fat (%)						
2.5	0.27	0.260	0.072	0.0024	0.0017	—
3.0	0.29	0.282	0.077	0.0025	0.0017	—
3.5	0.31	0.304	0.082	0.0026	0.0018	—
4.0	0.34	0.326	0.087	0.0027	0.0018	—
4.5	0.36	0.344	0.092	0.0028	0.0019	—
5.0	0.38	0.365	0.098	0.0029	0.0019	—
5.5	0.40	0.387	0.103	0.0030	0.0020	—
6.0	0.42	0.410	0.108	0.0031	0.0021	—

[a]To allow for growth of lactating heifers, increase the maintenance allowance for all nutrients except vitamin A by 20% during the first lactation and 10% during the second lactation.

Source: *Nutrient Requirements of Dairy Cattle,* Fifth Revised Edition, 1978, National Academy of Sciences—National Research Council, Washington, D.C.

TABLE III-C *Recommended Nutrient Content of Rations for Dairy Cattle*

	Lactating cow rations					
Nutrients (concentration in the feed dry matter)	*Cow wt (lb)*	*Daily milk yields (lb)*				*Maximum concentrations (all classes)*
	≤ 900	*< 18*	*18–29*	*29–40*	*> 40*	
	1100	*< 24*	*24–37*	*37–51*	*> 51*	
	1300	*< 31*	*31–46*	*46–64*	*> 64*	
	≥ 1550	*< 40*	*40–57*	*57–78*	*> 78*	
Ration No.		I	II	III	IV	Max.
Crude Protein, %		13.0	14.0	15.0	16.0	—
Energy						
Ne_1, Mcal/lb		0.64	0.69	0.73	0.78	—
NE_m, Mcal/lb		—	—	—	—	—
NE_g, Mcal/lb		—	—	—	—	—
ME, Mcal/lb		1.07	1.15	1.23	1.31	—
DE, Mcal/lb		1.26	1.34	1.42	1.50	—
TDN, %		63	67	71	75	—
Crude Fiber, %		17	17	17	17	—
Acid Detergent Fiber, %		21	21	21	21	—
Ether Extract, %		2	2	2	2	—
Minerals						
Calcium, %		0.43	0.48	0.54	0.60	—
Phosphorus, %		0.31	0.34	0.38	0.40	—
Magnesium, %		0.20	0.20	0.20	0.20	—
Potassium, %		0.80	0.80	0.80	0.80	—
Sodium, %		0.18	0.18	0.18	0.18	—
Sodium Chloride, %		0.46	0.46	0.46	0.46	5
Sulfur, %		0.20	0.20	0.20	0.20	0.35
Iron, ppm		50	50	50	50	1,000
Cobalt, ppm		0.10	0.10	0.10	0.10	10
Copper, ppm		10	10	10	10	80
Manganese, ppm		40	40	40	40	1,000
Zinc, ppm		40	40	40	40	500
Iodine, ppm		0.50	0.50	0.50	0.50	50
Molybdenum, ppm		—	—	—	—	6
Selenium, ppm		0.10	0.10	0.10	0.10	5
Fluorine, ppm		—	—	—	—	30
Vitamins						
Vit A, IU/lb		1,450	1,450	1,450	1,450	—
Vit D, IU/lb		140	140	140	140	—

Source: *Nutrient Requirements of Dairy Cattle,* Fifth Revised Edition, 1978, National Academy of Sciences—National Research Council, Washington, D.C.

TABLE III-C *Recommended Nutrient Content of Rations for Dairy Cattle (Continued)*

Nutrients (concentration in the feed dry matter)	*Non-lactating cattle rations*					*Maximum concen-trations (all classes)*
	Dry pregnant cows	*Mature bulls*	*Growing heifers and bulls*	*Calf starter concen-trate mix*	*Calf milk replacer*	
Ration No.	V	VI	VII	VIII	IX	Max.
Crude Protein, %	11.0	8.5	12.0	16.0	22.0	—
Energy						
Ne_1, Mcal/lb	0.61	—	—	—	—	—
NE_m, Mcal/lb	—	0.54	0.57	0.86	1.09	—
NE_g, Mcal/lb	—	—	0.27	0.54	0.70	—
ME, Mcal/lb	1.01	0.93	1.01	1.42	1.71	—
DE, Mcal/lb	1.20	1.12	1.20	1.60	1.90	—
TDN, %	60	56	60	80	95	—
Crude Fiber, %	17	15	15	—	—	—
Acid Detergent Fiber, %	21	19	19	—	—	—
Ether Extract, %	2	2	2	2	10	—
Minerals						
Calcium, %	0.37	0.24	0.40	0.60	0.70	—
Phosphorus, %	0.26	0.18	0.26	0.42	0.50	—
Magnesium, %	0.16	0.16	0.16	0.07	0.07	—
Potassium, %	0.80	0.80	0.80	0.80	0.80	—
Sodium, %	0.10	0.10	0.10	0.10	0.10	—
Sodium Chloride, %	0.25	0.25	0.25	0.25	0.25	5
Sulfur, %	0.17	0.11	0.16	0.21	0.29	0.35
Iron, ppm	50	50	50	100	100	1,000
Cobalt, ppm	0.10	0.10	0.10	0.10	0.10	10
Copper, ppm	10	10	10	10	10	80
Manganese, ppm	40	40	40	40	40	1,000
Zinc, ppm	40	40	40	40	40	500
Iodine, ppm	0.50	0.25	0.25	0.25	0.25	50
Molybdenum, ppm	—	—	—	—	—	6
Selenium, ppm	0.10	0.10	0.10	0.10	0.10	5
Fluorine, ppm	—	—	—	—	—	30
Vitamins						
Vit A, IU/lb	1,450	1,450	1,000	1,000	1,720	—
Vit D, IU/lb	140	140	140	140	270	—
Vit E, ppm	—	—	—	—	300	—

Source: Nutrient Requirements of Dairy Cattle, Fifth Revised Edition, 1978. National Academy of Sciences–National Research Council, Washington, D.C.

TABLE III-D *Nutrient Content of Some Common Dairy Feeds*

		Dry matter basis (moisture free)									
		Net energy									
	Dry matter	NE_l	NE_m	NE_g	TDN	Crude protein	Crude fiber	ADF	Calcium	Phosphorus	Vitamin A equivalent
	(%)	(Mcal/lb)			(%)	(%)	(%)	(%)	(%)	(%)	(1,000 IU/lb)
Alfalfa, fresh	27	0.621	0.594	0.268	61	19.0	28	35	1.72	0.31	36.1
Alfalfa hay, bud stage	89	0.669	0.640	0.372	65	21.7	24	31	2.12	0.30	90.9
Alfalfa hay, early bloom	90	0.590	0.562	0.268	58	17.2	31	38	1.25	0.23	15.4
Alfalfa hay, full bloom	88	0.544	0.522	0.200	54	15.0	35	42	1.28	0.20	2.2
Alfalfa dehydrated, 15% protein	93	0.621	0.594	0.313	61	16.3	33	41	1.32	0.24	19.9
Alfalfa silage, 25–40% DM (same values as for alfalfa hay of same maturity)											
Almond hulls, 13% crude fiber	91	0.579	0.561	0.358	57	4.4	14	27	0.23	0.11	—
Barley grain	89	0.866	0.889	0.594	83	13.9	6	7	0.05	0.37	—
Barley hay	87	0.576	0.558	0.249	57	8.9	26	—	0.21	0.30	—
Beet pulp, dried	91	0.812	0.812	0.540	78	8.0	22	34	0.75	0.11	—
Bermudagrass hay, coastal	91	0.476	0.457	0.086	48	6.0	34	35	0.46	0.18	—
Bonemeal, steamed	95	0.122	0.259	0	16	12.7	2	—	30.51	14.31	81.1
Brewers grains, dried	92	0.680	0.653	0.390	66	26.0	16	23	0.29	0.54	—
Bromegrass hay, late bloom	90	0.544	0.522	0.200	54	7.4	40	44	0.30	0.35	—
Citrus pulp, dried	90	0.798	0.798	0.526	77	6.9	14	23	2.07	0.13	—
Clover hay, ladino	91	0.621	0.594	0.313	61	23.0	19	32	1.38	0.24	29.3
Coconut meal, mech-extd	93	0.844	0.857	0.572	81	21.9	13	—	0.23	0.66	—
Coconut meal, solv-extd	92	0.767	0.753	0.490	74	23.1	16	—	0.18	0.66	—
Corn fodder	82	0.667	0.640	0.372	65	8.9	26	33	0.43	0.23	0.9
Corn stover	87	0.599	0.571	0.281	59	5.9	34	39	0.60	0.09	—
Corn cobs, ground	90	0.467	0.458	0.068	47	2.8	35	35	0.12	0.04	0.1
Corn ears, ground	87	0.835	0.843	0.563	80	9.3	9	—	0.05	0.26	—
Corn grain, ground	89	0.921	0.975	0.644	88	10.0	2	3	0.03	0.31	0.4
Corn silage, well-eared	25–40	0.721	0.699	0.440	70	8.0	24	31	0.27	0.20	8.2
Corn silage, average	25–40	0.694	0.670	0.406	68	8.0	24	31	0.27	0.20	8.2
Corn silage, not well-eared	25–40	0.667	0.640	0.372	65	8.4	32	41	0.34	0.20	2.3
Cottonseed hulls	90	0.367	0.390	0	38	4.3	50	71	0.16	0.73	—
Cottonseed, whole	93	1.034	1.175	0.730	98	24.9	18	29	0.15	0.73	—

TABLE III-D *Nutrient Content of Some Common Dairy Feeds (Continued)*

	Dry matter	Dry matter basis (moisture free): Net energy NE_1	NE_m	NE_g	TDN	Crude protein	Crude fiber	ADF	Calcium	Phosphorus	Vitamin A equivalent
	(%)	(Mcal/lb)	(Mcal/lb)	(Mcal/lb)	(%)	(%)	(%)	(%)	(%)	(%)	(1,000 IU/lb)
Cottonseed meal, mech-extd, 41% protein	94	0.798	0.798	0.526	77	43.6	13	20	0.17	1.28	—
Cottonseed meal, solv-extd, 41% protein	92	0.780	0.766	0.503	75	44.8	13	—	0.17	1.31	—
Dicalcium phosphate	96	0	0	0	0	0	0	0	23.70	18.84	0
Fat, animal (not over 3% of ration)	99	2.381	2.381	1.188	182	0	0	0	—	—	—
Hominy feed	91	0.966	1.052	0.680	92	11.8	6	12	0.06	0.58	1.8
Limestone, ground	100	0	0	0	0	0	0	0	36.07	0.02	0
Milo grain	88	0.835	0.844	0.563	80	11.7	2	9	0.03	0.33	0.2
Molasses, beet, 79 deg. brix	77	0.780	0.767	0.503	75	8.7	0	0	0.21	0.04	—
Molasses, citrus	65	0.798	0.798	0.599	77	10.9	0	0	2.01	0.14	—
Molasses, sugarcane, dehydrated	96	0.698	0.676	0.417	68	10.7	5	—	0.87	0.29	—
Molasses, sugarcane, 79.5 deg. brix	75	0.744	0.726	0.467	72	4.3	0	0	1.19	0.11	—
Monosodium phosphate	87	0	0	0	0	0	0	0	0	25.80	0
Oats, grain	89	0.789	0.785	0.517	76	13.6	12	17	0.07	0.39	—
Oat hay,	88	0.621	0.594	0.317	61	9.2	31	36	0.26	0.24	12.0
Oat silage, boot stage	20–30	0.635	0.603	0.331	62	—	—	—	—	—	—
Oat silage, dough stage	25–35	0.599	0.576	0.281	59	9.7	34	—	0.47	0.33	—

Oyster shells, ground	100	0	0	0	1.0	1.0	0	0	38.22	0.07	0
Phosphate rock, defluorinated	100	0	0	0	0	0	0	0	31.65	13.70	0
Pineapple bran	87	0.758	0.739	0.481	73	4.6	18	—	0.24	0.12	—
Rice bran	91	0.679	0.653	0.390	66	14.0	12	16	0.07	1.62	—
Safflower meal, solv-extd	92	0.558	0.531	0.218	55	23.9	34	—	0.37	0.80	—
Sodium tripolyphosphate	96	0	0	0	0	0	0	0	0	25.98	0
Sorghum silage, grain variety	25–40	0.558	0.531	0.218	55	8.3	26	—	0.32	0.18	—
Soybean meal, solv-extd, 44% protein	89	0.844	0.857	0.572	81	49.6	7	10	0.36	0.75	—
Soybean meal, solv-extd, 46% protein	89	0.844	0.857	0.572	81	51.8	5	8	0.36	0.75	—
Soybean meal, solv-extd, 48% protein	89	0.844	0.857	0.572	81	54.0	3	4	0.36	0.75	—
Sudangrass hay	89	0.599	0.571	0.281	59	11.0	29	42	0.56	0.31	10.7
Sudangrass silage	20–30	0.599	0.571	0.290	59	11.1	34	—	0.48	0.19	19.0
Timothy hay, early bloom	88	0.635	0.603	0.331	62	10.0	32	37	0.53	0.26	9.7
Timothy hay, late bloom	88	0.558	0.531	0.218	55	7.7	33	43	0.38	0.18	1.8
Timothy silage, 25–40% DM (same values as for Timothy hay of same maturity)											
Urea, 46% N	90	0	0	0	0	287.5	0	0	0	0	0
Vetch hay	88	0.635	0.603	0.331	62	19.0	31	43	1.18	0.34	79.1
Wheat bran	89	0.721	0.694	0.435	70	18.0	11	12	0.12	1.32	0.5
Wheat grain, soft	86	0.921	0.975	0.644	88	12.0	3	4	0.06	0.41	—
Wheat millrun	90	0.767	0.726	0.494	74	17.0	9	—	0.10	1.13	—

Source: *Nutrient Requirements of Dairy Cattle*, 5th Revised Edition, 1978, National Academy of Sciences—National Research Council, Washington, D.C.

Table IV-A *Composition of Various Diluents for Liquid Semen*[a]

Ingredients	*Original IVT*	*IVT 4G-IB*	*IVT 4G-2B*	*IVT 4G-4B*	*Arkansas No. 2*	*Tenn. YCCG*	*Germany EIBL*	*Cornell CUE*
Sodium bicarbonate	0.21	0.21	0.42	0.83	0.21	0.21	0.18	0.21
Sodium citrate	2.00	1.48	1.00	0.09	1.00	1.60	1.67	1.45
Potassium chloride	0.04	0.04	0.04	0.04	—	0.04	0.03	0.04
Glucose	0.30	1.20	1.20	1.20	0.30	1.00	0.25	0.30
Sulfanilamide	0.30	0.30	0.30	0.30	0.30	0.30	0.25	0.30
Egg yolk (%)	10	15	15	15	10	20	30	20
Penicillin (I.U./ml)	1,000	1,000	1,000	1,000	1,000	500	500	1,000
Streptomycin (μg/ml)	1,000	1,000	1,000	1,000	1,000	500	500	1,000
Glycine	—	—	—	—	1.00	—	—	0.937
Glutathione (reduced)	—	—	—	—	0.154	—	—	—
Catalase	—	0.01	0.01	0.01	—	—	—	—
Gassed with	CO_2	CO_2	CO_2	CO_2	CO_2	CO_2	CO_2	—[b]

[a]Grams per 100 milliliters unless otherwise noted.
[b]0.087 g Citric acid added per 100 ml to adjust pH, resulting in self-carbonation.
Source: Salisbury and Van Demark; *Physiology of Reproduction and Artificial Insemination of Cattle.* San Francisco: W. H. Freeman and Co., 1961.

Table IV-B *Example Method for Preparation of a Frozen Semen Extender*

a. Preparation of stock solution of sodium citrate•$2H_2O$. To make 100 ml of concentrate stock solution of sodium citrate, weigh 3.57 grams of sodium citrate into a 100-ml graduated cylinder; add double distilled water to about 70 ml total volume. Place this solution in a water bath and bring to just a boil. Cool the solution to room temperature and add double distilled water to 100 ml total volume. Store this concentrate stock solution at 5°C in an appropriately labeled glass-stoppered container for no more than 1 week.

b. Preparation of non-glycerol extender. Accurately pour 65 ml of the concentrated stock solution of sodium citrate prepared above (a) into a 100-ml glass-stoppered mixing cylinder. Add double distilled water to a total volume of 79 ml. Add 1 ml of a mixture of penicillin and streptomycin. (This antibiotic mixture is made by compositing a 1 million unit vial of penicillin with 1 gram vial of streptomycin and adding 10 ml of double distilled water. Thus, 1 ml of this mixture contains 100,000 units of penicillin and 100,000 micrograms of streptomycin.) Lastly, add 20 ml of egg yolk to give 100 ml of non-glycerol extender.

The antibiotic mixture should be stored at 5°C and may be used for no more than 1 week. The complete non-glycerol extender should be prepared fresh each day, although one may prepare it as much as 20 hours in advance provided it is stored at 5°C.

c. Preparation of glycerol extender. Accurately pour 66 ml of concentrated stock solution of sodium citrate prepared above (a) into a 100-ml glass-stoppered mixing cylinder. Add 14 ml of glycerol to a total volume of 80 ml and mix thoroughly. Add 20 ml of egg yolk to a total volume of 100 ml, and mix immediately to avoid packing of the yolk at the bottom of the cylinder.

Table IV-C *Example Method of Freezing Semen*

Fresh semen is diluted at 35°C to one-half the desired sperm concentration with the complete non-glycerolated extender (Table IV-B). Generally 60 million motile sperm per ml are used, but it may go to 40 million or less.

The partially diluted semen is gradually cooled to 5°C over a 4-hour period. The cooled semen is diluted slowly by adding an equal volume of complete glycerol extender (Table IV-B). To achieve the slow addition of glycerol extender add the extender in 4 equal portions at 15-minute intervals with mixing after each addition. The fully diluted semen is allowed to equilibrate for 18 hours at 5°C, packaged in ampules, straws, or pellets, and then frozen in a liquid nitrogen freezer. To freeze ampules, the temperature is lowered at about 1 to 3°C per minute to −15°C and at 4°C per minute until −150°C at which time the ampules are transferred to liquid nitrogen for storage at −196°C. Straws are usually frozen in nitrogen vapor and stored at −196°C. Pellets of semen are formed by dropping 0.1 ml drops of extended semen into depressions in blocks of dry ice. In a few minutes they achieve −79°C and may then be transferred to liquid nitrogen temperatures at −196°C.

Table IV-D *Causes of Common off-flavors of Milk and Methods of Prevention and Control*

1. Barny (cowy, musty, unclean)
   Causes—Foul odors inhaled by cows housed in dirty stables which are poorly ventilated travel by way of the lungs and circulatory system to the udder and enter the milk. Bedding and manure may enter the milk via the milking machine from dirty udders or by allowing the teat cups to touch the floor or bedding materials.
   Prevention—Clean cows, clean milkers, and well-ventilated, clean stables will eliminate this flavor which is more of a problem in conventional barns than in milking parlors.

2. Bitter (rancid, tainted)
   Cause—The enzyme lipase, which is always present in milk, when activated by any one of several factors hydrolyzes milkfat releasing fatty acids, some of which have a pronounced disagreeable flavor. Agitation of warm milk in the risers of pipeline milking systems causes rancidity. Temperature changes in milk may catalyze the lipase. These include slow cooling; the addition of warm milk to cold milk; warming milk to 80°F or above, then cooling to low temperatures; freezing and thawing milk. Cows in late lactation giving small amounts of milk often produce milk which is highly susceptible to rancidity.
   Prevention—Eliminate risers if possible and keep agitation at a minimum in handling warm milk. Cool milk as quickly as possible to 40°F or less. Avoid freezing milk, and do not permit cooled milk to be rewarmed. Check strippers and cows in late lactation for rancid milk. If present, dry off the cow. The bitter flavor may be judged to be unclean when the milk is first drawn. However, after several hours the milk will have a tainted taste that renders it undrinkable.

3. Feed (garlic, grassy, onion, silage, weedy)
   Causes—The most common off-flavor in milk is caused by various types of feeds consumed by cows. Any sudden change in the ration is likely to result in a corresponding change in the flavor of the milk. Feeding silage before milking, especially grass silage with a high butyric acid content, will cause a strong silage flavor in the milk. If cows breathe the odor from silage, it will also appear in the milk though less pronounced than if silage is fed prior to milking.
   If cows are pastured, off-flavors result from weeds eaten by the cows or from the herd being turned onto lush pasture. Wild garlic and wild onion are examples of weeds that cause objectionable flavors in milk.
   Prevention—Unless the herd has constant access to silage in loose-housing, plan to feed silage after milking. Bringing cows into the barn from pasture at least 3 hours before milking will reduce off-flavors in milk. Use herbicides to control noxious weeds.

4. Malty or high acid (grapenuts, sour)
   Causes—High bacteria counts are the cause of malty and high-acid milk. Malty flavor is present when the milk has a very high bacteria count but before it has had time to sour. The usual source of the bacteria is milk-handling equipment which is not properly cleaned and sanitized. Slow cooling or failure to lower the temperature of the milk to 40°F allows the bacteria to multiply rapidly.
   Prevention—Keep the cows clean. Clean milking equipment immediately after each use, and sanitize it prior to using. Cool milk rapidly to 40°F, and hold it at that temperature. The key words are *clean* and *cold*.

5. Oxidized (cardboard, tallowy)
   Causes—Since the flavor of oxidized milk resembles the taste of cardboard or tallow, consumers naturally object to it. There are 3 causes of oxidized flavor in milk: (1) Some cows produce milk with a cardboard flavor especially when fed in the barn or drylot. (2) When milk is exposed to sunlight for 15 minutes or more, it is quite likely to develop an oxidized flavor. (3) Dissolved copper or iron act as catalysts in milk. These may be in the water supply used for rinsing equipment or may be picked up by the milk from rusty or worn utensils.
   Prevention—Do not expose milk to sunlight. Use stainless steel equipment. Drain all chlorine from utensils and equipment after sanitizing. Although considerable research has been done with anti-oxidants and alpha tocopherol (vitamin E) as feed additives to prevent oxidized flavor in the milk of certain cows, a practical solution to this problem has not been achieved yet.

6. Miscellaneous
   a. *Cooked*—Pasteurization, especially the short-time high-temperature method, may impart a "cooked" flavor to milk as a result of the release of volatile sulfhydryl

Table IV-D *Causes of Common off-flavors of Milk and Methods of Prevention and Control (Continued)*

compounds from milk proteins. Milk dealers must be constantly alert in all phases of milk processing to avoid flavor defects.

b. *Flat*—Milk that is low in total solids has a flat taste. Milk low in lactose lacks the slightly sweet flavor of normal milk, and milk which is low in milkfat lacks the smooth flavor which is characteristic of higher testing milk.

c. *Salty*—Cows with mastitis, in late lactation, and occasionally individual cows in earlier stages of lactation may yield salty milk as a result of excessive amounts of sodium chloride from the blood plasma passing into the alveoli of the udder. The milk should be discarded, the mastitis treated, and cows in late lactation should be dried off.

d. *Physical defects*—There are 3 common physical defects which detract from the palatability of milk.

1. *Chalky*—Milk that is homogenized before UHT pasteurization may acquire a chalky texture. Homogenization after pasteurization will eliminate this problem.
2. *Ropy*—Ropy milk is quite viscous and has slimy threads on its surface. It is not to be confused with stringy milk from quarters with acute mastitis. Ropy milk does not develop ropiness until several hours after milking. Coliform bacteria and some streptococci form a gelatinous membrane around their bodies, and gums and mucins from it cause the ropiness in milk. Unless the bacteria have caused lactic acid fermentation, the flavor is normal and it is safe to drink. Obviously consumers will not appreciate its slimy texture. Proper washing of the cow's udders with a sanitizing solution before milking, sterile utensils, and pasteurization of milk will eliminate this defect.
3. *Sweet curd*—Some bacteria (*Bacillus subtilis, Bacillus cereus,* and *S. liquefaciens*) secrete a rennet-like enzyme which coagulates milk without an increase in acidity. The enzyme digests the phospholipid membrane surrounding the fat globules allowing them to coalesce and form a cream line. The enzyme is destroyed by pasteurization, but some bacteria may survive and if the milk reaches temperatures between 105 and 110°F the defect may occur.

7. Unnatural flavors

Careless or thoughtless use of disinfectants, medicines, or sprays on milking cows or in the dairy barn can result in unnatural flavors in milk which are highly objectionable to consumers.

a. *Disinfectants*—All sprays with a creosote base can impart an objectionable flavor to milk if ingested on forage or inhaled by cows.

b. *Insecticides*—Fly sprays and other insecticides can also give milk an unnatural flavor or odor.

c. *Medications*—Some preparations used for treating sore teats have pungent odors which will be picked up by milk passing through the teat cup.

d. *Pesticides*—Spray residues on the leaves of alfalfa or other forage crops consumed by milking cows can cause off-flavors in milk.

Most of these problems can be avoided by consulting your veterinarian relative to odorless medications and by good judgment in using disinfectants around the barn and in the choice of insecticides and pesticides.

TABLE IV-E *Dairy Farm Inspection Report*

(Inspecting Agency)

NAME

LOCATION

Permit No.

Pounds Sold Daily --------------------

Plant --------------------

SIR: An inspection of your dairy farm has this day been made, and you are notified of the violations marked below with a cross (X). Violation of the same requirement on two successive inspections calls for permit suspension and/or court action.

COWS

1. Abnormal Milk:

Cows secreting abnormal milk milked last or in separate equipment ---------- (a) ----

Abnormal milk properly handled and disposed of -- (b) ----

Proper care of abnormal milk handling equipment ---------- (c) ----

MILKING BARN, STABLE, OR PARLOR

2. Construction:

Floors, gutters, and feed troughs of concrete or equally impervious materials; in good repair -- (a) ----

Walls and ceilings smooth, painted or finished adequately; in good repair; ceiling dust-tight -- (b) ----

Separate stalls or pens for horses, calves, and bulls ---------- (c) ----

Adequate natural and/or artificial light; well distributed ---------- (d) ----

Proper feed storage facilities ---------- (e) ----

Properly ventilated; no overcrowding ---------- (f) ----

3. Cleanliness:

Clean and free of litter ---------- (a) ----

No swine or fowl ---------- (b) ----

4. Cowyard:

Graded to drain; no pooled water or wastes ---- (a) ----

Cowyard clean; cattle housing areas properly maintained ---------- (b) ----

No swine ---------- (c) ----

Manure stored inaccessible to cows ---------- (d) ----

MILKHOUSE OR ROOM

5. Construction and Facilities:

Floors

Smooth; concrete or other impervious material; in good repair ---------- (a) ----

Graded to drain ---------- (b) ----

Drains trapped, if connected to sanitary system -- (c) ----

Walls and Ceilings

Approved material and finish ---------- (a) ----

Good repair (windows, doors, and hose port included) ---------- (b) ----

Lighting and Ventilation

Adequate natural and/or artificial light; properly distributed ---------- (a) ----

Adequate ventilation ---------- (b) ----

Doors and windows closed during dusty weather -- (c) ----

Vents and lighting fixtures properly located ---- (d) ----

Miscellaneous Requirements

Used for milkhouse operations only; sufficient size ---------- (a) ----

No direct opening into living quarters or barn, except as permitted by Ordinance ---------- (b) ----

Liquid wastes properly disposed of ---------- (c) ----

Cleaning Facilities

Two-compartment wash and rinse vat of adequate size ---------- (a) ----

Suitable water heating facilities ---------- (b) ----

Water under pressure piped to milkhouse ------ (c) ----

6. Cleanliness:

Floors, walls, windows, tables, and similar non-product contact surfaces clean ---------- (a) ----

No trash, unnecessary articles, animals or fowl -- (b) ----

Pesticides and medicinals not stored in milkhouse ---------- (c) ----

TOILET AND WATER SUPPLY

7. Toilet:

Provided; conveniently located ---------- (a) ----

Constructed and operated according to Ordinance ---------- (b) ----

No evidence of human wastes about premises --- (c) ----

Toilet room in compliance with Ordinance ------ (d) ----

8. Water Supply:

Constructed and operated according to Ordinance ---------- (a) ----

Complies with bacteriological standards -------- (b) ----

No connection between safe and unsafe supplies; no improper submerged inlets ---------- (c) ----

UTENSILS AND EQUIPMENT

9. Construction:

Smooth, nonabsorbent, corrosion-resistant, non-toxic materials; easily cleanable; seamless hooded pails ---------- (a) ----

In good repair; accessible for inspection -------- (b) ----

Approved single-service articles; not reused ---- (c) ----

Strainers, approved design ---------- (d) ----

Approved CIP milk pipeline system ---------- (e) ----

10. Cleaning:

Utensils and equipment clean ---------- (a) ----

11. Sanitization:

All multi-use containers and equipment subjected to approved sanitization process (see Ordinance) ---------- (a) ----

12. Storage:

Left in treating chamber or sanitizing solution until used, or stored properly above floor ---- (a) ----

Stored to assure complete drainage, where applicable ---------- (b) ----

Single-service articles properly stored ---------- (c) ----

13. Handling:

Sanitized milk contact surfaces not exposed to contamination ---------- (a) ----

MILKING

14. Flanks, Udders, and Teats:

Milking done in barn, stable, or parlor ---------- (a) ----

MILKING—Continued

Brushing completed before milking begun ------ (b) ----

Flanks, bellies, udders, and tails of cows clean at time of milking; clipped when required ------ (c) ----

Udders and teats treated with sanitizing solution and dried, just prior to milking ---------- (d) ----

No wet hand milking ---------- (e) ----

15. Surcingles, Milk and Anti-Kickers:

Clean; stored above floor in clean place ---------- (a) ----

Stools, easily cleanable construction and not padded ---------- (b) ----

16. Transfer and Protection of Milk:

Immediate removal to milkhouse or room -------- (a) ----

Transfer, pouring, and/or straining facilities properly protected ---------- (b) ----

PERSONNEL

17. Hand-washing Facilities:

Soap, running water, and individual sanitary towels in milkroom and convenient to milking operations ---------- (a) ----

Wash and rinse vats not used as hand-washing facilities ---------- (b) ----

18. Personnel Cleanliness:

Hands washed clean and dried before milking, or performing milkhouse functions; rewashed when contaminated ---------- (a) ----

Clean outer garments worn ---------- (b) ----

COOLING

19. Cooling:

Milk cooled to 50° F. or less, within 2 hours after milking; maintained thereat until delivered -- (a) ----

VEHICLES

20. Vehicles:

Vehicles clean ---------- (a) ----

Constructed so as to protect milk ---------- (b) ----

No contaminating substances transported ------ (c) ----

INSECTS AND RODENTS

21. Insect and Rodent Control:

Fly breeding minimized by approved manure disposal methods (see Ordinance) ---------- (a) ----

Manure packs properly maintained ---------- (b) ----

All milkhouse openings effectively screened or otherwise protected; doors tight and self-closing; screen doors open outward ---------- (c) ----

Milkhouse free of insects and rodents ---------- (d) ----

Approved pesticides; used properly ---------- (e) ----

Equipment and utensils not exposed to pesticide contamination ---------- (f) ----

Surroundings neat and clean; free of harborages and breeding areas ---------- (g) ----

REMARKS:

DATE | SANITARIAN:

NOTE.—Item numbers correspond to required sanitation items for Grade A raw milk for pasteurization in the Grade A Pasteurized Milk Ordinance and Code—1965 Recommendations of the U.S. Public Health Service.

For sale by the Superintendent of Documents, U.S. Government Printing Office, Washington, D.C., 20402 U.S. GOVERNMENT PRINTING OFFICE: 1965—O-769-854

Source: Department of Health, Education, and Welfare, Public Health Service Publication, PHS-1783, January, 1965.

TABLE V-A *Interpretation of the California Mastitis Test (CMT)*

After adding the CMT reagent, rotate the plastic paddle for about 10 seconds in a circular pattern to mix the reagent and milk. The reaction is then scored and recorded as follows:

Symbol	*Meaning*	*Reaction*	*Estimated number of cells per ml*
—	negative	no evidence of precipitation	0–200,000
T	trace	a slight precipitate	150,000–500,000
1	weak positive	distinct precipitate, no gel formation	400,000–1,500,000
2	distinct positive	mixture thickens, some gel formation	800,000–5,000,000
3	strong positive	gel causes surface to become convex	over 5,000,000
+	alkaline milk	deep purple color	reduced secretory activity
y	acid milk	yellow color	[a]

[a]Acid milk in the udder is rare. A yellow color indicates a pH of 5.2 as a result of fermentation of lactose by bacteria after the milk was drawn from the cow and held for a period of time.

Source: Public Health Service: Screening Tests for the Detection of Abnormal Milk. P.H.S. Pub. no. 1306, Dept. of Health, Education, and Welfare, Washington, D.C., 1965.

Table V-B *Insulation "R" Values for Some Common Materials*

Material	*Insulation value*[a] *Per inch thickness*	*For thickness listed*
(1) Batt or blanket insulation		
Glass wool, mineral wool, or fiberglass	3.50 approx, read label	
(2) Fill-type insulation		
Glass or mineral wool	3.00	
Vermiculite (expanded)	2.13 to 2.27	
Shavings or sawdust	2.22	
Paper or wood pulp	3.70	
(3) Rigid insulation		
Wood fiber sheathing	2.27 to 2.63	
Expanded polystyrene,		
extruded	4.00 to 5.26	
molded	3.57	
Expanded polyurethane (aged)	6.25	
Glass fiber	4.00	
(4) Ordinary building materials		
Concrete, poured	0.08	
Plywood, 3/8″	1.25	0.47
Plywood, 1/2″	1.25	0.62
Hardboard, 1/4″	1.00 to 1.37	
Cement asbestos board, 1/8″		0.03
Lumber (fir, pine), 3/4″	1.25	0.94
Wood beveled siding 1/2″ × 8″		0.81
Asphalt shingles		0.44
Wood shingles		0.94
(5) Window glass, includes surface condition		
Single-glazed		0.88
Single-glazed + storm windows		1.79
Double-pane insulating glass		1.45 to 1.73
(6) Air space (3/4″ or larger)		0.90
(7) Surface conditions		
Inside surface		0.68
Outside surface (15 mph wind)		0.17

[a]Mean temperature of 75°F.
Source: ASHRAE Handbook of Fundamentals, 1972.

TABLE V-C *Water Absorption of Bedding*

Material	*Lb. water absorbed per lb. bedding*
Wood	
Tanning bark	4.0
Dry fine bark	2.5
Pine chips	3.0
sawdust	2.5
shavings	2.0
needles	1.0
Hardwood chips, shavings, or sawdust	1.5
Corn	
Shredded stover	2.5
Ground cobs	2.1
Straw, baled	
Flax	2.6
Oats	2.5
Wheat	2.2
Hay, chopped mature	3.0
Shells, Hulls	
Cocoa	2.7
Peanut	2.5
Cottonseed	2.5
Oats	2.0
Sand	0.25

Source: Dairy Housing and Equipment Handbook. Midwest Plan Service Publ. No. 7, 1976.

TABLE V-D *Space Required to Store Selected Bedding Materials*

	Wt. per cu. ft.		*Storage space per ton*	
Material	*Avg.*	*Range*	*Avg.*	*Range*
	lbs		cu. ft.	
Poor hay, baled	8.0	5–10[a]	250	200–400
Poor hay, chopped	7.5	6–9[b]	275	225–325
Poor hay, loose	4.0	3–5	500	400–600[c]
Sawdust, dry	10.0	8–12	200	175–250
Sawdust, green	20.0	18–22	100	90–110
Shavings, baled	20.0	16–24	100	85–125
Shavings, loose	10.0	8–12	200	175–250
Straw, baled	7.0	6–8[d]	300	250–350
Straw, chopped	5.0	4–6	400	350–500
Straw, loose	3.5	3–4	575	500–650

[a]Figures are for field baled hay. Hay baled in tight, wire bales from the mow or stack may weigh 10 to 20 pounds per cubic foot.
[b]Lower weight is for short (1.5″–2.0″) cut, and higher for long cut (3.0″) chopped hay.
[c]Lower figure is for shallow mows, higher for deep mows in which hay has settled.
[d]Field baled, tight wire bales may weigh 10–15 pounds.
Source: Figures compiled from various sources.

TABLE V-E *Estimated Daily Bedding Requirements for Dairy Cows*

Type of housing	*Type of bedding*		
	Straw	*Shavings*[a]	*Sawdust*[b]
	——lbs. per cow per day——		
Free stalls	3–5	3–6	8–10
Stanchion stalls with rubber mats	2–5	6–8	8–12
Stanchion barn–concrete floor	5–7	8–15	20–42
Bedded pack loose housing[c]	7–12	15–30	40–50

[a]Kiln dried 8% moisture, weight basis.
[b]Fifty percent moisture (green), weight basis.
[c]With separate lots for exercise.
Source: Ace, D. L., Dairy Reference Manual, College of Agric., Ext. Service, Pa. State U., 1970.

TABLE V-F *Volume of Milkhouse and Parlor Wastes*

Weekly operation	*Water volume*
Bulk tank, automatic	50 to 60 gal./wash
manual	30 to 40 gal./wash
Pipeline—volume increases with line diameter or length	75 to 125 gal./wash
Pail milkers	30 to 40 gal./wash
Miscellaneous equipment	30 gal./day
Cow prep, automatic	1 to 4½ gal./wash/cow (estimated average = 2 gal.)
manual	¼ to ½ gal./wash/cow
Parlor floor	40 to 75 gal. daily
Milkhouse floor	10 to 20 gal. daily

Source: Livestock Waste Facilities Handbook, MWPS-18. Midwest Plan Service, Iowa State Univ., Ames, Iowa.

Table V-G *Cost of Producing Milk at Two Different Production Levels in Two Different Regions of the United States*

	Minn.–Wis.		*Southwestern states*	
	12,000	*16,000*	*12,000*	*16,000*
Income				
Milk	$1,605	$2,140	$1,731	$2,308
Cull cows	160	184	142	156
Bull calves	43	46	29	29
Surplus heifer calves	62	80	108	72
Manure credit	15	15	—	—
TOTAL	1,885	2,465	2,010	2,565
Expenses				
Feed				
Grain mixture	333	447	408	586
Forages	396	418	611	561
Other	56	66	28	28
Total Feed	785	931	1,047	1,175
Breeding, vet., utilities, Accounting and supplies	191	211	143	151
Machinery and equipment costs	100	100	89	89
Milk hauling	29	38	68	92
TOTAL	1,105	1,280	1,347	1,507
Return to Capital, Labor, Overhead, Risk, & Management	780	1,185	663	1,058
Capital costs	444	545	330	450
Ownership costs	117	117	106	106
	561	662	436	556
Return to Labor, Overhead, Risk, & Management	219	523	227	502
Labor costs	313	364	260	318
Return to Overhead, Risk, & Management	−94	159	−33	184
Milk Break-Even Price	14.04	12.53	16.69	13.37
Price received: February 1982	13.38	13.38	14.42	14.42

Source: Adapted from February, 1982 Report on Costs and Returns of Milk Production study, Dairy Extension Office, Oklahoma State University, Stillwater, Oklahoma

Table V-H *Examples of the Annual Labor Requirements for a 40-Cow Stall Barn Herd and a 125-Cow Free Stall Herd*

The stall barn system chosen utilizes only minimal labor saving equipment, while the free-stall system has incorporated much equipment for this purpose. The input data used to derive these estimates were obtained from Appendix Table V-I.

The formula used is $a + bx$, where a = fixed time (regardless of herd size), b = added time per cow (or per ton, per acre), and x = number of cows in herd.

Example 1. 40 cow stanchion barn, milking 25 cows per hour, daily haul of manure, hand feeding of hay (3/8 of forage dry matter) and silage, average efficiency of youngstock care, raises 40 acres corn for grain, 35 acres of corn silage, and 30 acres of hay.

Chore routine		*Hr/yr*
Milking cows (25.2 × 40)	1,008	
Set up and clean up	273	
		1,281
Hauling manure (117 + (4.0 × 40))		277
Feeding and bedding		
Silage (10.9 + (.25 × 600T))	161	
Hay (76.4 + (0.5 × 120T))	136	
Bedding (57.2 + (1.5 × 40))	117	
		414
Other dairy		
Heat detection (50.0 + (1.1 × 40))	94	
Breeding (0.6 × 40)	24	
Turn in and out of stall barn (65.0 + (1.2 × 40))	113	
Youngstock care (180.0 + (9.1 × 40))	544	
Miscellaneous (50.0 + (3.2 × 40))	178	
		953
Total, dairy		2,925
Cropping		
Corn (5 × 40A.)	200	
Corn silage (6 × 35A.)	210	
Hay (8 × 30A.)	240	
		650
Total		3,575
Average		9.8 hr/day

Example 2. 125 cow free stall barn, milking 40 cows per hour, 6 month storage of liquid manure with daily scraping, forage is 3/8 hay and 5/8 corn silage (dry matter basis), minimal bedding used, efficient calf raising system, raises 120 acres of corn for grain, 110 acres of corn silage, and 100 acres of hay.

Chore routine		*Hr/yr*
Milking cows (15.8 × 125)	1,975	
Holding corral (122 + (0.4 × 125))	172	
Set up and clean up	365	
		2,512
Handling manure (257 + (3.7 × 125))		720
Feeding and bedding		
Silage (10.9 + (0.25 × 1,875T))	480	
Hay (30.5 + (0.7 × 375T))	293	
Bedding (73.3 + (0.9 × 125))	186	
		959

TABLE V-H *Examples of the Annual Labor Requirements for a 40-Cow Stall Barn Herd and a 125-Cow Free Stall Herd (Continued)*

Chore routine		Hr/yr
Other dairy		
Heat detection (50.0 + (1.1 × 125))	188	
Breeding (0.6 × 125)	75	
Youngstock care (120.0 + (5.5 × 125))	808	
Other (50.0 + (3.2 × 125))	450	
		1,521
Total, dairy		5,712
Cropping		
Corn (3.0 × 120A.)	360	
Corn silage (4.0 × 110A.)	440	
Hay (6.0 × 100A.)	600	
		1,400
Total		7,112
Average, 2 men		9.7 hr/day

TABLE V-I *Estimated Yearly Labor Requirement per Cow*[a]

Chore routine	Annual labor requirement[b]	
	Fixed	Added/cow
	(hr)	(hr)
I. Milking Cows		
Stall barn		
Set up and clean up	273	—
Milking (cows/man-hr): 25	—	25.2
30	—	21.0
35	—	18.0
40	—	15.8
Parlor milking		
Set up and clean up	365	—
Collect cows in holding corral	122	0.4
Milking (cows/man-hr): 30	—	21.0
35	—	18.0
40	—	15.8
60	—	10.5
II. Manure Handling		
Stall barn		
Gutter cleaner, daily haul	117	4.0
Gutter cleaner, stacker storage	102	3.2
Gutter cleaner, silo or earthen basin	113	3.0[c]
Basement storage	103	2.9
Free stall barn		
Tractor scraper, daily haul	252	4.4
Tractor scraper, 6-month storage	257	3.7
Basement storage, slotted floor	166	2.4
Any liquid storage, mechanical scraper	166	2.4[c]
III. Feeding and Bedding		
Stall barn		
Silage feeding (2× daily)	10.9	(0.25/T silage)
Hay feeding (2× daily)	76.4	(0.5T hay)

TABLE V-I *Estimated Yearly Labor Requirement per Cow*[a] *(Continued)*

Chore routine	*Annual labor requirement*[b] *Fixed*	*Added/cow*
	(hr)	(hr)
Bedding distribution, liberal amount	57.2	1.5
Bedding distribution, minimal amount	57.2	0.5
Free stall barn		
Silage feeding (2× daily)	10.9	(0.25/T silage)
Hay feeding (from adjacent storage)	30.5	(0.7/T hay)
Free-stall maintenance and bedding distribution		
liberal amount	73.3	1.8
minimal amount	73.3	0.9
IV. Other Dairy Activities		
Heat detection	50.0	1.1
Breeding chores	—	0.6
Turn in and out of stall barn	65.0	1.2
Youngstock care		
Labor efficient	120.0	5.5
Average	180.0	9.1
Other miscellaneous work	50.0	3.2

Chore routine	*Annual hours per acre*
V. Cropping Programs	
Corn	3–5
Oats	2–4
Corn silage	4–6
Hay	6–10
Hay crop silage	5–7

[a]Adapted from: Brown, L. H., and Speicher, J. A.: How to check your labor efficiency. Hoard's Dairyman *116:*1, 1971; Hoglund, C. R.: Dairy systems analysis handbook, Michigan State Univ. Agr. Econ. Report No. 300, 1976; and personal communication from E. I. Fuller, Dept. of Agr. Economics, Univ. of Minnesota, 1976.
[b]Estimated annual labor requirement is determined by the formula $a + bx$, where a = fixed hours, b = added hours per cow, x = number of cows in the herd. For example, the labor in hauling manure daily from a 40-cow stall barn is: $117 + (4.0 \times 40 \text{ cows}) = 277$ hours.
[c]Reduce by 0.6 hr/cow if distribution is by irrigation rather than liquid manure spreader.

Table V-J *Estimated Investment and Annual Operating Costs for Dairy Facilities and Equipment*[a]

Line	Item	*Investment*[b]		Assumed
		Basic	*Added cost*	*annual cost*[c]
		($)	($/cow)	(%)
	I.[d] Stall Barn, Housing and Milking (40 to 100 cows)			
1	Barn structure, including concrete	10,880	536.75	13.5
2	Stalls	—	118.40	17.0
3	Water, plumbing, and wiring	4,240	21.95	13.5
4	Insulation and ventilation	765	178.10	13.5
5	Milk room	3,700	66.60	13.5
6	Milking system, pipeline	4,195	135.65	24.0
7	Bulk tank	2,070	112.50	21.5
	II.[d] Free Stall Barn, Totally Enclosed System (75 to 250 cows)			
11	Barn structure, including concrete	2,015	555.00	13.5
12	Free stalls	—	78.45	23.0
13	Water, plumbing, and wiring	3,910	26.05	13.5
14[e]	Insulation and ventilation	1,480	222.00	13.5
	III.[d] Free Stall Housing, Outside Feeding (75 to 250 cows)			
21	Barn structure, including concrete	1,275	346.30	13.5
22	Concrete lot or slab	590	136.15	12.5
23	Free stalls	—	78.45	23.0
24	Water system and wiring	3,905	26.05	13.5
	IV. Milking Parlor and Milking System			
31	Milkroom and parlor	22,500	139.10	15.5
32	Milking system	5,475	72.50	24.0
33	Bulk tank	6,510	68.10	21.5
	Add-ons:			
34	Power gates	6,070	—	25.0
35	Feedbowl covers	—	203.50 ($/parlor stall)	25.0
36	Detaching unit	3,255	1,960.00 ($/parlor stall)	25.0
	V.[f] Waste Handling			
	Solid Manure			
41	Gutter cleaner	2,713	54.25	21.0
42	Stacker	3,955	29.10	22.5
43	Storage (6-month)	3,550	88.80	12.5
44	Manure spreader, daily haul	1,580	52.30	22.5
45	Manure spreader, seasonal hauling	3,670	59.95	25.8
	Liquid Manure			
	Basement storage (6-month)			
51	Stall barn, slotted gutter	8,683	611.70	12.5
52	Free stall barn, slotted floor	14,210	603.85	12.5
53	Free stall barn, scraper	11,840	580.15	12.5
54	Silo storage	6,215	182.05	14.5

Table V-J *Estimated Investment and Annual Operating Costs for Dairy Facilities and Equipment*[a] *(Continued)*

Line		Item	Investment[b] Basic	Added cost	Assumed annual cost[c]
			($)	($/cow)	(%)
55		Earthen basin, incl. fence and concrete	2,370	111.00	14.5
56		Gutter cleaner	2,713	54.25	21.0
57		Scraper, tractor	1,895	7.40	25.8
58		Scraper, mechanical	5,755	16.00	25.8
59		Pump and agitator, basement storage	3,020	11.25	26.8
60		Pump and agitator, silo storage	3,995	—	26.8
61		Pump and agitator, earthen basin	3,020	11.25	26.8
62		Underground pump system	6,070	2.95	26.8
63		Ventilation fan, basement storage	520	3.00	26.8
64		Liquid spreader, <100 cow herd	3,465	41.45	25.8
65		Liquid spreader, >100 cow herd	7,015	22.50	25.8
	VI.[g]	Feed Storage and Handling		($/unit)	
		Concrete upright silos, incl. unloaders			
71		Corn silage and low moisture silage	14,210	41.20/ton	15.6
72		High moisture corn	11,335	0.45/bu	16.8
		Sealed upright silos, incl. unloaders			
73		Corn silage and low moisture silage	18,620	82.90/ton	15.6
74		High moisture corn	15,755	1.05/bu	16.8
		Bunker silo, incl. tractor unloader			
75		Corn silage and low moisture silage	7,815	12.00/ton	12.5
76		Feed room, mix and weigh equipment	9,115	—	17.0
77		Mechanical conveying equipment	4,470	—	25.0
		Mechanical feeders			
78		Stall barn, auger or shuttle	590	162.80	25.0
79		Stall barn, belt	590	222.00	25.0
80		Free stall barns, auger or shuttle	590	36.25	25.0
81		Free stall barns, belt	590	65.85	25.0
		Mixing wagons or carts			
82		Stall barns, mechanical cart	1,275	—	20.5
83		Stall barns, electrical cart	5,180	—	25.0
		Free stall barns, mixing wagon			
84		with scales	10,210	48.85	22.8
85		without scales	5,030	48.85	20.5
86		(wider feed alley when feeding is inside)	60	23.70	13.5
87		Feed bunks, both sides	—	28.10	17.0
88		Feed bunks, single side	—	47.35	17.0
89		Hay and bedding storage, pole barn	—	71.05	13.5
	VII.	Miscellaneous Facilities and Equipment			
101		Holding corral, for 60 cows, cold barn	6,060	—	13.5
102		Holding corral, for 60 cows, warm barn	7,115	—	13.5
103		Powered crowd gate	1,775	—	21.0
104		Hospital area, cold barn	1,000	22.50	13.5
105		Hospital area, warm barn	1,255	28.20	13.5
106		2 box stalls for stall barn	3,640	—	13.5

Table V-J *Estimated Investment and Annual Operating Costs for Dairy Facilities and Equipment*[a] *(Continued)*

Line	*Item*	*Investment*[b] *Basic*	*Added cost*	*Assumed annual cost*[c]
		($)	($/cow)	(%)
	VIII.[h] Calf and Youngstock Facilities (70 to 250 cow herds)			
	Warm, insulated and ventilated, total confinement unit, birth to freshening			
111	with conventional manure handling	14,800	461.00	17.5
112	with liquid manure facilities	16,775	559.65	15.0
113	Cold, naturally ventilated calf barn, with open pole barn for older heifers	12,210	326.35	15.0
114	Calf hutches for milk fed calves, open pole barn for weaned calves, open lot with windbreak or shade for yearling heifers	5,180	137.65	17.5
	Partial facilities			
115	30 × 44-ft. cold calf barn for youngest 30 calves, conventional manure	13,615	—	17.5
116	30 × 44-ft. warm calf barn for youngest 30 calves, liquid manure	20,275	—	15.0
117	Calf hutches, each	220	—	21.0

[a]Adapted from: Hoglund, C. R.: Dairy systems analyses handbook. Michigan State Univ. Agr. Econ. Report No. 300, 1976; and adjusted to 1983 prices by correcting for inflation, using source data from: Agricultural Prices, Statistical Reporting Service, USDA, Publication Pr 1–3, 1983.

[b]Estimated investments are determined by applying the following formula: Investment = a + bx, where a = basic cost, b = added cost, x = number of cows (or units). For example, the estimated cost of a stall barn structure (line 1) for a 50 cow herd is: $10,880 + (536.75 × 50) = $37,717.50

[c]Assumed annual costs are based on expected life (including an obsolescence factor), interest at either 10 or 12% rate (depending on expected life) times 50% of value, depreciation, repairs, and insurance.

[d]Stall barn systems: choose from Sec. I, plus appropriate items from Sec. V and VI.
Free stall systems: choose from either Sec. II or III, plus appropriate items from Sec. IV, V, and VI.

[e]Line 14 to be included only in "warm" free stall systems.

[f]Stall barn, daily haul = lines 41 and 44.
Stall barn, seasonal haul = lines 41, 42, 43 and 45.
Stall barn, liquid manure = lines 51, 59, 63 and 64 or
lines (54 or 55), 56, (60 or 61), 62 and 64.
Free stall barn = lines 52, 59, 63, (64 or 65) or
lines 53, (57 or 58), 59, 63, (64 or 65) or
lines (54 or 55), (57 or 58), (60 or 61), 62, (64 or 65).

[g]Stall barn, hand feeding = lines (71 or 73), possibly (72 or 74), possibly 76, possibly 89.
Stall barn, mechanical feeding = lines (71 or 73), possibly (72 or 74), 76, 77, (78 or 79), possibly 89.
Free stall barn, mechanical feeders = choice of (71 through 75), possibly 77, (80 or 81), (87 or 88), possibly 89.
Free stall barn, mixer wagon = choice of (71 through 75), (84 or 85), possibly 86, (87 or 88), possibly 89.

[h]Cost estimates provided are only approximate guidelines, based on 1 replacement heifer for each 3 cows annually for herds exceeding 70 cows. If detailed cost estimates are desired, use Sec. II, III, V, and VI by making appropriate corrections for differences in size of animal (body weight).

Glossary

ABBERRANT. Abnormal.

ABORTION. Premature expulsion of the fetus from the uterus.

ACETIC ACID. CH_3COOH, the acid of vinegar and one of the volatile fatty acids produced by ruminal bacteria.

ACTIVE IMMUNITY. Immunity acquired when an individual produces immune products (e.g., antibodies) in response to an antigenic stimulus.

ADDITIVITY. Gene effect which causes a predictable phenotypic change to occur when one gene is substituted for another in its allelic series, no matter what other genes are present in the genotype.

ADENOHYPOPHYSIS. Anterior pituitary.

ADENOSINE TRIPHOSPHATE. Nucleotide containing 3 phosphoric acid groups, 2 of which yield a large amount of energy upon hydrolysis.

ADHESION. Fusion by formation of new tissue between surfaces normally separate.

ADRENALECTOMY. Removal of the adrenal gland.

AFFERENT. Leading toward a center; opposed to *efferent.*

AGE AND MONTH-OF-CALVING FACTORS. Factors used to eliminate the environmental effects of different ages and months of the year at calving in order to standardize lactation records for genetic evaluations.

AGGLUTINATION. Aggregation or clumping of cells in a fluid.

ALIMENTARY TRACT. Pathway of food and residues through the body.

ALLELES. Genes located at the same point (locus) on each of a pair of chromosomes.

ALTERNATE AM-PM SAMPLING PLAN. Dairy record-keeping plan in which milk weights and samples are taken at only one milking each test period, the milking alternating from AM to PM in succeeding test periods.

ALVEOLUS. Hollow ball of cells.

AMBIENT AIR TEMPERATURE. Temperature of the air encompassing an object on all sides.

AMINE. One of a class of compounds which contain nitrogen and are derived from ammonia by replacing hydrogen atoms with an organic radical.

AMINO ACID. One of a class of organic compounds containing the amino (NH_2) group and the carboxyl (COOH) group. They form the chief structure of proteins.

AMMONIA. NH_3, a colorless alkaline gas with a penetrating odor; soluble in water.

ANAEROBIC. Pertaining to or living in the absence of oxygen.

ANALOG. Having similar function but different structure.

ANASTOMOSIS. Union of a hollow organ or vessel with another.

ANCESTOR. Animal of a previous generation which has passed on genes through a line of descent.

ANEMIA. Deficiency in the quality or quantity of blood.

ANESTRUS. Period of sexual quiescence.

ANLAGE. First recognizable rudiment of a developing tissue or organ. Plural is *anlagen.*

ANOMALY. Deviation from normal.

ANTHELMINTIC. Agent which destroys worms in the digestive tract.

ANTIBIOTIC. Chemical substance produced by microorganisms which inhibit or destroy bacteria and other microorganisms.

ANTIDIURETIC. Suppressing urinary secretion.

ANTIDIURETIC HORMONE. Hormone of the posterior pituitary which promotes water resorption in the kidneys and increases blood pressure by constricting arterioles and coronary blood vessels.

ANTIGEN. Any substance, usually a protein, that stimulates production of a specific antibody in an animal.

ANTIOXIDANT. Compound which prevents oxidation.

ANTRUM. A cavity or chamber.

ARTERIOLE. Small branch of artery nearest a capillary.

ATAXIA. Lack of normal coordination of parts of the body, especially inability to coordinate movement of voluntary muscles.

ATHEROSCLEROSIS. Fatty deposits on the lining of the large and medium-sized arteries.

ASH. Residue left after burning; primarily composed of minerals.

ATRESIA. Degeneration of unruptured ovarian follicles.

AUTONOMIC. Self-controlling; pertaining to that portion of the nervous system which regulates involuntary processes.

BACKFLUSH. A system for sterilization of rubber inflations.

BACTERIA. Large group of widely distributed one-celled microorganisms, which may appear singly or in colonies as spherical, rod-shaped, or spiral threadlike cells.

BACTERIN. Vaccine containing attenuated or killed bacteria administered to increase the resistance of an animal to a specific infectious disease.

BACTERIOSTATIC. Inhibiting growth of bacteria without destroying them.

BAKEWELL, ROBERT. British livestock breeder who, during the period 1760 to 1795, was the first to use successfully estimates of breeding value, selection, and systems of mating to make genetic improvement.

BATESON, WILLIAM. British geneticist (1861–1926) who influenced the science of genetics by studying genetic variation as a basis for evolutionary change.

BIAS. Average error in estimating the true value of a trait from a sample of data that causes an overestimate (positive bias) or an underestimate (negative bias).

BIMONTHLY D.H.I. An unofficial dairy record-keeping plan in N.C.D.H.I.P. where the same procedures are usually followed as in Official D.H.I. except that the intervals between tests are 2 months long instead of approximately 1 month.

BIOMETRY. Study of variation in biological organisms.

BLASTULA. Stage in embryonic development in which cells are arranged to form a single-layered hollow sphere.

BLOOD SERUM ALBUMIN. Chief water-soluble protein in the serum of blood.

BOLUS. Mass of food ready to be swallowed or regurgitated.

BREED. Group of animals having a common origin and identifying traits (frequently color) which distinguish them as belonging to a certain group.

BREEDING VALUE. Genetic worth of an animal's genotype for a certain trait; twice the animal's genetic transmitting ability.

BUTTONS. See *cotyledons.*

BUTYRIC ACID. $CH_3CH_2CH_2COOH$, a rancid, sticky acid from the putrefaction of protein; one of the volatile fatty acids produced by ruminal bacteria.

CALORIE. Amount of heat required to raise the temperature of 1 gram of water from 14.5°C to 15.5°C. (Note: This term must be spelled with a small *c*.)

CALORIGENIC. Producing heat or energy.

CAPACITATION OF SPERM. Maturation of sperm within the reproductive tract of the female which permits fertilization.

CARBOHYDRATE. One of a class of compounds with the general chemical formula $(CH_2O)_n$. They include the sugars, starches, cellulose, and gums. They are formed in all green plants by photosynthesis.

CARBON DIOXIDE. CO_2, an odorless, colorless gas resulting from oxidation of carbon. It is formed in the tissues and excreted by the lungs.

CARCINOGEN. Any cancer-producing substance.

CAROTENE. Fat-soluble orange or red pigment of plants which may be transformed in the animal body into vitamin A.

CARRIER. Animal or person who carries disease organisms without showing symptoms of the disease.

CARUNCLE. Maternal cotyledon; a specialized area on uterine lining in ruminants where placenta attaches.

CASEIN. One of a group of several phosphoproteins which comprise the principal proteins in milk.

CATABOLISM. Metabolic breakdown of complex substances to simple substances.

CATALYZE. To change the velocity of a reaction. A catalyst is any substance that affects the rate of reaction but may be recovered unchanged at the end of the reaction.

CATHETER. Hollow tube designed to be inserted into various body cavities for the withdrawal of fluids or, in the case of artificial insemination catheters, deposition of semen.

CELLULOSE. Carbohydrate $(C_6H_{10}O_5)_n$ forming the skeleton of most plant structures and plant cells.

CESAREAN SECTION. Delivery of a fetus through a surgical incision of the abdominal and uterine walls.

CHROMOSOME. Bodies, occurring in pairs in the nuclei of cells, which carry the genetic material in the form of "genes" arranged linearly along the chromosome.

CLASSIFICATION SCORE. A final point rating placed on animals in the breed association conformation rating programs according to how closely each animal resembles the breed's ideal conformation.

CLAW (OF MILKING MACHINE). A chamber to which the teat cups are attached; delivers milk from the teat cups to the milk line.

CLOSEBREEDING. An intense form of inbreeding usually applied to the mating of closely related animals such as full sibs or parent-offspring.

COAGULATION. Process of clot formation.

CO-ENZYME. Chemical co-factor needed by an enzyme for efficient performance of catalytic function.

COITUS. Sexual union of two individuals of the opposite sex.

COLLATERAL RELATIVES. Animals which are related but not linearly.

COLLINGS BROTHERS. British cattle breeders who successfully applied many of Bakewell's principles plus their own insight in a manner which led to the formation of the Shorthorn breed and establishment of the first herdbook for cattle.

COLLOID. Any substance in a state of fine division with particles ranging from 100 millimicrons to 1 millimicron dispersed throughout the medium.

COMPLEMENTARY MILK. Portion of milk present in the udder that is not removed even by thorough milking. Also called *residual milk.*

COMPLEMENTARY SIRE SELECTION. Applying selection criteria to a group of bulls to meet herd breeding goals, rather than selecting bulls individually.

COMPONENT MILK PRICING. Payment for milk partially on the basis of composition in addition to milk fat, usually *solids-not-fat* or protein content.

CONCEPTION. Fertilization of the ovum.

CONFIDENCE INTERVAL. The interval above and below an estimated value within which we are confident that the true value lies with a specified level of probability.

CONJUNCTIVA. Membrane that lines the inner surface of the eyelids and covers the frontal margin of the eyeball.

CONTEMPORARY COMPARISON. Method for estimating the genetic transmitting ability of bulls and cows using information on contemporaries; i.e., cows that were born, raised, and completed their lactations during a certain time period, frequently limited to first lactation records.

CONTEMPORARY GROUP. A grouping of records into 2 groups, first lactations in one group and all other lactations in another group; used to increase accuracy in the U.S.D.A.-M.C.C. genetic evaluation procedure.

CONTINUOUS DISTRIBUTION. Method of depicting the frequency of occurrence of different levels of traits which do not occur in discrete classes (such as sex) but vary in increments too small to measure (such as body weight). Such traits are usually measured for convenience in arbitrary finite units such as pounds, kilograms, inches, or centimeters.

CONTOUR TILLAGE. Plowing a sloping field along contour lines in which all points have the same elevation for the purpose of controlling soil erosion.

COORDINATING GROUP. Abbreviated title of the Coordinating Group for the National Cooperative Dairy Herd Improvement Program, which is the representative body with authority delegated from the sponsors of N.C.D.H.I.P. to oversee the operation of N.C.D.H.I.P.

CORNIFICATION. Conversion of living cells lining tissues to a hard-packed layer of dead cells.

CORONARY. Pertaining to the blood vessels which supply blood to the heart muscle.

CORPUS LUTEUM. Temporary structure formed on the ovary after ovulation. During the time in which it secretes progesterone, the animal does not undergo estrous cycles.

CORRELATION. Tendency of two or more traits to vary in the same direction (positive correlation) or in opposite directions (negative correlation) due to common forces or influences. Mathematically it ranges from +1 to −1.

COTYLEDON. Area of fetal membranes which join the maternal caruncles of the uterus to form the placenta. Sometimes called *buttons*.

COUNTERVAILING DUTIES. Import assessments against government subsidized exports from other countries.

COVARIANCE. Variation which is common between two traits due to the same genetic and/or environmental effects acting on both traits.

COVER CROP. Crops, such as rye, planted between seasons for cultivated crops to protect the soil from erosion by wind or water.

COW INDEX. Name commonly given to values of genetic transmitting ability of cows calculated by the U.S. Department of Agriculture.

CROSSBREEDING. Mating of animals of different breeds.

CRYOSCOPE. Instrument used to determine the freezing point of liquids.

CRYPTORCHIDISM. Failure of one or both testes to descend into the scrotum.

CULLING. The removal from a herd of cows of lower genetic or phenotypic merit to increase production and profits.

CUSTOM FREEZING. The drawing and freezing of semen from a specified bull by someone not the owner of the bull on a contract basis.

CYANOSIS. Bluish discoloration of the skin or mucous membranes as a result of an excessive concentration of reduced hemoglobin in the blood.

CYST. Pouch or sac filled with fluid or semisolid material.

DAIRY RECORDS PROCESSING CENTERS. The computing facilities where information from the periodic tests in N.C.D.H.I.P. herds is summarized and analyzed and where information to be used in future management decisions is produced and returned to the dairyman.

DAM. Female parent.

DAUGHTER-DAM COMPARISON. Method of estimating genetic transmitting ability by expressing a cow's yield as a deviation from her dam's yield. This procedure has been superseded by several more accurate methods.

DAUGHTER-HERDMATE DEVIATION. Amount by which a cow differs in yield or other traits from other cows calving in the same herd during the same time period.

DE. Digestible energy.

DENATURE. Destruction of the native state of a substance.

DEPOT FAT. Fat stored in the body.

D.H.I.A. Dairy Herd Improvement Associations, an association which dairymen join to participate in the dairy record-keeping plan sanctioned under the National Cooperative Dairy Herd Improvement Program.

D.H.I.R. Dairy Herd Improvement Registry, a dairy record-keeping plan sponsored by the breed associations.

DIABETES MELLITUS. Disorder in carbohydrate metabolism caused by a lack of insulin. Characterized by excess glucose in blood and urine.

DIAPHRAGM. Musculomembranous partition separating the abdominal and thoracic cavities.

DICOUMAROL. Blood anticoagulant isolated originally from spoiled sweet clover and later made synthetically.

DIFFERENTIAL (IN MILK PRICING). An amount added to or subtracted from the price of milk at a specified base content of fat or other components based on the level of fat or components in the milk.

DIFFERENTIATION. Development of specialized form or function.

DIPLOID. Having two sets of homologous chromosomes. Somatic cells normally are diploid, whereas gametic cells have only one set of chromosomes, i.e., are haploid.

DIRECT COMPARISON METHOD. Procedure for estimating the relative genetic transmitting ability of a group of sires by comparing the relative yield of their daughters in the same herds and time periods; also referred to as the Linear Model Method.

DISACCHARIDE. One of a group of sugars which are composed of two monosaccharides.

DISCRETE DISTRIBUTION. Method of depicting the frequency of occurrence of different categories of traits which occur in discrete classes, such as sexes or months of calving.

DISPERSION. Distribution of particles of one substance in a medium of another substance, frequently a fluid.

DIURETIC. Increasing urinary secretion.

DM. Dry matter.

DOMINANCE. The effect where one gene of an allelic series masks the phenotypic expression of another gene in the same allelic series. The masking effect may be complete or incomplete.

DORSAL. Pertaining to the top surface of an animal.

D.R.P.C. Abbreviation commonly used for the Dairy Records Processing Centers.

DRY PERIOD. Period of non-lactation between two periods of lactation.

DUAL-PURPOSE. Those breeds of cattle, such as Milking Shorthorns, in which selection is practiced for both carcass and milking qualities.

DYSPNEA. Difficult or labored respiration.

DYSTOCIA. Abnormal or difficult labor and/or birth.

ECTODERM. Outermost of the three germ layers of the embryo.

EDEMA. Accumulation of fluids in the intercellular tissue spaces of the body.

EFFERENT. Leading away from a center; opposed to *afferent.*

EJACULATION. Sudden or rhythmic discharge of sperm and seminal fluid from the male.

EMBRYO. Period in the development of an individual between conception and the completion of organ formation. See *fetus.*

EMBRYO TRANSPLANT. The removal of an embryo in the very early stages of development from one cow and insertion of that embryo into another cow for development and birth. Transfer is usually made from a genetically superior cow to a genetically inferior one.

EMULSION. Dispersion of fine particles or globules of one liquid in another liquid.

ENDOCRINE. Pertaining to internal secretions elaborated directly into the blood or lymph that affect another organ or tissue in the body.

ENDODERM. Innermost of the three germ layers of the embryo.

ENDOGENOUS. Developing or originating within the organism.

ENDOMETRIUM. Inner lining of the uterine wall.

ENE. Estimated net energy.

ENTODERM. See *endoderm.*

ENVIRONMENT. All the external factors within which an animal's genotype acts to determine the animal's phenotypic traits.

ENVIRONMENTAL VARIANCE. That portion of the phenotypic variance which is caused by environmental effects.

ENZYME. Organic compound, frequently protein, capable of accelerating or producing by catalytic action some change in a substrate.

EOSINOPHIL. Type of white blood cell that has a particular affinity for the dye eosin.

EPINEPHRINE. Potent vasopressor hormone derived from the adrenal medulla.

EPISTASIS. Interaction of genes which are not in the same allelic series, to affect the phenotype of a trait.

EPITHELIUM. Tissue lining the internal cavities and covering the external surfaces of the body.

EQUAL PARENT INDEX. One of the earliest indexes of a sire's genetic transmitting ability. It assumed that the daughters' average yield was halfway between the sire's transmitting ability and the average yield of the dams of the sire's daughters.

EQUITY. Value of a property over and above any debts or liens against it.

ERUCTATION. Act of belching.

ERYTHROCYTE. Red blood cells which carry oxygen.

ESCUTCHEON. The area just above the rear udder.

ESOPHAGUS. Canal from the mouth to the stomach.

ESTIMATED PRODUCING ABILITY. An estimation of the amount of milk and/or components that a cow will yield based on pedigree information and including the cow's performance if available.

ESTIMATED TRANSMITTING ABILITY. An estimation of an animal's genetic transmitting ability based on pedigree information and the animal's performance if available.

ESTRADIOL-17β. Principal estrogenic steroid hormone secreted from the ovary.

ESTROUS. Pertaining to the entire cycle of reproductive changes in the nonpregnant female mammal.

ESTRUS. Period of sexual receptivity in female animals. Also called heat.

ETIOLOGY. Study of the causes of disease.

EXCRETION. Elimination of waste products from the body.

EXOCRINE. Pertaining to secretions elaborated into a duct and conveyed to other organs or to the outside of the body.

EXOGENOUS. Originating outside the organism; pertaining to the introduction of materials into the body from outside.

EXPORT SUBSIDIES. Government subsidies on exports to make them more competitive in international trade.

EXUDATION. Abnormal discharge of fluid.

FARM-RETAIL PRICE SPREAD. Difference between the price paid by consumers for one unit of a food product and the farm value of an equivalent amount of the product.

FAT-CORRECTED MILK (FCM). Adjustment of milk with different fat percentages to equivalent amounts on an energy basis. The two most commonly used formulas are:
4% FCM = (0.4 × pounds of milk) + (15 × pounds of fat)
3.5% FCM = (0.4324 × pounds of milk) + (16.218 × pounds of fat)

FATTY ACID. One of a group of compounds with the general chemical formula $CH_3 \cdot (CH_2)_nCOOH$ which usually combine with glycerol to form glycerides or fats.

FECES. Excrement discharged from the intestines.

FEDERAL MILK ORDER. U.S. government regulated farm milk pricing.

FERTILE. Capable of producing offspring.

FERTILIZATION. Union of sperm and egg cells.

FETUS. Period in the development of an individual between completion of organ formation and birth. See *embryo*.

FISHER, SIR RONALD A. British biometrician (1890–1962) who developed much of the underlying theory of modern biological statistics and statistical genetics.

FOLLICLE. Ovarian structure containing the egg and its accompanying cells.

FORMIC ACID. HCOOH, a colorless pungent liquid.

FREEMARTIN. An infertile female calf born co-twin to a bull; the abnormal genital tract development present in the female is the result of circulatory connections between the 2 sexes of calves in utero; approximately 90% of the female members of female/male twin sets are freemartins.

FREQUENCY DISTRIBUTION. Method of depicting the proportion of animals whose phenotypes occur at various points along a horizontal axis; the height of the line drawn over the scale representing the frequency of occurrence at each point on the scale.

FRESHEN. To give birth to a calf and simultaneously to begin a period of lactation.

FULL SIBS. Animals with the same sire and dam; for example, full brothers, full sisters, or a full brother-sister pair.

GALACTOPOIETIC. Pertaining to stimulation of an established lactation.

GALACTOSE. Carbohydrate which in combination with glucose forms lactose.

GALTON, SIR FRANCIS. British statistician (1822–1911) who is considered to be the father of the science of eugenics; i.e., the application of statistical theory for genetic improvement.

GALVANIZE. To coat with zinc.

GAMETE. A mature reproductive cell, either an ovum or sperm, containing genes representing a random sample of one-half an animal's breeding value or its transmitting ability.

GAMMA GLOBULIN. Protein fraction of blood serum especially high in antibodies.

GENE. Hereditary unit, located on a chromosome, which affects a specific trait and is not known to subdivide.

GENE FREQUENCY. Proportion (percentage) of a certain gene in its allelic series.

GENERATION LENGTH (INTERVAL). Average length of time from birth of one generation to birth of the succeeding generation.

GENETIC BASE. The average genetic merit of a population (usually a breed) at a specified time used as a reference point to express estimates of genetic transmitting ability.

GENETIC CORRELATION. Tendency of two traits to vary in the same direction (positive genetic correlation) or in the opposite direction (negative genetic correlation) due to common genetic influences.

GENETIC TRANSMITTING ABILITY. The average genetic merit for some specified trait that is passed on by a parent to its offspring; one-half breeding value.

GERM PLASM. The full complement or half complement of genetic material in the form of live animals, bull semen, cow ova, or embryos.

GESTATION. Pregnancy. The period of development of an individual between fertilization and birth.

GLAND. Any organ that secretes a specific product for use in or discharge from the body.

GLUCOCORTICOID. Steroid hormone from the adrenal cortex which stimulates the formation of carbohydrates by breaking down non-carbohydrate materials such as fat and protein.

GLUCONEOGENESIS. Formation of glucose from non-carbohydrates, such as fats or proteins.

GLUCOSE. Primary sugar in blood, $C_6H_{12}O_6$, used for energy, lactose synthesis, ribonucleic acid synthesis, and glycerol synthesis.

GLYCEROL. Three carbon molecule, $CH_2OH \cdot CHOH \cdot CH_2OH$, which normally combines with three fatty acids to form a triglyceride.

GLYCOGEN. Principal storage form of glucose primarily found in liver and muscle with the general chemical formula $(C_6H_{10}O_5)_n$.

GLYCOLYSIS. Metabolic breakdown of carbohydrates.

GOITER. Enlarged thyroid gland.

GONAD. Germ cell-producing tissue; an ovary or testis.

GONADOTROPHIN. See *gonadotropin.*

GONADOTROPIN. Hormone that stimulates the gonads. In cattle this would include FSH and LH.

GRADE CATTLE. Animals possessing the distinct characteristics of a particular breed but not registered with a breed association. Most grades have one registered parent, usually the sire.

GRADING UP. Use of sires (usually purebred) of superior genetic merit on cows (usually grades) of lesser genetic merit.

GRANULOSA. Cells lining the ovarian follicle.

GREEN MANURE CROPS. Crops, such as rye or clover, planted and plowed under to improve the fertility and texture of the soil by adding organic matter. They may also function as cover crops.

HALF SIB. One of a pair of animals with a common parent, meaning a half-brother or half-sister.

HAPLOID. Condition found in the sex cells or gametes in which half the normal (diploid) number of chromosomes are found. Condition necessitated by sexual reproduction and occurs during meiosis.

HEAT. An increase in temperature. Also, another name for estrus.

HEMOGLOBIN. Red pigment in red blood cells that transports oxygen to the tissues.

HEMORRHAGE. An excessive loss of blood from the blood vessels.

HEMORRHAGIC SEPTICEMIA. See *shipping fever.*

HERD BOOK. Permanent record of the identification and ancestry of livestock.

HERDMATE COMPARISON. Method of estimating genetic transmitting ability in which a cow's yield is compared to that of other cows which calved in the same herd during the same season.

HEREDITARY VARIANCE. That portion of the phenotypic variance that is due to genetic causes usually including additive, dominance, and epistatic effects.

HERITABILITY. That fraction or proportion of variation in a trait which is due to genetic effects (usually defined in terms of additive genetic effects for purposes of selection).

HERRINGBONE MILKING PARLOR. An overhanging, zigzag-designed milking stall which allows group milking of several cows at a time.

HETEROSIS. Hybrid vigor; the amount by which the F_1 generation exceeds the average of the parents (genetic heterosis) or the superior parent (economic heterosis).

HETEROZYGOUS. Pertaining to that condition where the two alleles at a given locus in an animal are not the same; i.e., the two chromosomes contain different alleles at that locus.

HISTOLOGY. Study of microscopic anatomy of tissues.

HISTOTROPH. The total of uterine secretions that nourish the developing embryo before the establishment of the placenta.

HOMEOSTASIS. Maintenance of a stable internal environment in the body.

HOMOZYGOUS. Pertaining to that condition where both genes at a particular locus are the same allele; i.e., the two chromosomes both contain the same allele at that locus.

HORMONE. Specific chemicals secreted from endocrine glands which influence the activities of other organs in the body.

HYBRID VIGOR. See *heterosis.*

HYDROCHLORIC ACID. HCl, a normal constituent of gastric juice.

HYDROGENATE. To combine with hydrogen; to reduce.

HYDROSALPINX. Distension of the oviduct with fluid.

β-HYDROXYBUTYRATE. Four-carbon ketone body used as a source of short-chain fatty acids in milk fat synthesis. Excess blood levels result in ketosis.

HYPERGLYCEMIA. Higher than normal concentration of glucose in the blood.

HYPOGLYCEMIA. Lower than normal concentration of glucose in the blood.

HYPOPHYSECTOMY. Removal of the pituitary gland.

HYPOPLASIA. Incomplete or reduced development.

HYPOTHALAMUS. Portion of the brain that regulates the anterior pituitary and many visceral activities.

HYSTERECTOMY. Removal of the uterus.

IMMUNE. Protected against a specific disease or poison.

IMMUNE GLOBULIN. Protein that contains a fraction of the serum antibodies.

IMPLANT. To insert or graft material into intact tissues.

IMPORT QUOTAS. Maximum legal imports permitted for various dairy products.

INBREEDING. Mating of animals which are more closely related than the average relationship in the population.

INBREEDING COEFFICIENT. Percentage which expresses the degree of extra homozygosity in an inbred animal due to the relationship between its parents.

INDEPENDENT CULLING LEVELS. Method of selection in which minimum acceptable phenotypic levels are assigned to several traits independently, and cows falling below any of these levels are culled regardless of their levels for the other traits.

INFERTILITY. Capacity for reproduction is reduced but not eliminated.

INFLAMMATION. Tissue response to injury characterized by pain, swelling, redness, and heat.

INORGANIC. Not of organic origin.

INSEMINATION. Deposition of semen in the female reproductive tract.

INTERACTION. Interacting of two sources of variation on a trait such that the effect caused by either source of variation is dependent in part on the influence of the other source.

INTER-HERD EFFECTS. Environmental effects which are different from herd to herd.

INTERPOLATION. Estimation of points falling between those given in a table.

INTERROGATOR. In electronic identification, the device that activates a passive transponder and receives and interprets an electronic signal from the transponder.

INTERSTITIAL CELLS. Cells located between the seminiferous tubules of the testis which secrete androgens. Also called *Leydig cells.*

INTRA-. A prefix meaning within.

INTRA-HERD EFFECTS. Environmental effects that affect all cows in a herd in much the same manner in the same time period.

INTRAMAMMARY PRESSURE. Pressure within the mammary gland.

INVOLUTION. Reduction in size and/or activity of an organ or tissue.

IOFC. The abbreviation for income-over-feed-cost; a figure on relative profitability of individual cows provided to dairymen in N.C.D.H.I.P.

IONIZE. To separate into ions, which are atoms or groups of atoms having either a positive (cation) or negative (anion) electrical charge.

ISLETS OF LANGERHANS. Irregular groups of cells in the pancreas which secrete insulin.

ISOTONIC. Having the property of equal tension, especially osmotic pressure.

KETONE BODY. Any compound containing the carbonyl group, CO.

KETONURIA. Presence of ketone (acetone) bodies in the urine.

KETOSIS. Metabolic disease characterized by excessive ketone body formation and high blood ketone level.

KILOCALORIE. 1000 calories; the amount of heat required to raise the temperature of one kilogram of water from 14.5°C to 15.5°C. Also called *large calorie.*

LABIA. Lips, or lip-shaped organ.

α-LACTALBUMIN. A milk protein; also a subunit of the enzyme lactose synthetase.

LACTEAL. Lymphatic vessel of the intestines which absorbs large amounts of fat, or chyle.

LACTOGEN. Prolactin.

β-LACTOGLOBULIN. A milk protein.

LACTOSE SYNTHETASE. Two-unit enzyme, composed of galactosyl transferase and α-lactalbumin. Catalyzes the formation of lactose from glucose and UDP-galactose.

LEUCOCYTE. See *leukocyte.*

LEUKOCYTE. White blood cells of which there are several types including neutrophils, basophils, eosinophils, monocytes, and lymphocytes.

LEYDIG CELLS. See *interstitial cells.*

LIBIDO. Sexual desire.

LINEAR MODEL PROCEDURE. A procedure for estimating the genetic transmitting ability of bulls by directly or indirectly comparing their daughters' performance in the same herds and seasons using complex matrix manipulations; sometimes referred to as the direct comparison procedure and also as Best Linear Unbiased Prediction (B.L.U.P.).

LINEBREEDING. Form of inbreeding in which an attempt is made to concentrate the genes of a superior ancestor in animals of later generations.

LOCUS. Region of a chromosome where a particular gene is located.

LONG-CHAIN FATTY ACIDS. Fatty acids containing more than 16 carbon atoms.

LUMEN. Cavity within any duct, tube, or alveolar organ.

LUSH, J. L. American animal geneticist who has developed many modern practices used in quantitative genetics as applied to livestock improvement.

LUTEINIZATION. Formation of the corpus luteum.

LUTEOLYTIC. Pertaining to the destruction of the corpus luteum.

LYMPHOCYTE. A type of white blood cell involved in the immune response.

MACROMOLECULES. Large molecules with molecular weights between a few thousand and hundreds of millions.

MATURE EQUIVALENT. Standardization of lactation records to the level of yield that would have been attained by each cow if she had been a mature cow and calved in the month of the year of highest calving frequency for her breed.

MCC. Abbreviation for the U.S.D.A.'s Modified Contemporary Comparison procedure of genetic evaluation.

ME. Metabolizable energy.

MEAN. Statistical and mathematical term for the average value of a set of values.

MEGACALORIE. 1,000,000 calories; equivalent to a Therm, an energy term used in older literature.

MEIOSIS. A type of cell division which produces the sex cells, or gametes, which contain half the number (haploid) of chromosomes found in the somatic cells of the same species.

MEMBRANE. Thin sheet of tissue which covers a surface or divides a space or organ.

MENOPAUSE. Cessation of menstruation in human females due to aging.

MESODERM. Middle layer of the three germ layers of the embryo.

MESSENGER (M) RNA. Ribonucleic acid formed on a template of deoxyribonucleic acid in the nucleus which subsequently moves to the ribosomes and specifies the sequence whereby amino acids are linked together to form a protein.

METABOLISM. Sum of the physical and chemical processes whereby the living organism is produced and maintained.

METABOLITE. Any substance produced in metabolism.

METHANE. CH_4, a colorless, odorless, inflammable gas produced by the decomposition of organic matter and excreted as one of the waste products of ruminal fermentation.

METRITIS. Uterine inflammation.

MICELLE. Colloidal particle composed of casein molecules in milk.

MICROFLORA. Plant life, visible only under a microscope, which is present in or characteristic of a special location, such as in the rumen.

MIGRATION. Introduction of new genotypes into a population by bringing in animals from outside the population.

MILK COMPONENTS. Usually refers to one or more of the 3 substances in milk that are considered to be (potential) bases for milk pricing: fat, protein, and solids-not-fat.

MILK FAT. A currently more popular and acceptable term that is synonymous with butterfat.

MILK ONLY RECORD (MOR). Dairy record-keeping plan similar to Official D.H.I. except that no milk fat samples are taken for determining fat content and the plan has "unofficial" status.

MILK WELL. Opening in the abdominal wall through which the milk vein (subcutaneous abdominal vein) enters to join the vena cava and return blood to the heart from the udder.

MINERALOCORTICOID. Steroid hormone from the adrenal cortex which promotes the retention of sodium and loss of potassium in the kidney.

MITOSIS. Cell division of somatic cells including longitudinal splitting of the chromosomes so that each daughter cell has the same number (diploid) of chromosomes as the parent cell before division.

MODIFIED CONTEMPORARY COMPARISON. A procedure adopted by U.S.D.A. in 1974 to eliminate assumptions inherent in the herdmate comparison and to provide highly accurate sire summaries and cow indexes.

MODIFYING GENES. Genes which cause minor changes in the phenotype by interacting with the gene(s) which are the major determiners of that phenotype.

MOLECULE. The smallest portion of an element or compound that retains chemical identity with the substance in mass.

MORPHOLOGY. Study of form and structure of plants and animals.

MOTOR NERVE. Nerve that conducts impulses away from the central nervous system to the periphery to cause some effect.

MUCOSA. Mucous membrane.

MUCOUS. Pertaining to mucus.

MUCUS. Slimy, sticky fluid produced by certain membranes.

MULTIPAROUS. Females having had two or more pregnancies which resulted in live offspring.

MULTIPLE ALLELES. Existence of 3 or more forms of a gene at a given locus in a population.

MULTIUNIT. In the dairy industry, the term *multi-unit* applies to cooperatives or corporations which own and operate several plants producing one or more manufactured dairy products in contrast to a single plant operated by an owner or partnership.

MUMMIFICATION. Drying up of a dead fetus.

MUTATION. Change in the genetic code at a locus which causes a different phenotypic effect and is transmissible to offspring.

MYOEPITHELIUM. Muscle-like cells that surround the alveoli of the mammary gland which in the presence of oxytocin contract to squeeze milk out of the udder.

MYOMETRIUM. Muscle layer of the uterus.

N.A.A.B. Abbreviation for the National Association of Animal Breeders, a trade association composed of the artificial insemination organizations.

NATIONAL D.H.I.A., INC. A national organization of dairymen whose herds are enrolled in N.C.D.H.I.P.

N.C.D.H.I.P. Abbreviation for the National Cooperative Dairy Herd Improvement Program.

NEGATIVE FEEDBACK. Mechanism whereby a hormone acts to inhibit the secretion of another hormone.

NE_g. Net energy for body weight gain.

NE_l. Net energy for lactating cows.

NE_m. Net energy for body maintenance.

NEUROHORMONE. Hormone which is synthesized in neurons.

NEUROHYPOPHYSIS. Posterior pituitary.

NEURON. A nerve cell.

NEUROTRANSMITTER. Chemicals associated with the nerves that are involved in transmission of impulses.

NONPROTEIN NITROGEN (NPN). Nitrogen in feed from substances such as urea, ammonium salts, amines, amino acids, etc., but not from preformed protein.

NORMAL DISTRIBUTION. Frequency distribution usually characteristic of biological traits and therefore the basis of many statistical procedures used in biological statistics and quantitative genetics.

NPN. Nonprotein nitrogen.

NYMPHOMANIA. Excessive sexual desire in the female. In dairy cows, the most common symptom is prolonged or constant estrus.

OFFICIAL D.H.I. Most widely used dairy record-keeping plan in the United States, records being recognized for "official" purposes such as sire summaries, cow indexes, awards, and advertising.

OOGENESIS. See *ovogenesis*.

ORGANELLE. Subcellular structures that make up a cell.

ORGANIC. Pertaining to substances derived from living organisms.

OSMOSIS. Differential passage of solute molecules across a semi-permeable membrane which is completely permeable to the solvent.

OSMOTIC EQUILIBRIUM. Balance in distribution of solutes achieved between two solutions separated by a semi-permeable membrane.

OUTBREEDING. System of mating animals which are less closely related than the average relationship in the population.

OUTCROSSING. System of mating in which several generations of inbreeding are broken by introducing animals not related to the inbred lines.

OVA. Plural of *ovum*.

OVA TRANSPLANT. The removal of an unfertilized female reproductive cell from one cow and insertion into another cow for fertilization and gestation.

OVOGENESIS. Formation and development of the ovum.

OVULATION. Release of a mature ovum from an ovarian follicle.

OVUM. Reproductive cell or gamete of the female.

OWNER-SAMPLER. Dairy record-keeping plan similar to Official D.H.I. except that milk weights and samples are taken by the dairyman instead of a D.H.I.A. supervisor; lactation records are "nonofficial."

OXIDATION. Loss of electrons from an atom or its gain in positive charges; for example, the chemical combination of oxygen with another atom.

PALPATION. Feeling by hand.

PANCREAS. Large, elongated gland which produces pancreatic juice involved in digestion, and secretion of insulin and glucagon for regulation of carbohydrate metabolism.

PAPILLAE. Small, nipple-shaped projections or elevations.

PARITY. Number of times a cow has calved.

PARTURITION. Act of giving birth to young.

PASSIVE IMMUNITY. Immunity acquired when an individual receives preformed immune products (e.g., antibodies) produced by another individual.

PATENT. Open or unobstructed.

PATERNAL HALF SIBS. Half-brothers or half-sisters having the same sire but different dams.

PATHOGEN. Any disease-producing organism.

P.D. Abbreviation for Predicted Difference.

P.D.C.A. Purebred Dairy Cattle Association, a group composed of the dairy cattle breed associations.

PEDIGREE. List or diagram of an animal's ancestors ideally including complete information on their performance and genetic merit.

PEPTIDE. One of a class of low-molecular weight compounds which yield 2 or more amino acids upon hydrolysis.

PERIPHERAL NERVOUS SYSTEM. Nerves outside the central nervous system.

PERISTALSIS. Rhythmic contractions by which the alimentary canal propels its contents.

PERITONEAL CAVITY. Cavity containing the digestive organs.

PERITONITIS. Inflammation of the peritoneum, the membrane lining the abdominal cavity and encasing the viscera.

PERMANENT ENVIRONMENTAL EFFECTS. Long-term environmental differences which have important effects on cows' phenotypes, such as inter-herd differences in housing, feeding, and milking practices.

PERMEABLE. Penetrable.

PERSISTENCY. Degree of maintenance of lactation in dairy cows.

PETECHIAE. Small, round, purplish-red spots which later turn blue or yellow caused by intradermal or submucous hemorrhage; often referred to as *pinpoint hemorrhage*.

PHAGOCYTOSIS. Ingestion of microorganisms, other cells, or foreign material by special cells of the body such as leukocytes.

PHENOL. Carbolic acid, C_6H_5OH.

PHENOTYPE. What an animal is and does; its physical characteristics and performance.

PHENOTYPIC CORRELATION. Degree to which phenotypic traits tend to vary in the same or opposite directions.

PHENOTYPIC VARIANCE. Variation in animal's physical traits and performance due to the action and interaction of genetic and environmental effects.

PHYSIOLOGY. Study of function of living plants and animals.

PINOCYTOSIS. Invagination of a surface membrane to enclose a particle.

PLACENTA. Vascular tissue in the uterus which allows exchange of nutrients and waste products between the mother and fetus.

PLEIOTROPY. Action of a single gene which affects 2 or more traits simultaneously.

POLLED. Naturally hornless, as opposed to having been dehorned.

POLYMORPHIC. Pertaining to existence of genetic variation in a population at a level higher than can be accounted for on the basis of genetic and environmental influences which have acted on the population.

POLYPEPTIDE. Combination of a few amino acids.

POPULATION. Group of animals which are considered genetically as a unit for purposes such as estimating gene frequencies, determining selection effects and systems of mating, and measuring genetic progress.

POPULATION GENETICS. Study of genetic changes in populations as a whole, due to forces which change gene and/or genotypic frequencies.

POSTNATAL. After birth.

POST PARTUM. After birth.

PPM. Parts per million.

PRECURSOR. Any compound from which another is formed.

PREDICTED DIFFERENCE. Estimate of genetic transmitting ability (i.e., one-half breeding value) of dairy bulls for performance traits in the United States, defined as the amount by which daughters of a bull will on the average differ in performance from the average breed performance in the genetic base period.

PREDISPOSITION. Latent susceptibility or tendency toward disease.

PREGNANT MARE'S SERUM GONADOTROPIN. A hormone, rich in FSH- and LH-like activity, secreted from the placenta into the serum of the mare.

PREHENSION. Act of seizing or grasping.

PRICE PARITY. Purchasing power for farmers equivalent to that which existed in 1910–1914.

PRICE SUPPORT PROGRAM. U.S. government dairy product purchase program designed to support farm milk prices at preannounced level.

PROBABILITY. The likelihood that some event or outcome will occur.

PROGENY. An animal's offspring.

PROGENY TEST. An evaluation of the transmitting ability of an individual by studying the performance of its offspring.

PROGESTATIONAL. Pertaining to the phase of the estrous cycle when the corpus luteum is actively secreting progesterone.

PROJECTION FACTORS. Factors used to extend lactation records that are terminated for reasons not associated with the genetic capability of cows and to extend records-in-progress to 305 days in length.

PROLAPSE. Abnormal protrusion of an organ, such as the uterus.

PROPIONIC ACID. CH_3CH_2COOH, an acid found in chyme and sweat; one of the volatile fatty acids produced by ruminal bacteria.

PROTEIN. One of a class of compounds composed of many amino acids which contain carbon, hydrogen, oxygen, nitrogen, and sometimes sulfur.

PROTOZOA. One-celled animals which are the lowest division of the animal kingdom.

PSYCHROPHILES. Cold-loving bacteria which grow rapidly at relatively low temperatures of about 50°F (10°C).

PUBERTY. Period of time during which the reproductive system acquires its mature form and function.

PUBIC SYMPHYSIS. Joint between the pubic bones.

PUBLIC LAW 480. The purpose of the Agricultural Trade Development and Assistance Act of 1954 as amended is "to increase the consumption of United States agricultural commodities in foreign countries, to improve the foreign relations of the United States, and for other purposes."

PUREBRED. An animal with two registered parents of the same breed.

PYOMETRA. Accumulation of pus in the uterus.

QUALITATIVE TRAIT. Trait whose measurement or description would fall into discrete classifications such as color, sex, or the presence of horns.

QUANTITATIVE TRAIT. Trait which is subject to more or less continuous variation and must be measured on a (pseudo) continuous scale such as body weight or milk yield.

R. Abbreviation for Repeatability of a sire summary or a cow index.

RANDOM DRIFT. Changes in gene frequency in a population which are due to chance or random causes rather than to identifiable causes.

REGISTERED. Recorded in the herdbook of the breed.

REGRESSION. Mathematical method for determining the amount of change in a trait that is due to change in another trait.

REGRESSION INDEX. Formerly used method for estimating breeding value, defined as one-half the sum of the equal parent index and the breed average.

REGURGITATION. Backward flowing, as the movement of undigested food up the esophagus.

RELEASING FACTOR. Small molecules found in the hypothalamus which stimulate synthesis and release of anterior pituitary hormones.

REPEATABILITY OF COW INDEX. A percentage that expresses the reliability of a cow index value.

REPEATABILITY OF PREDICTED DIFFERENCE. Percentage which expresses the reliability of a Predicted Difference; mathematically, the regression factor used to regress the daughter-herdmate deviation.

REPEATABILITY OF PRODUCTION RECORDS. Tendency of successive records by the same cow to be more alike than records by different cows due to similar genetic and environmental effects.

REPEATABILITY OF SIRE SUMMARY. A percentage that expresses the reliability of the Predicted Difference; a factor in the P.D. formula derived from the amount and distribution of daughter information that is used to weight the relative contributions of progeny and pedigree to the P.D. value.

RESIDUAL MILK. See *complementary milk*.

RICKETS. Disease, especially of young animals, caused by a deficiency of vitamin D and/or imbalances of calcium and phosphorus.

RODENT. Mammal belonging to the order Rodentia which have gnawing or biting habits; such as rats, squirrels, or beaver.

RUMINATING. Chewing the cud; remastication of a bolus of feed which has been regurgitated from the rumen of the animal.

SAGITTAL PLANE. Anteroposterior plane or section parallel to the long axis of the body.

SALPINGITIS. Inflammation of the oviduct.

SATURATED FATTY ACID. Fatty acid that has no double bonds; all carbon atoms are combined with the maximum number of hydrogen atoms.

SCOURS. Diarrhea, a profuse watery discharge from the intestines.

SCROTUM. An external pouch which contains the testes and their accessory ducts.

SCRUB CATTLE. Cattle of nondescript breed and breeding.

SCURS. Small rounded growths of horn tissue attached to the skin of polled or dehorned animals in the area of the horn pits.

SEBACEOUS GLANDS. Ductular glands located in the skin which secrete a fatty material, sebum.

SEBUM. Fatty secretion of the sebaceous glands.

SECONDARY SEX CHARACTERISTICS. Distinct anatomical traits which characterize a sex but are not directly related to reproduction.

SECRETION. Intracellular synthesis and expulsion of a specific substance from the cell.

SELECTION. The causing of a differential rate of reproduction with the goal of having the genetically superior animals produce the most offspring.

SELECTION DIFFERENTIAL. Amount by which those animals selected to be parents of the next generation surpass the average of the population from which they were selected.

SELECTION INDEX. Procedure in which selection can be made for several traits simultaneously in the most efficient manner by weighting the traits according to their heritability, genetic correlations, phenotypic variation, and economic worth.

SELECTION PRESSURE. The intensity of selection for or against a trait in the selection process.

SEMINAL PLASMA. Fluid from the accessory sex glands of the male.

SENESCENCE. Aging.

SENSES. Perception, especially of the five senses of sight, hearing, smell, touch, and taste.

SENSORY NERVE. Nerve which transmits impulses from sensory organs to the central nervous system.

SEROLOGICAL. Pertaining to use of blood serum of animals in certain tests such as *in vitro* study of antigen-antibody reactions.

SEX CELLS. Reproductive cells, ova in females and spermatozoa in males.

SHIPPING FEVER. Disease characterized by pneumonia or septicemia. Highest incidence occurs in animals subjected to stress.

SHORT-CHAIN FATTY ACIDS. Fatty acids containing 16 or fewer carbon atoms.

SHORT-LOOP FEEDBACK. Mechanism whereby a hormone acts to inhibit its own secretion.

SIB. A brother or sister.

SIBLING. See *sib*.

SILENT ESTRUS. Estrus which occurs without the typical behavioral signs of normal estrus.

SINGLE TRAIT SELECTION. Selecting for or against one trait ignoring all other traits.

SIRE SUMMARY. An estimate of the genetic transmitting ability of a bull.

SNF. See *solids-not-fat*.

SOLIDS-NOT-FAT. Total milk solids minus the fat. It includes lactose, protein, and minerals.

SOLUBLE (S) RNA. See *Transfer RNA*.

SOLUTION. In a true solution the molecules of one substance, the solute, are dispersed among those of the solvent to form a homogeneous mixture.

SOMATIC CELL COUNT. The number of somatic cells estimated to be in 1 ml of milk, expressed in the DHIA Program reports on a scale to the base $\log_2$.

SOMATIC CELLS. All cells which are part of an animal's body except the sex cells.

SOMATOTROPHIN. See *somatotropin*.

SOMATOTROPIN. Hormone of the anterior pituitary which causes growth in young animals; also called *growth hormone*.

SPERMATOGENESIS. Formation and development of sperm.

SPERMATOZOA. Mature male sex cells containing the haploid number of chromosomes.

SPHINCTER. A ring-like muscle which closes a natural opening.

STANDARD DEVIATION. In a normally distributed trait, the range above and below the mean within which approximately 68% of the measurements would fall.

STARCH. Carbohydrate with the general formula $(C_6H_{10}O_5)_n$ from various plant tissues.

STATE MILK ORDER. State government regulated farm, wholesale, and retail milk pricing.

STATISTICS. Mathematical procedures by which masses of data are reduced to a few meaningful numbers which can be readily interpreted in order to draw conclusions concerning the original data.

STERILITY. Inability to produce offspring.

STEROID. One of a class of compounds characterized by a 4-ring carbon structure resembling cholesterol. They are soluble in organic solvents.

STRAY VOLTAGES. Common terminology for electrical "neutral to earth" voltages in which an electrical current may pass through a cow's body.

STRESS. Subjection to external forces.

STRIPCROPPING. Planting alternating strips of grasses and/or clovers and cultivated crops along the contour lines on sloping fields to control soil erosion in the cultivated crop.

STROMA. Connective tissue which comprises the framework of an organ.

SUPEROVULATION. Shedding of an abnormally large number of ova.

SUPPOSITORY. An easily infusible medication for the urinary, reproductive, or lower intestinal tracts.

SUTURE. To stitch; or series of stitches to secure the edges of a wound.

SWEAT GLAND. Ductular glands located in the subcutaneous tissues which secrete a clear fluid onto the surface of the skin.

SYMBIOSIS. Living together or close association of two dissimilar organisms.

SYMPATHETIC NERVOUS SYSTEM. Part of the autonomic nervous system which, in general, has the opposite effects of the other part of the autonomic nervous system, the parasympathetic nervous system.

SYNDICATE. A group of breeders who have common (usually financial) interest in an animal frequently for purposes of getting the animal progeny tested.

SYNDROME. Group of symptoms which occur together and characterize a disease.

SYNERGISM. The combined action of two or more agents which produce a greater response than the sum of their individual effects.

SYNTHESIZE. To build up a compound by union of its elements.

SYNTHETIC. Artificial.

SYSTEM OF MATING. Procedure which causes the non-random combining of genotypes to change the genotypic frequencies in the next generation but not the gene frequencies; common examples are crossbreeding and various forms of inbreeding.

TANDEM METHOD OF SELECTION. Exertion of selection pressure first on one trait, then on a second and so on, rather than selection for all traits simultaneously.

TDN. Total digestible nutrients.

TEMPORARY ENVIRONMENTAL EFFECTS. Environmental effects which impinge on an animal for relatively short periods such as a sudden weather change or a minor injury.

TEST INTERVAL. Number of days between two consecutive test days when DHIA data are collected in a herd.

THERAPEUTIC. Curative; referring to the art of healing.

THORACIC DUCT. Major lymph collecting duct extending from the posterior region of the abdomen to the anterior vena cava in the thorax where lymph is returned to the blood.

TIE-IN-SALES. Marketing device whereby, in order to obtain a desired product, the purchaser must also order a quantity of a second item, usually one that is a slow seller.

TITLE I SALES. Title I Sales of U.S. agricultural commodities under Public Law 480 are either for local currency of the importing country deposited in a U.S. government account or may require a specified down payment in dollars upon delivery of the commodity with the balance to be repaid in dollars over a period of up to 20 years.

TITLE II FOREIGN DONATIONS. Title II of Public Law 480 authorizes the President to make donations up to $600 million annually to meet famine or other urgent relief requirements, to combat hunger and malnutrition, and to promote self-help activities designed to alleviate the causes of and need for such assistance.

TONUS. Tension caused by muscle contraction.

TRACHEA. Windpipe.

TRANSCRIPTION. Process whereby genetic information contained in DNA is used to specify the sequence of the molecules of an RNA chain.

TRANSFER (T) RNA. Any of at least 18 structurally similar RNA molecules which combine with a specific amino acid and deliver the amino acid to the mRNA on the ribosome.

TRANSLATION. Process whereby genetic information which has been transcribed onto an mRNA molecule is used to specify the order of amino acids during protein synthesis.

TRANSMITTING ABILITY. One-half an animal's breeding value; the average genetic superiority or inferiority which is transmitted by an animal to its offspring.

TRANSPONDER. In electronic identification, a passive device containing electronic circuitry permanently coded with identifying information unique to each animal; the device is attached to or implanted in the animal and activated by an interrogator.

TRIGLYCERIDE. Group of compounds consisting of a molecule of glycerol and 3 fatty acids.

TRI-IODOTHYRONINE. Potent naturally occurring hormone of the thyroid gland.

TRIMONTHLY D.H.I. An unofficial dairy record-keeping plan in which data are obtained and processed every third month.

TYPE. Physical conformation of an animal; may refer to a single trait or conformation in general.

UBIQUITOUS. Existing or being everywhere at the same time.

UNDULANT FEVER. Brucellosis in man.

UNIFORM SERIES EARTAG. Eartags used by state health authorities and the DHIA program for unique identification of animals. Eartag numbers are listed in the configuration SS-AAA-NNNN, in which SS is a 2-digit state code, AAA is a field of 3 letters, and NNNN is a field of 4 digits.

UNSATURATED. A fatty acid that contains double bonds; not all of the carbon atoms are combined with the maximum number of hydrogen atoms.

UREA. NH_2CONH_2, a white, crystalline substance found in urine, blood, and lymph, which is the final product of protein metabolism in the body. Synthetic urea can be used by ruminal bacteria as a nitrogen source for protein synthesis.

UTERINE MILK. See *histotroph.*

VACCINE. Suspension of attenuated or killed microorganisms (bacteria, rickettsiae, or viruses) which is injected into animals to prevent, ameliorate, or treat infectious diseases.

VAGINITIS. Inflammation of the vagina.

VALERIC ACID. $CH_3CH_2CH_2CH_2COOH$, one of the volatile fatty acids produced by ruminal bacteria.

VARIANCE. Statistical measure of the variation in a trait; defined as the average squared deviation of a single observed value from the average of all observed values.

VASOCONSTRICTION. Reduction in the diameter of blood vessels, especially arterioles.

VASOPRESSIN. Antidiuretic hormone.

VENEREAL. Caused or propagated by mating.

VENTRAL. Pertaining to the abdominal surface of an animal.

VENULE. A small vein.

VERIFIED IDENTIFICATION PROGRAM. A program sponsored by National D.H.I.A., Inc. in which a D.H.I.A. Supervisor verifies the required identification information so that the D.H.I.A. can issue an identification certificate that permanently identifies an animal and its parentage.

VESTIBULE. Space at the entrance of a canal; for example, at the vulva.

VILLI. Plural of *villus.*

VILLUS. A minute vascular protrusion on the surface of a mucous membrane.

V.I.P. Abbreviation for Verified Identification Program.

VITAMINS. Organic compounds that are necessary in small amounts for normal metabolism.

WADAM. Abbreviated name for the weigh-a-day-a-month dairy record-keeping plan.

WITHERS. Top of the shoulders.

WRIGHT, SEWALL. American geneticist renowned for his contributions to quantitative genetics.

ZEBU. Strain of cattle *(Bos indicus)* characterized by a hump on the shoulders and loose folds of skin especially on the dewlap, which is very well adapted to tropical areas. They are usually inferior to European breeds *(Bos taurus)* in milk yield.

ZYGOTE. Diploid cell formed by the union of a sperm and egg, each of which was haploid.

Index

Page numbers in *italics* refer to figures;
page numbers followed by t refer to tables.